Handbook of Communications System Engineering

Handbook of Communications System Engineering

Editor:

Nelson Bolívar, Ph.D

ArclerPress

www.arclerpress.com

Handbook of Communications System Engineering

Editor: Nelson Bolívar, Ph.D

708 3rd Avenue, 6th Floor
New York
NY 10017
United States of America
www.arclerpress.com

ISBN: 978-1-68094-402-0

Library of Congress Control Number: 2016951050

About the Editor

Nelson Bolívar, Ph.D

Nelson Bolivar has a PhD. in physics from the University of Lorraine in France finished in 2014. His expertise is in quantum systems and condensed matter. His interest includes spintronic devices and correspondences between general relativity and condensed matter. He is currently an associate professor at the Central University of Venezuela

Contents

List of Contributors ... xiii

Preface ... xvii

Introduction ... xix

Chapter 1 Modeling Components and Services of LTE Mobile Communications System ... 1

1. Introduction ... 1
2. Long Term Evolution System ... 2
3. Formal Specification ... 3
4. Model Analysis ... 6

Conclusions And Future Work ... 7

References ... 7

Chapter 2 Spatio-Temporal Channel Characterization:Theoretical Framework and Applications in Mimo System Design ... 9

1. Introduction ... 9
2. Background ... 10
3. Space, Wave Vector, and Compound Impulses ... 11
4. Spatial Kernel Functions ... 12
5. Spatio-Temporal Channel Characterization—Theoretical Framework ... 13
6. Spatio-Temporal Channel Characterization—Practical Framework ... 14
7. Mimo Antenna Array Design ... 15
8. Illustrative Examples: Aoa Estimation ... 16

Conclusions ... 18

Acknowledgements ... 18

References ... 18

Chapter 3 Advances in Mimo Techniques for Mobile Communications ... 21

1. Introduction ... 21
2. Current Implementation Status ... 21
3. Ieee 802.11n Wi-Fi ... 22
4. Ieee 802.16 Wimax ... 22
5. Ieee 802.20 Mbwa ... 23
6. Single-User Mimo Techniques ... 23
7. Multiuser Mimo ... 34
8. Convex Optimization ... 52

Conclusions and Future Research ... 52

References ... 53

Chapter 4 5g: Convergence of Wireless Communications ... 57

1. Introduction ... 57
2. Vision for 5g Systems ... 58
3. Modulation Techniques for Higher Spectral Efficiency ... 60
4. Mobile Traffic Offloading ... 63
5. Cognitive Radio ... 65
6. Software Defined Radio ... 65
7. Software Defined Networking ... 66
8. Blended Infrastructure ... 66

Conclusions 67
Acknowledgments 67
References 67

Chapter 5 Analysis of Privacy in Mobile Telephony Systems 71
1. Introduction 71
2. Background: Mobile Telephony Systems 73
3. Gsm/Umts Architecture 74
4. Analysis of Mobile Systems' Privacy: 2g/3g Weaknesses 76
5. Analysis of Mobile Systems' Privacy: Pure 3g Linkability Attack 83
6. Implementation of Some 3g Protocols Attacks 85
7. Privacy-Friendly Fixes 87
8. Automatic Verification of the Fixed Protocols 89
Conclusions 93
Acknowledgments 93
References 94

Chapter 6 Performance Improvement of Wireless Communications Using Frequency Hopping Spread Spectrum 99
1. Introduction 99
2 Review of Related Theories 99
3. Gilbert-Elliot Channel Model 100
4. Matlab Modelling 101
5. Evaluation of Frequency Hopping 102
Conclusions 103
Acknowledgement 104
References 104

Chapter 7 Wireless Communications on Linear Topologies 105
1. Introduction 105
2. Related Work 105
3. Wiwi Architecture and Communication Protocol 106
4 Testing and Prototyping 109
5. Designing a Fault-Tolerant Wiwi Node 110
Conclusions and Future Work 111
Acknowledgements 112
References 112

Chapter 8 A Random Access Protocol With Multi-Packet Reception for Infrastructure-Less Wireless Autonomic Networks 113
1. Introduction 113
2. System Model and Assumptions 114
3. Capacity of a Mobile Node With MPR Capability 114
4. The Optimum Throughput for Autonomic Networks 115
5. Maximizing the Gain Considering Performance and Cost 116
Conclusions 117
References 117

Chapter 9 Reconfigurable Microstrip Double-Dipole Antennas for Personal Wireless Communications 119
1. Introduction 119
2. Antennas' Geometries and Parametric Study 120
3. Results 121

Conclusions ... 125
Acknowledgements ... 127
References ... 128

Chapter 10 A Wideband Circularly Polarized Patch Antenna for 60 Ghz Wireless Communications ... 129
1. Introduction ... 129
2. Antenna Structure/Packaging ... 130
3. Detailed Antenna Design ... 130
4. Comparison of Measurement and Simulation ... 133
Conclusion ... 133
Acknowledgements ... 134
References ... 134

Chapter 11 Development of 16-Channels Compact Eeg System Using Real-Time High-Speed Wireless Transmission ... 137
1. Introduction ... 137
2. Materials ... 137
3. Methods ... 139
4. Results ... 142
Acknowledgements ... 142
References ... 143

Chapter 12 Real-Time Audio & Video Transmission System Based on Visible Light Communication ... 145
1. Introduction ... 145
2. Vlc System Description ... 145
3. Results and Analysis ... 147
4. Modeling and Designing of an Indoor VLC System ... 148
Conclusions ... 148
Acknowledgements ... 149
References ... 149

Chapter 13 Dc Programming in Communication Systems ... 151
1. Introduction ... 151
2. Dc Programming and DCA ... 152
3. Minimizing a Nonconvex Function Under a Convex Constraint Set ... 154
4. Nonconvex Constraint Set ... 157
5. Integer Variables ... 158
6. Another Issue: Solving Convex Programs By DCA ... 161
Conclusion ... 161
References ... 161

Chapter 14 Home Networking at 60 Ghz ... 165
1. Introduction ... 165
2. Motivation and Background ... 165
3. Use Case Scenario ... 166
4. Communication Infrastructures ... 167
5. Challenges and Research Issues ... 167
Conclusion ... 171
Acknowledgment ... 171
References ... 171

Chapter 15 Performance of Bluetooth 2.X + Edr Technology in Noisy Environments ... 173
1. Introduction ... 173
2. Characterization of the Delay for the Ideal Case ... 175
3. Characterization of the Delay with Packet Retransmissions ... 177
4. Discussion of the Performance of Bluetooth 2.X + EDR ... 180
5. Empirical Evaluation of the Model for the Basic Rate ... 182

Conclusions ... 183
Acknowledgments ... 184
References ... 184

Chapter 16 Constrained Pre-Equalization Accounting for Multi-Path Fading Emulated Using Large RC Networks: Applications to Wireless and Photonics Communications ... 187
1. Introduction ... 187
2. Channel Modelling Using RC Networks ... 188
3. Optimization of the Input Waveform ... 189
4. Conversion of the Optimization Problem to a Milp Problem ... 190
5. Simulation Results ... 191
6. Discussion ... 193

Conclusion ... 195
Acknowledgments ... 195
References ... 195

Chapter 17 Broadband Traffic in Next-Generation Networks ... 201
1. Introduction ... 201
2. Building Long-Term Traffic Models ... 202
3. Drivers of Future Applications And Services ... 203
4. Traffic Volumes of Video Streaming ... 205
5. 20+ Years Broadband Traffic Forecast ... 206

Conclusions ... 207
Acknowledgement ... 208
References ... 208

Chapter 18 TCP Performance in Mobile Environments ... 211
1. Introduction ... 211
2. Problems Statement ... 212
3. A Cross-Layer Approach for Improving TCP Performance Over Mobile Ipv6 ... 214
4 Analysis and Comparison ... 215

Simulation and Evaluation ... 218
Acknowledgements ... 222
References ... 222

Chapter 19 Detection of Turbo-Coded Ofdm Signals ... 225
1. Introduction ... 225
2. System Model ... 226
3. Enhanced Frame Structure ... 228
4. Receiver ... 229
5. The Channel Capacity ... 230
6. Simulation Results ... 231

Conclusions and Future Work ... 233
An Approximate and Simple Cramér–Rao Bound on the Variance of the Frequency Offset Estimation Error ... 233
References ... 234

Chapter 20 Acoustic Information Leakage Through Fiber Optic Communications ... **237**
1. Introduction ... 237
2. Physical Principles Eavesdropping ... 237
3. Script Eavesdropping Threats ... 238
4. Preventing Eavesdropping Threats ... 239
5. Practical Implementation Protection ... 240
6. Advice on Protection ... 241
Conclusion ... 242
References ... 242

Index ... **243**

List of Contributors

Ishtiaq Ahmed Choudhry
Department of Communication and Networking, King Faisal University, Al-Ahsa, KSA

Nazir Ahmad Zafar
Department of Computer Science, King Faisal University, Al-Ahsa, KSA

Viet-Ha Pham
Department of Electrical and Computer Engineering, Université Laval, Québec, QC, Canada G1V 0A6

Jean-Yves Chouinard
Department of Electrical and Computer Engineering, Université Laval, Québec, QC, Canada G1V 0A6

Dominic Grenier
Department of Electrical and Computer Engineering, Université Laval, Québec, QC, Canada G1V 0A6

Huu-Tue Huynh
Bac Ha International University, Hanoi, Vietnam

Xianbin Wang
Department of Electrical and Computer Engineering, University of Western Ontario, London, ON, Canada N6A 5B8

Yiyan Wu[4]
Communications Research Centre Canada, Ottawa, ON, Canada K2H 8S2

Farhan Khalid
Institute of Telecommunications, University of Stuttgart, Stuttgart, Germany

Joachim Speidel
Institute of Telecommunications, University of Stuttgart, Stuttgart, Germany

Raúl Chávez-Santiago
Norwegian University of Science and Technology, Trondheim, Norway
Intervention Center, Oslo University Hospital, Oslo, Norway
Institute of Clinical Medicine, University of Oslo, Oslo, Norway

Michał Szydełko
Huawei R&D Sweden, Stockholm, Sweden

Adrian Kliks
Poznan University of Technology, Poznan, Poland

Fotis Foukalas
Athena Research and Innovation Centre, Industrial Systems Institute, Patras, Greece

Yoram Haddad
Jerusalem College of Technology, Jerusalem, Israel

Keith E. Nolan
Intel Labs Europe, Intel, Leixlip, Ireland

Mark Y. Kelly
Intel Labs Europe, Intel, Leixlip, Ireland

Moshe T. Masonta
Department of Electrical Engineering, Tshwane University of Technology, Pretoria, South Africa
Meraka Institute, Council for Scientific and Industrial Research (CSIR), Pretoria, South Africa

Ilangko Balasingham
Norwegian University of Science and Technology, Trondheim, Norway
Intervention Center, Oslo University Hospital, Oslo, Norway
Institute of Clinical Medicine, University of Oslo, Oslo, Norway

Myrto Arapinis
University of Edinburgh, Edinburgh, UK

Loretta Ilaria Mancini
University of Birmingham, Birmingham, UK

Eike Ritter
University of Birmingham, Birmingham, UK

Mark Dermot Ryan
University of Birmingham, Birmingham, UK

Naser Hossein Motlagh
Department of Information Technology, Vaasa University of Applied Sciences, Vaasa, Finland

Daniele De Caneva
Pervasive Technologies Laboratory, Agemont, Italy

Pier Luca Montessoro
Department of Electrical, Managerial and Mechanical Engineering, University of Udine, Udine, Italy

Jahangir H. Sarker
School of Information Technology and Engineering (SITE), University of Ottawa, Ottawa K1N 6N5, Canada

Hussein T. Mouftah
School of Information Technology and Engineering (SITE), University of Ottawa, Ottawa K1N 6N5, Canada

Abdelnasser Eldek
Department of Computer Engineering, Jackson State University, Jackson, USA.

Abubaker Abdallah
Department of Computer Engineering, Jackson State University, Jackson, USA.

Mahmoud Manzoul
Department of Computer Engineering, Jackson State University, Jackson, USA.

Rongguo Zhou
Department of Electrical and Computer Engineering, the University of Arizona, Tucson, USA

Duixian Liu
IBM T. J. Watson Research Center, New York, USA.

Hao Xin
Department of Electrical and Computer Engineering, the University of Arizona, Tucson, USA

B. R. Myung
Graduate Programs of Biomedical Engineering, Yonsei University, Seoul, Korea

S. K. Yoo
Correspondence author, Department of medical Engineering, College of Medicine, Yonsei University, Seoul, Korea

Yingjie He
Institute of Communication Technology, Nanjing University of Posts and Telecommunications, Nanjing, China

Liwei Ding
Institute of Communication Technology, Nanjing University of Posts and Telecommunications, Nanjing, China

Yuxian Gong
Institute of Communication Technology, Nanjing University of Posts and Telecommunications, Nanjing, China

Yongjin Wang
Institute of Communication Technology, Nanjing University of Posts and Telecommunications, Nanjing, China

Hoai An Le Thi
Laboratory of Theoretical and Applied Computer Science, UFR MIM, University of Lorraine, Ile du Saulcy, 57045 Metz, France

Tao Pham Dinh
Laboratory of Mathematics, LMI, National Institute for Applied Sciences, Rouen, Avenue de l'Université, 76801 Saint-Etienne-du-Rouvray Cedex, France

Zulkuf Genc
Wireless and Mobile Communications (WMC) Group, Faculty of Electrical Engineering, Mathematics, and Computer Science (EEMCS) of Delft University of Technology, , 2600 GA Delft, The Netherlands

Bao Linh Dang
Wireless and Mobile Communications (WMC) Group, Faculty of Electrical Engineering, Mathematics, and Computer Science (EEMCS) of Delft University of Technology, , 2600 GA Delft, The Netherlands

Jing Wang
Wireless and Mobile Communications (WMC) Group, Faculty of Electrical Engineering, Mathematics, and Computer Science (EEMCS) of Delft University of Technology, , 2600 GA Delft, The Netherlands

Ignas Niemegeers
Wireless and Mobile Communications (WMC) Group, Faculty of Electrical Engineering, Mathematics, and Computer Science (EEMCS) of Delft University of Technology, , 2600 GA Delft, The Netherlands

Jose-Rafael Luque
Dpto. Tecnologı´a Electro´nica, Universidad de Ma´laga, 29071 Ma´laga, Spain

Maria-Jose Morón
Dpto. Tecnologı´a Electro´nica, Universidad de Ma´laga, 29071 Ma´laga, Spain

Eduardo Casilari
Dpto. Tecnologı´a Electro´nica, Universidad de Ma´laga, 29071 Ma´laga, Spain

Rubens Junqueira Magalhães Afonso
Divisa˜o de Engenharia Eletro^nica, Instituto Tecnolo´gico de Aerona´utica, Sa˜o Jose´ dos Campos, SP 12228-900, Brazil

Roberto Kawakami Harrop Galvão
Divisa˜o de Engenharia Eletro^nica, Instituto Tecnolo´gico de Aerona´utica, Sa˜o Jose´ dos Campos, SP 12228-900, Brazil

Karl Heinz Kienitz
Divisa˜o de Engenharia Eletro^nica, Instituto Tecnolo´gico de Aerona´utica, Sa˜o Jose´ dos Campos, SP 12228-900, Brazil

Hashem Alyami
School of Systems Engineering, University of Reading, Reading RG6 6AY, UK

Victor M. Becerra
School of Systems Engineering, University of Reading, Reading RG6 6AY, UK
School of Engineering, University of Portsmouth, Anglesea Building, Anglesea Road, Portsmouth PO1 3DJ, UK

Sillas Hadjiloucas
School of Systems Engineering, University of Reading, Reading RG6 6AY, UK

P. Loskot
College of Engineering, Swansea University, Singleton Park, Swansea SA2 8PP, UK

M. A. M. Hassanien
College of Engineering, Swansea University, Singleton Park, Swansea SA2 8PP, UK

F. Farjady
Aston Institute of Photonic Technologies, Aston University, Aston Triangle, Birmingham B4 7ET, UK

M. Ruffini
CTVR, Dunlop Oriel House, Trinity College, University of Dublin, Dublin, Ireland

D. Payne
CTVR, Dunlop Oriel House, Trinity College, University of Dublin, Dublin, Ireland

Deguang Le
Institute of Computer Science, University of Göttingen, Lotzestr. 16-18, 37083 Gottingen, Germany

Xiaoming Fu
Institute of Computer Science, University of Göttingen, Lotzestr. 16-18, 37083 Gottingen, Germany

Dieter Hogrefe
Institute of Computer Science, University of Göttingen, Lotzestr. 16-18, 37083 Gottingen, Germany

K. Vasudevan
Department of EE, IIT Kanpur, Kanpur, India

Vladimir V. Grishachev
Russian State Geological Prospecting University (RSGPU), Moscow, Russia

Preface

As the world continues to move, so does the advancement of technology. Every year sheds light on improvements to yesterday's devices and new inventions appearing on the horizon. At the center of it all is a desire to speed up communication across the globe, with the intention of making inconveniences a thing of the past. Modern civilization depends on advanced communication technologies. The application of electricity to communications with inventions such as the telephone and telegraph meant people could send information instantly over long distances. More recent advances such as satellites and the Internet have extended communications worldwide and made global news and information commonplace. Communications technology continues to improve with each passing year, bringing more information choices to you at lower costs.

Today's communication systems are increasingly complex. Communications system engineering may involve developing or maintaining cable-based or wireless telephone and voicemail systems, group scheduling programs, webinars, Voice over Internet Protocol or automatic call answering and routing systems. Handbook of Communications System Engineering aims to cover technical information relevant to those interested in handling and learning about communications systems to a medium-high level. It covers from on mobile communications, network and the protocols involved, to wireless and control. It compiles revised and extendsed research work written by prominent researchers and experts in their respective field. Students and engineers will find this book to be a good comprehensive resource that focus on specific examples that are also top research material in the area, suitable for the interested reader. A couple of centuries ago mankind did not even dream of communicating with the use of the sky, much less dream of something like satellite communication. A couple of decades ago we did not dream of what we could do with satellite communication. Today satellite communication has become one of most powerful methods of communication. Everything from phones to the internet to the TV and radio can be used through satellite communication. Through satellite communication it is possible for an individual to communicate with any one who has a similar communication device. This can be done from any where in the world. The latest GPS can show an individuals position with pinpoint accuracy. An individual can ask for directions from the system and it will show you the closest route that you can get there. If the road to the destination is under construction or if you cannot go in that path the GPS will choose a different route for you. The best and worst part about communication is that it evolves, develops and expands. The way we see the world today will be totally different to how we might see the world in another 10 years time. Communication methods and ways might change and expand in ways that we never thought possible.

Editor

Nelson Bolívar, Ph.D

INTRODUCTION

Every day, in our work and in our leisure time, we come in contact with and use a variety of modern communication systems and communication media, the most common being the telephone, radio, television, and the Internet. Through these media we are able to communicate (nearly) instantaneously with people on different continents, transact our daily business, and receive information about various developments and events of note that occur all around the world. Electronic mail and facsimile transmission have made it possible to rapidly communicate written messages across great distances. Can you imagine a world without telephones, radio, and TV? Yet, when you think about it, most of these modern-day communication systems were invented and developed during the past century. Here, we present a brief historical review of major developments within the last two hundred years that have had a major role in the development of modern communication systems

What Is Digital Signaling?

Digital signaling is the exchange of coded voltages or currents to effect control communications between two modules or equipment. It started in early telegraphy and telephony at very low speed. Later, digital signaling existed in parallel interfaces between equipment, and later as a serial data stream. Digital signaling has the potential to provide new solutions in all facets of daily living. There are many possibilities for new products and services in both communications and automation.

The early telegraphy that started in the late 1800s was a series of dots and dashes that represented coded characters, which could be considered as slow digital signaling. Later, telephone systems used another slow signaling for on/off hook and pulse dialing. By interrupting the line current, a phone user was able to signal the numbers being dialed. By the late 1900s, parallel digital signaling was very common between modules in computer equipment and accessories such as printers and hard disks. For instance, transistor-transistor logic (TTL) devices were very popular digital circuit family using 0 to 5 volts direct current (VDC) as the standard levels for digital signaling.

As of 2011, there were hundreds of variations in digital signaling used in computing and communications. The layered architecture of the open systems interconnection (OSI) has introduced signaling protocols embedded in other signaling protocols. Digital signal converters have taken many forms, such as converting wired Internet into wireless Internet services.

The digital television (TV) signal is now in broadcast over the air and as digital cable signal. Digital TV sets may receive the signals directly, while analog TV sets may use a converter. The digital TV signal results in a very clear picture free from any form of noise or distortion seen in analog TV systems.

Digital signal processing technology has drastically improved the way and the speed that analog is converted to digital. The digital signal processor (DSP) uses very fast built-in analog-to-digital converters, and the DSP units may be used in groups to compensate for any lack of speed. Digital signal will continue to be improved with the development of better and faster DSP units and components. In television, this not only gives a clearer picture but also allows for interactive TV and video on demand. The home viewer is able to signal the TV station on choices and responses, which is known as interactive mode.

What is Telephony?

Telephony is something that impacts the lives of just about everyone on a daily basis. Here are some basics about telephony, as well as some history about the development of voice communications.

In its simplest form, telephony is simply a collective term for all types of voice equipment that are designed to allow interactive communication between two points. The most commonly identified piece of telephony equipment is the telephone itself. Considered to be an essential in just about every residence and in all businesses, the telephone provides basic point to point audio communication that many of us have come to rely on extensively during the course of the day.

Along with telephones, telephone switching systems or telephone exchanges allow for the processing of all our telephone calls. The basic framework for a telephone calls involves a point of origin, which initiates a signal to a switch that the subscriber wishes to place a call. The switch accepts the inbound signal and delivers the call to the subscriber whom the original caller wishes to speak with, also called the point of termination.

If the call is for a subscriber outside area covered by the switch, then the connection will be routed to a remote switch or series of switches, until the connection is established at both ends. Once the original switch receives confirmation that the connection is established, then the verbal communication actually begins. In a similar manner, when the connection is terminated at either the point of origin or the point of termination, the switches involve interact to confirm the disconnect has occurred and free up the circuits for use by other subscribers.

Originally, telephony equipment involved nothing more than local point to point calls. In order to make these calls, the first telephone switch was created. Essentially, calls were switched manually by an operator. A caller would pick up the line, ring the operator and tell the operator whom he or she wished to speak with. Over time, the switches became more sophisticated with the assignment of an alphanumeric number or exchange for each subscriber.

Eventually, automated switching equipment and the issuing of numeric designations, or phone numbers, made the need for manual switching obsolete. The use of automated switches also paved the way for switching from an older analog phone signal to the more sensitive digital telephony technology that we enjoy today, both with land line service and cellular service.

As telecommunications equipment and solutions continue to evolve, telephony equipment will also continue to change with the times and redefine the nature of how we communicate with one another.

What is Internet Telephony?

Internet telephony refers to the science or technology of integrating telephone services into computer networks. In essence, it converts analog voice signals into digital signals, transmits them, then converts them back again. Voice over IP (VoIP) is a common form of this service.

With traditional telephone service, sometimes referred to as POTS (Plain Old Telephone Service), voice signals use telephone lines — copper wires — and circuit switches to communicate. Internet telephony eliminates the telco company all together by using computer networks to send voice signals. All information is transferred across the Internet in "data packets." For example, if someone sends a friend an email, the email is broken up into a series of data packets that each take their own route to the destination mail server. Once there, the packets reassemble themselves into the full email message.

Internet telephony also transmits using data packets. Analog voice signals are digitized, sent in discreet packets to the destination, reassembled and reverted back to analog signals. By using this system, a person can place long-distance calls free of telephone charges. The catch is that both parties must have the correct software. If this technology is used to call a land-line or cell phone, charges apply, though they are usually minimal.

Some online VoIP services provide free Internet telephony software and use a prepaid system to keep monies on account for calling from it to a land line. The charge is a small and made per minute, per call. The rate does not change, whether calling someone local or in another country. Again, if both parties use VoIP software, there is no charge at all.

Internet telephony has drastically improved since its first incarnations. Initial VoIP was very poor quality, but now many users report land line-like quality. There are many advantages to using the Internet to make calls, not just for family members and friends to stay in touch free of charge, but for multi-state or multi-national corporate PBXs where routine long distance calls between offices are significant. A potential disadvantage of using Internet telephony for corporate environments is that VoIP tends to have more downtime than POTS. Computer or network problems can interfere with the calls, though many VoIP programs kick calls to POTS if there is a problem.

Voice mail and other telephone services are often available from these services, and installation of simple VoIP software is easy for anyone with a small amount of skill. Industry insiders predict that, in time, Internet telephony technologies and services will supplant much of the workload currently handled by plain old telephone service.

What is a Cable Telephony?

Cable telephony is the use of a cable television connection as a means of transmitting telephone calls. It uses the local cable network to send and receive data in a similar manner to a cable Internet connection, though cable telephony does not itself use the Internet and so is distinct from Internet telephony. Due to the ability to compress data sent over cable lines, cable telephony can provide more efficient use of bandwidth than conventional telephones, though it also suffers from reliability issues not faced by conventional land-line connections.

The principal benefit of cable telephony is that it requires much less bandwidth to transmit the same amount of data as a conventional phone. Similarly to digital cable television, this compression of data moving over the cable network makes it possible for the network to carry more data at once than would be possible with older technology. This more efficient use of bandwidth allows for higher sound quality. As cable telephony uses the same network as cable internet connections and cable television, some cable companies offer these services bundled together, which can be cheaper than receiving telephone, Internet, and television services separately. This is commonly called a triple play service.

The principal downside of cable telephony is that users of the service share bandwidth with other cable telephone users and with other uses of the cable network, such as television signals and cable Internet connections. This means that, when the cable network is being heavily used, there is a risk that some users' calls will be dropped or delayed due to insufficient bandwidth to meet the demand. In addition to the possible inconveniences it can cause, this also poses a potential safety risk if someone attempts to make an emergency call, such as a call to the fire department or hospital, during peak usage hours and is unable to connect immediately.

Another disadvantage of cable telephony is that it requires an electricity source in the user's home to make or receive calls. Conventional telephones receive a supply of electrical current sent over the phone lines by the phone company, called phantom power, that allows them to send out calls even from a building that has lost electricity. This cannot be done over a cable telephone connection, though some cable telephones are sold with emergency battery backups to deal with this issue.

What are Wireless Telecommunications?

Wireless telecommunications is any means of electronic data transfer using electromagnetic means, most commonly radio waves. The data transferred is often in the form of digital information, which include voice transmission. Wireless telecommunications can also include analog signals, though these are being phased out in many industries. Telecommunication is also usually thought of as being two-way, with both sides having receiving and transmission capabilities.

Most types of wireless telecommunications work through a base station. This station receives information both from fixed lines and wireless sources, such as radio waves. The base station then takes those signals and converts them from one type of medium to the other. It is possible the radio waves would be forwarded to a satellite or other type of system and then transferred to another base station before going to a fixed land line system.

One major exception to this process is the use of satellite phones, which are still quite expensive, and generally not used by the average consumer. These phones can forgo the use of a base station and transmit directly to overhead satellites. This capability greatly enhances their functionality to receive and send signals in locations where traditional cell phones cannot. The drawbacks are the price of such phones, and the power it takes to run them, both of which are substantially more than with cell phones.

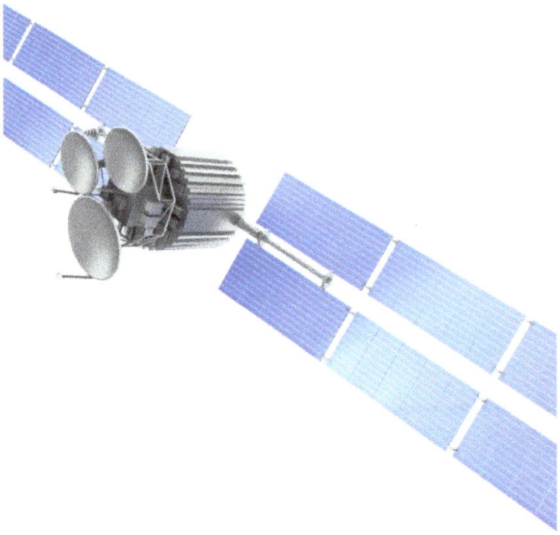

Though they have been around a long time, wireless telecommunications have only recently begun dominating the marketplace, especially for personal phone calls. Popularity since the beginning of the 21st century has ballooned. The use of systems such as cellular telephones enable users to take their phones with them anywhere, which greatly enhances their on-demand communication opportunities. While this may offer many users a greater convenience, there have been some drawbacks to wireless telecommunications as well, including more government regulation, and disputes over who has the right to what airwaves.

The prevalent use of cell phones and other radio transmission devices has prompted the Federal Communications Commission in the United States and other governing bodies around the world to work quickly to improve regulations. As the system becomes increasingly crowded with signals, the FCC has worked to implement changes, such as moving television signals from an analog-based system to a digitized one and allowing certain cellular networks to only transmit on certain frequencies. This frees up more bandwidth across the spectrum.

The switch in reliance from land line telecommunications to wireless telecommunications has concerned some. Some see wireless systems as less reliable, especially in emergency situations. Others argue that when storms knock down phone lines, cell towers usually make it through and provide sometimes the only means of communication. In general, the safest bet is to have both types of systems available to you in an emergency.

What does a Communication Technician do?

People who need to troubleshoot a computer or have their telephone systems repaired often turn to a communication technician. Communication technicians install and maintain electronic communication equipment, such as telephone systems and computer modems. They are also responsible for building and maintaining electronic circuits and data networks.

Humans rely very heavily upon complex communication networks. Communication technicians are needed to maintain and repair all of the devices within these networks. They must be able to properly utilize diagnostic and measuring tools to test, alter, and repair any equipment necessary.

A communication technician might be responsible for working on cellular phones, wireless Internet services, two-way radios, handheld computers, and other electronic equipment. A technician may also work on important law enforcement or military devices, such as vehicle location equipment or marine radios. Satellite systems and their related equipment are also included.

To become a communication technician, a person should be very familiar with digital radio technology, cellular systems, and electronics in general. He or she will need good reading skills to be able to read work orders depicting equipment needs. Good communication skills are also necessary to be able to converse with equipment operators and clients in order to assess problems and equipment needs. Sometimes schematic drawings and other written materials are needed in order to locate and rectify problems.

When wireless communications break down, a field communication technician is typically dispatched to repair the damage. This may happen on a work site or wireless network relay sites. Much of the work of a communication technician is not emergency, however, but standard maintenance. Many corporations and other users of wireless networks hire communication techs in order to maintain their systems periodically.

Many communications techs work in stores, repair shops, or service centers. At these locations, they fix wireless communications devices usually brought in by customers. They may also help advice customers about replacement parts and other devices they may be interested in purchasing or upgrading.

Large telecommunications companies typically employ many communication technicians. In this field, they are responsible for both addressing the needs of the telecommunications companies themselves as well as the customers using their services. Such careers may also include duties in sales, purchasing office equipment, and executing office administration.

Formal training in electronics is generally preferred in a communications technician. This can be acquired through community and technical colleges, as well as universities. Many communications technicians receive their training and experience while serving in the armed forces.

What is a Wireless DMX System?

A wireless DMX system is used to transmit lighting data without the use of traditional wireline DMX cables. The systems are used by theme parks, theatrical companies and event companies that rely on large scale lighting effects. Data is sent across Wi-Fi radio links to control when certain lights are powered on and off. In this way, a wireless DMX system helps to create the illusion of lighting animation.

There are several methods in which a wireless DMX system might transmit data to lighting fixtures. Some of the technology behind these methods is remarkably similar to the use of wireless communications in the cell phone industry. For example, time division multiple access (TDMA) is one of the ways in which communication occurs between a wireless DMX receiver and transmitter. TDMA technology allows a portion of the radio spectrum to be used simultaneously by multiple devices, which increases capacity and functionality.

These systems are largely designed by the user since each DMX unit has the capability of being set up as either a transmitter or a receiver. "Point-to-point," "point-to-multipoint," or "multipoint-to-multipoint" configurations are possible. For example, a row of stage lights might be set up to receive data transmissions from a single receiver in a "point-to-point" design. A second row of lights above the stage may be set up to be controlled by the same transmitter that controls the first row in a "point-to-multipoint" configuration.

Wireless DMX systems operate using a computerized communication network. Multiple networks might be set up and designed for larger scale lighting displays or a single intranet design could work better in a smaller display situation. Data is transmitted quickly to lighting that is in poorly accessible locations such as the top of tall buildings. The technology makes entertaining and decorative lighting more convenient with mobile platforms like carnival floats or moving stage sets.

Concert halls have found wireless DMX systems to be more cost-efficient since they eliminate labor and cable breakage. It is possible to install receivers on the roof of the building that are in direct communication with a receiver installed on the operating box. This technology allows operators to control the lighting from any position in the arena.

Interactive outdoor lighting displays are also made possible by wireless DMX systems. In this scenario, a control box is set up to receive text message phrase commands from the general public. Each command triggers the control box to send a different sequence of data to the light fixtures, resulting in a unique animation-type display.

CHAPTER 1

MODELING COMPONENTS AND SERVICES OF LTE MOBILE COMMUNICATIONS SYSTEM

Ishtiaq Ahmed Choudhry, and Nazir Ahmad Zafar

Keywords: LTE Communications System, Formal Methods, Emerging Software, Mathematical Language, Z Notation

INTRODUCTION

Long Term Evolution (LTE), a wireless broadband technology, is designed to support roaming Internet use and access via mobiles phones and other such handset devices. LTE is also referred as fourth generation (4G) technology and provides significant improvements over previous cellular communication networks. LTE is the latest wireless standard for International Mobile Telecommunication (IMT) system which has been deployed in many countries and few others are in pipeline for its deployment. LTE can operate in different bandwidths ranging from 1.4 MHz to 20 MHz and will provide a data rate of 100 Mbps to 1 Gbps depending upon the mobility of the user equipment. LTE—Advanced Release 10 is set to provide higher bit rate (DL 3Gbps, UL 1.5 Gbps). The IMT system is a complex one due to its different Radio Access Networks (RAN) and protocols. International Mobile Telecommunication systems allow users to access services and information at any time and at any place. However, these telecommunication systems provide mobility services with different requirements, resources and constraints. For example, poor quality of service, service interruptions due to mobility, seamless handover, interruption time at handovers and downward compatibility to other Radio Access Networks are some of the common key issues that are addressed in the scientific literature [1-7]. There are many other issues which need to be addressed due to complexity and dynamic nature of wireless systems, new information technology infrastructure, multimedia applications and software requirements

LTE offers many real time multimedia applications by integrating with the existing technologies. Although there exists a lot of research in this area but still there is an uncertainty and sufficient improvement are required in it due to complexity and flexibility of the system. The telecom industry is still discovering the system, network approaches and optimization techniques with the support and cooperation of academics and researches around the globe. Because LTE will integrate many heterogeneous networks and services, the seamless implementation for a complete development of LTE communications system is an open research problem.

Formal methods are mathematical techniques used to model software systems. It enables us to write system specification in such a way that it helps us understanding states of the system and its properties in a rigorous way. Formal methods are effective abstraction technique for managing complexity of large scale, dynamic and distributed systems. By use of formal methods, systems requirements, specification and limitations are addressed at early stages of systems and software development. In this way, a thorough development processes leads to a complete, consistent and robust model. The most important benefit of using formal methods is that we can move gradually from informal requirements to a detailed model maintaining the system properties and constraints.

In the existing work [8], Milner extended the system from Calculus of Communication System to pie-calculus developing a dynamic communication model which can be used to describe the mobility issues. In this work, mobility is assumed as movement of links between the connected components. Acharya et al., [9] have used formal techniques to overcome the consistency issue at various phases of development process in wireless mobile communication networks. Formal models are applied in [10], [11] using Z and object Z, however, few discrepancies are identified and the systems are not addressed completely. Duke et al., [12] have presented several case studies to illustrate application of formal methods. One of his study regarding mobile phone system which keeps track of several states of mobiles with limited number of operations, is an interesting example. The same case study is extended by Battaz in [13] to address the mobility issues explicitly using

Citation: I. Ahmed Choudhry and N. Ahmad Zafar, "Modeling Components and Services of LTE Mobile Communications System," International Journal of Communications, Network and System Sciences, Vol. 5 No. 12, 2012, pp. 815-824. doi: 10.4236/ijcns.2012.512086.

Z specification language. To further extend the same system, Battaz has described handover procedure among two objects in more details. The requirements capturing and analysis of LTE system is presented in [14]. In our paper, LTE is developed by the 3G Partnership Group (3GPP) [15] and its specifications is laid out mainly focusing on model abstraction.

In this paper, formal specification of static structures of the LTE system is presented using Z notation. First of all, fundamental definitions used in the model are presented. Then formal specification of few important services, phone book, messages and call record is described. In the next, formal description of user equipment is given based on the above definitions. Further, formal description of core components, namely, evolved Node B, Home Subscriber Server, Serving Gateway and Mobility Management Entity is described. Finally, LTE system is presented after composition of the above structures. The relationships among the components are defined for well definedness. An integration of the components is described in terms of properties. The description of the formal specification is provided by Z notation. The specification of the system is analyzed using Z/Eves toolset.

Rest of the paper is organized as: In Section 2, basic model of the LTE communications system is introduced. Formal model using Z notation is provided in Section 3. Model analysis is presented in Section 4. Finally, conclusions and future work are addressed in Section 5.

LONG TERM EVOLUTION SYSTEM

The previous cellular systems were unable to provide connectivity to packet switched networks and hence were not able to share services. The LTE has been designed to provide seamless Internet Protocol (IP) connectivity between user equipment (UE) and the packet data network (PDN), without any disruption to the applications run by end user during mobility. While the term "LTE" encompasses the evolution of the Universal Mobile Telecommunications System (UMTS) radio access through the Evolved UTRAN (E-UTRAN), it is accompanied by an evolution of the non-radio aspects under the term "System Architecture Evolution" (SAE), which includes the Evolved Packet Core (EPC) network. Together LTE and SAE comprise the Evolved Packet System (EPS) [16]. The LTE architecture is composed of separate components that includes UE (User Equipment), eNB (evolved Node B—more commonly known as base station in UMTS), SGW (Serving Gateway), MME (Mobility Management Entity), HSS (Home Subscriber Server), and PDN-GW (Packet Data Network Gateway). We have dedicated this section to define LTE components. After defining the system and its functional components formal description of the system will be provided.

Evolved Radio Access Network

The evolved RAN for LTE consists of a single node, i.e., the eNodeB (eNB) that interfaces with the user equipment (UE). The architecture of all eNBs is called a flat architecture. The eNB is responsible for all radio-related functions. It offers Radio Resource Control (RRC) functionality corresponding to the control plane. It performs many functions including Radio Resource Management (RRM), Admission Control (AC), radio mobility control, scheduling and dynamic allocation of resources to UEs in both uplink and downlink. Another function performed by eNBs is header compression to ensure efficient use of radio interface by compressing the IP packet headers that could otherwise represent a significant overhead, especially for small packets such as VoIP. When UE and eNB is connected, then all data sent over the radio interface is encrypted

On the network side, all of these functions reside in the eNBs, each of which can be responsible for managing multiple cells. Unlike some of the previous second- and third-generation technologies, LTE integrates the radio controller function into the eNB. This allows tight interaction between the different protocol layers of the radio access network (RAN), thus reducing latency and improving efficiency. Such distributed control eliminates the need for a high-availability, processing-intensive controller, which in turn has the potential to reduce costs and avoid "single points of failure." Furthermore, as LTE does not support soft handover there is no need for a centralized data-combining function in the network [16].

Serving Gateway

All user data packets (IP packets) are transferred through Serving Gateway, while also acting as the mobility anchor for the user plane during inter-eNB handovers. It also serves as the mobility anchor for interworking with other 3GPP technologies such as general packet radio service (GPRS) and UMTS. For idle state of user equipment, the serving gateway terminates the download link data path and triggers paging when the data arrives for the UE. It manages and stores UE contexts, e.g. parameters of the IP bearer service, and network internal routing information. It performs replication of the user traffic in case of lawful interception. In addition, SGW performs some administrative functions in the visited network such as collecting information for charging (for example, the volume of data sent to or received from user).

Mobility Management Entity

The MME is the key control-node for the LTE accessnetwork. MME is the control node that processes the signaling between the UE and the Core Network (CN). The protocols running between the UE and the CN are known as the Non Access Stratum (NAS) protocols. It is also responsible for idle mode UE tracking and paging procedure including retransmissions. It is involved in the carrier activation/deactivation process and is also responsible for choosing the SGW for a UE at the initial attach and at time of intra-LTE handover involving CN node relocation. It is responsible for authenticating the user (by interacting with the HSS). The Non-Access Stratum (NAS) signaling terminates at the MME and it is also responsible for generation and allocation of temporary identities to UEs. It checks the authorization of the UE to camp on the service termination point in the network for ciphering/integrity protection for NAS signaling and handles the security key management. Lawful interception of signaling is also supported by the MME. The MME also provides the control plane function for mobility between LTE and 2G/3G access networks with the S3 interface terminating at the MME from the SGSN. The MME also terminates the S6a interface towards the home HSS for roaming UEs.

Home Subscriber Server

The Home Subscribe Server is the main Internet Multimedia Server IMS database which also acts as database in evolved packet core EPC networks. The HSS is a super HLR that combined legacy HLR and AUC (Authentication Center) functions together for circuit switched (CS) and packet switched (PS) domains. The AUC generates the vectors for authentication and security keys. In the IMS architecture, the HSS connects to application servers as well as the Call Session Control Function (CSCF) using the DIAMETER protocol. HSS is the master repository for subscriber profiles, device profiles, and state information. As a mandatory control plane function in the LTE Evolved Packet Core (EPC), the HSS manages subscriber identities, service profiles, authentication, authorization, and quality of service (QoS) for LTE and IP Multimedia Subsystem (IMS) networks. An architecture of the LTE System, its components and relationship among the components is presented in the Figure 1 (reproduced from www.lte.org).

Packet Data Network Gateway

Another very important component is the PDN gateway. The PDN Gateway is responsible for IP address allocation for the UE, as well as QoS enforcement and flowbased charging according to rules from the PCRF (Policy Control and Charging Rules Function). It is responsible for the filtering of downlink user IP packets into the different QoS-based carriers. This is performed based on Traffic Flow Templates (TFTs). The P-GW performs QoS enforcement for guaranteed bit rate (GBR) carriers. It also serves as the mobility anchor for interworking with non-3GPP technologies such as CDMA2000 and WiMAX® networks. This component is partially modeled in this paper, however, a more completed function will be modeled later.

From the above model it is clear that this system is very much complex in terms of hardware equipment, communication protocols, and multimedia applications.

FORMAL SPECIFICATION

Formal specification of the LTE system is provided in this section. Z is used for the formal specification and Z/Eves tool is used for the model analysis. Schema is a powerful structure in Z used for variables definitions, components encapsulation and defining properties in terms of invariants. In the formal model, first of all, fundamental definitions used in the LTE system are presented. In these definitions, phone book, messages and call record facilities are considered for services description. Then major components namely user equipment, eNB, HSS and SGW are defined. In the next section, formal definition of an important component, MME is described. Finally, LTE communications system is presented based on composition of the above structures.

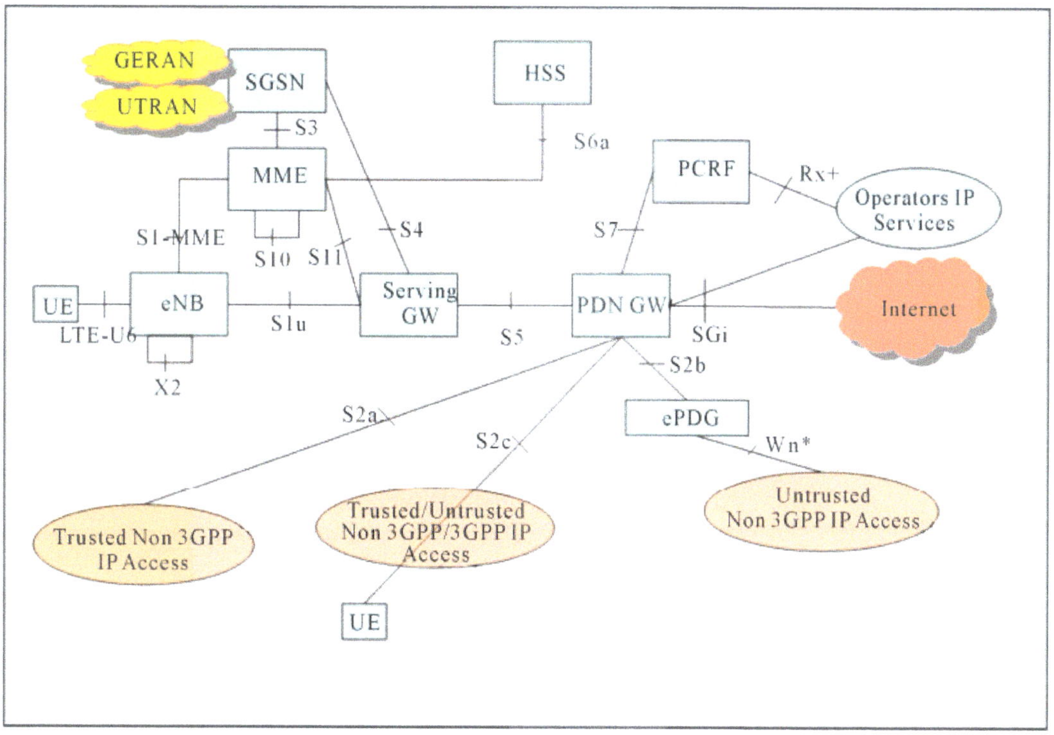

Figure 1: An architecture of LTE system.

Services

Formal definition of phone book is described below by using schema PhoneBook which consists of four components, namely, mobiles, officials, personal and type. The first three variables are defined as functions of type equipment identifier to Person. The schema person contains two variables, that is, name and address. The last one variable in definition of PhoneBook is used to represent type of the phone number, i.e., home, official or mobile. The schema consists of two parts in addition to the name of the schema written in the first horizontal line. Definitions of variables used are given in first part of the schema and invariants are defined in the second part.

PhoneBook

$mobiles, officials, personals: EquipmentId \rightarrow Person$
$type: PType$

$\forall eid: EquipmentId; person: Person | eid \in \mathrm{dom}\ mobiles \wedge person \in \mathrm{ran}\ mobiles$
$\bullet\ (eid, person) \in mobiles \Rightarrow type = MOBILE$
$\forall eid: EquipmentId; person: Person$
$|\ eid \in \mathrm{dom}\ officials \wedge person \in \mathrm{ran}\ officials$
$\bullet\ (eid, person) \in officials \Rightarrow type = OFFICE$
$\forall eid: EquipmentId; person: Person$
$|\ eid \in \mathrm{dom}\ personals \wedge person \in \mathrm{ran}\ personals$
$\bullet\ (eid, person) \in personals \Rightarrow type = HOME$

Invariants: 1) For each equipment identifier in do- main of function mobiles and person in range of mobiles, it holds that type of mobile is MOBILE; 2) For each equipment identifier in domain of function officials and person in range of officials, the type of mobile is OFFICIAL; 3) For each equipment identifier in domain of function personals and person in range of the personals, the type of mobile is described as PERSONAL

$[String, EquipmentId]; PType ::= MOBILE \mid OFFICE \mid HOME$

Person

$name: \mathrm{seq}\ String$
$address: \mathrm{seq}\ String$

Formal definition of messages s described below by using the schema Messages. The schema consists of four components, inbox, sent, outbox and type. The variables inbox, sent and outbox are defined as functions of type equipment identifier to Message. The schema Message consists of three variables, namely, equipment identifier, data stored and sending or receiving time. The data sent or received is defined as a sequence of strings. The sending or receiving times are denoted by Time.

$[Time]$

Message

$ueid: EquipmentId$
$data: \mathrm{seq}\ String$
$time: Time$

$MType ::= RECEIVED \mid DELIVERED \mid FAILED \mid DEFFERED$

Messages

$inbox, sent, outbox: EquipmentId \rightarrow Message; type: MType$

$\forall eid: EquipmentId; message: Message \mid eid \in \mathrm{dom}\ inbox \wedge message \in \mathrm{ran}\ inbox$
$\bullet\ (eid, message) \in inbox \Rightarrow type = RECEIVED$
$\forall eid: EquipmentId; message: Message \mid eid \in \mathrm{dom}\ sent \wedge message \in \mathrm{ran}\ sent$
$\bullet\ (eid, message) \in sent \Rightarrow type = DELIVERED$
$\forall eid: EquipmentId; message: Message \mid eid \in \mathrm{dom}\ outbox \wedge message \in \mathrm{ran}\ outbox$
$\bullet\ (eid, message) \in outbox \Rightarrow type = FAILED \vee type = DEFFERED$

Invariants: 1) For each equipment identifier in domain of function inbox and message in range of the inbox, it holds that status of message is RECEIVED; 2) For each equipment identifier in domain of function sent and message in range of the sent, the status of message is DELIVERED; 3) For each equipment identifier in domain of function outbox and message in range of outbox the status of the message is represented as DEFFERED. Formal specification of calls record is described by the schema CallsRecord which consists of outgoing, incoming, missed and type. The first three variables are described by the functions of type equipment identifier to Call. The schema Call consists of two variables, that is, equipment identifier and sending or receiving time.

CallsRecord

$outgoing: EquipmentId \rightarrow Call$
$incoming: EquipmentId \rightarrow Call$
$missed: EquipmentId \rightarrow Call$
$type: CType$

$\forall eid: EquipmentId; call: Call \mid eid \in \mathrm{dom}\ outgoing \wedge call \in \mathrm{ran}\ outgoing$
$\bullet\ (eid, call) \in outgoing \Rightarrow type = OUTGOING$
$\forall eid: EquipmentId; call: Call \mid eid \in \mathrm{dom}\ incoming \wedge call \in \mathrm{ran}\ incoming$
$\bullet\ (eid, call) \in incoming \Rightarrow type = INCOMING$
$\forall eid: EquipmentId; call: Call \mid eid \in \mathrm{dom}\ missed \wedge call \in \mathrm{ran}\ missed \bullet (eid, call) \in missed \Rightarrow type = MISSED$

Call

$ueid: EquipmentId$
$time: Time$

$CType ::= INCOMING \mid OUTGOING \mid MISSED$

Invariants: 1) For each equipment identifier in domain and call in range of the outgoing function, it is satisfied that status of the call is OUTGOING; 2) For each equipment identifier in domain and call in range of the incoming, it is satisfied that status of the call is INCOMING; 3) For each equipment identifier in domain and call in range of the missed function, the property holds that status of the call is MISSED.

LTE Components

The formal description of User equipment, eNB, HSS and SGW are presented here. User equipment is a device used directly by an end-user to communicate with other connected users. It can be a telephone, laptop or any other such device connected to the base station. The user equipment is defined by the schema UserEquipment given below. The equipment identifier is denoted by the EquipmentId. The equipment has three states that is active, idle or detached. The Phone Book, Messages and Call Record schemas are called to define the user equipment. The equipment has information about its position which is defined in terms of latitude and longitude. Each variable has two parts that is an integral and fractional. The user equipment contains information about its state, i.e., active, idle or detached. Further, it contains sim identifier and key. The equipment has other two variables called control and voice frequencies. The value of frequency will be either allocated or null. In predicate part of the schema, it is described that control frequency has allocated status if and only if the equipment status is active or idle. The control frequency is null if and only if the equipment is detached.

$Limit: \mathbb{N}$

Position

$north, nfrac: \mathbb{N}; west, wfrac: \mathbb{N}$

$0 \leq north \wedge north \leq 180 \wedge 0 \leq west \wedge west \leq 360$
$0 \leq nfrac \wedge nfrac \leq Limit \wedge 0 \leq wfrac \wedge wfrac \leq Limit$

$[SimKey, SimId, Frequency]$
$UEState ::= ACTIVE \mid IDLE \mid DETACHED$
$FStatus ::= NULL \mid ALLOCATED$

UserEquipment

$ueid: EquipmentId; PhoneBook$
$messages: Messages; calls: CallsRecord$
$position: Position; uestate: UEState$
$simid: SimId; simkey: SimKey$
$cfstatus, vfstatus: FStatus; cfreq, vfreq: Frequency$

$cfstatus = ALLOCATED \Leftrightarrow uestate \in \{IDLE, ACTIVE\}$
$cfstatus = NULL \Leftrightarrow uestate = DETACHED$

An evolved Node B is the radio access part of LTE which contains at least one radio transmitter, receiver and control section. The receiver contain resource management and logic control functions. The evolved Node B is denoted by the schema eNB given below. The schema consists of seven components. The first one node is identifier denoted by NodeId. The next three components are frequencies, namely, total set of frequencies, control and voice frequencies. The next component is a set of equipment connected to the evolved Node B. Finally, two functions for control frequencies and voice frequencies are described from a set of frequency to equipment identifier. The node identifier is of type Node.

$[Node]$

eNB

$enb: Node; uequipments: \mathbb{F}\ UserEquipment$
$frequencies; cfrequencies ;vfrequencies: \mathbb{F}\ Frequency$
$calloc; valloc: Frequency \rightarrow EquipmentId$

$cfrequencies \neq \emptyset \wedge vfrequencies \neq \emptyset$
$cfrequencies \cap vfrequencies = \emptyset$
$frequencies = cfrequencies \cup vfrequencies$
$\forall ue: UserEquipment \mid uequipments \neq \emptyset$
$\quad \bullet\ ue \in uequipments \Rightarrow ue\ .\ cfreq \in cfrequencies \wedge ue\ .\ vfreq \in vfrequencies$
$\forall f1, f2: Frequency; ei: EquipmentId \mid (f1, ei) \in calloc \wedge (f2, ei) \in calloc \quad \bullet\ f1 = f2$
$\forall f1, f2: Frequency; ei: EquipmentId \mid (f1, ei) \in valloc \wedge (f2, ei) \in valloc \quad \bullet\ f1 = f2$
$\forall fr: Frequency \mid fr \in \text{dom}\ calloc \bullet fr \in cfrequencies$
$\forall fr: Frequency \mid fr \in \text{dom}\ valloc \bullet fr \in vfrequencies$
$\forall ue1: EquipmentId \mid ue1 \in \text{ran}\ calloc$
$\quad \bullet\ \exists ue2: UserEquipment \mid ue2 \in uequipments \bullet ue1 = ue2\ .\ ueid$
$\forall ue1: EquipmentId \mid ue1 \in \text{ran}\ calloc$
$\quad \bullet\ \exists ue2: UserEquipment \mid ue2 \in uequipments \bullet ue1 = ue2\ .\ ueid$

Invariants: 1) Control and voice frequencies are nonempty; 2) Intersection of control and voice frequencies is an empty set; 3) The union of control and voice frequencies is equal to set of total frequencies; 4) The control and voice frequencies of each equipment connected to evolved node are in the set of total frequencies in the node; 5) The control frequency is a one to one function; 6) The voice frequency is a one to one function that is a voice frequency can be assigned to only one equipment; 7) The control frequency in the domain of frequency allocation function to each user equipment, is in the set of control frequencies of evolved node; 8) The voice frequency in the domain of frequency allocation function to each user equipment, is in the set of voice frequencies of evolved node; 9) The range of control frequency is subset of set of equipments of evolved node; 10) The range of voice frequency function is sub-set of set of equipments of evolved node.

The HSS manages subscription related information and supports the network control layer. It consists of various components some of the important information is given below in schema HSS. The first one component is its identifier. The second one is access having value granted or not denied. The next three components are for storing record about sim identifiers, sim keys and equipment identifiers. The relationship between sim identifiers with keys and equipment are also defined. Home location and visitor location registers are represented by hlrs and vlrs respectively which are power set of location register (LR). Three types of networks, GERAN, UTRAN and 2G/3G, are considered in the schema. The sim is described by the schema sim which consists of sim key, sim's owner, balance to call, render and other information. The last one component is added because this schema can be extended in our future work to have more information needed for the definition of sim.

$[Render]$

Sim

$simkey: SimKey; owner: \text{seq}\ String$
$balance: \mathbb{N}; render: Render; otherinfo: \text{seq}\ String$

$[LR, HId]; Access ::= GRANTED \mid DENIED$

$RAN ::= GERAN \mid UTRAN \mid G23\ ;\ BalLimit: \mathbb{N}$

HSS

$hid: Hid\ ; access: Access; sims: SimId \rightarrow Sim$
$srecord: SimId \rightarrow EquipmentId; simkeys: \mathbb{F}\ SimKey$
$hlrs, vlrs: \mathbb{F}\ LR\ ; networktype: RAN$

$\forall sid: SimId; sim: Sim \mid sid \in \text{dom}\ sims \wedge sim \in \text{ran}\ sims \wedge (sid, sim) \in sims$
$\quad \bullet\ sim\ .\ balance \geq BalLimit \Rightarrow access = GRANTED$
$\forall sid: SimId; sim: Sim \mid sid \in \text{dom}\ sims \wedge sim \in \text{ran}\ sims \wedge (sid, sim) \in sims \quad \bullet\ sim\ .\ balance < BallLimit \Rightarrow access = DENIED$
$\forall sid: SimId \mid sid \in \text{dom}\ sims \bullet sid \in \text{dom}\ srecord$
$\forall sid: SimId \mid sid \in \text{dom}\ srecord \bullet sid \in \text{dom}\ sims$
$\forall sid: SimId; sim: Sim \mid sid \in \text{dom}\ sims \wedge sim \in \text{ran}\ sims \wedge (sid, sim) \in sims \quad \bullet\ sim\ .\ simkey \in simkeys$
$\forall lr: LR \bullet lr \in hlrs \Rightarrow lr \notin vlrs \wedge lr \in vlrs \Rightarrow lr \notin hlrs$

Invariants: 1) Each sim identifier in the HSS is in the domain of sims function. If the sim balance is more than the minimum limit than permission for call is guaranteed; 2) If the sim balance is less than the minimum required limit then permission for call is not guaranteed; 3) Each sim in the domain of sim function is in the domain of sim-equipment function; 4) Each sim in the domain of sim-equipment function is in the domain of sims function; 5) Each sim key in the sim record function is in the HSS; 6) The intersection of sets, home location and visitors location registers, is empty.

The primary functionality of Serving Gateway (SGW) is to manage user mobility and act as a segregation point between the radio access network and the other core networks. The SGW maintains data movement between evolved nodes and the PDN Gateway. In a functional point of view, SGW is a termination point of the packet data network interface. Considering functionality of SGW, formal specification of SGW is given below by using the schema SGW. The schema consists of four components, namely, identifier, internet protocol, data and networks. The data is defined as a basic set type. The radio access network types are same as defined above in the definition of HSS schema. In the predicate part of the schema, it is stated that the data packet is sent towards a network based on the request of sender.

$[Data]; IP ::= IP1 \mid IP2 \mid IP3$

SGW

$sgw: \text{seq}\ \mathbb{N};\ iprotocol: IP;\ data: Data;\ network: RAN$

$iprotocol = IP1 \Rightarrow network = GERAN$
$iprotocol = IP2 \Rightarrow network = UTRAN$
$iprotocol = IP3 \Rightarrow network = G23$

The Mobility Management Entity (MME) is the key control node for the Long Term Evolution network. The primary functionality of MME is monitoring tracking and paging procedure for idle mode user equipment. It is involved in activation and deactivation processes and is responsible for choosing the SGW for user equipment at the initial process. It is responsible for security issues including authentication of the user by interacting with the HSS. It verifies authorization of user equipment to site on the service provider and enforces the equipment roaming restrictions. Furthermore, the MME provides the control function for mobility between Long Term Evolution and other access networks. Formal specification of Mobility Management Entity is given below using the schema MME. The schema consists of five components, namely, evolved nodes, home subscriber servers, serving gateways, verification and request. The formal definitions of the components are already given in terms of schemas. In MME schema, evolved nodes, home subscriber servers and serving gateways are assumed as power sets. Similar to above, all sets are assumed as finite collections of the components. Request has two values either success or failure. Similarly verification is either done or rejected.

$Request ::= SUCCESS \mid FAILURE$
$Verification ::= VERIFIED \mid REJECTED$

MME

$enbs: \mathbb{F}\ eNB;\ hsss: \mathbb{F}\ HSS;\ sgws: \mathbb{F}\ SGW$
$verification: Verification;\ request: Request$

$\forall enb: eNB \mid enb \in enbs$
$\bullet\ \forall ue: UserEquipment \mid ue \in enb\ .\ uequipments$
$\bullet\ \exists hss: HSS \mid hss \in hsss$
$\bullet\ \exists sid: SimId;\ uei: EquipmentId \mid (sid, uei) \in hss\ .\ srecord$
$\bullet\ ue\ .\ ueid = uei \wedge ue\ .\ simid = sid$
$\forall enb: eNB \mid enb \in enbs$
$\bullet\ \forall ue: UserEquipment \mid ue \in enb\ .\ uequipments$
$\bullet\ \exists hss: HSS \mid hss \in hsss$
$\bullet\ \exists sk: SimKey \mid sk \in hss\ .\ simkeys \bullet\ ue\ .\ simkey = sk$
$\forall enb: eNB \mid enb \in enbs$
$\bullet\ \forall ue: UserEquipment \mid ue \in enb\ .\ uequipments$
$\bullet\ verification = VERIFIED \Rightarrow request = SUCCESS$
$\forall enb: eNB \mid enb \in enbs$
$\bullet\ \forall ue: UserEquipment \mid ue \in enb\ .\ uequipments$
$\bullet\ verification = REJECTED \Rightarrow request = FAILURE$

Invariants: 1) For each evolved node and for each user equipment in the node there exists home subscriber server which contains the sim identifier of the equipment; 2) For an evolved node and for each equipment in the node there exists home subscriber server containing sim identifier of the equipment; 3) For each user equipment, if identity is verified, it is successful; 4) For an equipment, if the identity is not verified then it is a failure. Finally, formal specification of LTE is given below using LTE schema which consists of five components, namely, set of evolved nodes, set of home subscriber servers, set of serving gateways, set of mobile management entities and network. The components definitions are given in first part and invariants are defined in the second part of the schema.

LTE

$eNBS: \mathbb{F}\ eNB;\ hSSS: \mathbb{F}\ HSS$
$sGWS: \mathbb{F}\ SGW;\ mMES: \mathbb{F}\ MME;\ networks: RAN$

$\forall mm: MME \mid mm \in mMES \bullet \forall enb: eNB \mid enb \in mm\ .\ enbs$
$\bullet\ enb \in eNBS$
$\forall mm: MME \mid mm \in mMES \bullet \forall hs: HSS \mid hs \in mm\ .\ hsss$
$\bullet\ hs \in hSSS$
$\forall mm: MME \mid mm \in mMES \bullet \forall sg: SGW \mid sg \in mm\ .\ sgws$
$\bullet\ sg \in sGWS$

Invariants: 1) For each mobile management entity and for every evolved node, the node is in the LTE system; 2) For each mobile management entity and for every home subscriber server, the server is in the LTE system; 3) For each mobile management entity and for every serving gateway, the gateway is in the LTE system.

MODEL ANALYSIS

Formal analysis of the specification is provided here in this section. Since there does not exist any computer tool which may assure about complete correctness of a formal model, therefore, even the specification is well-written, it may contain potential errors. Hence, an art of writing specification does not provide guarantee about correctness of a system. However, if the specification is analyzed and validated with the rigorous computer tools, it certainly improves confidence by identifying errors if exists. There exist various tools for analyzing Z specification, Z/Eves is one of the powerful tools used in this research for analyzing the formal specification of long term evolution communications system. A snapshot of the part of the specification analysis is provided in Figure 2. First column on the left of the figure shows status of the syntax checking and the second column represents the proof correctness of the specification. The symbol "Y" stands that specification is correct syntactically and proof is also correct. The symbol "N" is used that errors exist which needs to be fixed by further analysis. Formal specification of our systems is fully analyzed to be correct and hence "N" does not appear in the figure. All the schemas are checked to be correct in syntax and has a correct proof.

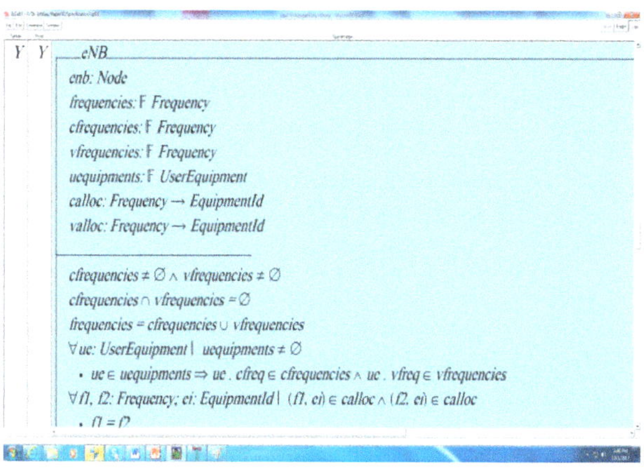

Figure 2: Model analysis using Z/Eves tool.

Table 1: Results of model analysis.

Schema Name	Syntax Type Check	Domain Check	Reduction	Proof
Person	Y	Y	NA	Y
PhoneBook	Y	Y	NA	Y
Message	Y	Y	NA	Y
Messages	Y	Y	NA	Y
Call	Y	Y	NA	Y
CallsRecord	Y	Y	NA	Y
Position	Y	Y	NA	Y
UserEquipment	Y	Y	NA	Y
eNB	Y	Y	NA	Y
Sim	Y	Y	NA	Y
HSS	Y	Y	NA	Y
SGW	Y	Y	NA	Y
MME	Y	Y	NA	Y
LTE	Y	Y	NA	Y

Summary of the results is presented in Table 1 given below. Name of the schema is given in first column of the table. The symbol "Y" in second column indicates that all schemas are proved automatically. Similarly, domain checking, reduction and proof by reduction are represented in columns 3, 4 & 5 respectively. The "NA" in column four is used that reduction was not required on the predicates and, hence, the formal specification is wellwritten and meaningful.

CONCLUSIONS AND FUTURE WORK

Modeling of complex systems has become a challenging task due to specifications and integration of their components. The Long Term Evolution (LTE) is a complex mobile communications system providing real time multimedia applications. Literature review reveals that some research work on addressing various issues of this system has been considered, however, LTE needs much more research contributions for its modeling, specifications, optimization and maintenance due to its complexity and distributed nature [17,18]. The telecom industry with the support of academia is discovering networks modeling and integration techniques for the systems and their implementation. Since LTE will integrate many heterogeneous networks and services, the seamless implementation for a complete development of LTE communications system is an open research problem [19-22]. In this paper, formal methods [23] in terms of Z notation are applied for modeling, specification and analysis of the LTE system. It is observed that the use of Z notation was effecttive to model and verify properties of the system because of its abstraction and expressive power. Further, the Z notation has a rigorous computer tool support addressing complexity of the system. At first, formal specification of services including messages and call record was provided. Then description of LTE components was presented. In the next section, formal description of the LTE system was described after composition of the schemas describing services and LTE components. Formal specification is analyzed using Z/Eves toolset. Based on our experience, it is observed that Z notation was useful at requirements analysis for further modeling and development of the system. The Z/Eves toolset facilitated and boosted our confidence for a complete and consistent description of the LTE model.

The preliminary results of this research were presented in [24] where LTE system was described at higher level of abstraction. In this paper, a detailed model is presented for addressing various issues including services, components and interaction protocols at software level of the LTE system. The other challenges, for example, quality of service, security issues and optimization techniques in this heterogeneous environment will be addressed in our future work.

REFERENCES

1. D. Singhal, M. Kunapareddy, V. Chetlapalli, V. B. James and N. Akhtar, "LTE-Advanced: Handover Interruption Time Analysis for IMT-A Evaluation," Proceedings of International Conference on Signal Processing, Communication, Computing and Networking Technologies, Thuckafay, 21-22 July 2011, pp. 81-85. doi:10.1109/ICSCCN.2011.6024519
2. Z. Bai, C. Spiegel, G. H. Bruck, P. Jung, M. Horvat, J. Berkmann, C. Drewes and B. Gunzelmann, "System Performance of UTRA LTE and LTE-Advanced," International Conference on Communications, Computing and Control Applications (CCCA), Hammamet, 3-5 March 2011, pp. 1-5.
3. Y. Wang, K. I. Pedersen, P. E. Mogensen and T. B. Sørensen, "Resource Allocation Considerations for Multi-Carrier LTE-Advanced Systems Operating in Backward Compatible Mode," IEEE 20th International Symposium on Personal, Indoor, and Mobile Radio Communications, Tokyo, 13-16 September 2009, pp. 370-374.
4. N. A. Zafar, U. Tajammul and N. Sabir, "Formal Specification of Call Setup Procedure of UMTS Communications System," Proceedings of International

Conference on Communication Software and Networks (ICCSN09), Macau, 27-28 February 2009.

5. N. A. Zafar and U. Tajammul, "Handover Based Formal Dynamic Model of UMTS Communication System," Proceedings of 4th International Computer Engineering Conference, Cairo, 29-30 December 2008, pp. 1-6.
6. S. Tariq and N. A. Zafar, "Modeling of Wireless Mobile Communication System Using Z Notation," Proceedings of the 1st International Conference on Computer, Control and Communication (IC4), Karachi, 12-13 November 2007.
7. S. Ahmed, A. K. Ramani and N. A. Zafar, "Formal Analysis of Mobility Management for Ad Hoc Networks," Proceedings of the International Conference on Wireless Networks (ICWN'11), Las Vegas, 18-21 July 2011, pp. 173-179.
8. R. Milner, "Communicating and Mobile Systems: The π- Calculus," Cambridge University Press, Cambridge, 1999.
9. S. Acharya, C. George and H. Mohanty, "Domain Consistency in Requirements Specification," Proceedings of the 5th International Conference on Quality Software, Melbourne, 19-20 September 2005, pp. 231-238. doi:10.1109/QSIC.2005.24
10. K. Taguchi and J. S. Dong, "An Overview of Mobile Object-Z," Springer-Verlag, New York, 2002, pp. 144-155.
11. M. Bettaz and M. Maouche, "Towards Mobile Z Schemas," International Journal of Computer Science and Application, Vol. 11, No. 11, 2005, pp. 101-117.
12. R. Duke, P. King, G. Rose and G. Smith, "The Object-Z Specification Language," University of Queensland, Brisbane, 1991.
13. M. Bettaz and M. Maouche, "Mobile Z Notation," Research Report, Philadelphia University, Philadelphia, 2002.
14. 3GPP TR 25.913, "Requirements for Evolved UTRA and Evolved UTRAN," V7.3.0 2006.
15. 3GPP TS 23.401, "GPRS Enhancements on EUTRAN Access," 2007. http://www.3gpp.org
16. Alcatel Lucent, "The LTE Network Architecture—A Comprehensive Tutorial," 2009. www.alcatel-lucent.com
17. M. M. A. Airchinnigh and M. Butler, "Service Specification Using Z," Technical Report, University of Southampton, Southampton, 1993. http://eprints.soton.ac.uk/253693
18. R. Cam and S. Vuong, "A Formal Specification in LOTOS, of a Simplified Cellular Mobile Communication System," Formal Description Techniques-II (FORTE'89), NorthHolland Publishing Co., Amsterdam, 1990, pp. 485-499.
19. L. Chen, et al., "System Level Simulation Methodology and Platform for Mobile Cellular Systems," IEEE Communication Magazine, Vol. 49, No. 7, pp. 148-155. 2011.
20. K. Doppler, "Device-To-Device Communication as an Underlay to LTE Advanced Networks," IEEE Communication Magazine, Vol. 47, No. 12, pp. 42-49.
21. Y. Sun, "Editorial Advanced Signal Processing for Wireless and Mobile Communications," IET Signal Processing, Vol. 3, No. 6, 2009, pp. 431-432. 109.
22. J. M. Wing, "A Specifier's Introduction to Formal Methods," IEEE Computer, Vol. 23, No. 9, 1990, pp. 8-24.
23. H. Luo, et al., "Quality Driven Cross Layer Optimized Video Delivery over LTE," IEEE Communication Magazine, Vol. 48, No. 2, 2010, pp. 102-
24. I. A. Choudhry, N. A. Zafar and M. Zahrani, "Validating Statistics of Long Term Evolution Mobile Communications System," Proceedings of the 2012 International Conference on Modeling, Simulation and Visualization Methods, Las Vegas, 16-19 July 2012, pp. 201-206.

CHAPTER 2

SPATIO-TEMPORAL CHANNEL CHARACTERIZATION: THEORETICAL FRAMEWORK AND APPLICATIONS IN MIMO SYSTEM DESIGN

Viet-Ha Pham, Jean-Yves Chouinard, Dominic Grenier, Huu-Tue Huynh, Xianbin Wang, and Yiyan Wu

Keywords: Spatio-temporal characterization, MIMO system design, Angle of arrival estimation, Multidimensional Fourier transform, MIMO channel characterization and modeling

INTRODUCTION

Fast growing demands of multimedia applications over mobile wireless transmission channels have raised the requirements of high transmission throughput at minimal transmission delays through fast time-varying wireless channels. The fourth generation (4G) communications technologies are designed in order to response to these requirements. These communications technologies, including LTE-Advanced [1] and mobile WiMAX [2], are expected to provide the peak downlink and uplink transmission rates of 1 Gbps and 500 Mbps, respectively, and to maintain connections in high mobility scenarios, e.g., at 100 km/h on highways and at 350 km/h on bullet trains.

The multiple-input and multiple-output (MIMO) is the enabling technology for the next 4G broadband wireless communications systems. The key idea of the MIMO technology is to maximize the number of independent parallel sub-channels in multi-antenna communication systems by optimally decomposing the spatial structure of transmission channels, i.e., by space–time channel coding [3–7]. As a result, the channel capacity linearly increases, without consuming additional power and bandwidth, as the number of antenna elements increases [8, 9]. This also puts channel characterization and modeling in a new context, in which rich multipath environments, e.g., Rayleigh fading channels, are preferable to communication systems. In order to facilitate the MIMO research, the first MIMO channel models were introduced as direct extensions of single antenna, or single-input and singleoutput, channels: the sub-channels were assumed to be uncorrelated and identical–independent Rayleigh fading [10–12]. While these MIMO channel models are simple, they do not reflect the interdependent nature between the sub-channels of a MIMO channel. In an attempt to improve this drawback, the sub-channel fading correlation was modeled and its effects on channel capacity were studied [13–17].

In order to build more realistic MIMO channel models, single-directional MIMO channel models were presented. In refs. [18–20], each multipath component was represented in angle of arrival (φ), delay (τ) and observation time (t), or Doppler shift (ν). Two channel impulse responses, including the angle-dependent time-varying channel impulse response (h(φ, t,τ)) and the multipath-Doppler spreading function (H(φ, ν, τ)) were introduced. In another approach, Durgin [21] proposed to use channel impulse responses, represented in time, frequency, and space ($\tilde{h}(t, f, r)$)and in Doppler shift, delay, and wave number ($\tilde{H}(\nu, \tau, k)$) to model MIMO channels. The receiver displacement (r) can be replaced by location vector (r) and the wave number (k) can be replaced by wave vector (k), in order to extend the results to three-dimensional (3D) space and 3D wave vector domains. The angle of arrival can be extracted from the received wave vector spectrum.

While the single-directional MIMO channel models characterize only the angle of arrival at the receiver, double-directional MIMO channel models include directional information at both the receiver and transmitter in order to facilitate the beamforming processing for both signal transmission and reception. In refs. [22–25], double direction implies angle of departure (AoD) and angle of arrival (AoA) while in ref. [26], it implies transmit spatial angle

Citation: Viet-Ha Pham, Jean-Yves Chouinard, Dominic Grenier, Huu-Tue Huynh, Xianbin Wang, Yiyan Wu, Spatio-temporal channel characterization: theoretical framework and applications in MIMO system design, DOI: 10.1007/s12243-011-0248-9.

and receive spatial angle. The spatial angle is the antenna element separation normalized by the signal wavelength and projected on the propagation direction. Double-directional channel representation includes time-varying double-directional channel impulse response h (t,τ,ϕ_R, ϕ_T) [22, 23], spatial spreading function $G(\theta_R, \theta_T)$ [26], and kernel function $h(k_o, k_i)$ and system function $h(k, k_d)$ [24, 25]. ϕ_T and ϕ_R are angle of departure and angle of arrival. θ_T and θ_R are transmit spatial angle and receive spatial angle. k_o and k_i are receive and transmit wave vector, while k and k_d are receive wave vector and Doppler wave vector shift. Dual MIMO channel representations in the space domain are related to double-directional MIMO channel representations through the space–wave vector duality and the 3D Fourier transform [24, 25].

The existing MIMO channel characterization and modeling techniques mentioned above have two major limitations. Firstly, channels and systems are represented in either the time domain or the space domain for simplicity and the inclusion of the other domain is assumed to be straight forward. However, there exists a relationship, and therefore an interdependence, between frequency and wave vector variables and between time and space variables. These relationships need to be taken into account in channel description and in multidimensional Fourier transform calculation. Secondly, the 3D Fourier transform is only valid for uniform plane waves. Propagating waves around the receiver can be approximated as uniform plane waves if the receiver is in the far field of the transmitter and scatterers, if any. Unfortunately, this approximation does not apply to the transmitter side. The propagating waves around the transmitter are spherical and the signal attenuation is not uniform.

In this paper, a novel MIMO channel characterization and modeling technique is proposed to overcome the two major issues of the existing characterization and modeling techniques. Firstly, a theoretical framework for MIMO channel characterization and modeling in the space, time, wave vector, and frequency domains is proposed. The space–time and wave vector–frequency interdependences are included in the framework description and the 4D Fourier transform calculation. Secondly, a practical framework, which is a realization of the theoretical framework in realistic scenarios, is proposed for MIMO channel characterization and modeling at both the transmitter and the receiver. The applications of the practical framework in designing transmit and receive MIMO antenna arrays and estimating angles of arrival are described. In addition, the presented work also finds applications in improving the 3D propagation channel simulation and modeling techniques, such as the one proposed in ref. [27].

The remainder of this paper is organized as follows. Section 2 recalls the fundamental concepts required for the development of the proposed framework. Section 3 discusses the wave vector–frequency spectrum estimation technique while taking into consideration the wave vector–frequency and space–time relationships. Section 4 introduces the space–time and wave vector– frequency compound impulses by applying the multidimensional Dirac impulse concept in the 4D space–time and 4D wave vector–frequency domains. Section 5 recalls the kernel function definitions. Sections 6 and 7 present the proposed theoretical and practical frameworks. Sections 8 and 9 discuss the framework applications in MIMO antenna array design and AoA estimation. Section 10 concludes the paper.

BACKGROUND

Wave vector

Wave vector is a vector representation of radio waves. The direction of a wave vector indicates the wave propagation direction and the magnitude is the angular wave number. The magnitude of the wave vector K is:

$$|\mathbf{k}| = \frac{2\pi}{\lambda} = \frac{\omega}{v_p}, \tag{1}$$

where λ is the wavelength, ω is the angular frequency, and v_p is the propagation velocity.

Let k_x, k_y, and k_z be the three components of vector k, then:

$$k_x = \frac{2\pi}{\lambda_x} = \frac{\omega}{v_{px}}$$

$$k_y = \frac{2\pi}{\lambda_y} = \frac{\omega}{v_{py}}$$

$$k_z = \frac{2\pi}{\lambda_z} = \frac{\omega}{v_{pz}}, \tag{2}$$

where λ_x, λ_y, and λ_z are the wavelengths and v_{px}, v_{py}, and v_{pz} are the phase velocities along the$x$-, y-, and z-axes, respectively.[1]

Plane Waves

A wave is said to be planar if its wave fronts are plane in the space domain. If the wave amplitudes on a wave plane are constants, then the wave is a uniform plane wave. An example of a uniform plane wave is the electromagnetic wave radiated from an infinite current sheet. Plane waves do not exist in practice, but they provide a simple framework for the study of electromagnetic wave radiation and propagation. Furthermore, plane waves can be used as an approximation of propagation waves in real physical scenarios.

3D Fourier Transform

A location in the three-dimensional (3D) space is represented by a vector, called space vector, or location/position vector (r). In the Cartesian coordinates system, space vector r has three components, denoted as r_x,r_y, and r_z. A duality exists between time and frequency [28]. For a uniform plane wave, the duality exists also between space vector and wave vector [24] through the 3D Fourier transform. The 3D Fourier transform converts a signal in the spatial domain into the wave vector domain. Alternatively, the 3D inverse Fourier transform converts a signal in the wave vector domain into the spatial domain. The continuous 3D Fourier and inverse Fourier transforms are:

$$S(\mathbf{k}) = \int s(\mathbf{r}) e^{j\mathbf{r}\cdot\mathbf{k}} d\mathbf{r} \tag{3}$$

$$s(\mathbf{r}) = \frac{1}{(2\pi)^3} \int S(\mathbf{k}) e^{-j\mathbf{k}\cdot\mathbf{r}} d\mathbf{k}, \tag{4}$$

where dk represents $dk_x dk_y dk_z$ and dr represents $dr_x dr_y dr_z$.

Wave Vector–Frequency and Space–Time Relationships

Wave vector and frequency have a direct relationship: $|\mathbf{k}| = \frac{2\pi}{\lambda} = \frac{\omega}{v_p} = \frac{2\pi f}{v_p}$. In Eq. 4, the signal wave vector spectrum component at wave vector k transmitted from the space origin is S (k). As the signal reaches location r, the signal component at wave vector k is S (k) $e^{-jk \cdot r}$. One observes that no attenuation affects the signal and that no changes occurred to wave vector k in the calculation of the three-dimensional Fourier transform as a lossless and scatter-free environment is assumed. In this environment, [2] the time variable (t) is the ratio between the magnitude of location vector |r| and the propagation velocity (v_p): $t = \frac{|\mathbf{r}|}{v_p}$.

Wave vector–frequency spectrum estimation

Four-dimensional compound Fourier transform

The wave vector–frequency spectrum is the compound transform of a 3D Fourier transform with respect to the space variable, $F_r(\cdot)$, and a Fourier transform with respect to the time variable, $F_t(\cdot)$:

$$S(f;\mathbf{k}) = \int\int s(t;\mathbf{r})e^{-j2\pi ft}e^{j\mathbf{k}\cdot\mathbf{r}}\mathrm{d}t\mathrm{d}\mathbf{r} \tag{5}$$

$$s(t;\mathbf{r}) = \frac{1}{(2\pi)^3}\int\int S(f;\mathbf{k})e^{j2\pi ft}e^{-j\mathbf{k}\cdot\mathbf{r}}\mathrm{d}f\mathrm{d}\mathbf{k}. \tag{6}$$

Frequency and wave vectors are related to each other through the wave number. Also, the location vector and the time are related to each other through the propagation velocity, as discussed in Section 2.4. This means that the integrals in Eqs. 5 and 6 are not performed all over the 4D *hyperspace*, i.e., 3D space and time in Eq. 5 and 3D wave vector and frequency in Eq. 6, but rather on a 4D *hypersurface*. Equations 5 and 6 can be rewritten as:

$$S(f;\mathbf{k}) = \int_{H_{s\_t}} s(t;\mathbf{r})e^{-j2\pi ft}e^{j\mathbf{k}\cdot\mathbf{r}}\mathrm{d}h_{s\_t} \tag{7}$$

$$s(t;\mathbf{r}) = \frac{1}{(2\pi)^3}\int_{H_{w\_f}} S(f;\mathbf{k})e^{j2\pi ft}e^{-j\mathbf{k}\cdot\mathbf{r}}\mathrm{d}h_{w\_f}. \tag{8}$$

where H_{s_t} is the integration hypersurface, or integration domain, defined by the space–time relationship, $t = \frac{|\mathbf{r}|}{v_p}$, and H_{w_f} is the integration hypersurface defined by the wave vector–frequency relationship, $|\mathbf{k}| = \frac{2\pi f}{v_p}$. Hypersurfaces H_{w_f} and H_{s_t} are dependent on the transmission media characteristics, i.e., propagation velocity v_p. From Eqs. 7 and 8, we have Eqs. 9 and 10.

Equations 9 and 10 are convenient in the mathematical sense as they describe explicitly the space–time and wave vector–frequency relationships in a general form. If one needs to modify the space–time and/or wave vector–frequency relationships, Eqs. 9 and 10 still remain unchanged.

Another approach to the compound Fourier transform calculation is to represent the time variable from the space vector, i.e. $t = \frac{|\mathbf{r}|}{v_p}$, and the frequency variable from the wave vector, i.e., $f = \frac{|\mathbf{k}|v_p}{2\pi}$. Assuming that $\frac{1}{v_p^2} \ll 1$ and $\frac{v_p^2}{4\pi^2} \gg 1$, Eqs. 7 and 8 become Eqs. 11 and 12.

$$S(f;\mathbf{k}) = \int s(t;\mathbf{r}(t))e^{-j2\pi ft}e^{j\mathbf{k}\cdot\mathbf{r}(t)} \times\left[1+\left(\frac{\mathrm{d}r_x}{\mathrm{d}t}\right)^2+\left(\frac{\mathrm{d}r_y}{\mathrm{d}t}\right)^2+\left(\frac{\mathrm{d}r_z}{\mathrm{d}t}\right)^2\right]^{1/2}dt \tag{9}$$

$$s(t;\mathbf{r}) = \frac{1}{(2\pi)^3}\int S(f;\mathbf{k})e^{j2\pi ft}e^{-j\mathbf{k}\cdot\mathbf{r}} \times\left[1+\left(\frac{\mathrm{d}k_x}{\mathrm{d}f}\right)^2+\left(\frac{\mathrm{d}k_y}{\mathrm{d}f}\right)^2+\left(\frac{\mathrm{d}k_z}{\mathrm{d}f}\right)^2\right]^{1/2}\mathrm{d}f \tag{10}$$

$$\begin{aligned} S(f;\mathbf{k})\Big|_{f=\frac{|\mathbf{k}|v_p}{2\pi}} &= \int s\left(t=\frac{|\mathbf{r}|}{v_p};\mathbf{r}\right)e^{-j2\pi f\frac{|\mathbf{r}|}{v_p}}e^{j\mathbf{k}\cdot\mathbf{r}} \\ &\quad\times\left[1+\left(\frac{\partial t}{\partial r_x}\right)^2+\left(\frac{\partial t}{\partial r_y}\right)^2+\left(\frac{\partial t}{\partial r_z}\right)^2\right]^{1/2}\mathrm{d}\mathbf{r} \\ &= \int s\left(t=\frac{|\mathbf{r}|}{v_p};\mathbf{r}\right)e^{-j2\pi f\frac{|\mathbf{r}|}{v_p}}e^{j\mathbf{k}\cdot\mathbf{r}} \\ &\quad\times\left[1+\frac{1}{v_p^2}\frac{r_x^2+r_y^2+r_z^2}{r_x^2+r_y^2+r_z^2}\right]^{1/2}\mathrm{d}\mathbf{r} \\ &= \int s\left(t=\frac{|\mathbf{r}|}{v_p};\mathbf{r}\right)e^{-j2\pi f\frac{|\mathbf{r}|}{v_p}}e^{j\mathbf{k}\cdot\mathbf{r}} \\ &\quad\times\left[1+\frac{1}{v_p^2}\right]^{1/2}\mathrm{d}\mathbf{r} \\ &\approx \int s\left(t=\frac{|\mathbf{r}|}{v_p};\mathbf{r}\right)e^{-j2\pi f\frac{|\mathbf{r}|}{v_p}}e^{j\mathbf{k}\cdot\mathbf{r}}\mathrm{d}\mathbf{r} \end{aligned} \tag{11}$$

$$\begin{aligned} s(t;\mathbf{r})\Big|_{t=\frac{|\mathbf{r}|}{v_p}} &= \frac{1}{(2\pi)^3}\int S\left(f=\frac{|\mathbf{k}|\,v_p}{2\pi};\mathbf{k}\right)e^{j|\mathbf{k}|v_pt}e^{-j\mathbf{k}\cdot\mathbf{r}} \\ &\quad\times\left[1+\left(\frac{\partial f}{\partial k_x}\right)^2+\left(\frac{\partial f}{\partial k_y}\right)^2+\left(\frac{\partial f}{\partial k_z}\right)^2\right]^{1/2}\mathrm{d}\mathbf{k} \\ &= \frac{1}{(2\pi)^3}\int S\left(f=\frac{|\mathbf{k}|\,v_p}{2\pi};\mathbf{k}\right)e^{j|\mathbf{k}|v_pt}e^{-j\mathbf{k}\cdot\mathbf{r}} \\ &\quad\times\left[1+\frac{v_p^2}{4\pi^2}\frac{k_x^2+k_y^2+k_z^2}{k_x^2+k_y^2+k_z^2}\right]^{1/2}\mathrm{d}\mathbf{k} \\ &= \frac{1}{(2\pi)^3}\int S\left(f=\frac{|\mathbf{k}|\,v_p}{2\pi};\mathbf{k}\right)e^{j|\mathbf{k}|v_pt}e^{-j\mathbf{k}\cdot\mathbf{r}} \\ &\quad\times\left[1+\frac{v_p^2}{4\pi^2}\right]^{1/2}\mathrm{d}\mathbf{k} \\ &\approx \frac{v_p}{(2\pi)^4}\int S\left(f=\frac{|\mathbf{k}|\,v_p}{2\pi};\mathbf{k}\right)e^{j|\mathbf{k}|v_pt}e^{-j\mathbf{k}\cdot\mathbf{r}}\mathrm{d}\mathbf{k} \end{aligned} \tag{12}$$

SPACE, WAVE VECTOR, AND COMPOUND IMPULSES

Impulse function definitions

In order to characterize the channel and system in any N-dimensional space, we need to use multidimensional Dirac impulses [29, pp. 23–24], [30, pp. 26–29]. Let x be an N-dimensional vector defined in $\mathbb{R}^N$. We generalize an N-dimensional Dirac impulse as follows:

$$\underbrace{\int\cdots\int}_{N} f(\mathbf{x})\delta(\mathbf{x}-\mathbf{x}_0)\mathrm{d}x_1\mathrm{d}x_2\cdots\mathrm{d}x_N = f(\mathbf{x}_0), \tag{13}$$

where f(x) is a function defined over $\mathbb{R}^N$, and x_1, x_2, ..., x_N are the components of vector x .

In the following, x might be a combination of vectors and variables r, k, f and t. δ(r), δ(k), δ(r, t), δ(r, f), δ(k, t), and δ(k, f) are the multidimensional Dirac impulses in space, wave vector, space–time, space–frequency, wave vector–time, and wave vector– frequency, respectively.

Space vector–wave vector duality

Applying the 3D Fourier transform on the space impulse function, δ(r), gives $F_r\{\delta(r)\} = 1$. Therefore, the wave vector spectrum of the space impulse is equal to unity and is independent of the wave vector. Similarly, the inverse 3D Fourier transform of the wave vector impulse function at wave vector k_0 is a complex exponential with equal unity magnitude: $\mathcal{F}_{\mathbf{k}}^{-1}\{\delta(\mathbf{k}-\mathbf{k}_0)\} = e^{-j\mathbf{k}_0\cdot\mathbf{r}}$. This means that an impulse at frequency f_0 emitted in the direction of k0 will propagate all over the space with equal (unit) magnitude. This is due to the assumption of uniform plane wave propagation

A frequency impulse, δ(f − f0), is white in time domain and a time impulse, δ(t), is white in frequency domain. This duality can be transposed to space and wave vector domains through the 3D Fourier transform. The wave vector impulse is white in the tri-dimensional space domain and the spatial impulse is white in the 3D wave vector domain. Due to the existing relationships between wave vector and frequency, and between location vector and time (Eqs. 11 and 6), the wave vector impulse, δ(k), is white in space and also white in time; the space impulse, δ(r), is white in wave vector and also in frequency. Therefore, a four-dimensional wave vector–time impulse, δ(k, t), and a 4D space–frequency impulse, δ(r, f), do not have physical sense.

Space–time relationship

Here, we review the space–time relationship using the multidimensional impulse functions. The space impulse at location r_0, $\delta(r - r_0)$, is white in wave vector and in frequency. A signal that is white in frequency implies that it is a time impulse. We need to determine the time instant that the impulse occurred in order to find the space–time relationship. The wave vector spectrum of the space impulse at location r_0, $\delta(r - r_0)$, is:

$$\mathcal{F}_{\mathbf{r}}\{\delta(\mathbf{r}-\mathbf{r}_0)\} = e^{j\mathbf{r}_0\cdot\mathbf{k}} = e^{j|\mathbf{r}_0||\mathbf{k}|\cos\angle\mathbf{r}_0.\mathbf{k}}. \quad (14)$$

Representing the wave number, |k|, through frequency, the wave vector spectrum becomes:

$$\mathcal{F}_{\mathbf{r}}\{\delta(\mathbf{r}-\mathbf{r}_0)\} = e^{j\frac{2\pi f}{v_p}|\mathbf{r}_0|\cos\angle\mathbf{r}_0.\mathbf{k}} = e^{j2\pi f\cos\angle\mathbf{r}_0.\mathbf{k}\frac{|\mathbf{r}_0|}{v_p}}. \quad (15)$$

Examining Eq. 15 along a given propagation direction, i.e., angle $\angle r_0$, k is set, the spectrum in the frequency domain of a spatial impulse is $e^{j2\pi f' t0}$, with f' = f cos $\angle r_0$, k and $t_0 = \frac{|\mathbf{r}_0|}{v_p}$. This is the spectrum of a time impulse at time instant t = t0. This shows clearly the space–time relationship referred to in Section 2.4. Similarly, assuming space–time relationship $t = \frac{|\mathbf{r}|}{v_p}$, then the wave vector–frequency relationship $|\mathbf{k}| = \frac{2\pi f}{v_p}$ is derived. This is another form of duality in space–time and wave vector–frequency domains: the wave vector– frequency relationship assumes the space–time relationship and vice versa

Space–time impulse functions

The space–time impulse function, δ(r, t), exists at the spatial origin, r = 0, and at the time origin, t = 0. Due to the space–time relationship mentioned above, a space impulse is also a space–time compound impulse. Applying Eq. 11, the 4D spectrum of a space–time compound impulse at position r_0 and time t_0 is:

$$\begin{aligned}
&\mathcal{F}_{\mathbf{r},t}\{\delta(\mathbf{r}-\mathbf{r}_0, t-t_0)\}\Big|_{t=\frac{|\mathbf{r}|}{v_p}, t_0=\frac{|\mathbf{r}_0|}{v_p}} \\
&= \int \delta(\mathbf{r}-\mathbf{r}_0, t-t_0)e^{j\mathbf{k}\cdot\mathbf{r}}e^{-j2\pi ft}d\mathbf{r}\Big|_{|\mathbf{k}|=\frac{2\pi f}{v_p}} \\
&= e^{j\mathbf{k}\cdot\mathbf{r}_0}e^{-j2\pi ft_0}\Big|_{|\mathbf{k}|=\frac{2\pi f}{v_p}}
\end{aligned} \quad (16)$$

Wave vector–frequency impulse functions

The wave vector impulse function (δ(k)) occurs only when the wave vector equals to wave vector zero (k = 0). In practice, we usually refer to an impulse at wave vector k_0 rather than at wave vector zero, as for a frequency impulse at frequency f_0, δ(f − f0). Due to the wave vector–frequency relationship, a wave vector impulse $\delta(k - k_0)$ is also a wave vector–frequency compound impulse at wave vector k0 and frequency f0. In space and time, a wave vector–frequency impulse is a complex exponential at frequency f0 propagating in the direction of k_0. Applying Eq. 12, the space–time representation of a wave vector–frequency compound impulse is found:

$$\begin{aligned}
&\mathcal{F}_{\mathbf{k},f}^{-1}\{\delta(\mathbf{k}-\mathbf{k}_0, f-f_0)\}\Big|_{|\mathbf{k}|=\frac{2\pi f}{v_p}, |\mathbf{k}_0|=\frac{2\pi f_0}{v_p}} \\
&= \frac{v_p}{(2\pi)^4}\int \delta(\mathbf{k}-\mathbf{k}_0, f-f_0)e^{-j\mathbf{k}\cdot\mathbf{r}}e^{j2\pi ft}d\mathbf{k}\Big|_{t=\frac{|\mathbf{r}|}{v_p}} \\
&= \frac{v_p}{(2\pi)^4}e^{-j\mathbf{k}_0\cdot\mathbf{r}}e^{j2\pi f_0 t}\Big|_{t=\frac{|\mathbf{r}|}{v_p}}
\end{aligned} \quad (17)$$

SPATIAL KERNEL FUNCTIONS

Definition of the spatial kernel functions

Here, for simplicity, we consider only the space and wave vector variables in signal and system representation. Time and frequency variables will be included, when it is necessary, in order to derive the complete signal and system characterization. In a communications system, the transmitted and received signals can be represented in either space or wave vector domains. A communications system transforms the transmitted signals represented in either space or wave vector, into received signals, either in space or wave vector domains. Four operators are defined to form the kernel functions: $K_{ss}(r_o,r_i)$, $K_{ww}(k_o, k_i)$, $K_{sw}(r_o, k_i)$, and $K_{ws}(k_o,r_i)$. r_i and r_o are the positions of the transmitter and the receiver, respectively, while k_i and k_o are the transmit and the receive wave vectors, respectively. These kernels functions are derived directly from Bello's results [31] in time and frequency domains, and developed by Guguen and Zein [24, 25] for spatial channel impulse responses. The relationships between the transmitted and received signals using these four kernel functions are:

$$s_o(\mathbf{r}_o) = \int s_i(\mathbf{r}_i)K_{ss}(\mathbf{r}_o, \mathbf{r}_i)d\mathbf{r}_i \quad (18)$$

$$S_o(\mathbf{k}_o) = \int S_i(\mathbf{k}_i) K_{ww}(\mathbf{k}_o, \mathbf{k}_i) d\mathbf{k}_i \quad (19)$$

$$s_o(\mathbf{r}_o) = \int S_i(\mathbf{k}_i) K_{sw}(\mathbf{r}_o, \mathbf{k}_i) d\mathbf{k}_i \quad (20)$$

$$S_o(\mathbf{k}_o) = \int s_i(\mathbf{r}_i) K_{ws}(\mathbf{k}_o, \mathbf{r}_i) d\mathbf{r}_i, \quad (21)$$

where $d\mathbf{r}_i = dr_{ix}dr_{iy}dr_{iz}$ and $d\mathbf{k}_i = dk_{ix}dk_{iy}dk_{iz}$. $s(\mathbf{r}_i)$ and $S(k_i)$ are the transmitted signal in 3D spatial domain and transmitted signal spectrum in wave vector domain respectively, $s(r_o)$ and $S(k_o)$ are the received signal in the 3D spatial domain and the received signal spectrum in wave vector domain

Spatial kernel function relationships

Applying the Fourier transform on Eq. 20, the equality becomes:

$$S_o(\mathbf{k}_o) = \int s_i(\mathbf{r}_i) \mathcal{F}_{\mathbf{r}_o} \{K_{ss}(\mathbf{r}_o, \mathbf{r}_i)\} d\mathbf{r}_i, \quad (22)$$

where F_{ro} $\{K_{ss}(r_o,r_i)\}$ is the Fourier transform of function $K_{ss}(r_o,r_i)$, with respect to space variable ro. Comparing this equation and Eq. 21, one obtains the following relationships:

$$K_{ws}(\mathbf{k}_o, \mathbf{r}_i) = \mathcal{F}_{\mathbf{r}_o} \{K_{ss}(\mathbf{r}_o, \mathbf{r}_i)\} \quad (23)$$

$$K_{ss}(\mathbf{r}_o, \mathbf{r}_i) = \mathcal{F}_{\mathbf{k}_o}^{-1} \{K_{ws}(\mathbf{k}_o, \mathbf{r}_i)\}. \quad (24)$$

Similarly, applying the Fourier transform on equation Eq. 20 and comparing with Eq. 19, the relationship between kernel function $K_{ww}(\cdot)$ and kernel function $K_{sw}(\cdot)$ is obtained:

$$K_{ww}(\mathbf{k}_o, \mathbf{k}_i) = \mathcal{F}_{\mathbf{r}_o} \{K_{sw}(\mathbf{r}_o, \mathbf{k}_i)\} \quad (25)$$

$$K_{sw}(\mathbf{r}_o, \mathbf{k}_i) = \mathcal{F}_{\mathbf{k}_o}^{-1} \{K_{ww}(\mathbf{k}_o, \mathbf{k}_i)\}. \quad (26)$$

In order to determine the relationship between kernel functions $K_{ws}(k_o,r_i)$ and $K_{ww}(k_o, k_i)$, the input signal $S_i(k_i)$ in Eq. 19 is replaced by the Fourier transform of its dual function, i.e., $s_i(r_i)$:

$$\begin{aligned} S_o(\mathbf{k}_o) &= \int \left\{ \int s_i(\mathbf{r}_i) e^{j\mathbf{r}_i \cdot \mathbf{k}_i} d\mathbf{r}_i \right\} K_{ww}(\mathbf{k}_o, \mathbf{k}_i) d\mathbf{k}_i \\ &= \int s_i(\mathbf{r}_i) \left\{ \int K_{ww}(\mathbf{k}_o, \mathbf{k}_i) e^{j\mathbf{r}_i \cdot \mathbf{k}_i} d\mathbf{k}_i \right\} d\mathbf{r}_i. \end{aligned} \quad (27)$$

Comparing this equation and Eq. 21, leads to:

$$\begin{aligned} K_{ws}(\mathbf{k}_o, \mathbf{r}_i) &= \int K_{ww}(\mathbf{k}_o, \mathbf{k}_i) e^{j\mathbf{r}_i \cdot \mathbf{k}_i} d\mathbf{k}_i \\ &= (2\pi)^3 \mathcal{F}_{\mathbf{k}_i}^{-1} \{K_{ww}(\mathbf{k}_o, -\mathbf{k}_i)\} \end{aligned} \quad (28)$$

Or

$$K_{ww}(\mathbf{k}_o, -\mathbf{k}_i) = \left(\frac{1}{2\pi}\right)^3 \mathcal{F}_{\mathbf{r}_i} \{K_{ws}(\mathbf{k}_o, \mathbf{r}_i)\}. \quad (29)$$

Applying similar reasoning to Eqs. 20 and 18, the relationships between kernel functions $K_{ss}(r_o,r_i)$ and $K_{sw}(r_o, k_i)$ are obtained:

$$K_{sw}(\mathbf{r}_o, -\mathbf{k}_i) = \left(\frac{1}{2\pi}\right)^3 \mathcal{F}_{\mathbf{r}_i} \{K_{ss}(\mathbf{r}_o, \mathbf{r}_i)\}$$

(30)

$$K_{ss}(\mathbf{r}_o, \mathbf{r}_i) = (2\pi)^3 \mathcal{F}_{\mathbf{k}_i}^{-1} \{K_{sw}(\mathbf{r}_o, -\mathbf{k}_i)\}. \quad (31)$$

The above relationships are summarized in Fig. 1.

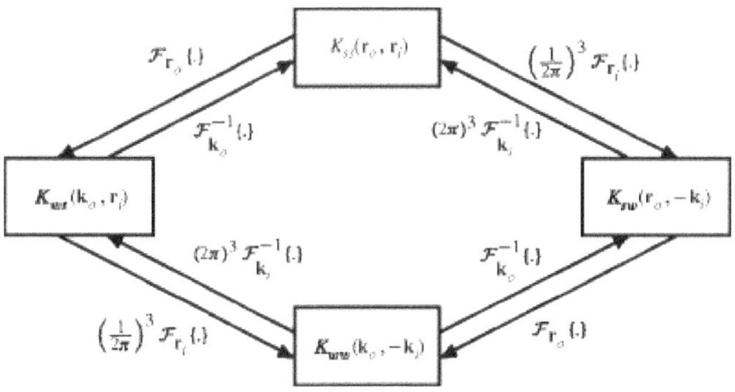

Figure 1: Relationships between the kernel functions

SPATIO-TEMPORAL CHANNEL CHARACTERIZATION—THEORETICAL FRAMEWORK

Space–Time Kernel Functions

In Section 5, we omitted the time and frequency variables from the signal and system representation. A complete system representation includes the variables in time, frequency, space and wave vector domains. Due to the time–frequency duality and the space–wave vector duality, the input and output signals of a communications system are represented as the functions of two variables, i.e., time or frequency variable, and space or wave vector variable. There are four combinations of these two variables: space–time, space–frequency, wave vector–time and wave vector-frequency. Following the same approach in Section 5, we defined a set of 16 kernel functions, as listed in Table 1. This set of kernel functions is complete in the sense that it characterizes systems and channels in all the related domains and the space–time and wave vector–frequency relationships are included implicitly. The relationships between the space–time kernel functions and their applications in practical contexts are discussed in subsequent paragraphs.

Table 1: Ensemble of Kernel functions in space–time–wave vector–frequency

No.	Kernel function	No.	Kernel function
1	$K_1(\mathbf{r_o}, \mathbf{r_i}; t_o, t_i)$	9	$K_9(\mathbf{r_o}, \mathbf{r_i}; f_o, f_i)$
2	$K_2(\mathbf{r_o}, \mathbf{k_i}; t_o, t_i)$	10	$K_{10}(\mathbf{r_o}, \mathbf{k_i}; f_o, f_i)$
3	$K_3(\mathbf{k_o}, \mathbf{k_i}; t_o, t_i)$	11	$K_{11}(\mathbf{k_o}, \mathbf{k_i}; f_o, f_i)$
4	$K_4(\mathbf{k_o}, \mathbf{r_i}; t_o, t_i)$	12	$K_{12}(\mathbf{k_o}, \mathbf{r_i}; f_o, f_i)$
5	$K_5(\mathbf{r_o}, \mathbf{r_i}; t_o, f_i)$	13	$K_{13}(\mathbf{r_o}, \mathbf{r_i}; f_o, t_i)$
6	$K_6(\mathbf{r_o}, \mathbf{k_i}; t_o, f_i)$	14	$K_{14}(\mathbf{r_o}, \mathbf{k_i}; f_o, t_i)$
7	$K_7(\mathbf{k_o}, \mathbf{k_i}; t_o, f_i)$	15	$K_{15}(\mathbf{k_o}, \mathbf{k_i}; f_o, t_i)$
8	$K_8(\mathbf{k_o}, \mathbf{r_i}; t_o, f_i)$	16	$K_{16}(\mathbf{k_o}, \mathbf{r_i}; f_o, t_i)$

Space–time kernel function relationships

Following the same reasoning as in Section 5, kernel functions in space–time–wave vector–frequency are also interconnected to each other by the multidimensional Fourier transform. These relationships are summarized in Fig. 2. Note that, in this figure, the Fourier transform on the input spatial vector (r_i)is multiplied by $\left(\frac{1}{2\pi}\right)^3$ while the inverse Fourier transform on the input wave vector $(\mathbf{k}_i)$ is multiplied by $(2\pi)^3$. In Fig. 2, we also need to assign a negative sign to either the input time (t_i) or the input location vector (r_i) or the input frequency (f_i) and the input wave vector (k_i). The 16 kernel functions are represented by the 16 vertices of two cubes: the vertex label numbers are the labels of the corresponding kernel functions defined in Table 1. The bold lines represent the 3D space–wave vector Fourier transform and inverse 3D Fourier transform pair while the thin lines represent the time–frequency Fourier transform and inverse Fourier transform pair. One observes that the 16 kernel functions are grouped into four groups of four functions: these are (K_1, K_2, K_3, K_4), (K_5,K_6, K_7, K_8), (K_9, K_{10}, K_{11}, K_{12}), and (K_{13}, K_{14}, K_{15}, K_{16}). The functions from the same group have a common time–frequency variable pair and lie on the same surface, which is parallel to the surface formed by four points (1, 2, 3, 4). They are interconnected by the 3D Fourier transform. These four groups are interconnected by the time–frequency Fourier transform pairs as shown by the thin lines at the edge of the cubes. Solid and dashed lines are used to show the external and the internal views of the kernel function relationship cube.

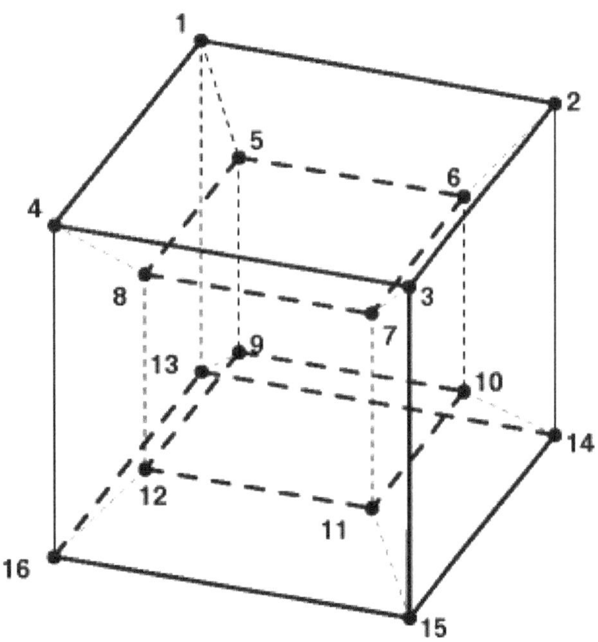

Figure 2: Complete kernel function relationship

SPATIO-TEMPORAL CHANNEL CHARACTERIZATION—PRACTICAL FRAMEWORK

The transposition of the spatio-temporal theoretical framework to practical wireless communication scenarios is not straightforward since propagating waves have some degree of curvature. In the far field [32] of the transmitter, the curvature becomes less significant and, within a bounded region, the propagating waves can be regarded approximatively as plane waves. In ref. [25], Guguen and Zein presented the utilization of the spatial kernel functions in characterizing MIMO channel for indoor WLAN applications. Unfortunately, the plane wave approximation is not always valid in indoor scenarios, since the receivers are not always in the far field of scatterers, e.g., walls and furniture. In this work, the plane wave approximation is made to find applications of the proposed framework to realistic scenarios at receiver side. However, this approximation cannot be used at transmitter side since the space region around the transmitter does not satisfy the far field criterion. The approach used for applying the framework at transmitter side will be discussed in Section 7.2.

Application of the Modeling Framework at the Receiver Side

When one considers uniform plane wave propagation, it is usually assumed that the receiver is located very far from the transmitter. In order to apply the proposed framework at the receiver in practical contexts, two constraints are introduced to approximate the propagating waves to uniform plane waves. Firstly, the receiver must be located in the far field regions of the transmitter and the scatterers, if any. Secondly, the receiving waves are considered planar in the vicinity of the current receiver position, which is called the *local planar zone* of the current receiver location. The arriving waves are considered virtually emitted from an infinite source, whose first, or the reference, wave front is found at the transmitter location and the wave vectors point in the transmitter–receiver direction. Once the receiver moves out of the current local planar zone, the arriving wave vector direction changes accordingly to the new transmitter–receiver axis. The received signals at two locations are regarded as coming from the same uniform plane wave if a location is found in the local planar zone of the other location, i.e., if the angle between the two transmitter–receiver axes is small. Whenever the arriving waves at the receiver can be approximated as uniform plane waves, one can apply directly the kernel functions represented in arrival time t_o, received frequency f_o, receiver location r_o and receive wave vector k_o to characterize the transmission channel, seen from receivers.

Application of the Modeling Framework at the Transmitter Side

As mentioned above, the spatio-temporal modeling framework cannot be used to characterize the transmitted signals and the channels, as seen from the transmitters, since the space regions around the transmitters do not satisfy the plane wave approximation criteria. This also means that the multidimensional Fourier transform does not provide the duality between transmitter location r_i and transmit wave vector k_i. The transmitted wave vector spectrum cannot be estimated from the transmitted signal in space domain, as was the case at the receiver side. In order to characterize fully systems and channels at transmitter side, the framework is modified by adding the AoD to the representation of kernel functions. The set of eight modified kernel functions, identified as $O(\cdot)$, and the input–output relationships based on these functions are listed in Table 2. The transmitted signal representation is also modified accordingly.

Table 2: Ensemble of modified Kernel functions in space–time–wave vector–frequency

No.	Kernel function	Input–output relationship
1	$O_1(\mathbf{r_o}, t_o; \mathbf{r_i}, \psi_i, t_i)$	$s_o(\mathbf{r}_o, t_o) = \int\int\int s_i(\mathbf{r}_i, \psi_i, t_i) O_1(\mathbf{r}_o, t_o; \mathbf{r}_i, \psi_i, t_i) d\mathbf{r}_i d\psi_i dt_i$
2	$O_2(\mathbf{r_o}, t_o; \mathbf{r_i}, \psi_i, f_i)$	$s_o(\mathbf{r}_o, t_o) = \int\int\int S_i(\mathbf{r}_i, \psi_i, f_i) O_2(\mathbf{r}_o, t_o; \mathbf{r}_i, \psi_i, f_i) d\mathbf{r}_i d\psi_i df_i$
3	$O_3(\mathbf{r_o}, f_o; \mathbf{r_i}, \psi_i, f_i)$	$S_o(\mathbf{r}_o, f_o) = \int\int\int S_i(\mathbf{r}_i, \psi_i, f_i) O_3(\mathbf{r}_o, f_o; \mathbf{r}_i, \psi_i, f_i) d\mathbf{r}_i d\psi_i df_i$
4	$O_4(\mathbf{r_o}, f_o; \mathbf{r_i}, \psi_i, t_i)$	$S_o(\mathbf{r}_o, f_o) = \int\int\int s_i(\mathbf{r}_i, \psi_i, t_i) O_4(\mathbf{r}_o, f_o; \mathbf{r}_i, \psi_i, t_i) d\mathbf{r}_i d\psi_i dt_i$
5	$O_5(\mathbf{k_o}, t_o; \mathbf{r_i}, \psi_i, t_i)$	$S_o(\mathbf{k}_o, t_o) = \int\int\int s_i(\mathbf{r}_i, \psi_i, t_i) O_5(\mathbf{k}_o, t_o; \mathbf{r}_i, \psi_i, t_i) d\mathbf{r}_i d\psi_i dt_i$
6	$O_6(\mathbf{k_o}, t_o; \mathbf{r_i}, \psi_i, f_i)$	$S_o(\mathbf{k}_o, t_o) = \int\int\int S_i(\mathbf{r}_i, \psi_i, f_i) O_6(\mathbf{k}_o, t_o; \mathbf{r}_i, \psi_i, f_i) d\mathbf{r}_i d\psi_i df_i$
7	$O_7(\mathbf{k_o}, f_o; \mathbf{r_i}, \psi_i, f_i)$	$S_o(\mathbf{k}_o, f_o) = \int\int\int S_i(\mathbf{r}_i, \psi_i, f_i) O_7(\mathbf{k}_o, f_o; \mathbf{r}_i, \psi_i, f_i) d\mathbf{r}_i d\psi_i df_i$
8	$O_8(\mathbf{k_o}, f_o; \mathbf{r_i}, \psi_i, t_i)$	$S_o(\mathbf{k}_o, f_o) = \int\int\int s_i(\mathbf{r}_i, \psi_i, t_i) O_8(\mathbf{k}_o, f_o; \mathbf{r}_i, \psi_i, t_i) d\mathbf{r}_i d\psi_i dt_i$

One observes that, at the transmitter side, the Fourier transform applies only on kernel functions with respect to transmit time variable (t_i) to provide the kernel functions in frequency domain, i.e., fi. Spatial variable r_i represents the transmitter location and it is independent from other variables, i.e., AoD ψ_i and transmit time t_i. Variable ψ_i represents the angle of departure or the transmit direction from the transmit antenna. In two-dimensional space, ψ_i is the azimuth angle, whereas in three-dimensional space, ψ_i is a pair of azimuth and elevation angles. t_i is the time variable at transmitter. Marginalization integral can be performed to remove some of the dependent variables. For instance, using marginalization integrals on kernel functions with respect to AoD variable, ψ_i, leads to the kernel function presentation in space and time domains, or in space and frequency domains, at transmitter side.

The relationships between the set of kernel functions $O(\cdot)$ is depicted in Fig. 3. The kernel function relationship cube is built similarly to that in Fig. 2. Kernel functions $O_1(\cdot)$, $O_2(\cdot)$, $O_3(\cdot)$ and $O_4(\cdot)$ are interconnected by Fourier transform on time and frequency variable pairs t_i–f_i and t_o–f_o. Kernel functions $O_5(\cdot)$, $O_6(\cdot)$, $O_7(\cdot)$ and $O_8(\cdot)$ are also interconnected by Fourier transform on time and frequency variable pairs t_i–f_i and t_o–f_o. Kernel function pairs $O_1(\cdot)$–$O_5(\cdot)$, $O_2(\cdot)$–$O_6(\cdot)$, $O_3(\cdot)$–$O_7(\cdot)$ and $O_4(\cdot)$–$O_8(\cdot)$ are interconnected by three-dimensional Fourier transform on variable pair r_o–k_o.

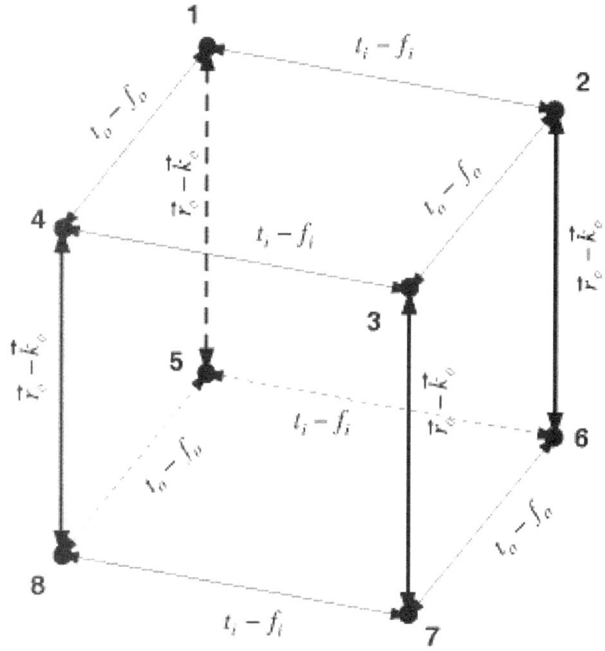

Figure 3: Reduced kernel function relationship

MIMO ANTENNA ARRAY DESIGN

In this section, we discuss the application of the reduced kernel functions, $O(\cdot)$, in order to design the transmit and receive antenna arrays for MIMO communications systems. Without loss of generality, we consider uniform linear array (ULA) antennas with N_T elements at the transmitter and N_R antenna elements at the receiver. The extension to other types of antenna arrays is straightforward. To maximize the channel capacity of a MIMO communications system, the sub-channels needs to be uncorrelated and the channel matrix needs to have high rank.

Transmit Antenna Array Design

Antenna Element Separation

To determine the transmit antenna element separation, the transmit coherence spatial vector has to be estimated. For ULA antenna, the coherence spatial vector becomes the coherence distance. The estimation is based on the kernel correlation functions. The correlation function of kernel function $O_1(\cdot)$ is presented in Eq. 32:

$$R_{O_1O_1}\left(\mathbf{r}_{o,1}, \mathbf{r}_{o,2}, t_{o,1}, t_{o,2}; \mathbf{r}_{i,1}, \mathbf{r}_{i,2}, \psi_{i,1}, \psi_{i,2}, t_{i,1}, t_{i,2}\right)$$
$$= E\left\{O_1\left(\mathbf{r}_{o,1}, t_{o,1}; \mathbf{r}_{i,1}, \psi_{i,1}, t_{i,1}\right) O_1^*\left(\mathbf{r}_{o,2}, t_{o,2}; \mathbf{r}_{i,2}, \psi_{i,2}, t_{i,2}\right)\right\} \quad (32)$$

where $E\{\cdot\}$ is the ensemble average and * is the conjugate operator. If the channel is wide sense stationary (WSS) in space, AoD and time domains, the correlation function becomes Eq. 33.

Marginalization integrals can be performed on the kernel correlation functions with respect to some variables in order to obtain compact kernel correlation functions [33]. Without loss of generality, $R_{OO}(r^i, \psi^i, t^i)$ denotes the transmit correlation function of a WSS channel, as it can be estimated from any kernel function O(·) that contains variables r^i, ψ^i, and t^i. The transmit coherence spatial vector, $r^{i,c}$, or the transmit coherence distance in the case of ULA antenna, $d^{i,c}$, is the distance over which the transmit correlation function $R_{OO}(r^i, \psi^i, t^i)$ drops below a given threshold, e.g., 50% of the maximum correlation peak. The transmit antenna elements must be separated by a distance greater than the transmit coherence distance $d^{i,c}$ in order to obtain sufficient sub-channel decorrelation.

$$R_{O_1O_1}(\Delta\mathbf{r}_o, \Delta t_o; \Delta\mathbf{r}_i, \Delta\psi_i, \Delta t_i) = E\{O_1(\mathbf{r}_o, t_o; \mathbf{r}_i, \psi_i, t_i) \cdot O_1^*(\mathbf{r}_o - \Delta\mathbf{r}_o, t_o - \Delta t_o; \mathbf{r}_i - \Delta\mathbf{r}_i, \psi_i - \Delta\psi_i, t_i - \Delta t_i)\} \quad (33)$$

$$R_{O_5O_5}(\Delta\mathbf{k}_o, \Delta t_o; \Delta\mathbf{r}_i, \Delta\psi_i, \Delta t_i) = E\{O_5(\mathbf{k}_o, t_o; \mathbf{r}_i, \psi_i, t_i) \cdot O_5^*(\mathbf{k}_o - \Delta\mathbf{k}_o, t_o - \Delta t_o; \mathbf{r}_i - \Delta\mathbf{r}_i, \psi_i - \Delta\psi_i, t_i - \Delta t_i)\} \quad (34)$$

$$R_{OO}(\Delta\mathbf{k}_o, \Delta t_o) = \int\int\int R_{O_5O_5}(\Delta\mathbf{k}_o, \Delta t_o; \Delta\mathbf{r}_i, \Delta\psi_i, \Delta t_i) \cdot d\Delta\mathbf{r}_i d\Delta\psi_i d\Delta t_i \quad (35)$$

$$R_{OO}(\Delta\mathbf{k}_o) = \int R_{OO}(\Delta\mathbf{k}_o, \Delta t_o)\, d\Delta t_o \quad (36)$$

Antenna Element Beam Separation

To further decrease the sub-channel correlation, directional antenna elements can be used. In this scenario, each transmit antenna element emits signal in a given direction. The element beams are separated also by an angle which is larger than the coherence angle of departure, $\psi_{i,c}$, defined as the angle over which the transmit correlation function, $R_{OO}(r_i, \psi_i, t_i)$, is below a specified threshold.

Transmit correlation function $R_{OO}(r_i,m,r_i,n, \psi_i, t_i)$ represents the kernel correlation function between the mth and nth transmit antenna elements. The coherence angle of departure estimated by this correlation function, $\psi_{i,c}(m, n)$, depends on these antenna elements. Taking into account the AoD coherence in designing the transmit beam of antenna elements further reduces the sub-channel correlation

Functions $R_{OO}(r_i, \psi_i, t)$ and $R_{OO}(r_i,m,r_i,n, \psi_i, t)$ are estimated at time instant t. This implies that the coherence angles of departure estimated from these functions are time-varying. A transmitter using these coherence angles, ψ_i,c(t) and ψ_i,c(m, n, t), in order to adaptively orient its beams obtains stable performance in fast time-varying channels.

Spatial-Frequency Multiplexing

Transmit correlation functions $R_{OO}(r_i, \psi_i, f_i)$ describe the transmit correlation in space, in terms of angle of departure and frequency separation. The coherence frequency estimated from correlation function $R_{OO}(r_i, \psi_i, f_i)$, i.e., $f_{i,c}$, represents the minimum frequency separation of transmit antenna elements to reduce the sub-channel correlation.

Similarly to coherence angle of departure, the coherence frequency $f_{i,c}(m, n)$, estimated from transmit correlation function $R_{OO}(r_i,m,r_{i,n}, f_i)$, represents the minimum frequency separation that the mth and nth transmit antenna elements needed to ensure uncorrelated sub-channels. More specifically, the coherence frequency $f_{i,c}(m, n, \psi_{i,m}, \psi_{i,n})$, estimated from transmit correlation function $R_{OO}(r_i,m,r_{i,n}, \psi_{i,m}, \psi_{i,n}, f_i)$, reflects the minimum frequency separation that the mth and the nth antenna elements must follow when their beams orient in directions $\psi_{i,m}$ and $\psi_{i,n}$, respectively. A system built based on these coherence values is capable of performing spatial-frequency multiplexing.

Receive antenna array design

The application of the proposed framework in designing receive antenna arrays is also based on the second order statistics of the kernel functions, as was the case for the transmit antenna array design discussed in Section 8.1. The generally receive correlation function of a space and time WSS channel is:

$$R_{OO}(\Delta\mathbf{r}_o, \Delta t_o) = \int\int\int R_{O_1O_1}(\Delta\mathbf{r}_o, \Delta t_o; \Delta\mathbf{r}_i, \Delta\psi_i, \Delta t_i)\, d\Delta\mathbf{r}_i d\Delta\psi_i d\Delta t_i \quad (37)$$

Performing marginalization over t_i variable, results in the receive correlation function is space, $R_{OO}(r_o)$. The receive coherence distance ($d_{o,c}$), estimated from the receive correlation function in space $(R_{OO}(\Delta\mathbf{r}_o))$represent the minimum antenna element separation in the receive antenna array.

The receive correlation functions, $R_{O5O5}(\cdot)$ and $R_{OO}(\cdot)$, in wave vector domain are presented in Eqs. 34–36. Differently from the situation observed at the transmitter side, the coherence receive wave vector, $\Delta\mathbf{k}_{o,c}$ contains both the minimum angle of arrival separation and receive frequency separation that the receive antenna elements needs to follow in order to guarantee low sub-channel correlation. This is because of the assumption made that the arrival waves are considered uniform plane waves.

ILLUSTRATIVE EXAMPLES: AOA ESTIMATION

In this section, we use some field test data to illustrate the utilization of the practical modeling framework in estimating the angle of arrival. In our measurements, the sounding signal is the ATSC DTV signal. The transmitter with one transmit antenna is mounted on top of a television tower. The receiver with two receive antennas, separated by 2.1 m or 5.5λ (wavelengths), is installed in a mobile laboratory unit. The spatial kernel functions,[3] $O(r_o;r_i)|_{ri=constant}$, were estimated by using the pseudorandom number sequence PN511, embedded in the ATSC signal (part 2 in [34]), and a sliding correlator.

AoA Estimation at Rural Site

The measurement site of this example is shown in Fig. 4. In order to perform spatial sampling, a mobile receiver with two receive antennas is moving at a speed of 60 km/h from point 1 to point 2 and then from point 2 to point 1. The antenna separation is 2.1 m, or 5.5λ, at a carrier frequency of 788 MHz, and the sampling interval is 0.4 m. The spatial kernel function estimated on the straight segment between the intersection and point 2 is used to estimate the angle of arrival. We assume that the channel is sufficiently stable during the measurement such that the samples are aligned in space within reasonable limits. Axis *x* points from the intersection to point 2, while axis *y* is perpendicular to axis *x* and points to the upper of Fig. 4.

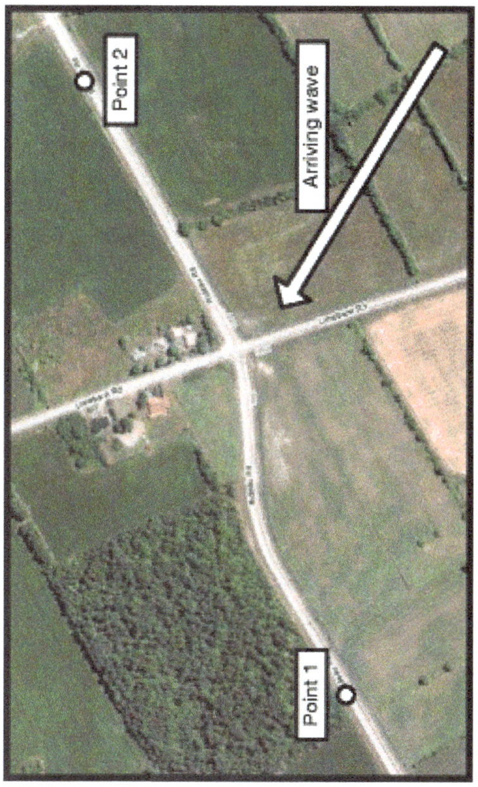

Figure 4: Satellite photo of downlink rural measurement sites

Figure 5a shows the LOS components of the spatial kernel function, $O(r_o;r_i)|_{ri=constant}$, represented by a 4 × 1, 000 matrix, for a downlink rural measurement. The sampling interval is 0.4 m along the x-axis and 2.1 m along the y-axis. Applying a two-dimensional FFT on kernel function $O(r_o;r_i)|_{ri=constant}$, with respect to the receiver location vector, ro, gives us its dual kernel function, $O(r_o;r_i)|_{ri=constant}$.

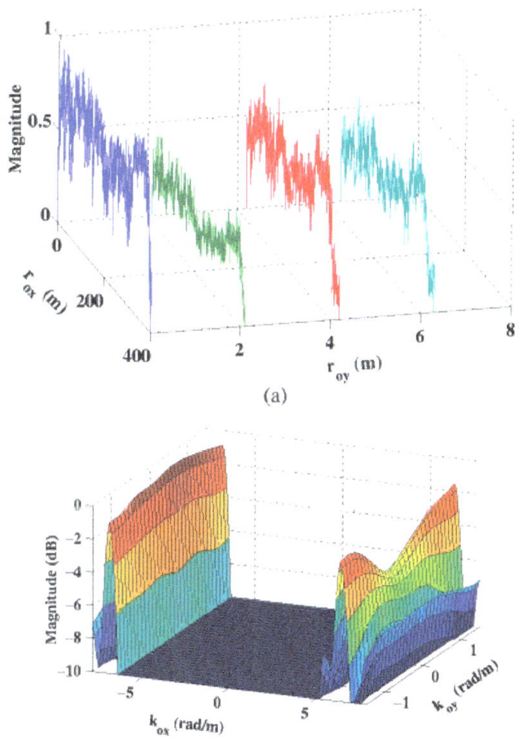

Figure 5: Kernel functions (**a**) $O(r_o;r_i)|_{ri=constant}$ and (b) $O(r_o;r_i)|_{ri=constant}$ for downlink measurement at rural site

Figure 5b shows the resulting 2D wave vector spectrum. In this figure, the principal, or main, spectrum component is found at wave vector component k_{ox} between −8 and −7 rad/m. There are two other strong spectrum components found at $k_{ox} \approx 6.5$ rad/m, $k_{oy} \approx -1$ rad/m, and at $k_{ox} \approx 6.5$ rad/m, $k_{oy} \approx 1$ rad/m. Note that for our measurements, the spatial resolution in the *y*-axis is limited since only four sampling points at a sampling step of 2.1 m are collected in the *y*direction at each sampling instant. Therefore, the information along the k_y-axis in the wave vector domain is not as precise as for the wave vector component along the k_x-axis. For this measurement, the carrier frequency is 788.31 MHz, the AoA of the LOS component is 120°. Therefore, the k_x component of the LOS component is −8.25 rad/m, which is very close to the value estimated from the kernel function $O(r_o;r_i)|_{ri=constant}$. The spectrum component found at $k_{ox} = -8\sim -7$ rad/m corresponds to the LOS component, while the components found at $k_{ox} \approx 6.5$ rad/m are the reflected components.

AoA Estimation at Suburban Site

The measurement site of this example is depicted in Fig. 6. Similarly to the rural site measurement, the mobile unit is traveling from point 1 to point 2 and then from point 2 to point 1 to perform the spatial sampling process. Kernel function $O(r_o;r_i)|_{ri=constant}$ estimated on the straight segment from point 1 to the intersection, is used in this example. Axis*x* point from point 1 to the intersection. Axis *y* is perpendicular to axis *x* and point to the lower right corner of Fig. 6. The speed of the mobile unit is 30 km/h. The spatial sampling in axis *x* is 0.2 m and in axis *y* is 2.1 m.

Figure 6: Satellite photo of downlink suburban measurement site

Figure 7a shows the LOS components of spatial kernel function $O\ (r_o;r_i)|_{ri=constant}$ while Fig. 7b shows the dual kernel function $O(r_o;r_i)|_{ri=constant}$. One observes clearly that there are three spectrum components found at $\mathbf{k}_{o_1} = (k_{o_1x}, k_{o_1y}) \approx (11, -1.2)$ rad/m, at $\mathbf{k}_{o_2} = (k_{o_2x}, k_{o_2y}) \approx (11, -0.1)$ rad/m and at $k_{o3} = (k_{o3\,x}, k_{o3\,y}) \approx (11, 1)$ rad/m. For this example, the carrier frequency is 788.31 MHz, the AoA of the LOS component is 50°. Therefore, the k_x component of the LOS component is 10.6 rad/m. This is very close to the values extracted from the received wave vector spectrum $O\ (r_o;r_i)|_{ri=constant}$, mentioned above. The three wave vector spectrum components found in Fig. 7b at positive k_{ox} correspond to the LOS component. At negative values of k_{ox}, we find some low-magnitude spectrum components, which likely correspond to the reflected waves.

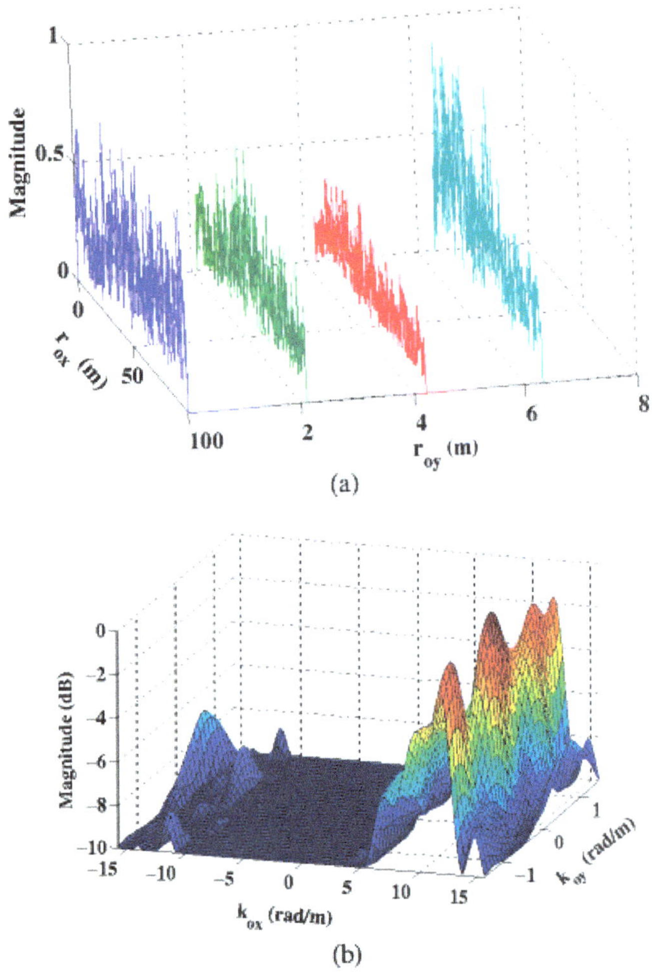

Figure 7: Kernel functions $O\ (r_o;r_i)|_{ri=constant}$ and (**b**) $O(r_o;r_i)|_{ri=constant}$ for downlink measurement at suburban site

CONCLUSIONS

In this paper, a new theoretical framework is proposed for MIMO channel characterization and modeling. The proposed spatio-temporal modeling framework allows us to represent MIMO channels and systems in space, time, wave vector, and frequency domains. The relationships between space and time and between wave vector and frequency are explicitly described. The proposed theoretical framework consists of sixteen kernel functions. The kernel functions are interconnected by Fourier transforms. Since the theoretical framework is based on uniform plane wave propagation, the application of the spatio-temporal framework to realistic propagation situations is not straightforward. For this reason, a practical framework is proposed to approximate the theoretical framework in realistic scenarios. The practical framework consists of eight kernel functions, which represent channels in space, time, receive wave vector, frequency, and direction of departure. Time–frequency duality exists at both the receiver and the transmitter sides, whereas the wave vector–location vector duality exists only at the receiver side. Applications of the practical framework in designing transmit and receive antenna arrays and in estimating angle of arrival using field test measurement data are presented.

ACKNOWLEDGEMENTS

The authors are greatly indebted to Mr. Sébastien Laflèche, Mr. Robert Gagnon, Mr. Benoit Ledoux, Mr. Charles Nadeau, Mr. Gilles Gagnon, and Mr. Bernard Caron from the Television Broadcast Technologies Research Group at the Communication Research Center in Ottawa, who very kindly provided the mobile channel measurement facilities essential to the field experiments conducted for the data analysis presented in this paper.

REFERENCES

1. Ghosh A, Ratasuk R, Mondal B, Mangalvedhe N, Thomas T (2010) LTE-advanced: next-generation wireless broadband technology [Invited Paper]. IEEE Wirel Commun Mag 17(3):10–22
2. Ahmadi S (2009) An overview of next-generation mobile WiMAX technology. IEEE Commun Mag 47(6):84–98
3. Foschini GJ (1996) Layered space–time architecture for wireless communication in a fading environment when using multi-element antennas. Bell Labs Tech J 1(2):41–59
4. Alamouti SM (1998) A Simple Transmit Diversity Technique for Wireless Communications. IEEE J Sel Areas Commun 16(8):10
5. Paulraj A, Nabar R, Gore D (2003) Introduction to space–time wireless communications. Cambridge University Press, Cambridge
6. Sterian CED, Wu Y, Pätzold M (2008) Super-quasiorthogonal space–time trellis codes for four transmit antennas with rectangular signal constellations. Ann Telecommun 63(5–6):331–350
7. Bouvet P-J, Hélard M, Nir VL (2006) Simple iterative receivers for MIMO LP-OFDM systems. Ann Telecommun 61(5–6):578–601
8. Naguib A, Paulraj A, Kailath T (1994) Capacity improvement with base-station antenna arrays in cellular CDMA. IEEE Trans Veh Technol 43(3):691–698
9. Molisch A, Win M, Choi Y, Winters J (2005) Capacity of MIMO systems with antenna selection. IEEE Trans Wirel Commun 4(4):1759–1772
10. Winters J (1987) On the capacity of radio communication systems with diversity in a Rayleigh fading environment. IEEE J Sel Areas Commun 5(5):871–878
11. Winters J, Salz J, Gitlin R (1994) The impact of antenna diversity on the capacity of wireless communication

systems. IEEE Trans Commun 42(2):1740–1751

12. Foschini G, Gans M (1998) On limits of wireless communications in a fading environment when using multiple antennas. Wirel Pers Commun 6(3):311–335
13. Shiu D, Foschini G, Gans M, Kahn J et al (2000) Fading correlation and its effect on the capacity of multielement antenna systems. IEEE Trans Commun 48(3):502–513
14. Chuah C, Tse D, Kahn J, Valenzuela R (2002) Capacity scaling in MIMO wireless systems under correlated fading. IEEE Trans Inf Theory 48(3):637–650MathSciNetMATH
15. Chizhik D, Foschini GJ, Gans MJ, Valenzuela RA (2002) Keyholes, correlations, and capacities of multielement transmit and receive antennas. IEEE Trans Wirel Commun 1(2):361–368
16. Gesbert D, Bolcskei H, Gore DA, Paulraj AJ (2002) Outdoor MIMO wireless channels: models and performance prediction IEEE Trans Commun 50(12):1926–1934
17. Almers P, Tufvesson F, Molisch AF (2003) Measurement of keyhole effect in a wireless multiple-input multiple-output (MIMO) channel. IEEE Commun Lett 7(8):373–375
18. Sayeed A, Sendonaris A, Aazhang B (1998) Multiuser detection in fast-fading multipath environments. IEEE J Sel Areas Commun 16(9):1691–1701
19. Bhashyam S, Sayeed AM, Aazhang B (2000) Time-selective signaling and reception for communication over multipath fading channels. IEEE Trans Commun 48(1):83–94
20. Onggosanusi E, Sayeed A, Van Veen B (2000) Canonical space–time processing for wireless communications. IEEE Trans Commun 48(10):1669–1680
21. Durgin GD (2003) Space–time wireless channels. Prentice Hall
22. Molisch A, Asplund H, Heddergott R, Steinbauer M, Zwick T (2006) The COST259 directional channel model—part I: overview and methodology. IEEE Trans Wirel Commun 5(12):3421–3433
23. Asplund H, Glazunov A, Molisch A, Pedersen K, Steinbauer M (2006) The COST 259 directional channel model—part II: macrocells. IEEE Trans Wirel Commun 5(12):3434–3450
24. Guguen P, Zein GE (2004) Les techniques multi-antennes pour les Réseaux Sans Fil (in French). Hermes Science, Paris
25. Chandran S, Guguen P, Zein GE (editors) (2004) Adaptive antenna arrays, trends and applications. Springer, Berlin. pp 360–376
26. Sayeed A (2002) Deconstructing multiantenna fading channels. IEEE Trans Signal Process 50(10):2563–2579
27. Fluerasu A, Letrou C (2009) Gaussian beam launching for 3D physical modeling of propagation channels. Anna Telecommun 64(11–12):763–776
28. Bello PA (1964) Time–frequency duality. IEEE Trans Inf Theory 10(1):18–33, 1
29. Zemanian AH (1965) Distribution theory and transform analysis: an introduction to generalized functions, with applications. McGraw-Hill, New YorkMATH
30. Dearholt DW, McSpadden WR (1973) Electromagnetic wave propagation. McGraw-Hill, New York
31. Bello PA (1963) Characterization of randomly time-variant linear channels. IEEE Trans Commun Syst 11(4):360–393, 12
32. Balanis CA (2005) Antenna theory analysis and design, 3rd edn. Wiley, New York
33. Papoulis A, Pillai SU (2001) Probability, random variables and stochastic processes, 4th edn. McGraw-Hill, New York
34. A/53: ATSC Digital Television Standard, parts 1–6 advanced television systems committee 2007

CHAPTER 3

ADVANCES IN MIMO TECHNIQUES FOR MOBILE COMMUNICATIONS

Farhan Khalid, and Joachim Speidel

Keywords: Multiple-Input Multiple-Output (MIMO), Multiuser MIMO, Wireless Communications, Beamforming, Diversity, Precoding, Capacity

INTRODUCTION

Multiple-input multiple-output (MIMO) wireless systems employ multiple transmit and receive antennas to increase the transmission data rate through spatial multiplexing or to improve system reliability in terms of bit error rate (BER) performance using space-time codes (STCs) for diversity maximization [1]. MIMO systems exploit multipath propagation to achieve these benefits, without the expense of additional bandwidth. More recent MIMO techniques like the geometric mean decomposition (GMD) technique proposed in [2] aim at combining the diversity and data rate maximization aspects of MIMO in an optimal manner. These advantages make MIMO a very attractive and promising option for future mobile communication systems especially when combined with the benefits of orthogonal frequency-division multiplexing (OFDM) [3,4].

The capacity of an M ′ N single-user MIMO (SUMIMO) system with M transmit and N receive antennas, in terms of the spectral efficiency i.e. bits per second per Hz, is given by [1]

$$C = \log_2\left[\det\left(\mathbf{I}_N + \frac{\rho}{M}\mathbf{H}\mathbf{H}^H\right)\right] \quad (1)$$

where H is the N ′ M MIMO channel matrix and ρ is the signal to noise ratio (SNR) at any receive antenna. Equation (1) assumes that the M information sources are uncorrelated and have equal power. Expressed in terms of the eigenvalues, Equation (1) can be written as [1]

$$C = \sum_{i=1}^{m} \log_2\left(1 + \frac{\rho}{M}\lambda_i\right) \quad (2)$$

where λ_i represent the nonzero eigenvalues of HH^H or H^HH for $N \leq M$ and $M < N$ respectively and m = min(M, N). Therefore, MIMO systems are capable of achieving several-fold increase in system capacity as compared to single-input single-output (SISO) systems by transmitting on the spatial eigenmodes of the MIMO channel.

Equation (2) also shows that the performance of MIMO systems is dependent on the channel eigenvalues. Very low eigenvalues indicate weak transmission channels which may make it difficult to recover the information from the received signals. Optimal power allocation based on the water-filling algorithm can be used to maximize the system capacity subject to a total transmit power constraint. Water-filling provides substantial capacity gain when the eigenvalue spread, i.e., the condition number $\lambda_{max}/\lambda_{min}$ is sufficiently large.

The MIMO concept becomes even more attractive in multiuser scenarios where the network capacity can be increased by simultaneously accommodating several users without the expense of valuable frequency resources.

This paper is arranged as follows: Section 2 provides an overview of the current wireless standards which support MIMO technologies. Sections 3 and 4 include detailed discussion and performance analysis of various important SU-MIMO and multiuser MIMO (MU-MIMO) techniques respectively that are proposed for the next generation wireless communication systems. In-depth description of several MU-MIMO uplink and downlink schemes is given in Section 4 followed by a brief discussion of the MU-MIMO capacity. Section 5 provides an overview of convex optimization which has become an important tool for designing optimal MIMO beamforming systems. Section 6 concludes this work and identifies the areas for future research.

CURRENT IMPLEMENTATION STATUS

There has been a lot of research on MIMO systems and

Citation: F. Khalid and J. Speidel, "Advances in MIMO Techniques for Mobile Communications," International Journal of Communications, Network and System Sciences, Vol. 3 No. 3, 2010, pp. 213-252. doi: 10.4236/ijcns.2010.33031.

techniques. MIMO-OFDM WLAN products based on the IEEE 802.11n standard are already available. The IEEE 802.16 wireless MAN standard known as WiMAX also includes MIMO features. Fixed WiMAX services are being offered by operators worldwide. Mobile WiMAX networks based on 802.16e are also being deployed while 802.16m is under development. IEEE 802.20 mobile broadband wireless access (MBWA) standard is also being formulated which will have complete support for mobility including high-speed mobile users e.g., on train networks. For other applications like cellular mobile communications which supports both voice and data traffic, MIMO systems are yet to be deployed. However, the 3GPP's long term evolution (LTE) is under development and adopts MIMO-OFDM, orthogonal frequency-division multiple access (OFDMA) and single-carrier frequency-division multiple access (SC-FDMA) transmission schemes. The following text presents a more detailed discussion of the various technical aspects of these standards and technologies.

IEEE 802.11N WI-FI

The IEEE 802.11n WLAN standard incorporates MIMOOFDM as a compulsory feature to enhance data rate. Initial target was to achieve data rates in excess of 100 Mb/s [5]. However, current WLAN devices based on 802.11n Draft 2.0 are capable of achieving throughput up to 300 Mb/s utilizing two spatial streams in a 40 MHz channel in the 5 GHz band [6].

Initially, there were two main proposals one form the WWiSE consortium and the other from the TGnSync consortium competing for adoption by the IEEE 802.11 TGn. However, another proposal by the Enhanced Wireless Consortium (EWC) was finally accepted as the first draft for IEEE 802.11n [7].

The IEEE 802.11n standard proposes the use of the legacy 20 MHz channel and also an optional 40 MHz channel. The available modulation schemes include BPSK, QPSK, 16-QAM and 64-QAM [5,6]. Convolutional coding with different code rates is specified and use of low-density parity-check (LDPC) codes is also supported [5,8]. The MIMO techniques adopted include both spatial multiplexing and diversity techniques. Open-loop MIMO (OL-MIMO) techniques which do not require channel state information (CSI) at the transmitter seem to have been preferred [9]. Non-iterative linear minimum mean square error (LMMSE) detection has primarily been considered so as to minimize the complexity associated with MIMO detection while ensuring reasonably good performance [10]. Spatial spreading mentioned in [11] is an open-loop MIMO spatial multiplexing technique where multiple data streams are transmitted such that the diversity is maximized for each of the streams. The MIMO diversity techniques introduced in the standard include space-time block coding (STBC) and cyclic shift diversity (CSD) which extend the range and reception of 802.11n devices. In addition, conventional receiver spatial diversity techniques like maximum ratio combining (MRC) are also specified. Transmit beamforming is also specified as an optional feature [6]. The Cisco Aironet 1250 series access point based on 802.11n draft 2.0 supports open-loop transmit beamforming [12]. 802.11n draft 2.0 specifies a maximum of 4 spatial streams per channel. Thus, a maximum throughput of 600 Mb/s can be achieved by using 4 spatial streams in a 40 MHz channel. In addition to spatial multiplexing and doubled channel bandwidth, more efficient OFDM with shorter guard interval (GI) and new medium access control (MAC) layer enhancements (e.g. closed-loop rate adaptation [13]) have also contributed to the increased throughput of 802.11n [6].

IEEE 802.16 WIMAX

The IEEE 802.16 worldwide interoperability for microwave access (WiMAX) is a recently developed wireless MAN standard that employs MIMO spatial multiplexing and diversity techniques. In addition to fixed WiMAX, the IEEE 802.16e Mobile WiMAX standard has also been developed and was approved in December 2005 [14]. Fixed WiMAX networks have already been deployed around the world and Mobile WiMAX deployments have also started.

802.16e-2005 is basically an amendment to the 802.16-2004 standard for fixed WiMAX with addition of new features to support mobility. 802.16e specifies the 2–6 GHz frequency band for mobile applications and the 2–11 GHz band for fixed applications (The single-carrier WirelessMAN-SC PHY specification for fixed wireless access however specifies the 10–66 GHz frequency band [15].). It also specifies a license-exempt band between 5–6 GHz. A cellular network structure is specified with support for handoffs and mobile users moving at vehicular speeds are also supported, thus enabling mobile wireless internet access [14,16]. In addition to single carrier transmission, the standard specifies OFDM transmission scheme with 128, 256, 512, 1024 or 2048 subcarriers. Both TDD and FDD duplexing is specified while the multiplexing/ multiple access schemes include OFDMA in addition to burst TDM/ TDMA. However, scalable OFDMA is specified in all mobile WiMAX profiles as the physical layer multiple access technique. The various channel bandwidths specified in the standard include 1.25, 1.75, 3.5, 5, 7, 10, 8.75, 10, 14 and 15 MHz. WiMAX supports adaptive modulation and coding schemes. The supported modulation schemes include BPSK, QPSK, 16-QAM and 64-QAM [14,16]. Optional 256-QAM support is provided in the WirelessMAN-SCa PHY [15]. Convolutional codes at rate1/2, 2/3, 3/4 or 5/6 are specified as mandatory for both uplink and downlink. In addition, convolutional turbo codes, repetition codes, LDPC and concatenated Reed-Solomon convolutional code (RS-CC) are specified as optional. The supported data rates range from 1 Mb/s to 75 Mb/s [14,16].

IEEE 802.16e supports both open-loop and closedloop MIMO. Open-loop MIMO techniques include spatial multiplexing (SM) and space-time coding (STC) [14,17,18]. 802.16e includes support for up to four spatial streams and therefore a maximum of 4 ´ 4 MIMO configuration [14,18]. STC is based on the Alamouti scheme (also STBC) and is also called space-time transmit diversity (STTD). It is an optional feature and may be used to provide higher order transmit diversity on the downlink [14]. In closed-loop MIMO, full or partial CSI is available at the transmitter through feedback. Eigenvector steering is employed to approach full capacity of the MIMO channel and water filling can be used to maximize throughput by allocating power in an optimal manner [9,19]. IEEE 802.16e supports closed-loop MIMO precoding for SM and also closed-loop STC [14,17]. However, closed-loop MIMO is not yet supported in the latest WiMAX Forum Wave 2 profiles [18]. Another MIMO mode called "collaborative spatial multiplexing" is also specified where two subscriber stations (SS), each having a single antenna, use the same subchannel for uplink transmission in order to increase the throughput [14,15, 17,20]. The adaptive antenna systems (AAS) supported in 802.16e also include closed-loop adaptive beamforming, which uses feedback from the SS to the base station (BS) to optimize the downlink transmission [14,15,18]. IEEE 802.16 Task Group m (TGm) has also been set up to develop the IEEE 802.16m standard which will enable interoperability between WiMAX and 3GPP's Long Term Evolution (LTE) standard for next generation mobile communications [21,22]. 802.16m is expected to support high-speed mobile wireless access (up to 350 km/h) and peak data rates of over 300 Mb/s using 4 ´ 4 MIMO [22]

IEEE 802.20 MBWA

The IEEE 802.20 working group was established to draft the IEEE 802.20 Mobile Broadband Wireless Access (MBWA) standard which is also nicknamed as MobileFi. IEEE 802.20 proposes a complete cellular structure and is designed and optimized for mobile data services at speeds up to 250 km/h. However, it can also support voice services due to very low transmission latency of 10–30 ms (better than the 25–40 ms for 802.16e). User data rates in excess of 1 Mb/s can be supported at 250 km/h [23–25]. MBWA is designed to operate in the licensed bands below 3.5 GHz [24,25]. 2.5 MHz to 20 MHz of uplink/downlink transmission bandwidth can be allocated per cell [25]. For a bandwidth of 5 MHz, peak aggregate data rate of around 16 Mb/s can be supported in the downlink [23,24] which obviously would be much greater for higher bandwidths. The transmission scheme is based on OFDM, with OFDMA used for downlink transmission while both OFDMA and code-division multiple access (CDMA) are specified for the uplink. Rotational OFDM is specified as an optional scheme. The standard supports both FDD and TDD operation. The supported modulation schemes include QPSK, 8-PSK, 16-QAM and 64-QAM. Support of hierarchical (layered) modulation involving the superposition of two modulation schemes is also included for broadcast and multicast services. The specified FEC coding schemes include convolutional codes, turbo codes and LDPC codes [25].

Various MIMO schemes are also supported. STTD (based on STBC) and SM are specified for SU-MIMO transmission, utilizing up to 4 transmit antennas. STTD is particularly important for high speed mobile access. Two different stream multiplexing schemes namely single codeword (SCW) and multiple code word (MCW) may be employed for MIMO transmission. These schemes also support closed-loop MIMO downlink transmission with rank adaptation. Both schemes utilize linear precoding at the BS for transmit beamforming based on the feedback of a suitable precoding matrix from the user equipment's (UE's) codebook to the BS. The standard also supports MU-MIMO or space-division multiple access (SDMA) transmission in the downlink which involves multiuser scheduling and precoding at the BS depending upon the feedback of the preferred precoding matrix index and differential channel quality indicator (CQI) reports from the UEs [25].

The IEEE 802.20 standard was supposed to be available in 2006 but was delayed due to lack of support from some of the key vendors and the political turmoil within the standards forum [23]. However, it was finally approved in June 2008 and made available by the end of August 2008 [25].

3gpp LTE

The 3rd generation partnership project's (3GPP) long term evolution (LTE) project is aimed at developing a new mobile communications standard for gradual migration from 3G to 4G. LTE physical layer is almost near completion. It specifies an OFDM based system with support for MIMO. Downlink transmission is based on OFDMA while SC-FDMA is used for the uplink due to its low PAPR characteristics. It supports both TDD and FDD operation. A packet switching architecture is specified for LTE [26,27].

LTE supports scalable bandwidths of 1.25, 2.5, 5, 10 and 20 MHz. Peak data rates of 100 Mb/s and 50 Mb/s are supported in the downlink and the uplink respectively, in 20 MHz channel. The standard specifies full performance within a cell up to 5 km radius and slight degradation from 5–30 km. Operation up to 100 km may be possible. It also supports high-speed mobility with high performance at speeds up to 120 km/h while the E-UTRAN (Evolved Universal Terrestrial Radio Access Network i.e., LTE's RAN) should be able to maintain the connection up to 350 km/h, or even up to 500 km/h. LTE also specifies very low latency operation with control plane (C-plane) latency of < 50-100ms and user plane (U-plane) latency of < 10 ms [27,28].

The single-user MIMO techniques supported include STBC and SM. Closed-loop multiple codeword (MCW) SM with codebook based precoding and with support for cyclic delay diversity (CDD) is specified. A maximum of two downlink spatial streams are specified. LTE also supports MU-MIMO in the downlink as well as in the uplink. Closed-loop transmit diversity using MIMO beamforming with rank adaptation is also supported. The supported antenna configurations for the downlink include 4 ´ 2, 2 ´ 2, 1 ´ 2 and 1 ´ 1 whereas 1 ´ 2 and 1 ´ 1 configurations are supported in the uplink [27,29,30]. However, multiple UE antennas in the uplink may be supported in future.

SINGLE-USER MIMO TECHNIQUES

Various open-loop and closed-loop SU-MIMO techniques are discussed in the following text along with performance analysis and comparison. Some of the techniques mentioned herein have already been adopted for the current standards while other advanced methods are likely candidates for the next generation wireless systems.

V-Blast

The vertical Bell Laboratories Layered Space-Time (VBLAST) [31] is one of the very first open-loop spatial multiplexing MIMO systems which has been practically demonstrated to achieve much higher spectral efficiencies than SISO systems, in rich scattering environments. In V-BLAST, a single data stream is demultiplexed into multiple sub streams which are mapped on to symbols and then transmitted through multiple antennas. Intersubstream coding is not employed in V-BLAST, however channel coding can be applied to the individual sub streams for reduction of bit error rate (BER). CSI in a

V-BLAST system is available at the receiver only by means of channel estimation. Figure 1 shows the simple block diagram of a V-BLAST system.

V-BLAST detection can be accomplished by using linear detectors like zero-forcing (ZF) or minimum mean square error (MMSE) detector along with symbol cancellation (also called successive interference cancellation). Symbol cancellation is a nonlinear technique which enhances the detection performance by subtracting the detected components of the transmit vector from the received symbol vector [31]. This technique, however, is prone to error propagation

The QR decomposition of the MIMO channel matrix can be used to represent the ZF nulling in V-BLAST [2]. Assuming a frequency-flat fading MIMO channel, the corresponding sampled baseband received signal for a V-BLAST system with M transmit and N receive antennas (M ≤ N) is therefore given by

$$\begin{aligned} \mathbf{y} &= \mathbf{Hx} + \mathbf{n} \\ &= \mathbf{QRx} + \mathbf{n} \end{aligned} \qquad (3)$$

where Q is an N ´ M unitary matrix with orthonormal columns, R is a M ´ M upper triangular matrix, x is the transmitted signal and n represents the noise vector. The discrete-time index is dropped to simplify notation. Multiplying both sides of Equation (3) by Q^H gives

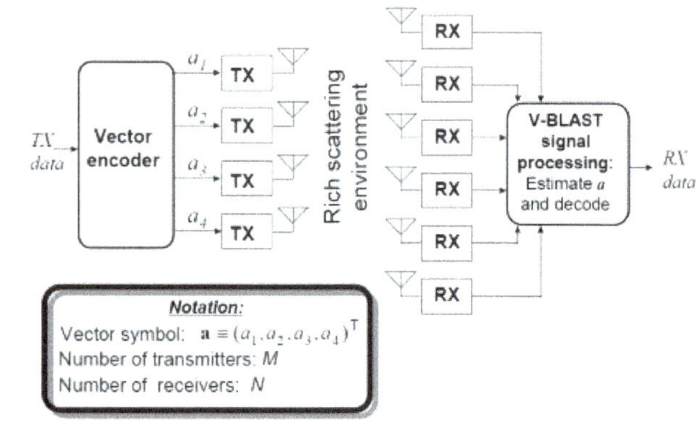

Figure 1: V-BLAST system block diagram [31].

The sequential signal detection in V-BLAST can be accomplished as follows [2]: for i M= -: 1:1

$$\hat{x}_i = C\left[\left(\tilde{y}_i - \sum_{j=i+1}^{M} r_{ij}\hat{x}_j\right) / r_{ii}\right]$$

End

where C represents mapping to the nearest modulation symbol.

The results for an initial V-BLAST prototype mentioned in [31] yielded spectral efficiencies of 20–40 bps/Hz in indoor scenarios which is quite impressive. However, later has shown that V-BLAST also performs reasonably well in mobile scenarios and can be employed for MIMO.

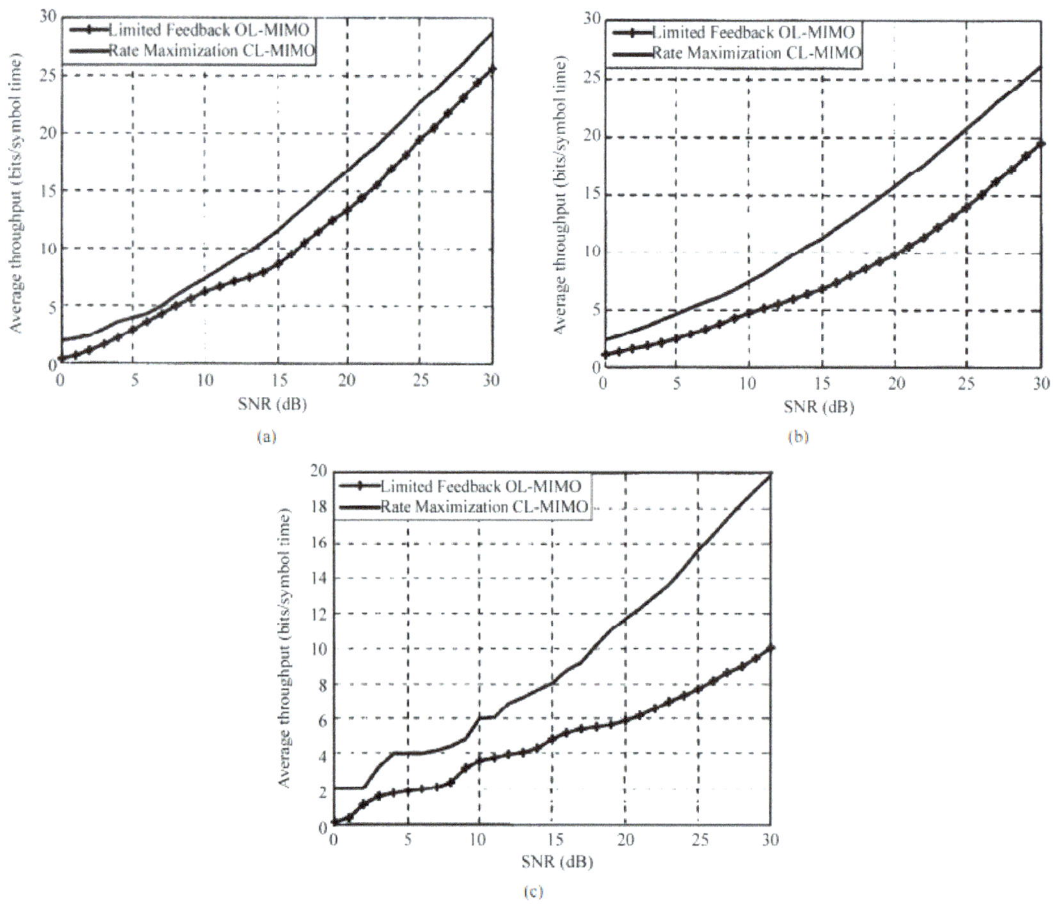

Figure 2: Throughput for (a) Rayleigh fading channel, (b) Spatial correlation channel and (c) Realistic channel model [34].

OFDM systems as well and further improvements have been suggested in the literature. [32] Proposes an extension of V-BLAST incorporating power and rate feedback which approaches closed-loop MIMO capacities. Equal power allocation with per-antenna rate control (PARC) produces the best results for the proposed system. PARC enables the transmitter to select the appropriate data rate and the associated modulation and coding scheme (MCS) for each transmit antenna based on the feedback of channel quality information from the receiver [33].

It presents a comparison between a modified V-BLAST system with limited feedback (including the modulation index and the number of streams to be used) and closed-loop MIMO (CL-MIMO) in [34]. CL-MIMO shows 15.1% throughput improvement for Rayleigh fading channel, 48.1% for spatially correlated channel and 104% for the case of a realistic channel model, at SNR of 25 dB. Figure 2 shows these results.

Spatial Multiplexing with Cyclic Delay Diversity

Spatial multiplexing (SM) can be combined with a simple diversity technique such as cyclic delay diversity (CDD) to obtain much better performance as compared to regular SM systems like V-BLAST. Such a system which combines SM and MIMO diversity is referred to as a joint diversity and multiplexing (JDM) system [35]. SM with CDD is also specified in the 3GPP LTE standard [30]. It proposes a cyclic delay assisted SM-OFDM (CDA-SM-OFDM) system which does not require any CSI at the transmitter, however complete CSI is required at the receiver [35]. Figure 3 shows the transmitter and receiver block diagram.

The blocks denoted $\Omega_1,\Omega_2,\ldots,\Omega_\Delta$ perform the cyclic delay operation which involves cyclic shifting of the signal within each group of D transmit antennas per SM branch. If there are P SM branches then the total number of transmit antennas is PD . The receiver for CDA-SM-OFDM system is similar to V BLAST.

CDD increases the channel frequency-selectivity since cyclic shifting of the OFDM signal and then adding those shifted signals linearly at the receiver inserts virtual echoes on the channel response. The resulting higher order frequency diversity can be exploited by any coded OFDM (COFDM) system [35].

Figure 4 shows a comparison of the CDA-SM-OFDM system capacity with 2 ′ 2 and 4 ′ 2 SM-OFDM systems. Here it can be seen that the capacity of the CDA-SM- OFDM system lies between that for the two SM-OFDM systems. However, the capacities for the SM-OFDM systems are plotted for the ideal case i.e. with the best possible STC and channel coding schemes. It can also be seen that the outage capacity i.e. the capacity obtained below 10% of the times, for the CDA-SM-OFDM system is much higher than the 2 ′ 2 SM-OFDM and closer to the 4 ′ 2 SM-OFDM. Thus the system performance for the CDA-SM-OFDM system shows a significant increase just by employing a simple STC i.e. CDD.

It has also be shown in [35] that the eigenvalue spread for the CDA-SM-OFDM system is generally higher than both the SM-OFDM schemes and this means that eigen-beamfoming can be employed for CDD based SM systems. In fact, the 3GPP LTE standard incorporates CDD based SM with precoding and specifies precoding matrices for small and large delay CDD [30].

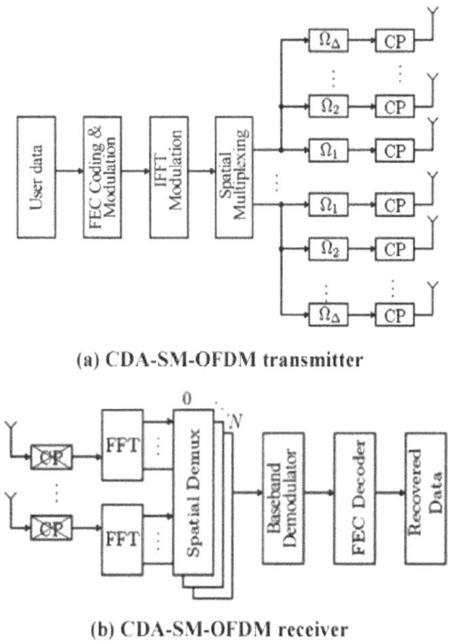

Figure 3: CDA-SM-OFDM system transmitter and receiver [35].

Figure 5 provides a comparison of the average spectral

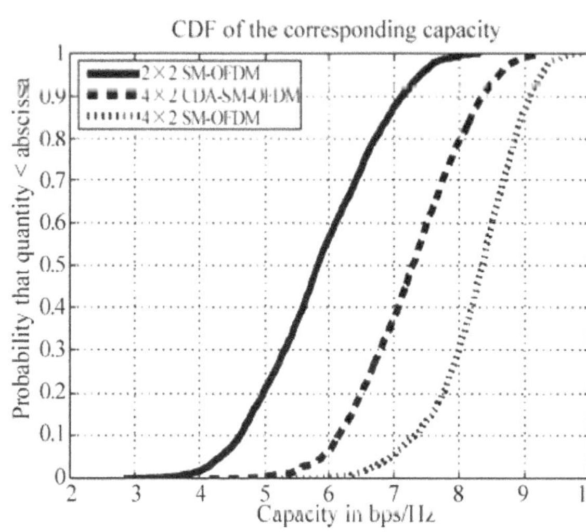

Figure 4: Comparison of system capacity for 4 ′ 2 CDASM-OFDM system with 2 ′ 2 and 4 ′ 2 SM-OFDM system [35].

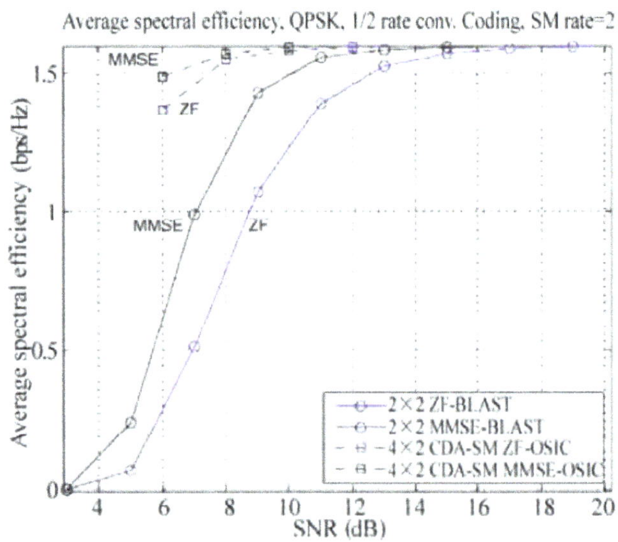

Figure 5: Average spectral efficiencies in bps/Hz [35].

Efficiencies of 2 ´ 2 SM-OFDM systems and 4 ´ 2 CDA-SM-OFDM systems for a low user mobility indoor WLAN scenario. Here it can be seen that the 4 ´ 2 CDA-SM-OFDM systems provide much higher spectral efficiencies at low SNR values.

Singular Value Decomposition Based MIMO Precoding

Singular value decomposition (SVD) based MIMO precoding is a closed-loop MIMO scheme where the precoding filter at the transmitter is designed by taking the SVD of the MIMO channel matrix H.

[36] Provides an analysis of the classical SVD based MIMO precoding scheme, SVD based precoding with ZF equalization, SVD based precoding with MMSE equalization and also an improved SVD based precoding technique. All of these schemes are analyzed with realistic channel knowledge at the transmitter. Figure 6 shows the block diagram of the SVD based MIMO-OFDM transmitter and receiver.

Classical SVD Precoding and Equalization

In SVD based techniques, the channel matrix H of a MIMO system with N_t transmit antennas and N_r receive antennas, is decomposed as

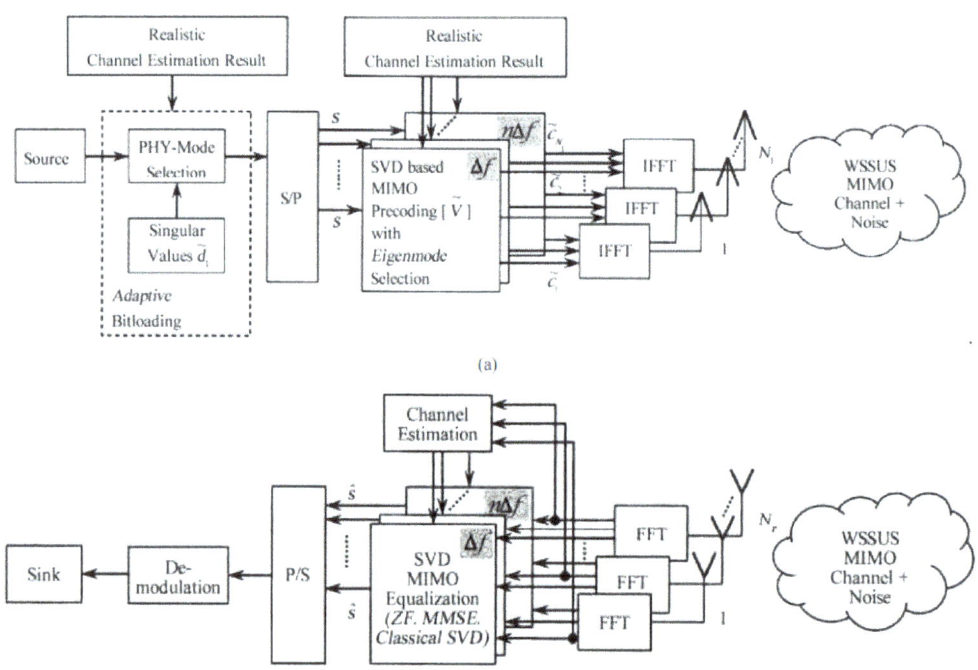

Figure 6: SVD based MIMO-OFDM system (a) Transmitter and (b) Receiver [36].

$$\mathbf{H} = \mathbf{U}\mathbf{D}\mathbf{V}^H \quad (5)$$

where and $\mathbf{U} \in \mathbb{C}^{N_r \times N_r}$ and $\mathbf{V} \in \mathbb{C}^{N_t \times N_t}$ are unitary matrices while $\mathbf{D} \in \mathbb{R}^{N_r \times N_t}$ is a diagonal matrix consisting of the ordered singular values d_i

The classical SVD approach utilizes matrix V for precoding at the transmitter. The columns of matrix are V the eigenvectors of HH^H. The received signal is given by

$$\mathbf{r} = \mathbf{H}\mathbf{V}\mathbf{s} + \mathbf{n} \quad (6)$$

where s is a vector of information symbols s_i and n is the noise vector corresponding to an additive white Gaussian noise (AWGN) process with variance σ_n^2 for each element. At the receiver, matrix U^H is employed for equalization and the detected signal vector is given by

$$\begin{aligned} \mathbf{y} &= \mathbf{U}^H\mathbf{r} = \mathbf{U}^H\mathbf{H}\mathbf{V}\mathbf{s} + \mathbf{U}^H\mathbf{n} \\ &= \mathbf{U}^H\mathbf{U}\mathbf{D}\mathbf{V}^H\mathbf{V}\mathbf{s} + \mathbf{U}^H\mathbf{n} \\ \mathbf{y} &= \mathbf{D}\mathbf{s} + \mathbf{U}^H\mathbf{n} \end{aligned} \quad (7)$$

Each individual received signal can be written as

$$y_i = d_i s_i + n_i' \tag{8}$$

where n_i' represents the i-th element of $U^H n$. The corresponding SISO SNR values are then given by

$$SNR_i = d_i^2 \frac{|s_i|^2}{2\sigma_n^2} \tag{9}$$

Equations (8) and (9) show that the singular values represent the MIMO processing gain for each of the eigenmodes. Therefore, SVD based MIMO precoding requires adaptive modulation and bit loading techniques for capacity maximization [36].

SVD Precoding with ZF Equalization

Linear ZF equalization can also be used at the receiver which is based on the inversion of the estimated MIMO channel matrix. ZF equalization requires the estimation of the product ofHV at the receiver. Assuming ideal channel knowledge at the receiver, the detected signal can then be given by

$$\begin{aligned} y &= (HV)^+ r \\ &= (HV)^+ HVs + (HV)^+ n \end{aligned} \tag{10}$$

$$y = s + (UD)^{-1} n \tag{11}$$

where $(\,)^+$ represents the pseudo inverse.

SVD Precoding with MMSE Equalization

MMSE equalization is based on minimizing the mean square error (MSE) between the transmitted and detected symbols. The minimum mean square error is given by

$$e = \min\left(\left\{|\hat{s}_i - s_i|^2\right\}\right) \tag{12}$$

where S_i is the transmitted symbol and $\hat{s}_i$ represents the received symbol. The detected signal for SVD based MIMO MMSE equalization is given by

$$y = \left[K\sigma_n^2 I + (HV)^H (HV)\right]^{-1} (HV)^H r \tag{13}$$

with $K = N_t$ utilized eigenmodes. As seen from Equation (13), MMSE MIMO equalization also requires the precoding matrix V at the receiver.

Improved SVD Precoding Technique with Realistic Channel Knowledge

An improved SVD based MIMO precoding technique is also proposed in [36] which maximizes MIMO capacity while considering realistic channel knowledge at the transmitter rather than the ideal one. The MIMO capacity for realistic channel knowledge is given by

$$C_H = \sum_{i=1}^{N_t} \log_2\left(1 + \tilde{d}_i^2 \frac{|s_i|^2}{2\sigma_n^2}\right) \tag{14}$$

where $\tilde{d}_i$ represent the singular values for the case of realistic channel knowledge. The improved technique considers the k strongest Eigen modes for transmission with $K < N_t$ if the following statement is fulfilled.

$$\sum_{i=1}^{N_t} \log_2\left(1 + \tilde{d}_i^2 \frac{|s_i|^2}{2\sigma_n^2}\right) \le \sum_{i=1}^{K} \log_2\left(1 + \frac{N_t}{K} \tilde{d}_i^2 \frac{|s_i|^2}{2\sigma_n^2}\right) \tag{15}$$

The remaining eigenmodes which correspond to the N_t - K unused eigenvectors of the precoding matrix $\tilde{V}$ are not utilized

Performance Comparison

Figure 7 shows the BER performance comparison of the classical SVD, ZF and MMSE equalization schemes for a 4 ′ 4 MIMO-OFDM system. A curve for ideal ZF equalization is also provided for reference. It is clear from the comparison that MMSE equalization provides the best results with realistic channel estimation.

Figure 8 shows the performance comparison of the MMSE equalization scheme with different values of K (utilized eigenmodes) for an uncoded 4 ′ 4 MIMO system with realistic channel knowledge at the transmitter. The BER curve for the case of ideal channel knowledge is also provided. The MMSE equalization scheme provides the best performance for = 2 utilized eigenmodes selected according to Equation (15).

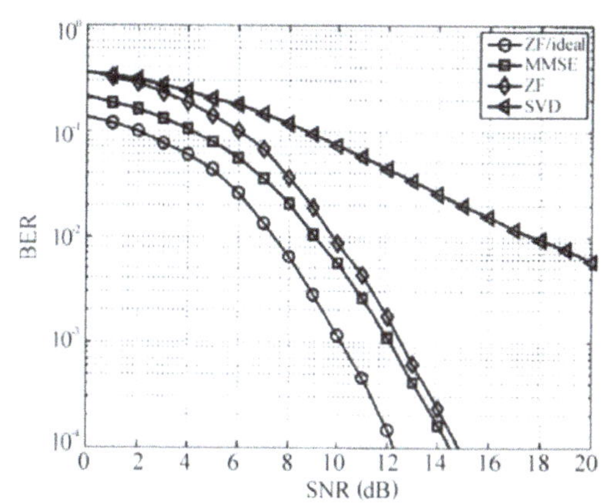

Figure 7: BER performance of uncoded 4 ′ 4 SVD based MIMO systems [36].

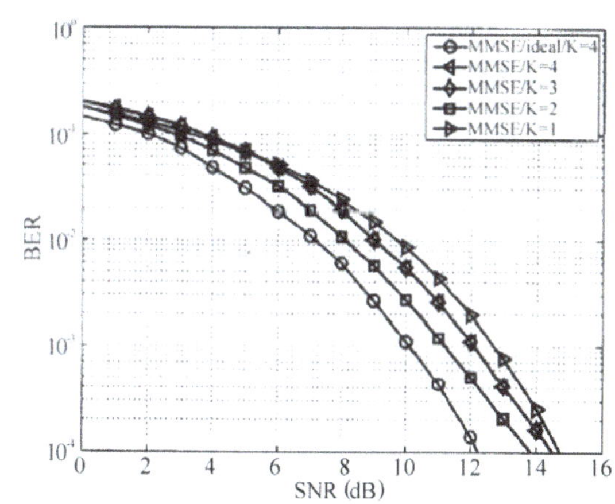

Figure 8: BER performance of an uncoded 4 ′ 4 SVD based MIMO system with MMSE equalization [36].

Geometric Mean Decomposition (GMD) Based MIMO

GMD based MIMO [2] is also a closed-loop joint transceiver design scheme which aims at optimally combining the benefits of MIMO diversity and spatial multiplexing. This technique utilizes the GMD of the MIMO channel matrix for precoder and equalizer design when the CSI is available at both the transmitter and the receiver. It is also applicable to MIMO-OFDM systems.

GMD calculation algorithm in [2] starts from the SVD of the channel matrix H which is given according to Equation (5)

$$\mathbf{H} = \mathbf{U}\mathbf{D}\mathbf{V}^H$$

The GMD is then given by

$$\begin{aligned}\mathbf{H} &= \mathbf{U}\mathbf{U}_R^H\mathbf{R}\mathbf{V}_R\mathbf{V}^H \\ &\triangleq \mathbf{Q}\mathbf{R}\mathbf{P}^H \end{aligned} \qquad (16)$$

where and are semi-unitary matrices, P being the linear precoder at the transmitter. $\mathbf{R} \in \mathbb{R}^{K\times K}$ is an upper triangular matrix whose diagonal elements are the geometric mean of the K nonzero singular values of H. The GMD scheme thus decomposes the MIMO channel into identical parallel subchannels which makes the symbol constellation selection and the overall system design much simpler. GMD can also be seen as an extended QR decomposition GMD MIMO can be implemented with the V-BLAST receiver and also with the zero-forcing dirty paper precoder (ZFDP). The V-BLAST technique has been discussed earlier in the text. The ZFDP technique also involves sequential nulling and cancellation but at the transmitter and utilizes CSI at the transmitter only. The ZFDP scheme combines QR decomposition and "dirty paper" precoding. The QR decomposition for ZFDP is given by

$$\mathbf{H}^H = \breve{\mathbf{Q}}\breve{\mathbf{R}} \qquad (17)$$

The sampled baseband received signal is then given by

$$\mathbf{y} = \breve{\mathbf{R}}^H\breve{\mathbf{Q}}^H\mathbf{x} + \mathbf{n} \qquad (18)$$

Substituting $\mathbf{x} = \breve{\mathbf{Q}}\breve{\mathbf{x}}$ we have

$$\mathbf{y} = \breve{\mathbf{R}}^H\breve{\mathbf{x}} + \mathbf{n} \qquad (19)$$

Let $\mathbf{s} \in \mathbb{C}^{K\times 1}$ be the transmitted symbol vector then $\breve{\mathbf{x}}$ should satisfy

$$\operatorname{diag}\{\breve{\mathbf{R}}\}\mathbf{s} = \breve{\mathbf{R}}^H\breve{\mathbf{x}} \qquad (20)$$

where the left-hand side represents the element-wise multiplication of the diagonal elements of $\breve{\mathbf{R}}$ with the elements of s. The solution to Equation (20) is then

$$\breve{\mathbf{x}} = \left(\breve{\mathbf{R}}^H\right)^{-1}\operatorname{diag}\{\breve{\mathbf{R}}\}\mathbf{s} \qquad (21)$$

The ZFDP scheme, unlike V-BLAST, does not suffer from the error propagation problem. However, due to the matrix inversion in Equation (21) the norm $\breve{\mathbf{x}}$ of can be significantly amplified resulting in increased transmitter power consumption. This problem can be resolved by using the Tomlinson-Harashima precoder to restrict the transmit signal level within acceptable limits [2]

Combining GMD with V-BLAST and ZFDP

The GMD-VBLAST scheme can be implemented beginning with the GMD of the channel matrix, H = QRP^H. The information symbol vector s is then encoded by the linear precoder P resulting in the transmit signal x = Ps. The resulting signal at the receiver is then given by

$$\mathbf{y} = \mathbf{Q}\mathbf{R}\mathbf{s} + \mathbf{n} \qquad (22)$$

which can be decoded simply by using the V-BLAST receiver. GMD-ZFDP scheme can be also be implemented in a similar way. The resulting K independent and identical subchannels are given by

$$y_i = \bar{\lambda}_H x_i + n_i \quad ; i = 1, \ldots, K \qquad (23)$$

where $\bar{\lambda}_H$ represent the subchannel gain and are in fact the identical diagonal elements of the matrix R [2].

Performance

Some simulation results from [2] depicting the performance of GMD based MIMO schemes are presented in the following text, assuming independent identically distributed (i.i.d) Rayleigh flat fading channels. Figure 9 shows a comparison of the capacity of GMD-MIMO with other schemes for 4 ´ 4 MIMO configuration. The informed transmitter (IT) curve corresponds to the Shannon channel capacity when CSI is available at both the transmitter and the receiver while the uninformed transmitter (UT) curve corresponds to the channel capacity when CSI is not available at the transmitter. MTM and MMD are both linear precoder design schemes for linear transceivers. MTM is based on the minimization of the trace of the MSE matrix while MMD minimizes the maximum diagonal elements of the MSE matrix resulting in nearoptimal performance. Clearly, GMD outperforms both MTM and MMD at high SNR and approaches optimal capacity. The capacity loss of GMD at low SNR is due to the ZF receiver. Based on GMD, the authors of [2] have also proposed another scheme called uniform channel decomposition (UCD) which can decompose a MIMO channel into identical subchannels in a strictly capacity lossless manner [37].

Figures 10 and 11 show the BER performance comparison of GMD-MIMO with ordered MMSE-VBLAST, MTM and MMD for 2 ´ 4 and 4 ´ 4 MIMO configurations respectively. GMD achieves much higher performance particularly at high SNR.

Figure 12 shows a performance comparison of GMDVLBAST and GMD-ZFDP when combined with OFDM for ISI suppression. GMD-VBLAST results in performance loss of about 2 dB because of error propagation.

Turbo-MIMO Systems

Turbo-MIMO systems represent a class of MIMO communication systems that combine the turbo-processing principle used in turbo coding with MIMO. These systems aim at attaining channel capacity close to the Shannon limit for MIMO channels with manageable complexity and can be implemented from diversity maximization or SM aspects [38].

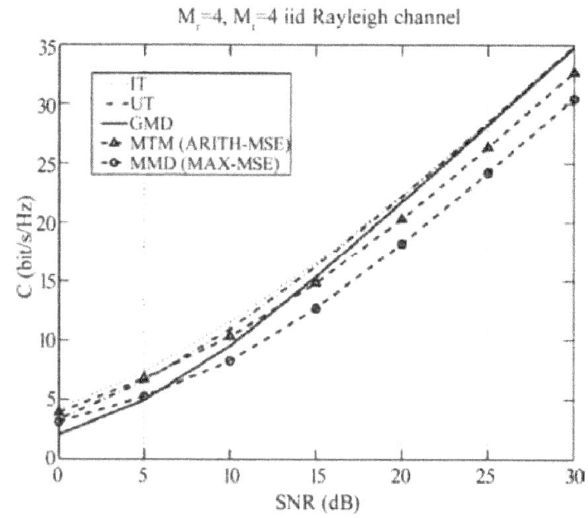

Figure 9: Average capacity for 4 ′ 4 MIMO configuration [2].

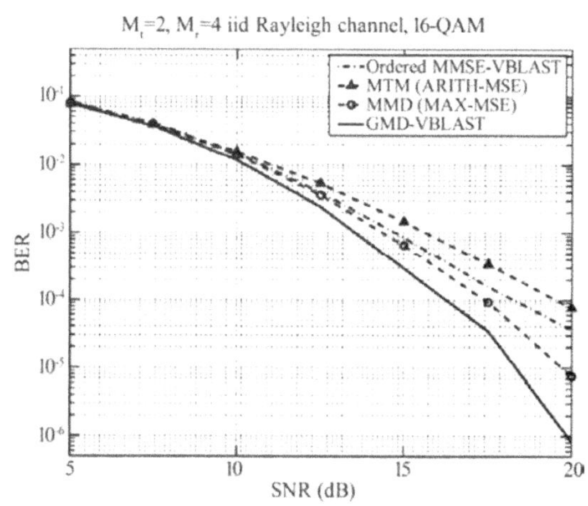

Figure 10: BER performance for 2 ′ 4 MIMO configuration [2].

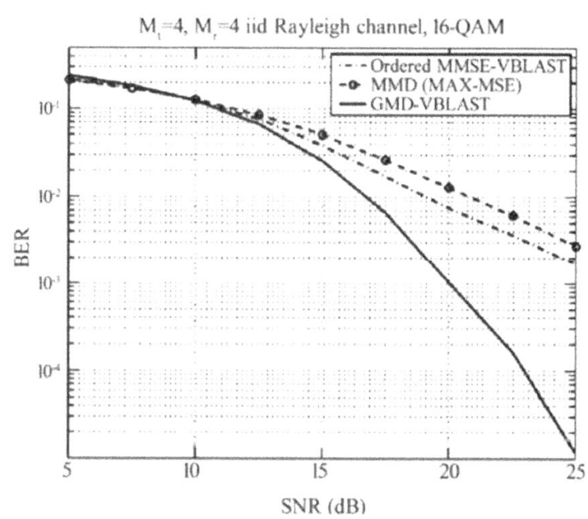

Figure 11: BER performance for 4 ′ 4 MIMO configuration [2].

A turbo-MIMO architecture known as Turbo BLAST is presented in [39]. This MIMO system is based on random layered space-time (RLST) coding which is a combination of independent block-time coding and space-time interleaving. The receiver uses iterative turbo-processing for RLST decoding and estimation of the flat fading MIMO channel matrix. A similar turbo-MIMO system based on space-time bit-interleaved coded modulation (ST-BICM) is presented in [38]. ST-BICM codes are formed by concatenation of a turbo encoded sequence and ST interleaving.

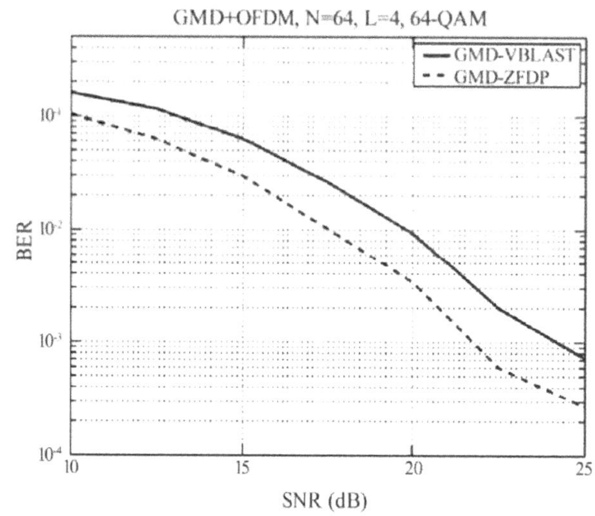

Figure 12: BER performance of GMD based MIMO-OFDM systems [2].

Figure 13 shows the block diagram of a ST-BICM MIMO system transmitter. The information bits are turbo encoded based on a linear forward error correction (FEC) code represented as the outer code. The encoded sequence is then bit-interleaved using a space-time pseudorandom interleaver denoted by P in the figure. Each interleaved substream is then independently mapped onto M-ary PSK or QAM symbols and transmitted using a separate antenna. The inner code basically represents a linear space-time mapper which allows for a flexible MIMO design with optimal diversity order and multiplexing gain or a desired tradeoff between the two. STBCs can be used to obtain the maximum diversity order while a symbol multiplexer can be used if full multiplexing gain is desired [38].

Figure 14 shows a double iterative decoding receiver for the ST-BICM MIMO system. It operates in two stages consisting of inner and outer iterative decoding loops. The inner and outer decoders are separated by an interleaver and a deinterleaver represented by P and P^{-1} respectively. This arrangement decorrelates the correlated outputs between the two stages. The decorrelator compensates for the interleaving operation at the transmitter. The two stages iteratively exchange information, producing a better estimate of the transmitted symbols after each iteration, until the receiver converges [38]. The inner decoder is in fact a MIMO detector, the optimal choice being the maximum a posteriori probability (MAP or APP) detector/decoder. However, due to the excessive computational complexity of APP detection, reduced-complexity near-optimal detectors like MMSESIC or reduced-complexity APP detectors e.g. the listsphere detector (LSD), iterative tree search (ITS) and multilevel bit mapping ITS (MLM-ITS) detectors can be used.

The outer decoder consists of a channel turbo decoder with two decoding stages separated by an interleaver and a deinterleaver denoted by α and α^{-1} respectively in Figure 14. This arrangement forms the outer iterative decoding loop of the ST-BICM MIMO receiver [38].

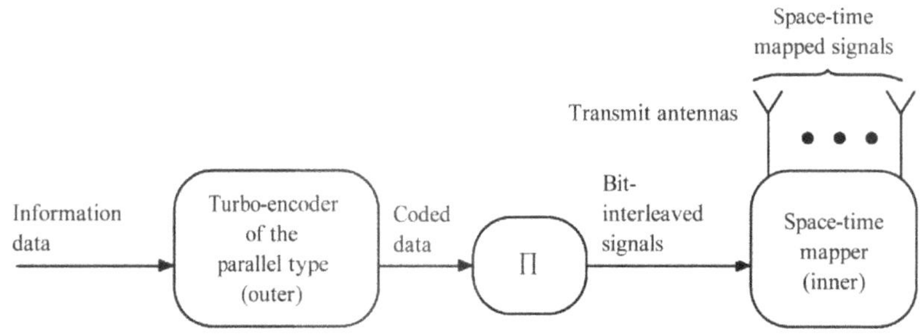

Figure 13: ST-BICM MIMO system transmitter [38].

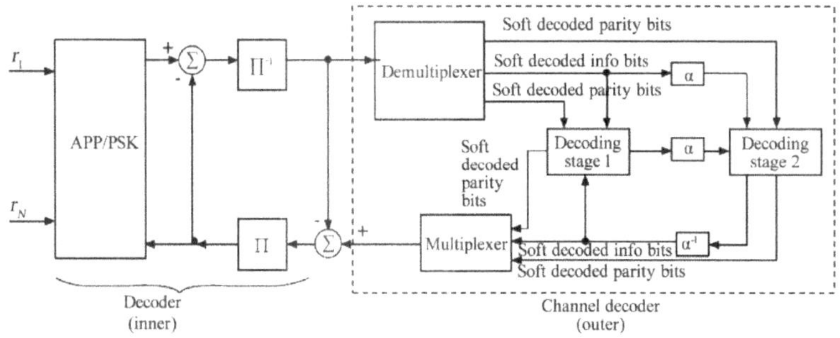

Figure 14: Receiver structure for the ST-BICM MIMO system [38].

Performance

Figure 15 shows the BER performance of a simulated 8 ´ 8 ST-BICM MIMO system using a rate-1/2, memory 2 turbo code as the outer channel code, with feed-forward and feedback generators 5 and 7 (octal) respectively. A block fading channel is assumed for the inner encoder which remains constant for a block size of 192 information bits with each block representing a statistically independent channel realization. A rich scattering Rayleigh MIMO model is used to select the elements of the MIMO channel matrix. 4 iterations are used in the inner decoder loop while 8 iterations are used in the outer channel decoder loop. The figure shows a comparison for different modulation schemes with MMSE-SIC and MLM-ITS inner detectors. The performance of MLMITS detection increases with larger list size M however, at the cost of increased complexity. The respective capacity limits for QPSK, 16-QAM and 64-QAM are also shown. At BER = 10-5 and M = 64, The ST-BICM systems using QPSK, 16-QAM and 64-QAM operate 1, 4 and 6 dB away from their respective capacity limits [38].

Figure 16 shows the BER performance of the STBICM system using MLM-ITS detector as the no. of iterations in the outer decoder increase from 1 to 5. Clearly the performance improves with the no. of iterations which pertains to only a linear increase in complexity. However, it can also be seen that the performance gain between successive iterations diminishes somewhat due to the feedback of correlated noise. Further increase in iterative gain can be achieved by using larger interleavers [38].

Limited Feedback Strategies for Closed-loop MIMO Systems

Certain closed-loop MIMO systems like the SVD and GMD based systems assume the availability of full CSI at the transmitter. Full CSI is available at the transmitter in a TDD system with duplex time less than the channel coherence time due to the reciprocity of the channel while in a FDD system a feedback channel for CSI is required thus consuming additional bandwidth. However, in practical scenarios the extra load resulting from large CSI feedback is not desirable and may not even be possible e.g. in case of rapidly varying mobile channels. Furthermore, results have shown that performance close to that with full CSI can be achieved by using limited feedback strategies utilizing only a few bits of feedback. Figure 17 shows the block diagram of a limited feedback MIMO system [40].

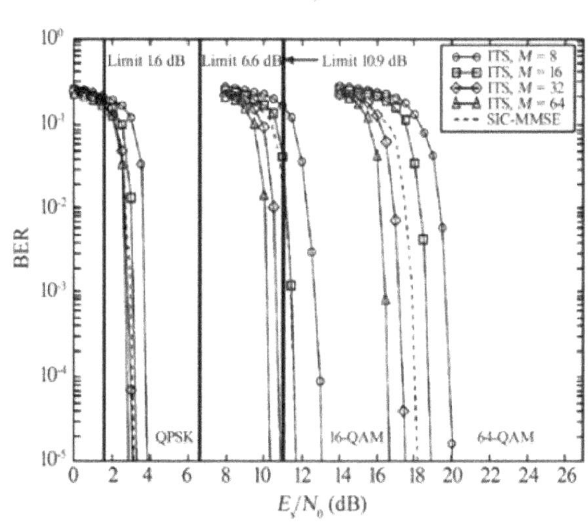

Figure 15: BER performance of 8 ´ 8 ST-BICM MIMO system with different modulation and inner detection schemes [38].

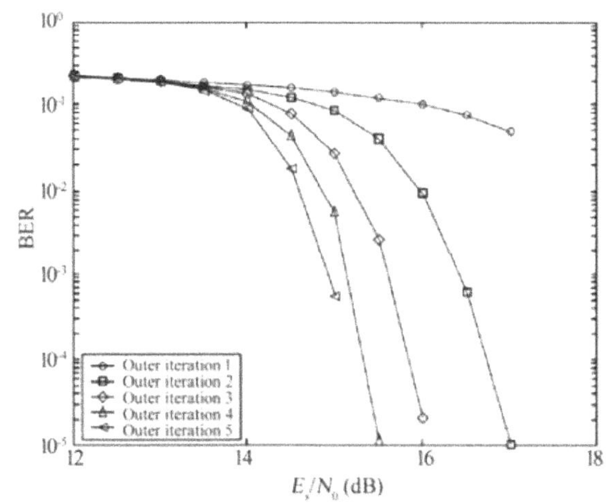

Figure 16: BER performance of 8 ´ 8 ST-BICM MIMO system with different no. of iteration in the outer decoder [38].

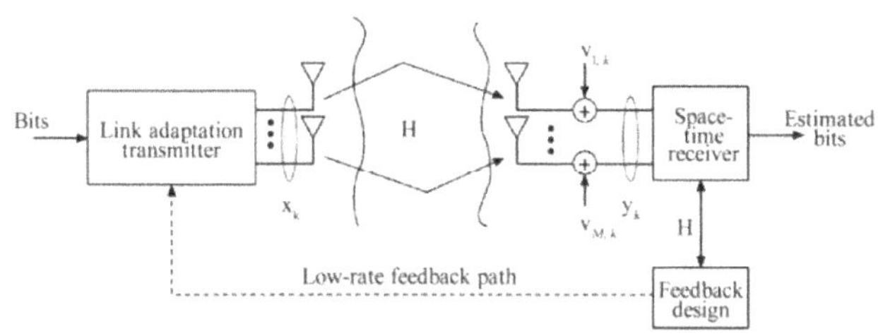

Figure 17: Limited feedback closed-loop MIMO system [40].

The feedback may be based on channel quantization or quantization of some properties of the transmitted signal. Channel quantization involves vector quantization (VQ) of the channel matrix H as depicted in Figure 18. The quantized version of the MIMO channel can then be fed back to the transmitter. However, it has been observed that quantization of the entire channel may not be necessary and it may be sufficient to include only some part of the channel structure like the channel singular vectors. For example, the optimal precoding matrix for an i.i.d MIMO channel consists of the eigenvectors of the channel covariance matrix, as columns. The feedback overhead may be further reduced by using only a limited number of quantized weighting vectors or matrices for precoding. This collection of precoding matrices is known as a precoding codebook and is shared by the transmitter and the receiver. The feedback consists of bits representing a particular precoding matrix within the codebook [40,41].

A large codebook length for vector quantization schemes results in increased complexity at the receiver due to the exhaustive search required for selecting a precoding matrix. In such cases, when the codebook length and therefore the corresponding no. of feedback bits B is large, scalar quantization of the elements of the precoding matrix can be employed instead. However, for small values of B, scalar quantization may become too inaccurate. In such cases, performance of scalar quantization can be improved by using the reduced rank approach where the columns of the precoding matrix are constrained to lie within a subspace of dimension less than the no. of transmit antennas N_t [40].

Figure 19 shows the symbol error rate (SER) performance of a simulated 4 ´ 5 limited feedback MIMO beamformer with different feedback strategies. The system uses 16-QAM modulation for transmission and MRC at the receiver. Optimal BF in the figure represents the optimal beamformer with unquantized feeback and full CSI at the transmitter. Grassmannian BF (6-bit) represents signal adaptive beamforming using a 6-bit feedback VQ codebook and results in the best performance, lying within 0.7 dB of the optimal BF and approximately 1dB better than the 40-bit channel quantization which suffers from large quantization error. The 6-bit quantized reduced rank (RR) beamformer with dimension D = 3, performs close to the 40-bit channel quantization [40].

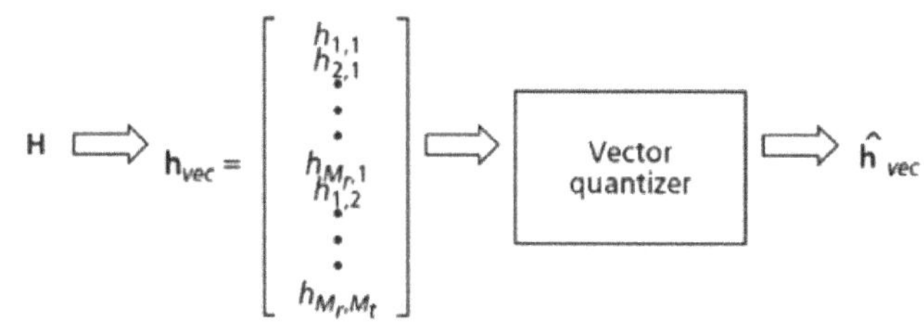

Figure 18: Channel Quantization [40].

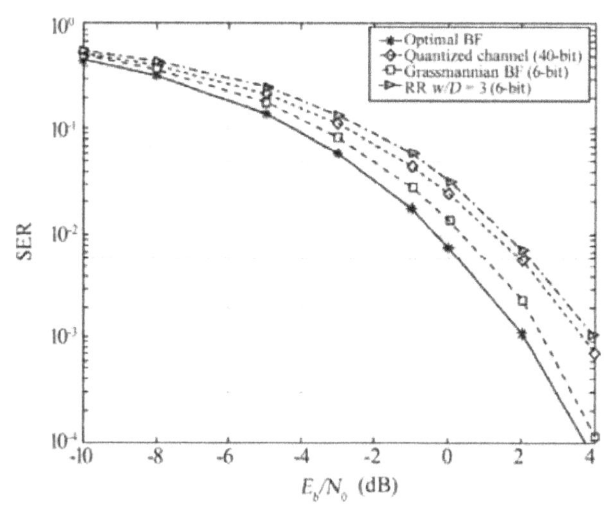

Figure 19: Limited feedback beamformer performance for 4 ´ 5 MIMO configuration [40].

Link Adaptation without Precoding

In addition to the precoding matrix, other information e.g. the received signal to interference and noise ratio (SINR) may also be included in the feedback for link adaptation. However, some MIMO schemes like the modified VBLAST schemes in [32,34] rely solely on this type of feedback without any precoding information.

Another example is the 2-codeword multiple codewords (2CW-MCW) scheme for FDD MIMO-OFDM cellular systems proposed in [41] that uses SINR feedback for each stream to select a suitable modulation and coding scheme (MCS) for each of the two simultaneously transmitted codewords. The two codewords are mapped onto 2 and 4 streams respectively for 2 ´ 2 and 4 ´ 4 antenna configurations. The mapping may either be fixed or adaptive. Adaptive mapping also makes use of the SINR feedback. Precoding is not used in this scheme resulting in reduced feedback overhead.

Partial Feedback Schemes

Partial feedback schemes for MIMO systems are based on the feedback of statistical channel information along with some instantaneous channel quality indicator (CQI) e.g. SNR, SINR etc. to the transmitter. A partial feedback scheme for MIMO-OFDM systems involving the decomposition of MIMO channel covariance matrix is presented in [42].

The covariance matrix R is calculated from the estimated MIMO channel matrix H (for the k-th subcarrier) at the receiver and is given by

$$\mathbf{R} = \mathrm{E}\{\mathbf{H}^H\mathbf{H}\} \qquad (24)$$

The matrix R is then decomposed using SVD which is given by

$$\mathbf{R} = \mathbf{U}_{Stat}\mathbf{\Lambda}_{Stat}\mathbf{V}_{Stat}^{H} \qquad (25)$$

Where $\mathbf{\Lambda}_{Stat}$ is a diagonal matrix containing the singular values while $\mathbf{U}_{Stat}$ and are $\mathbf{V}_{Stat}$ unitary matrices. The feedback includes the matrix $\mathbf{\Lambda}_{Stat}$ and the column vectors of $\mathbf{V}_{Stat}$ for power allocation and spatial processing (precoding) at the transmitter [42].

Figure 20 shows the block diagram of a $N_T \times N_R$ MIMO-OFDM system based on this partial feedback scheme which transmits N_S spatial streams using N_C OFDM subcarriers. The received signal vector for the k-th subcarrier is given by

$$\mathbf{y} = \mathbf{HQAx} \qquad (26)$$

where x is the transmit data vector, H is the MIMO channel matrix for the k-th subcarrier, A is a diagonal matrix with N_S diagonal elements determined by the matrix Λ_{Stat} feedback for power allocation to the active spatial streams, and the $N_T \times N_S$ matrix Q represents a spatial processing transformation which maps the spatial streams to the transmit antennas. The Q matrix is constructed from the vectors of V_{Stat} received at the transmitter via feedback and is used to maximize the received energy for each transmitted spatial stream. This enables maximum ratio transmission (MRT) and SVD beamforming along with tracking of spatial variations of the MIMO channel [42].

The MIMO channel covariance and the corresponding channel singular values do not vary rapidly with time even at vehicular speeds around 100 km/h [42]. This greatly reduces the feedback load on the system and makes this closed-loop MIMO-OFDM system suitable for mobile environments.

Figures 21 and 22 show the simulated frame error rate (FER) performance of the proposed MIMO-OFDM system in comparison with open-loop SM and perfect CSI feedback MIMO-OFDM systems, for 2 ´ 2 and 4 ´ 4 MIMO configurations respectively. The figures include FER performance curves for QPSK and 64-QAM modulation in a low speed mobile scenario using the ITU PB channel profile with vehicular speeds of 3 km/h. The OFDM scheme is based on 512-point FFT with 15 subchannels for data transmission each consisting of 20 continuous subcarriers, for a total bandwidth of 5 MHz. The frame duration is about 0.5ms. Turbo coding is employed for FEC and MMSE detection is used at the receiver.

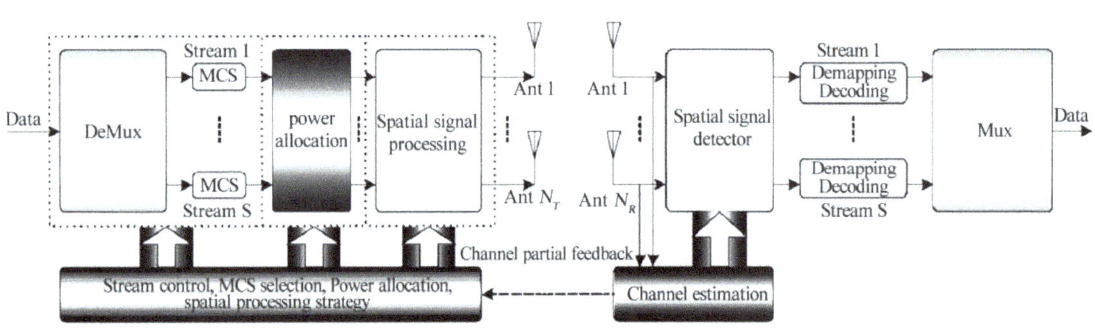

Figure 20: MIMO-OFDM system with partial feedback [42].

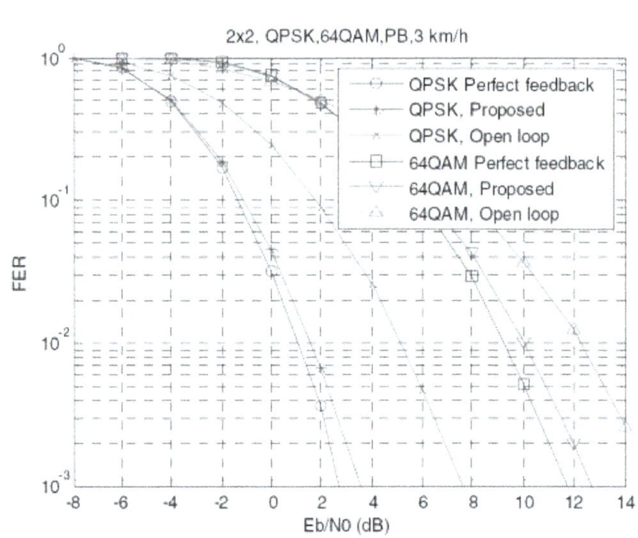

Figure 21: FER performance for coded 2 ´ 2 MIMO configuration [42].

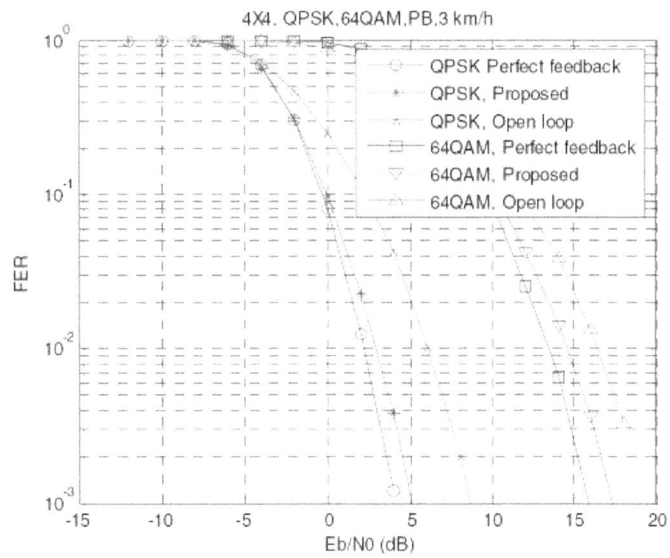

Figure 22: FER performance for coded 4 ´ 4 MIMO configuration [42].

Equal power allocation is used for the open-loop SM system while water-filling is used for the closed-loop systems. Ideal channel knowledge is assumed at the receiver for all systems and antenna correlations are not considered [42]. As seen from the results, the proposed system operates quite close to the perfect CSI feedback system and shows substantial performance gain over the open-loop system. The small performance loss in comparison with the perfect feedback system is primarily due to the quantization error associated with limited feedback [42]. Efficient feedback reconstruction algorithms for improvement of the closed-loop transmit diversity scheme constituting the mode 1 of 3GPP's wideband code-division multiple access (WCDMA) 3G standard are presented in [43]. These algorithms efficiently reconstruct the beamforming weights at the transmitter while considering the effect of feedback error. Performance results for vehicular speeds up to 100 km/h are provided. The proposed techniques are applicable to closed-loop MIMO diversity systems and may possibly be extended to 4G systems.

In [44], the optimal MIMO precede designs for frequency-flat and frequency-selective fading channels are presented, assuming partial CSI at the transmitter consisting of transmit and receive correlation matrices. The elements of transmit and receive correlation matrices are determined from the respective transmit and receive antenna spacing and angular spread. It is shown that from the capacity maximization perspective, the optimal precede for a frequency-flat fading channel is an eigen-beamformer. On the other hand, the optimal precede for a frequency-selective fading channel represented by L uncorrelated effective paths consists of P + L parallel eigen beamformers where P is an arbitrary value depending on the no. of vectors in a transmission data block [44]. A closed-loop limited feedback MIMO scheme called multi-beam MIMO (MB-MIMO) is proposed in [45] for 3GPP LTE E-UTRA downlink. MB-MIMO employs multiple fixed beams at the base station (Node B) to transmit multiple data streams. The no. of beams and data streams to be used are adaptively selected using a codebook at the UE. The selected precoding vectors or beam indices constituting a precoding matrix are then fed back to the Node B. The MB-MIMO scheme can adaptively switch between MIMO SM and transmit beamforming (Tx-BF) modes. Tx-BF is used if a single beam is selected and SM is used if multiple beams are selected. The proposed scheme eliminates the need for a hardware calibrator (HW-CAL) at Node B that was required for a previously proposed MB-MIMO implementation. HW-CAL compensates the phase variations caused by RF components and was needed to align the phase condition of each transmit antenna element for maximizing the transmit beamforming gain. The proposed scheme uses a larger codebook based on an extended precoding matrix which includes phase terms that can be controlled to align the phase condition of the 4 node B antenna elements. This results in high beamforming gain even without HW-CAL. However, 4 additional bits or a total of 8 bits are required for feedback, which is still a small number.

MIMO over High-Speed Mobile Channels

Open-loop MIMO diversity techniques like STC and space frequency coding (SFC) are appropriate choices for high-speed mobile channels that vary rapidly with time. In such scenarios, maintaining a reliable link becomes the foremost priority rather than maximizing system throughput. High-speed mobile channels undergo fast fading which may cause time variation of the fading channel within an OFDM symbol period. This results in the loss of subchannel orthogonality and leads to interchannel interference (ICI) due to the distribution of leakage signals over other OFDM subcarriers. The error floor associated with ICI increases with the speed of the mobile terminal [46]. An improved MIMO-OFDM technique for high-speed mobile access in cellular environments is proposed in [46]. This technique reduces ICI and provides diversity gain as well as noise averaging even for highly correlated channels. ICI is reduced by transmitting weighted data on adjacent subcarriers. The weights are selected such that the mean ICI power is minimized. The adopted weight selection procedure however results in suboptimal weights. Diversity gain in [46] is achieved by using space-frequency block coding (SFBC) which is based on Alamouti code but the coding is applied in frequency domain i.e. to OFDM subcarriers rather than to

OFDM symbols in time domain [47]. Figure 23 shows the data assignment scheme for the 2 ′ 1 SFBC-OFDM system, without the weighting factors. The transmit data is assigned to subcarrier groups each consisting of two adjacent subcarriers, as shown in the figure. Instead of SFBC, other diversity techniques such as STBC, space-frequency trellis coding (SFTC), maximal

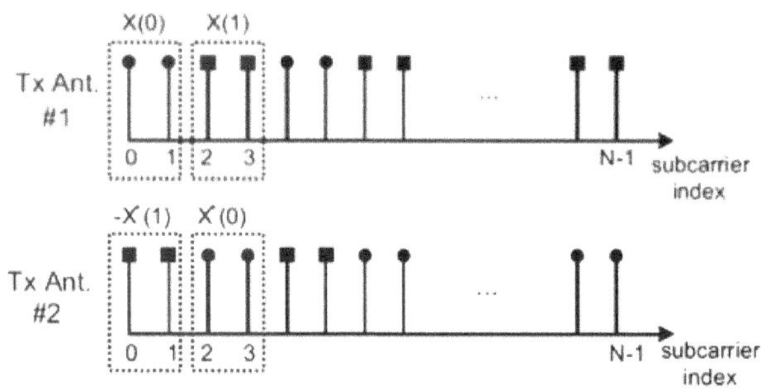

Figure 23: Data assignment scheme for ICI reduction in 2 ′ 1 SFBC-OFDM system [46].

Ratio receive combining (MRRC) etc. can also be used. The proposed technique operates without CSI at the transmitter and does not require any pilot signals for channel tracking. However, it is suitable for OFDM systems with subcarrier group spacing less than the channel coherence bandwidth because the channel coefficients are assumed to be identical for adjacent subcarriers.

Figure 24 shows the simulated BER performance of the proposed SFBC-OFDM scheme in comparison with conventional SFBC MIMO-OFDM schemes for 2 ′ 1 antenna configuration using I-METRA MIMO channel model Case A for downlink transmission with mobile speed of 250 km/h. Case A corresponds to a frequencyflat Rayleigh fading channel with uncorrelated antennas. 25 MHz of downlink channel bandwidth is used at 2 GHz with 2048 OFDM subcarriers. Performance results for QPSK and 16-QAM are provided.

Figure 25 shows the performance comparison using

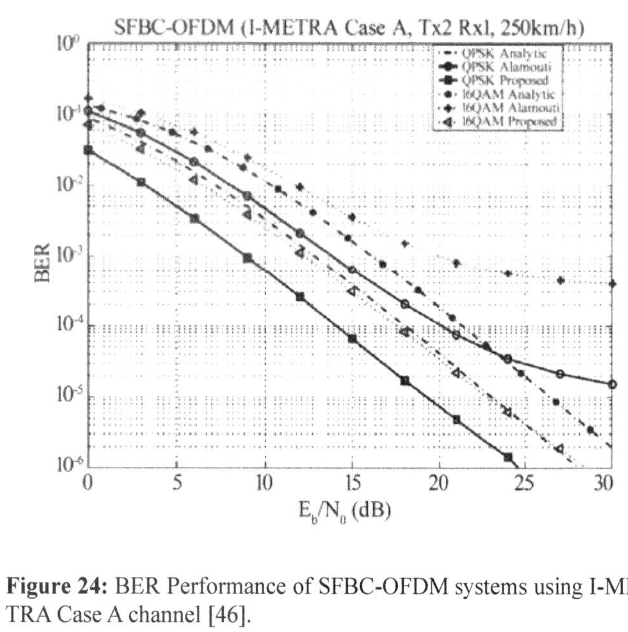

Figure 24: BER Performance of SFBC-OFDM systems using I-METRA Case A channel [46].

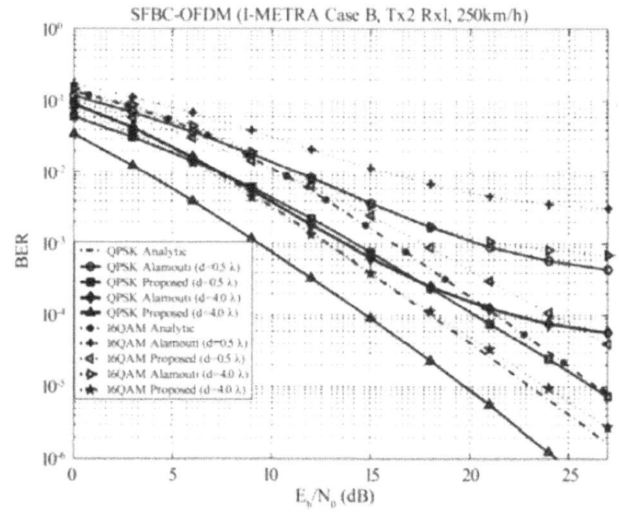

Figure 25: BER Performance of SFBC-OFDM systems using I-METRA Case B channel [46].

I-METRACaseBwhichcorrespondstoafrequencyselective fading channel with correlated transmit antennas in an urban macro cellular environment. The proposed SFBC-OFDM scheme clearly outperforms the conventional SFBC-OFDM schemes in both cases. The conventional SFBC-OFDM scheme referred to as Alamouti in the figures is severely performance limited due to the error floor phenomenon resulting from ICI introduced by the high-speed mobile user at 250 km/h.

MULTIUSER MIMO

Multiuser MIMO (MU-MIMO) systems consist of multiple antennas at the BS and a single or multiple antennas at each UE. MU-MIMO enables space-division multiple access (SDMA) in cellular systems which increases the system capacity by exploiting the spatial dimension (i.e. the location of UEs) to accommodate more users within a cell. It also provides beamforming or array gain as well as diversity gain due to the use of multiple antennas. In case of multiple antennas at the UE, spatial multiplexing can also be employed to further enhance the spectral efficiency [48]. The uplink and the downlink of a MU-MIMO system represent two different problems which are discussed in the following text.

The MU-MIMO Uplink

The MU-MIMO uplink channel is a MIMO multiple access channel (MIMO-MAC) [49] where the users simultaneously transmit data over the same frequency channel to the BS equipped with multiple antennas. The BS must separate the received user signals by means of array processing, multiuser detection (MUD), or some other method [48]. Figure 26 shows various linear and nonlinear MUD schemes for MIMO-OFDM systems, some of which are discussed in the later sections.

Classic SDMA-OFDM MUDs

An overview of some classic MUDs for MU-MIMOOFDM is presented in [3]. The discussion is based on the

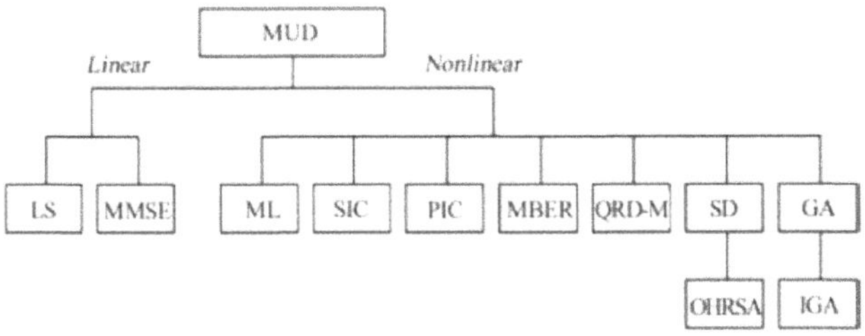

Figure 26: Various multiuser detectors (MUDs) for MIMOOFDM systems [3].

uplink MIMO SDMA-OFDM system model of Figure 27 where each of the L UEs uses a single transmit antenna while the BS is equipped with P antennas.

The complex-valued P ´ 1 received signal vector at the BS antenna array for the k-th subcarrier of the n-th OFDM symbol is given by

$$\mathbf{x} = \mathbf{H}\mathbf{s} + \mathbf{n} \tag{27}$$

where s is the L ´ 1 transmitted signal vector, n is the P ´ 1 AWGN noise vector and H is the P ´ L channel transfer function matrix consisting of L column vectors, each containing the transfer functions for a particular UE. Therefore, H can be represented as

$$\mathbf{H} = \left[\mathbf{H}^{(1)}, \mathbf{H}^{(2)}, \cdots, \mathbf{H}^{(L)}\right] \tag{28}$$

where

$$\mathbf{H}^{(l)} = \left[H_1^{(l)}, H_2^{(l)}, \cdots, H_P^{(l)}\right]^T, \quad l = 1, \cdots, L \tag{29}$$

is a P ´ 1 vector whose elements are the channel transfer functions for the transmission paths between the transmit antenna of the l-th UE and the P BS antennas

It is assumed that the complex signal s $^{(l)}$ transmitted by the l-th user has zero mean and variance σ_l^2 while the AWGN signal n_p has zero mean and variance σ_n^2 .

The channel transfer functions $H_p^{(l)}$ are assumed to be independent, stationary, complex Gaussian distributed processes with zero mean and unit variance.

The classic MUD schemes [3] are discussed in the following text

1) MMSE MUD

Figure 28 shows the schematic diagram of a MMSE SDMA-OFDM MUD. The multiuser signals received at each BS antenna are multiplied by a complex-valued array weight $w_p^{(l)}$ and then summed up. The superscript l represents a particular user which means that a separate set of weights is used for detection of each user's signal. The combiner output y $^{(t)}$ is subtracted from a user specific reference signal r (t) known at the BS and the UE, resulting in an error signal ϵ(t) The error signal is used for weight estimation according to the MMSE criterion. The steepest descent algorithm can be used in this regard for stepwise weight adjustment for each subcarrier of each user. The performance of the MMSE MUD improves as the no. of antennas P in the BS antenna array is increased and degrades when the no. of users increase.

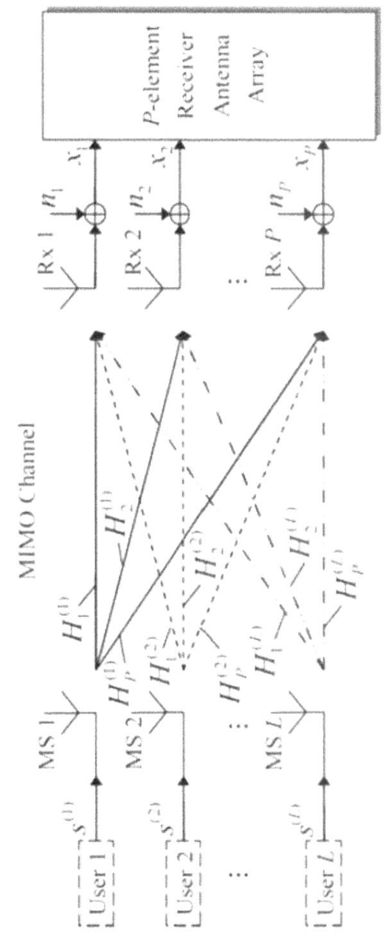

Figure 27: Uplink MIMO SDMA-OFDM system model with single antenna at each UE [3].

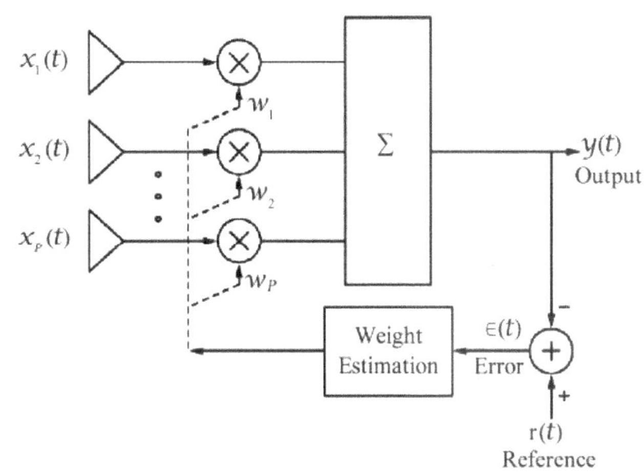

Figure 28: MMSE SDMA-OFDM MUD [3].

2) Successive Interference Cancellation (SIC) MUD:

The successive interference cancellation (SIC) MUD enhances the MMSE MUD using SIC. For each subcarrier, the detection order of the users is arranged according to their estimated total received signal power at the BS antenna array and the strongest user's signal with the least multiuser interference (MUI) is detected using the MMSE MUD. The detected user signal is then subtracted from the composite multiuser signal and the next strongest user is detected by the same procedure. This process continues till the detection is completed for all users. SIC results in high diversity gain at the MMSE combiner, which mitigates the effects of MUI as well as channel fading. The SIC MUD is also effective in near-far scenarios that result from inaccurate power control. However, it is prone to errors in power classification of user signals and also to interuser error propagation. Figure 29 shows the BER performance comparison of MMSE MUD and SIC MUD (M-SIC with M = 2) for an SDMAOFDM scenario with four single-antenna UEs and a four-antenna BS antenna array using QPSK modulation. The indoor short wireless asynchronous transfer mode (SWATM) channel model is used. Copyright © 2010 SciRes. IJCNS

3) Parallel Interference Cancellation (PIC) MUD:

The PIC MUD does not require any power classification of the received user signals. The detection procedure consists of two iterations for all subcarriers. In the first iteration, MMSE detection is used to estimate all user signals y $^{(l)}$ from the received composite multiuser signal vector x. In case of channel encoded transmission, all user signals must be decoded, sliced, channel encoded

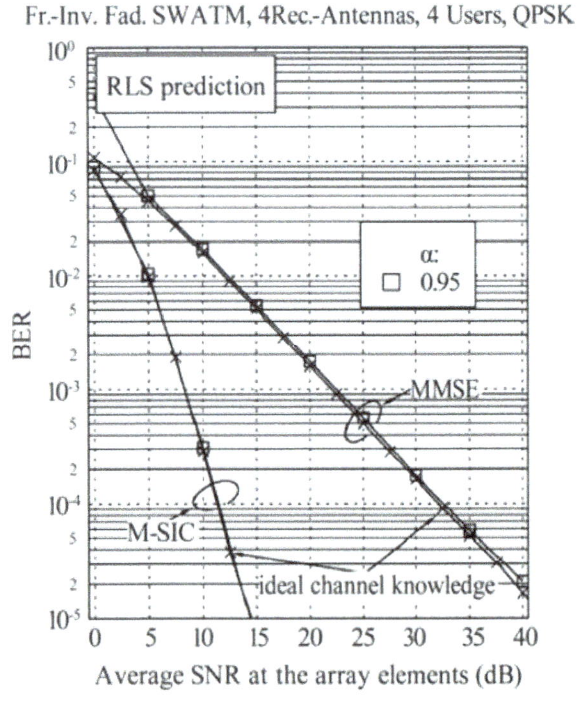

Figure 29: BER performance of MMSE MUD and SIC MUD [3].

Again and also remodulated onto subcarriers. In the second detection iteration, signal vectors $\tilde{\mathbf{x}}^{(l)}$ for all L users are reconstructed and an estimate $\tilde{y}^{(k)}$ of each user signal is generated by subtracting the signal vectors $\tilde{\mathbf{x}}^{(l)}, l \neq k$ of all other users followed by MMSE combining. The estimated user signals are then channel decoded and sliced. The PIC MUD scheme is also vulnerable to interuser error propagation.

4) Maximum Likelihood (ML) MUD:

The ML MUD employs the ML detection principle to find the most likely transmitted user signals through an exhaustive search. It provides the optimal detection performance but also has the highest complexity of any other MUD. For an OFDM-SDMA system with L simultaneous users, the ML MUD produces the estimated L ′ 1 symbol vector $\hat{\mathbf{s}}_{ML}$ consisting of the most likely transmitted symbols of the L users for a particular OFDM subcarrier, as given by

$$\hat{\mathbf{s}}_{\mathrm{ML}} = \arg\min_{\check{\mathbf{s}} \in \mathcal{M}^L} \|\mathbf{x} - \mathbf{H}\check{\mathbf{s}}\|^2 \tag{30}$$

where M ^{L} is a set containing 2^{mL} trial vectors, m being the no. of bits per symbol depending on the modulation scheme used resulting in 2^m constellation points. Therefore, the computational complexity of the ML MUD increases exponentially with the no. of users L thus making it prohibitive for practical implementation

5) Sphere Decoding (SD) aided MUD:

SD-aided MUDs use SD for reduced-complexity ML multiuser detection with near-optimal performance. SD reduces the ML search to within a hypersphere of a certain radius around the received signal. The radius of this search sphere determines the complexity of the MUD. Various SD algorithms have been proposed in literature like the complex-valued SD (CSD) and multistage SD (MSD) which significantly reduce the complexity by reducing the required search radius [3].

Layered Space-Time MUD

A V-BLAST based MUD scheme referred to as layered space-time MUD (LAST-MUD) is presented in [50] for CDMA uplink. This scheme is somewhat similar to the SIC MUD since V-BLAST detection also incorporates SIC.

Figure 30 shows the layered space-time MU-MIMO system block diagram. Here the single antenna users are arranged in G groups each containing M users for a total of K = G M′ . The UEs within each group are treated as the multiple transmit antennas of a V-BLAST system. The users within each group share the same unique spreading code which distinguishes the groups from one another. Therefore, out of the K total spreading codes, only G are unique. The N ′ K random spreading matrix consisting of K length N code vectors is denoted by $\mathbf{S} = [\mathbf{S}_1, \cdots, \mathbf{S}_1, \mathbf{S}_2, \cdots, \mathbf{S}_2, \cdots, \mathbf{S}_G, \cdots, \mathbf{S}_G]$ The proposed system can also accommodate users with multiple antennas thus enabling spatial multiplexing for achieving high data rates. In that case each user with multiple transmit antennas will be considered as one group. The K ′ 1 transmitted symbol vector is represented as $\mathbf{b} = [b_1, \cdots, b_K]^T$ where each element represents the bit transmitted by a particular user. The channel between the users and the BS is considered to be a frequency-flat fading MIMO channel and is denoted by the channel matrix $\mathbf{H} = [\mathbf{h}_1, \cdots, \mathbf{h}_P]^T$ where h_p is the K ′ 1 channel coefficient vector between all K users and the p-th BS antenna. The BS is equipped with a total of P antennas

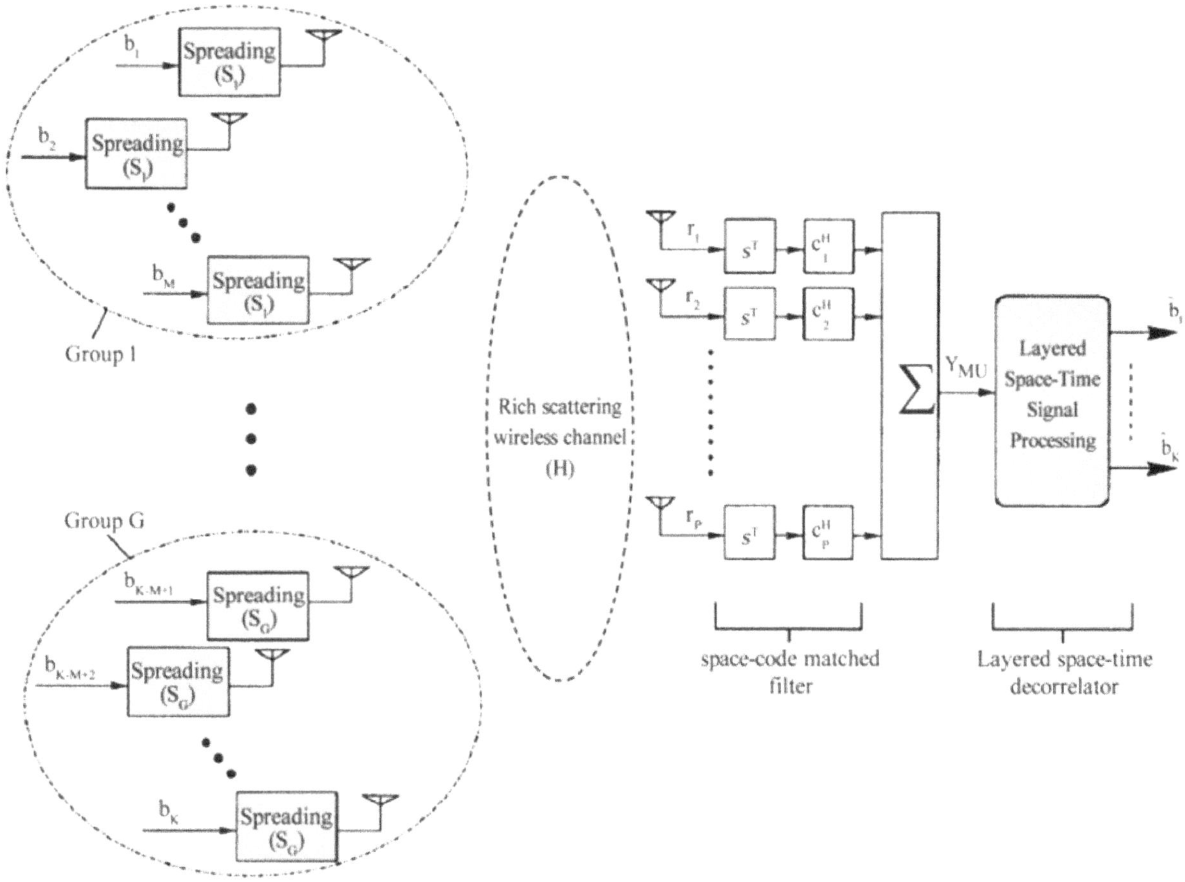

Figure 30: LAST MU-MIMO system [50].

The N ′ 1 received baseband signal at the p-th BS antenna for a certain symbol period after chip-matched filtering is given by

$$\mathbf{r}_p = \mathbf{S}\mathbf{C}_p\mathbf{b} + \mathbf{n}_p \tag{31}$$

Where $\mathbf{C}_p = \text{diag}(\mathbf{h}_p)$ is the complex diagonal channel matrix for the p-th BS antenna and Cp = diag( ) np is the corresponding complex-valued AWGN noise vector with zero mean and variance 2 $_s$. A frequency-flat fading MIMO channel is assumed as well as perfect channel estimation and symbol synchronization. The users are assumed to be separated by a considerable distance so that the antennas of different users are not correlated. The channel estimation and symbol synchronization at the BS is also assumed to be ideal.

The BS employs space-code matched filtering to separate the different user groups. The K ′ 1 sufficient statistic vector Y_{MU} is then fed to the layered spacetime decorrelator which eliminates the remaining interuser interference to produce the estimated symbol vector $\hat{\mathbf{b}} = [\hat{b}_1, \cdots, \hat{b}_K]$. The vector Y_{MU} is given by

$$\mathbf{Y}_{MU} = \sum_{p=1}^{P} \mathbf{C}_p^H \mathbf{S}^T \mathbf{r}_p = \tilde{\mathbf{R}}_{MU}\mathbf{b} + \tilde{\mathbf{n}} \tag{32}$$

Where $\tilde{\mathbf{R}}_{MU}$ is the K ′ K space-code cross-correlation matrix,

$$\tilde{\mathbf{R}}_{MU} = \sum_{p=1}^{P} \mathbf{X}_p^H \mathbf{X}_p \tag{33}$$

with $\mathbf{X}_p = \mathbf{S}\mathbf{C}_p$. The K ′ 1 real Gaussian noise vector n% with covariance matrix $\sigma^2 \mathbf{R}_{MU}$ is given by

$$\tilde{\mathbf{n}} = \sum_{p=1}^{P} \mathbf{X}_p^H \mathbf{n}_p \tag{34}$$

The detection algorithm is iterative and consists of three steps: 1) computation of the nulling vector, 2) user signal estimation and 3) interference cancellation (SIC).

For the i-th iteration, the first step consists of calculating the pseudoinverse $[\tilde{\mathbf{R}}_{MU}(i)]^+$ of $\tilde{\mathbf{R}}_{MU}(i)$. The user signals are then ranked according to their post detection SNRs (following the space-code matched filtering) and the user having the highest SNR, given by

$$k_i = \arg \max_{j \notin \{k_1, \cdots, k_{i-1}\}} \frac{\|b_j\|^2}{\sigma^2 \left[\tilde{\mathbf{R}}_{MU}(i)\right]^+_{(j,j)}} \tag{35}$$

is selected. The subscripts i and j in this equation denote the elements of an array, vector or matrix. The nulling vector for the selected user is $\mathbf{w}_{k_i} = [\tilde{\mathbf{R}}_{MU}(i)]^+_{k_i}$ i.e., the k_i-th column of $[\tilde{\mathbf{R}}_{MU}(i)]^+$ The slicer output is then given by

$$z_{k_i} = \mathbf{w}_{k_i}^T \mathbf{Y}_{MU}(i) \tag{36}$$

resulting in the estimated symbol $\hat{b}_{k_i}$ The final step is interference cancellation where the detected symbol is subtracted from the received signal vector resulting in the symbol vector for the next iteration i +1, given by

$$\mathbf{r}_p(i+1) = \mathbf{r}_p(i) - (\mathbf{X}_p(i))_{k_i} \hat{b}_{k_i} \tag{37}$$

where $(\mathbf{X}_p(i))_{k_i}$ is the k_i-th column of X_p (i). Similarly, $\mathbf{X}_p(i+1)$ and $\tilde{\mathbf{R}}_{MU}(i+1)$ are obtained by striking out the k_i-th column of $\mathbf{X}_p(i)$and the k_i-th row and column of $\tilde{\mathbf{R}}_{MU}(i)$ respectively $\mathbf{Y}_{MU}(i+1)$ is then given by

$$\mathbf{Y}_{MU}(i+1) = \sum_{p=1}^{P} \mathbf{X}_p(i+1)^H \mathbf{r}_p(i+1) \tag{38}$$

This process is repeated until all K user signals are detected. Two reduced complexity versions of LAST-MUD called serial layered space-time group multiuser detector (LASTG-MUD) and parallel LASTG-MUD are also presented in [50].

Figure 31 shows the SER performance of the LASTMUD scheme using 4-QAM modulation with 12 simultaneous users (single-antenna) and 6 BS antennas, as the no. of user groups is increased. Fixed spreading factor of N = 15 is used. The performance improves as the users are distributed into more (smaller) groups since the no. of unique spreading codes also increases. For G = 1, the performance is equivalent to V-BLAST and represents the worst case.

Figure 32 shows the SER performance as the no. of users is increased by adding more user groups with M = 4 users per group. The LAST-MUD scheme provides substantial increase in network capacity by accommodating a large no. of simultaneous users with good SER performance.

Smmse Sic Mud

A MUD scheme for MIMO-OFDM systems referred to as successive MMSE receive filtering with SIC (SMMSE SIC) is presented in [51]. The MMSE SIC MUD suffers from performance loss in scenarios with multiple closely spaced antennas located at the same UE. The proposed scheme tackles this problem by successively calculating the rows of the receive matrix at the BS for each of the UE transmit antennas, followed by SIC thus transforming the uplink MU-MIMO channel into a set of parallel SU-MIMO channels.

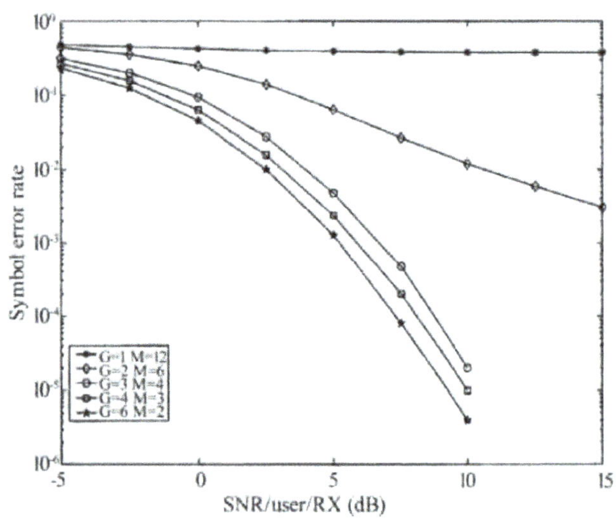

Figure 31: Grouping effect on SER performance of LAST-MUD [50].

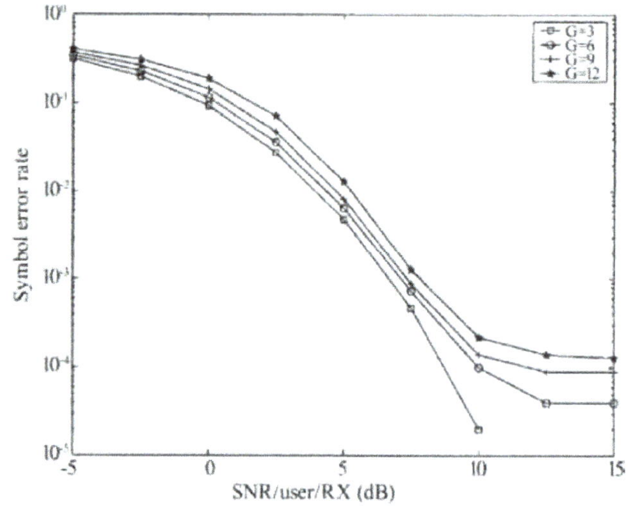

Figure 32: SER performance of LAST-MUD with increasing number of users [50].

Figure 33 shows the MU-MIMO uplink employing SMMSE SIC detection. The system consists of K simultaneous users, each equipped with M_{T_i} transmit antennas for i = 1, k, and therefore, a total of $M_T = \sum_{i=1}^{K} M_{T_i}$ transmit antennas. The BS has M_R receive antennas. s_i and r_i represent the i-th user data vector and the receive vector respectively whereas D_i and F_i are the respective UE transmit matrix and the BS receive matrix. The MIMO channel matrix is represented as

$$\mathbf{H} = [\mathbf{H}_1\ \mathbf{H}_2 \cdots \mathbf{H}_K] \tag{39}$$

where $\mathbf{H}_i \in \mathbb{C}^{M_R \times M_{T_i}}$ is the MIMO channel matrix between user i and the BS for $i = 1, \cdots, K$. The $M_R \times M_T$ receive filter matrix at the BS is given by

$$\mathbf{F} = [\mathbf{F}_1\ \mathbf{F}_2 \cdots \mathbf{F}_K] \tag{40}$$

where $\mathbf{F}_i \in \mathbb{C}^{M_{T_i} \times M_R}$ corresponds to the i-th user. Each row of F_i corresponds to one of the M_{Ti} transmit antennas at the UE.

The proposed algorithm successively calculates the rows of each F_i using the MMSE criterion such that the users are ordered according to their respective total MSE in ascending order (starting with the minimum MSE). The total MSE of user i is obtained by summing up the MSE corresponding to each of its individual transmit antennas. Following the receive filtering by the receive filter matrix F, SIC is performed to eliminate the MUI. This process has the effect of transforming the multiuser uplink channel into a set of parallel M_{Ti} M_{Ti} ´ SU MIMO channels represented as $\mathbf{F}_i\mathbf{H}_i$

The SMMSE SIC detection in [51] has been considered for STBC based MIMO transmission i.e. the Alamouti scheme and also for dominant eigenmode transmission (DET). The open-loop Alamouti scheme provides MIMO diversity without the need for any CSI at the UEs. On the other hand, DET requires full CSI at both the transmitter and the receiver, which means that the UE transmit matrices need D_i to be computed at the BS and then fed forward to the UEs. However, beside the full diversity gain equal to that of STBC, DET also provides the maximum array gain by transmitting over the strongest eigenmode of the MIMO channel [49,51,52]

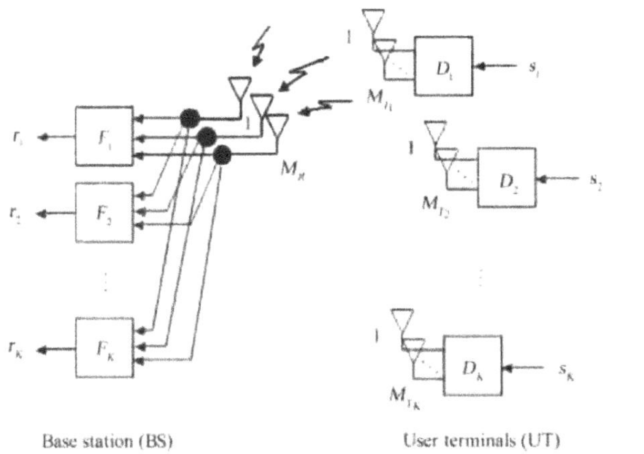

Figure 33: Block diagram of SMMSE SIC MUD based MUMIMO system [51].

Figure 34 shows the BER performance of SMMSE SIC Alamouti in comparison with V-BLAST. The impact of channel estimation errors is also shown. The simulated MIMO-OFDM system consists of 6 BS antennas and 3 UEs equipped with 2 antennas each, denoted as 6 ´ {2, 2, 2} MU-MIMO configuration. The MIMO channel H is assumed to be frequency selective with the power delay profile defined by IEEE 802.11n - D for non-line-of-sight (NLOS) conditions. The channel model for each user's channel H_i takes into account the antenna correlation at the BS. However, the antennas at each UE are considered to have low spatial correlation assuming large angular spread at the user. A total of 64 OFDM subcarriers are used with subcarrier spacing of 150 kHz. The user data is encoded using a 1/2 rate convolutional code. 4-QAM modulation is used for each subcarrier of SMMSE SIC Alamouti system while BPSK is used for V-BLAST so that the data rate remains the same for both systems.

Clearly, the SMMSE SIC Alamouti system outperforms V-BLAST by a large margin particularly when the channel estimation errors are taken into account

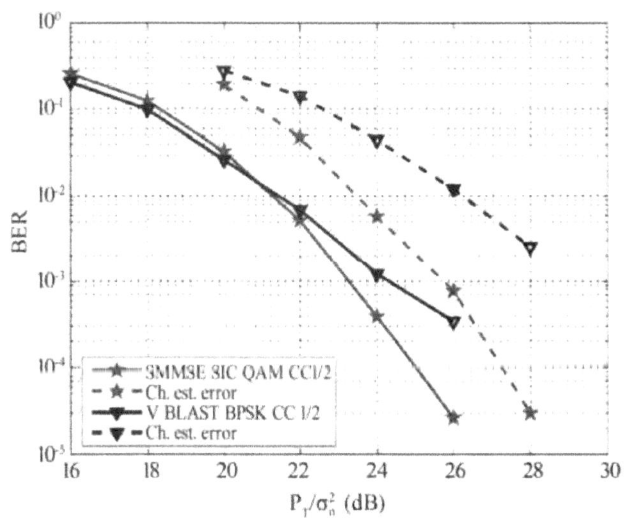

Figure 34: BER performance of SMMSE SIC Alamouti and V-BLAST for 6 ´ {2, 2, 2} configuration [51].

Figure 35 shows the performance gain of SMMSE SIC DET over SMMSE SIC Alamouti. SMMSE SIC DET shows substantial gains in the high SNR region and these become more significant as the no. of users in the system increases.

Turbo MUD

An iterative Turbo MMSE MUD scheme is presented in [53] for single-carrier (SC) space-time trellis-coded (STTC) SDMA MIMO systems in frequency-selective fading channels. This scheme can jointly detects multiple UE transmit antennas while cancelling the MUI from undetected users along with co-channel interference (CCI) and ISI through soft cancellation. Unknown co-channel interference (UCCI) from other interferers not known to the system is also considered. It is also shown that the no. of BS antennas required to achieve the corresponding lower performance bound of single-user detection is equal to the no. of users rather than the total no. of transmit antennas. The receiver derivations in [53] are provided for M-PSK modulation but can be extended to QAM as well.

Figure 36 shows the system model while the UE transmitter block diagram is given in Figure 37. The system has a total of K + K_I simultaneous users, each indexed by $k = 1,\cdots K, K+1, \ldots K_I$ where the first K are the users to be detected at the BS while the remaining K_I are unknown interfering users representing the source of UCCI. Each UE is equipped with N_T transmit antennas. However, the system can also support unequal no. of antennas at the UEs. Each UE encodes the bit sequence $c_k(i)$ for $i = 1,\cdots, BN_T$ using a rate k_0 / N_T STTC code, where B represents the frame length in symbols. The encoded sequences $b_k(i) \in \mathbb{Q}$, $i = 1,\cdots, BN_T$ are

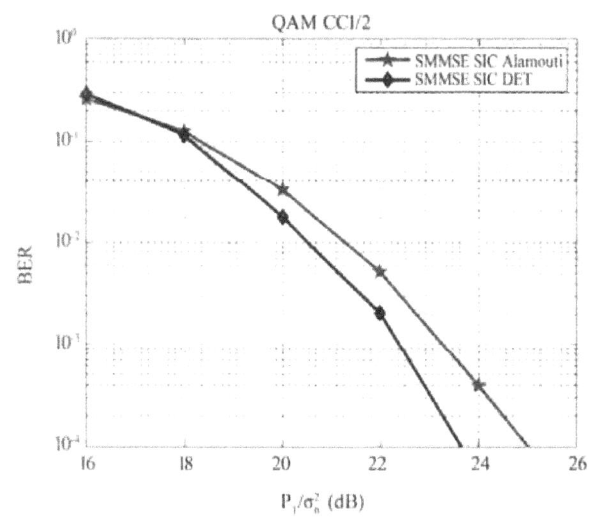

Figure 35: BER performance of SMMSE SIC Alamouti and SMMSE SIC DET for 6 ´ {2, 2, 2} configuration [51].

Grouped into B blocks of N_T symbols, where $\mathbb{Q} = \{\alpha_1, \cdots, \alpha_{2^{k_0}}\}$ represents the modulation alphabet of M-PSK, and then interleaved by user-specific permutation of blocks of length N_T within a frame, such that the positions within each block remain unchanged. This interleaving process preserves the rank properties of the STTC code. User-specific training sequences of length T_{NT} are then attached at the start of the interleaved sequences. After serial-to-parallel (S/P) conversion of the entire frame, the resulting sequences $b_k^{(n)}(i)$ for $n = 1,\cdots, N_T$, $i = 1,\cdots, B+T$ are transmitted over the frequency-selective MIMO channel using the N_T transmit antennas.

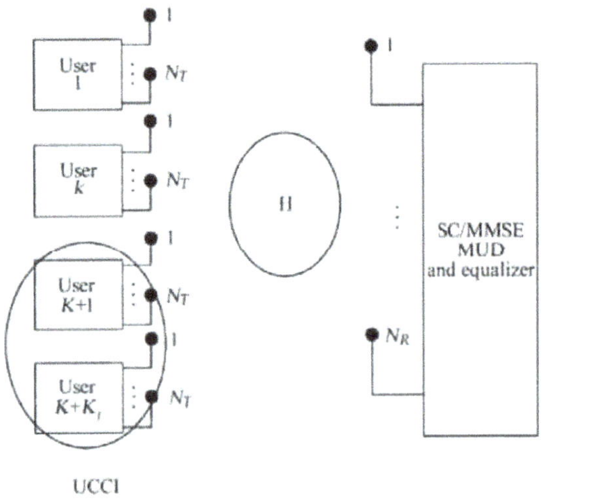

Figure 36: System model of Turbo MUD based STTC MU-MIMO system [53].

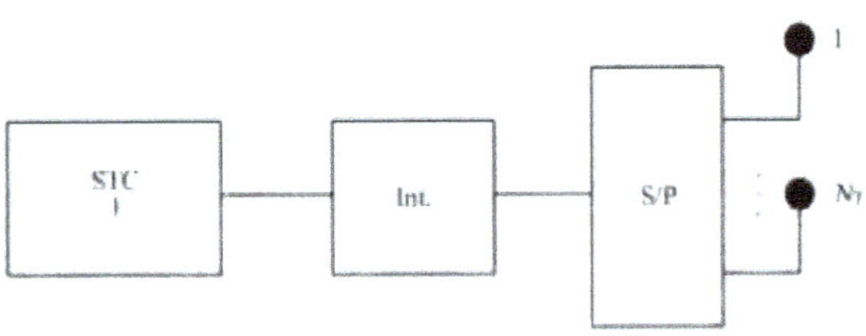

Figure 37: UE transmitter block diagram [53].

The transmit signals from all users are received at the N_R BS receive antennas. The space-time sampled received signal vector $\mathbf{y}(i)\in\mathbb{C}^{LN_R\times1}$ at time instant i is given by

$$\mathbf{y}(i)=\underbrace{\mathbf{H}\mathbf{u}(i)}_{\text{desired}}+\underbrace{\mathbf{H}_I\mathbf{u}_I(i)}_{\text{UCCI}}+\underbrace{\mathbf{n}(i)}_{\text{noise}},\quad i=1,\cdots,T+B \tag{41}$$

which can also be represented as

$$\mathbf{y}(i)=\left[\mathbf{r}^T(i+L-1),\cdots,\mathbf{r}^T(i)\right]^T \tag{42}$$

where each vector $\mathbf{r}(i)\in\mathbb{C}^{N_R\times1}$ consists of N_R elements $r_m(i),\ m=1,\cdots,N_R$ representing the received signal sample after matched filtering at the m-th receive antenna and L is the no. of paths of the frequency-selective MIMO channel. The channel matrix $\mathbf{H}\in\mathbb{C}^{LN_R\times KN_T(2L-1)}$ has the structure

$$\mathbf{H}=\begin{bmatrix}\mathbf{H}(0) & \cdots & \mathbf{H}(L-1) & \cdots & \mathbf{0}\\ \vdots & \ddots & & \ddots & \vdots\\ \mathbf{0} & \cdots & \mathbf{H}(0) & \cdots & \mathbf{H}(L-1)\end{bmatrix} \tag{43}$$

Each element $\mathbf{H}(l)\in\mathbb{C}^{N_R\times KN_T}$ is given by

$$\mathbf{H}(l)=\begin{bmatrix}h_{1,1}^{(1)}(l) & \cdots & h_{1,1}^{(N_T)}(l) & \cdots & h_{K,1}^{(1)}(l) & \cdots & h_{K,1}^{(N_T)}(l)\\ \vdots & \ddots & \vdots & \ddots & \vdots & \ddots & \vdots\\ h_{1,N_R}^{(1)}(l) & \cdots & h_{1,N_R}^{(N_T)}(l) & \cdots & h_{K,N_R}^{(1)}(l) & \cdots & h_{K,N_R}^{(N_T)}(l)\end{bmatrix} \tag{44}$$

where $h_{k,m}^{n}(l)$ represents the complex gain for the l-th path between the n-th transmit antenna of user k and the m-th BS receive antenna. The channel matrix $\mathbf{H}_I\in\mathbb{C}^{LN_R\times K_IN_T(2L-1)}$ corresponding to UCCI has a similar structure as H, and consists of matrices $\mathbf{H}_I(l)\in\mathbb{C}^{N_R\times K_IN_T}$.

The vectors $\mathbf{u}(i)\in\mathbb{Q}^{KN_T(2L-1)\times1}$ and $\mathbf{u}_I(i)\in\mathbb{Q}^{K_IN_T(2L-1)\times1}$ consist of the respective transmit sequences $b_k^{(n)}(i)$ of the desired and the unknown users, and are expressed as

$$\begin{aligned}\mathbf{u}(i)&=\left[\mathbf{b}^T(i+L-1),\cdots,\mathbf{b}^T(i),\cdots,\mathbf{b}^T(i-L+1)\right]^T\\ \mathbf{u}_I(i)&=\left[\mathbf{b}_I^T(i+L-1),\cdots,\mathbf{b}_I^T(i),\cdots,\mathbf{b}_I^T(i-L+1)\right]^T\end{aligned} \tag{45}$$

where the vectors $\mathbf{b}(i)\in\mathbb{Q}^{KN_T\times1}$ and $\mathbf{b}_I(i)\in\mathbb{Q}^{K_IN_T\times1}$ are given by

$$\begin{aligned}\mathbf{b}(i)&=\left[b_1^{(1)}(i),\cdots,b_1^{(N_T)}(i),\cdots b_K^{(1)}(i),\cdots b_K^{(N_T)}(i)\right]^T\\ \mathbf{b}_I(i)&=\left[b_{K+1}^{(1)}(i),\cdots,b_{K+1}^{(N_T)}(i),\cdots b_{K+K_I}^{(1)}(i),\cdots b_{K+K_I}^{(N_T)}(i)\right]^T\end{aligned} \tag{46}$$

$\mathbf{n}(i)\in\mathbb{C}^{LN_R\times1}$ is the AWGN vector with covariance matrix σ^2 I

The Turbo receiver block diagram is shown in Figure 38. For the k-th user, the receiver first associates the signals from the user's N_T transmit antennas to N_T/n_0 sets of equal size n_0 such that the antennas, represent the first set and so on. The training sequence $\mathbf{u}(i),\ i=1,\cdots,T$ is used to obtain an estimate $\hat{\mathbf{H}}$ of the channel matrix H. The UCCI-plus-noise covariance matrix R is then estimated. The estimate for the first iteration is given by

$$\hat{\mathbf{R}}=\frac{1}{T}\sum_{i=1}^{T}\left(\mathbf{y}(i)-\hat{\mathbf{H}}\mathbf{u}(i)\right)\left(\mathbf{y}(i)-\hat{\mathbf{H}}\mathbf{u}(i)\right)^H \tag{57}$$

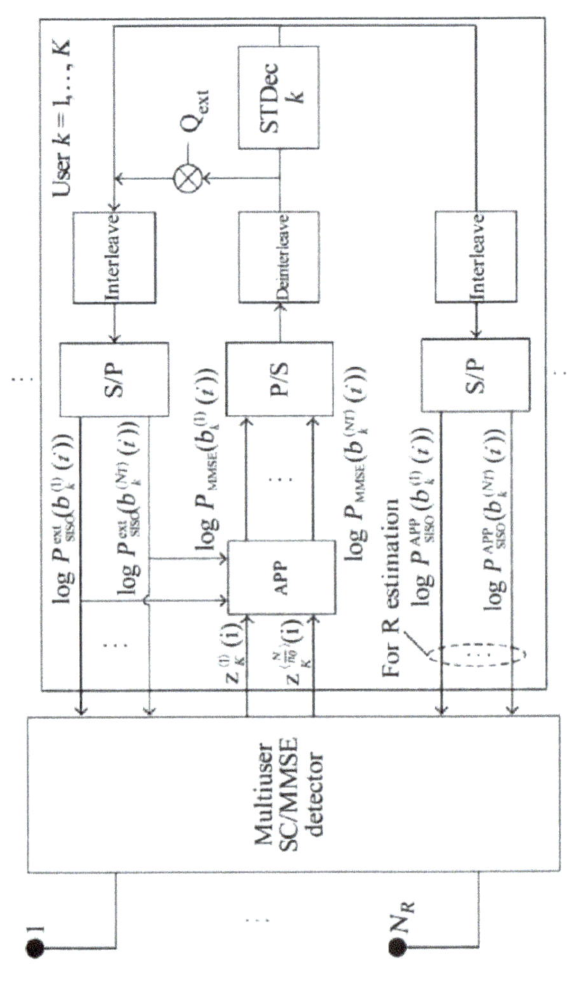

Figure 38: Block diagram of the iterative Turbo multiuser receiver [53].

From the second iteration onward, the soft feedback vector $\bar{\mathbf{u}}(i)$ from the APP (also MAP) SISO decoding is used to obtain the covariance matrix estimate

$$\hat{\mathbf{R}} = \frac{1}{T}\sum_{i=1}^{T}\left(\mathbf{y}(i)-\hat{\mathbf{H}}\mathbf{u}(i)\right)\left(\mathbf{y}(i)-\hat{\mathbf{H}}\mathbf{u}(i)\right)^{H} + \frac{1}{B}\sum_{i=T+1}^{T+B}\left(\mathbf{y}(i)-\hat{\mathbf{H}}\bar{\mathbf{u}}(i)\right)\left(\mathbf{y}(i)-\hat{\mathbf{H}}\bar{\mathbf{u}}(i)\right)^{H}. \tag{48}$$

Vector $\bar{\mathbf{u}}(i)$ consists of the sequences $\bar{b}_k^{(n)}(i)$ calculated using the a posteriori probability $P_{\text{SISO}}^{\text{APP}}$. The signals $b_k^{(n)}(i),\ n=1,\cdots,n_0$ for user k are y detected using linear MMSE MUD which filters the signal vector

$$\mathbf{y}_k^{(1)} = \mathbf{y}(i) - \mathbf{H}\mathbf{u}_k^{(1)}(i), \quad i = T+1,\cdots,B+T \tag{49}$$

where the vector $\mathbf{u}_k^{(1)}(i)$ is calculated using the extrinsic probability $P_{\text{SISO}}^{\text{est}}$ obtained after APP SISO decoding and $(\cdot)^{(1)}$ deno he first set of n_0 antennas for user k. The weighting matrix $\mathbf{W}_k^{(1)}(i)$ f the MMSE MUD satisfies the criterion,

$$\left[\mathbf{W}_k^{(1)}(i),\mathbf{A}_k^{(1)}(i)\right] = \arg\min_{\mathbf{W},\mathbf{A}}\left\|\mathbf{W}^{H}\mathbf{y}_k^{(1)}(i) - \mathbf{A}^{H}\boldsymbol{\beta}_k^{(1)}(i)\right\|^2$$

(50)

subject to the constraint $[\mathbf{A}]_{j,j}=1,\ j=1,\cdots,n_0$ to avoid the trivial solution $\left[\mathbf{W}_k^{(1)}(i),\mathbf{A}_k^{(1)}(i)\right]=[\mathbf{0},\mathbf{0}]$.

The vector $\boldsymbol{\beta}_k^{(1)}(i) \in \mathbb{C}^{n_0\times 1}$ is given by

$$\boldsymbol{\beta}_k^{(1)}(i) = \left[b_k^{(1)}(i),\cdots,b_k^{(n_0)}(i)\right]^{T}. \tag{51}$$

The corresponding output $\mathbf{z}_k^{(1)}(i) \in \mathbb{C}^{n_0\times 1}$, of the MMSE MUD is given by

$$\begin{aligned}\mathbf{z}_k^{(1)}(i) &= \mathbf{W}_k^{(1)}(i)\mathbf{y}_k^{(1)}(i)\\ &= \boldsymbol{\Omega}_k^{(1)}(i)\boldsymbol{\beta}_k^{(1)}(i) + \boldsymbol{\psi}_k^{(1)}(i)\end{aligned} \tag{52}$$

where the matrix $\boldsymbol{\Omega}_k^{(1)}(i) \in \mathbb{C}^{n_0\times n_0}$ consists of the equivalent channel gains after filtering and $\boldsymbol{\psi}_k^{(1)}(i) \in \mathbb{C}^{n_0\times 1}$ is the filtered AWGN vector The MMSE MUD outputs $\mathbf{z}_k^{(\gamma)}(i)$ along with the parameters $\boldsymbol{\Omega}_k^{(\gamma)}(i)$ and $\boldsymbol{\psi}_k^{(\gamma)}(i)$ for all antenna sets $\gamma = 1,\cdots,N_T/n_0$ of user k, are passed on to the APP SISO detector which calculates the extrinsic probabilities for SISO decoding. This iterative procedure is continued for all antenna sets of the remaining users until all users are detected. The complexity of this receiver is primarily associated with the MMSE and APP blocks and is on the order of $O\{\max(L^3N_R^3, 2^{k_m n_0})\}$

Two special cases of the proposed receiver architecture are considered in [53]. Receiver 1, with n_0 = 1, detects the transmit antennas of user k one by one s receiver type has the lowest complexity since the complexity depends exponentially on n_0 Receiver 2, with no= N_T jointly detects all N_T transmit antennas of user k and has the highest complexity

The simulated SER and FER performance of the two receiver types vs. per-antenna E_s/N_0 is shown in Figure 39 and Figure 40, for a par user referred to as user 1. Performance comparison with optimal joint ML

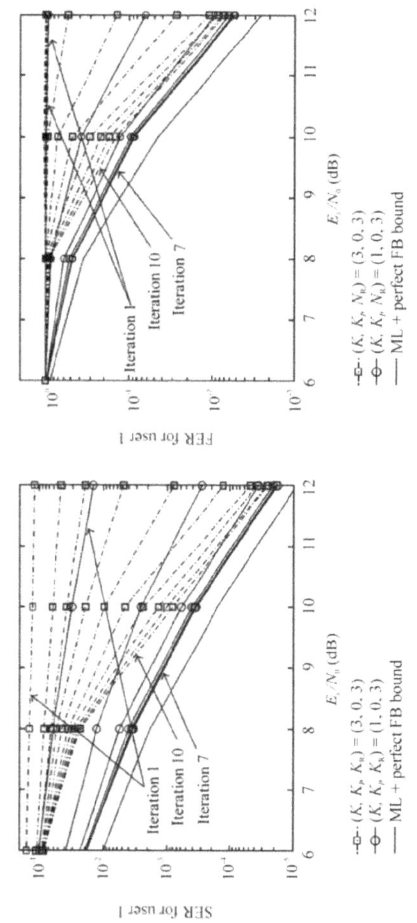

Figure 39: Receiver 1's (a) SER and (b) FER performance, (K, KI, N_R) = (3, 0, 3) and (1, 0, 3), NT = 2 [53].

Detection of all N_T antennas of user 1 followed by MAP SISO decodi assuming perfect feedback (FB), is also provided. The results are provided for 1 and 10 iterations for $(K,K_I,N_R)=(3,0,3)$ and for 1 and 7 iterations for $(K,K_I,N_R)=(1,0,3)$ with NT = 2 transmit antennas per user K, K_I and N_R represent the no. of desired users, unknown interferer (UCCI) and BS receive antennas. QPSK modulation is used for transmission and frequency-selective fading is assumed with L = 5 uncorrelated Rayleigh distributed paths. For a sufficient no. of iterations, both receivers perform reasonably close to the ML receiver, particularly in the high SNR region. Receiver 2 shows slightly better performance than receiver 1.

Figure 41 shows the SER and FER performance of the two receivers for $(K,K_I,N_R)=(2,1,3)$ with $N_T=2$ transmit antennas per desired user and a single transmit antenna for the unknown interferer (UCCI) i.e $N_I=1$ Two cases of signal-to-UCCI interference ratio (S e considered. For SIR = 3 dB, the signal transmitted from UCCI's antenna is assumed to have the same power as that of the signal from one antenna of the desired user whereas, for SIR = 0 dB, UCCI's antenna transmits at twice the transmit power of a desired user's antenna. The performance of both receivers is obviously better for the 3 dB SIR case. However, the UCCI has a considerable impact on performance and more iterations are needed to achieve a reasonably low SER or FER. The performance will degrade further in case of multiple UCCI antennas and also as the UCCI sources i.e. the no. of unknown interferers increase

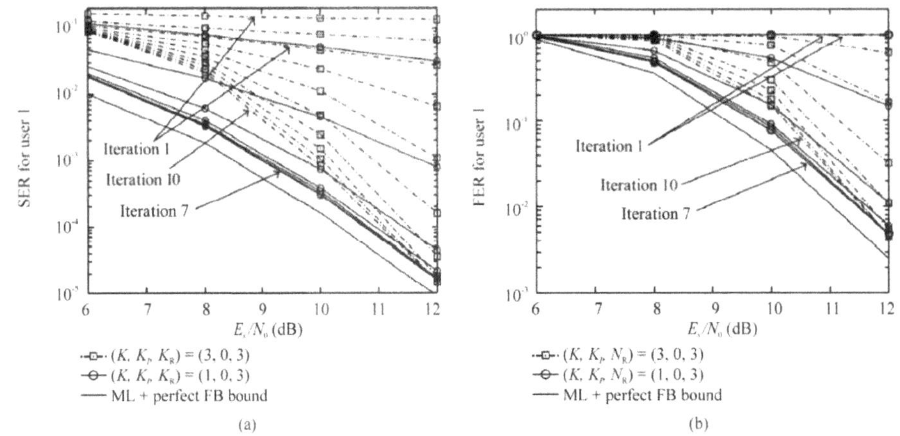

Figure 40: Receiver 2's (a) SER and (b) FER performance, (K, KI, N_R) = (3, 0, 3) and (1, 0, 3), N_T = 2 [53].

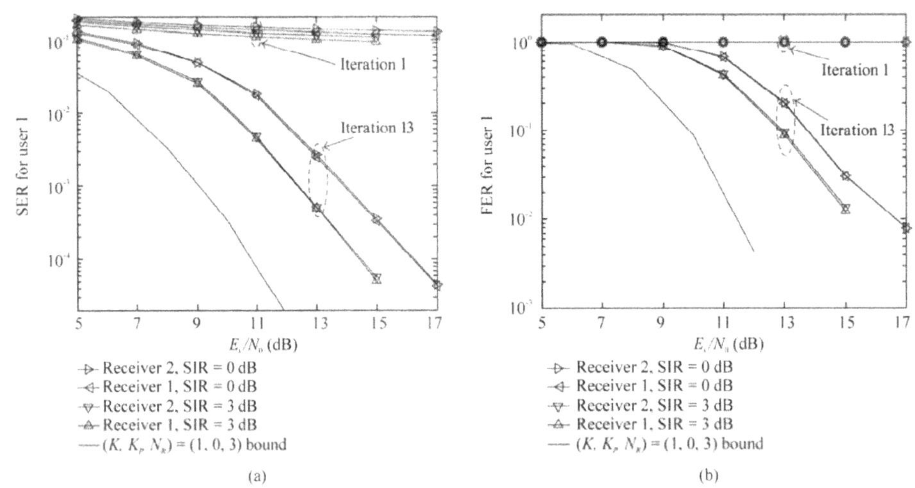

Figure 41: Receiver 1's and 2's (a) SER and (b) FER performance **$(K, K_I, N_R) = (2, 1, 3), N_T = 2, N_I = 1$ [53].**

Genetic Algorithm (GA) Assisted MUDs

Genetic algorithms (GAs) are based on the theory of evolution's concept of survival of the fittest, where the genes from the fittest individuals of a species are passed on to the next generation through the process of "natural selection". When applied to MUDs, an "individual" represents the L-dimensional MUD weight vector corresponding to the L users. These MUD weights are then optimized using GA by genetic operations of "mating" and "mutation" to get a new generation of individuals i.e. the MUD weights. The initial "population" (MUD weights) is typically obtained from the MMSE solution which is retained throughout the GA search process as an alternate solution in case of poor convergence [3].

Considering the SDMA-OFDM system model of Figure 27 with a single transmit antenna at each UE and P receive antennas at the BS, the ML-based decision metric or objective function (OF) for a GA-assisted MUD corresponding to the p-th receive antenna can be written as

$$\Omega_p(\tilde{\mathbf{s}}) = \left|x_p - \mathbf{H}_p\tilde{\mathbf{s}}\right|^2 \tag{53}$$

where p_x is the received symbol corresponding to the p-th BS antenna for a specific OFDM subcarrier and H_p is the p-th row of the channel transfer function matrix H. The estimated symbol vector of the L users corresponding to the p-th BS antenna is then given by

$$\hat{\mathbf{s}}_{GA_p} = \arg\left\{\min_{\tilde{\mathbf{s}}}\left[\Omega_p(\tilde{\mathbf{s}})\right]\right\} \tag{54}$$

The combined decision metric for the P receive antennas can therefore be written as

$$\Omega(\tilde{\mathbf{s}}) = \sum_{p=1}^{P}\Omega_p(\tilde{\mathbf{s}}) = \left\|\mathbf{x} - \mathbf{H}\tilde{\mathbf{s}}\right\|^2 \tag{55}$$

Therefore, the decision rule for the GA-assisted MUD is to find an estimate $\hat{\mathbf{s}}_{GA}$ of the L ′ 1 transmitted symbol vector such that $\Omega(\tilde{\mathbf{s}})$ (is minimized [3].

Review and analysis of various GA-assisted MUDs is provided in [3] for the SDMA-OFDM uplink consisting of a single transmit antenna at each UE. Figure 42 shows the schematic diagram of the SDMA-OFDM uplink system based on the concatenated MMSE-GA MUD. The concatenated MMSE-GA MUD uses the MMSE estimate 1 MMSE $\hat{\mathbf{s}}_{MMSE} \in \mathbb{C}^{L\times 1}$, of the transmitted symbol vector of the L users as initial information for the GA. $\hat{\mathbf{s}}_{\text{MMSE}}$ is given by

$$\hat{\mathbf{s}}_{\text{MMSE}} = \mathbf{W}^{H}_{\text{MMSE}}\mathbf{x} \quad (56)$$

Where $\mathbf{W}_{\text{MMSE}} \in \mathbb{C}^{P\times L}$ is the MMSE MUD weight matrix, expressed as

$$\mathbf{W}_{\text{MMSE}} = \left(\mathbf{H}\mathbf{H}^{H} + \sigma_n^2\mathbf{I}\right)^{-1}\mathbf{H} \quad (57)$$

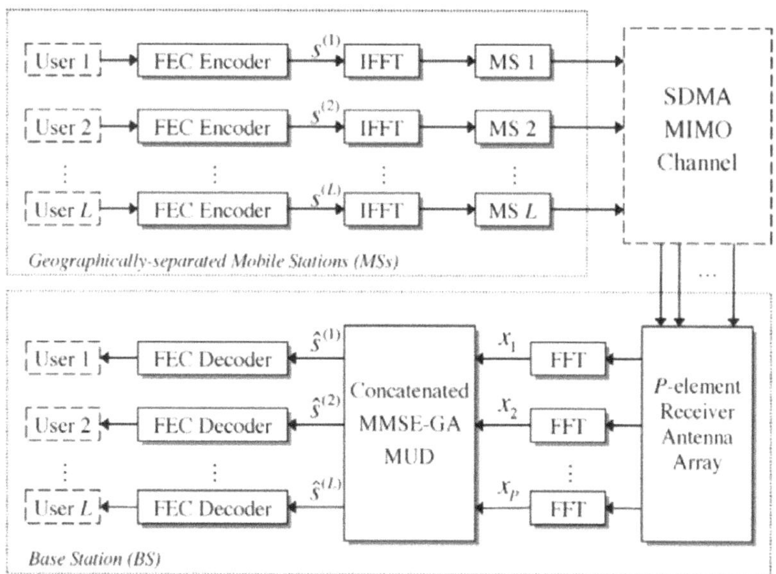

Figure 42: SDMA-OFDM uplink system based on concatenated MMSE-GA MUD [3].

Using this MMS 1 E estimate, the 1st GA generation, y = containing a population of X individuals, is created. The x-th individual is a symbol vector denoted as

$$\tilde{\mathbf{s}}_{(y,x)} = \left[\tilde{s}^{(1)}_{(y,x)}, \tilde{s}^{(2)}_{(y,x)}, \cdots, \tilde{s}^{(L)}_{(y,x)}\right], \quad x = 1,\cdots,X \quad y = 1,\cdots,Y \quad (58)$$

where each element $\tilde{s}^{(l)}_{(y,x)}$ called a gene belongs to the set of complex-value modulation symbols corresponding to the particular modulation scheme used.

The GA search procedure shown in Figure 43 is then initiated, which involves several GA operations like mating, mutation, elitism etc. leading to the next generation. This process is repeated for Y generations and the individual with the highest fitness value is considered to be the detected L ´ 1 multiuser symbol vector for the corresponding OFDM subcarrier. All users are jointly detected by the concatenated MMSE-GA MUD and therefore, no error propagation exists between the detected users

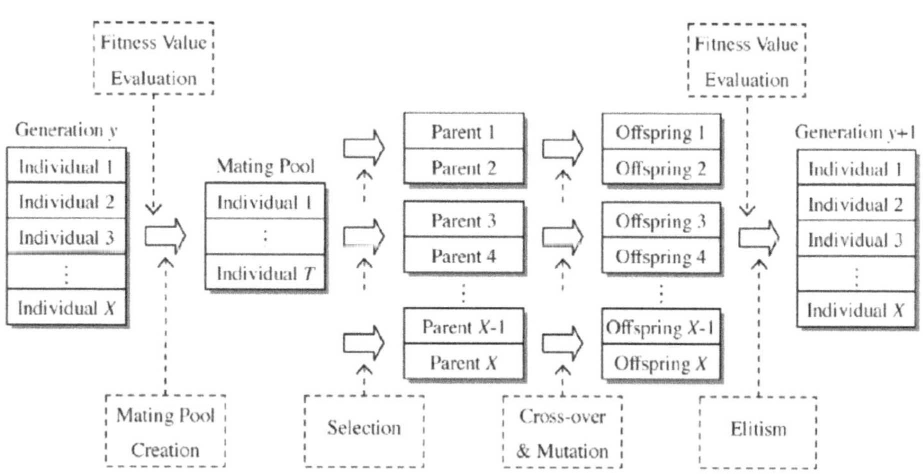

Figure 43: GA search procedure for one generation [3].

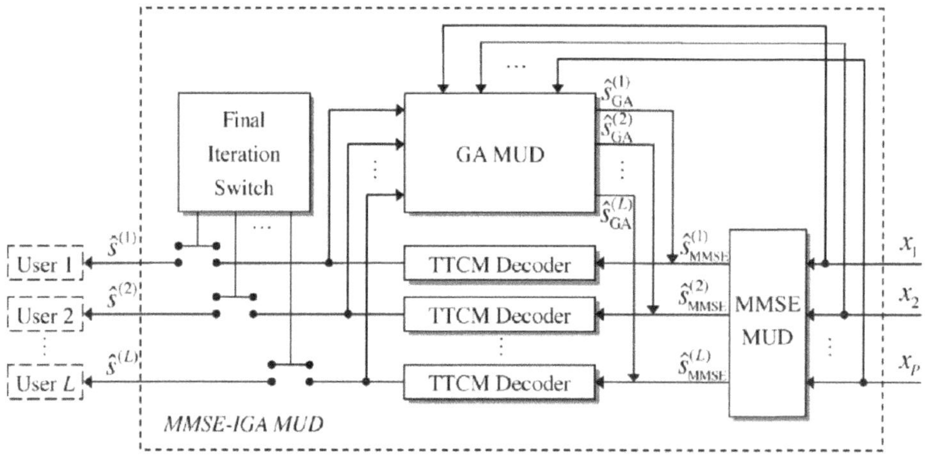

Figure 44: Schematic diagram of an IGA MUD [3].

Enhanced GA MUDs, utilizing an advanced mutation technique called biased Q-function-based mutation (BQM) instead of the conventional uniform mutation (UM), and incorporating the iterative turbo trellis coded modulation (TTCM) scheme for FEC decoding, are also discussed in [3]. Figure 44 shows the schematic diagram of an MMSE-initialized iterative GA (IGA) MUD incorporating TTCM. The P ´ 1 received symbol vector x is detected by the MMSE MUD to get the estimated symbol vector $\hat{s}_{MMSE}$ of the L users consisting of the symbols $\hat{s}_{MMSE}^{(l)}$, $l = 1, \cdots, L$. Each of these symbols are then decoded by a TTCM decoder to get a more reliable estimate. The resulting symbol vector is then used as the initial information for the GA MUD. The GA-estimated symbol vector $\hat{s}_{GA}$ is then fed back to the TTCM decoders for further provident of the estimate. This optimization process involving the GA MUD and the TTCM decoders is continued for a desired no. of iterations. The final estimates $\hat{s}^{(l)}$ of the L users' symbols are then obtained at the output after the final iteration.

Figure 45 shows the BER performance comparison of various MMSE-initialized TTCM-assisted GA and IGA MUD based SDMA-OFDM systems consisting of L = 6 (single-antenna) users and P = 6 BS antennas. 4-QAM modulation and SWATM channel model is used. GA population size of X = 20 (also X = 10 for TTCM, MMSE-IGA (2)) and a total of Y = 5 generations are considered. UM and BQM mutation schemes are employed. Performance curves for 1 ´ 1 SISO AWGN, 1 ´ 6 MRC AWGN, TTCM-MMSE SDMA-OFDM and TTCM-ML SDMA-OFDM systems are also provided for reference. The TTCM-MMSE-GA MUDs and the IGA MUDs in particular provide exceptionally good BER performance. The performance of the TTCM-MMSEIGA scheme with 2 iterations and X = 20 (represented as TTCM, MMSE-IGA (2) in the figure), is in fact identical to the optimum TTCM-ML MUD

The BER performance comparison for rank-deficient scenarios where the no. of users exceeds the no. of BS antennas resulting in insufficient degrees of freedom for separating the users, is given in Figure 46. Performance curves for L = 6, 7 and 8 users and P = 6 BS antennas are provided. The IGA MUD schemes still perform reasonably well with relatively small performance degradation.

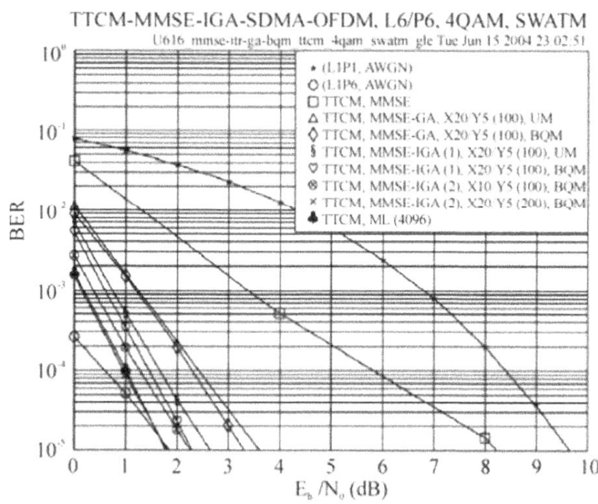

Figure 45: BER performance of TTCM-MMSE-GA/IGA SDMA-OFDM systems, L = 6, P = 6 [3].

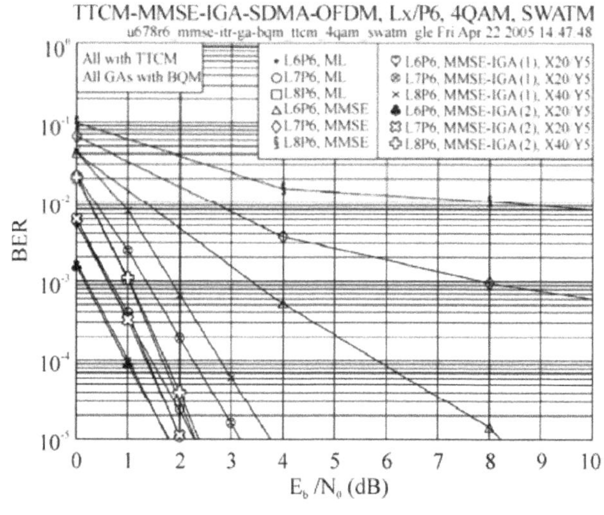

Figure 46: BER performance of TTCM-MMSE-GA/IGA SDMA-OFDM systems, L = 6, 7, 8 and P = 6 [3].

Figure 47 compares the complexity of the TTCMMSE-GA and TTCM-ML MUDs in terms of the no. of calculations, as the no. of users increase. The no of BS antennas is kept equal to the no. of users i.e. L=P. The complexity of the GA MUD increases very slowly with the number of users as compared to the ML MUD, resulting in a huge difference as more users are added to the system.

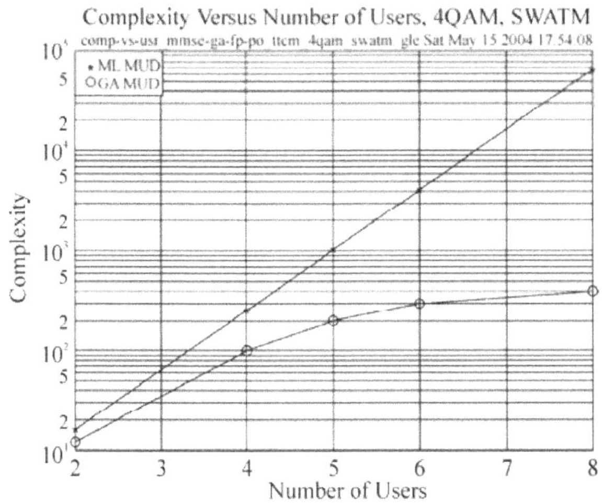

Figure 47: Complexity of TTCM-MMSE-GA and TTCMML SDMA-OFDM systems vs. no. of users, L = P [3].

The MU-MIMO Downlink

The MU-MIMO downlink channel is referred to as the MIMO broadcast channel (MIMO-BC) [49] where the BS equipped with multiple antennas, simultaneously transmits data to multiple UEs consisting of one or more antennas each, as shown in Figure 48. The multiuser interference (MUI) (also called multiple access interference, MAI) can be suppressed by means of transmit beamforming or "dirty paper" coding. Therefore, CSI feedback from each user is required for precoding at the BS. Various linear and nonlinear precoding techniques for MU-MIMO downlink systems are discussed in the following text.

Channel Inversion

Channel inversion is a linear precoding technique for MU-MIMO downlink systems where each UE is equipped with a single receive antenna. As the name suggests, channel inversion uses the inverse of the channel matrix for precoding to remove the MUI, as illustrated in Figure 49.

Assuming that the no. of receive antennas M_R £ M_T, the no. of transmit antennas, ZF precoding can be used for this purpose. The $M_T \times 1$ transmitted signal vector is then given by

$$\mathbf{x} = \mathbf{H}^{+}\mathbf{d} = \mathbf{H}^{H}\left(\mathbf{H}\mathbf{H}^{H}\right)^{-1}\mathbf{d} \qquad (59)$$

where d is the vector of data symbols to be precoded and H^+ is the pseudoinverse of the M_R′ M_T channel matrix H. Vector d can have any dimension up to the rank of H [48]. The i-th column of the prefiltering or precoding matrix P_{ZF} is given by [49]

$$\mathbf{p}_{ZF,i} = \frac{\mathbf{h}_i^{(+)}}{\sqrt{\left\|\mathbf{h}_i^{(+)}\right\|^2}} \qquad (60)$$

where $\mathbf{h}_i^{(+)}$ is the i-th column of H^+ . The combined received signal vector can be expressed as

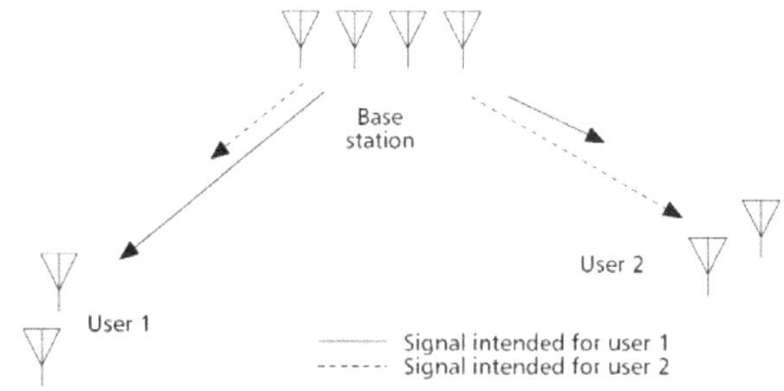

Figure 48: The MU-MIMO downlink [48].

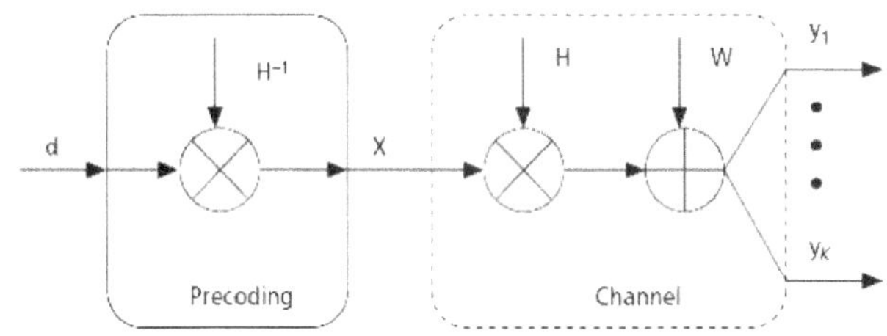

Figure 49: Channel inversion [48].

$$\mathbf{y} = \mathbf{d} + \mathbf{w} \qquad (61)$$

where w is the noise vector. Therefore, ZF precoding is only suitable for low-noise or high transmit power scenarios [48].

MMSE precoding, also called "regularized" channel inversion provides a better alternative. In this case, the transmitted signal vector is given by

$$\mathbf{x} = \mathbf{H}^{H}\left(\mathbf{H}\mathbf{H}^{H} + \alpha\mathbf{I}\right)^{-1}\mathbf{d} \qquad (62)$$

where a is the loading factor. For a MU-MIMO downlink system with total transmit power P_T and K simultaneous users, $\alpha = K / P_T$ maximizes the SINR at the receivers [48].

Block Diagonalization

Block diagonalization (BD) or block channel inversion, which was first proposed in [54], is a generalization of channel inversion to multi-antenna UEs [48]. BD also requires the total no. of receive antennas M_R to be less than or equal to the no. of transmit antennas M_T i.e. M_R £ M_T.

Consider the system model of Figure 50. The system consists of K simultaneous users each having re- M_{Ri} receive antennas for $i = 1,\cdots,K$ such that the total no. of receive antennas $M_R = \sum_{i=1}^{K} M_{R_i}$. The combined channel matrix $\mathbf{H} \in \mathbb{C}^{M_R \times M_T}$ is given by

$$\mathbf{H} = \left[\mathbf{H}_1^T \;\; \mathbf{H}_2^T \;\; \cdots \;\; \mathbf{H}_K^T\right]^T \qquad (63)$$

where $\mathbf{H}_i \in \mathbb{C}^{M_{R_i} \times M_T}$ represents the MIMO channel from the M_T BS antennas to user i. The combined precoding matrix $\mathbf{P} \in \mathbb{C}^{M_T \times S}$ can be expressed as

$$\mathbf{P} = \left[\mathbf{P}_1 \;\; \mathbf{P}_2 \;\; \cdots \;\; \mathbf{P}_K\right] \qquad (64)$$

Where $\mathbf{P}_i \in \mathbb{C}^{M_T \times S_i}$ is the precoding matrix for the, i-th user $S \le M_R$ is the total no. of transmitted data streams and $S_i \le M_{R_i}$ is the no. of data streams transmitted to user i. P ne to be selected in such a way that HP becomes block diagonal is end, a matrix $\tilde{H}_i$ is defined as

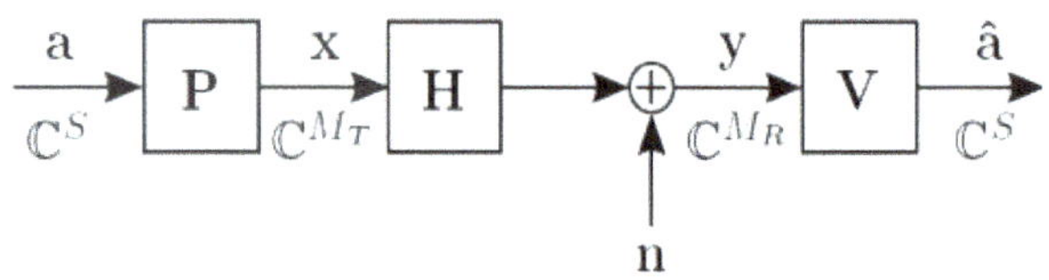

Figure 50: System model for MU-MIMO downlink transmission [55].

$$\tilde{H}_i = \left[H_1^T \cdots H_{i-1}^T \; H_{i+1}^T \cdots H_K^T\right]^T \tag{65}$$

which contains all but the i-th user's channel matrix. Therefore P_i lies in the null space of $\tilde{H}_i$ and consists of unitary column vectors which are obtained by the SVD of $\tilde{H}_i$ given by

$$\tilde{H}_i = \tilde{U}_i\tilde{D}_i\left[\tilde{V}_i^{(1)} \;\; \tilde{V}_i^{(0)}\right]^H \tag{66}$$

The rightmost $M_T - \tilde{L}_i$ singular vectors $\tilde{V}_i^{(0)} \in \mathbb{C}^{M_T \times M_T - \tilde{L}_i}$ form an orthogonal basis for the null space of $\tilde{H}_i$ where $\tilde{L}_i$ is the rank of $\tilde{H}_i$. The product $H_i\tilde{V}_i^{(0)}$ with dimensions $M_{R_i} \times (M_T - \tilde{L}_i)$ represents e equivalent channel matrix for user i after eliminating the MUI. Thus, BD transforms the MU-MIMO downlink system into a set of K parallel $(M_T - \tilde{L}_i) \times M_{R_i}$ SU-MIMO systems Using SVD $H_i\tilde{V}_i^{(0)}$ can be expressed as

$$H_i\tilde{V}_i^{(0)} = U_i\begin{bmatrix} D_i & 0 \\ 0 & 0 \end{bmatrix}\left[V_i^{(1)} \;\; V_i^{(0)}\right]^H \tag{67}$$

where D_i is an $L_i \times L_i$ diagonal matrix, assuming Li to be the rank of $H_i\tilde{V}_i^{(0)}$ The product of $\tilde{V}_i^{(0)}$ and the first L_i singular vectors $V_i^{(1)}$ produces an orthogonal basis of dimension L_i and represent the transmission vectors which maximize the information rate for the i-th user while elimination the MUI. Therefore, the precoding matrix P_i for user i consists of $\tilde{V}_i^{(0)}V_i^{(1)}$ with appropriate power scaling Optimal power allocation is achieved by water-filling, using the diagonal element of the matrices D_i and can either be implemented globally to maximize the overall information rate of the system or on a per-user basis ,57].

Successive Optimization

Successive optimization (SO) [54,56,57] is a successive precoding algorithm which addresses the power control problem in BD where capacity loss occurs due to the nulling of overlapping subspaces of different users. First, an optimum ordering of the users is determined like in case of V-BLAST detection. The precoding matrix for each user is then designed in a successive manner so that it lies in the null space of the channel matrices of the previous users only. In other words the transmit power of user i is optimized in such a manner that it does not interfere with users $1,\cdots,i-1$. However interference with the successive users is allowed. The combined channel matrix for the previous i –1 users can be written as

$$\hat{H}_i = \left[H_1^T \;\; H_2^T \;\cdots\; H_{i-1}^T\right]^T \tag{68}$$

and its SVD is given by

$$\hat{H}_i = \hat{U}_i\hat{D}_i\left[\hat{V}_i^{(1)} \;\; \hat{V}_i^{(0)}\right]^H \tag{69}$$

$\hat{V}_i^{(0)}$ contains the $M_T - \hat{L}_i$ rightmost singular vectors where $\hat{L}_i$ is the rank of $\hat{H}_i$ The precoding matrix P_i that lies in the null space of $\hat{H}_i$ is then determined as $P_i = \hat{V}_i^{(0)}P_i'$ for some choice of P_i'

Dirty Paper Coding

Dirty paper coding is a nonlinear precoding technique and is based on the concept introduced by Costa [58] where the AWGN channel is modified by adding interference which is known at the transmitter. This concept is analogous to "writing on dirty paper" where the writing is the desired signal and the dirt represents the interference. Since the transmitter knows where the "dirt" or interference is, writing on di paper is the same as writing on clean paper [48].

In addition to SU-MIMO system e.g. the GMDZFDP scheme mentioned in Subsection 3.4, dirty paper coding is also applicable to MU-MIMO downlink transmission. In a MU-MIMO downlink system, CSI feedback from the users is available at the BS and it can figure out the interference produced at a particular user by the signals meant for other users. Therefore, dirty paper coding can be applied to each user's signal at the BS so that the known interference from other users is avoided.

Various dirty paper coding techniques for the MUMIMO downlink are discussed in [48]. A well-known approach is to use QR decomposition of the channel matrix, given by H =LQ, where L is a lower triangular matrix and Q is a unitary matrix Q^H is then used for transmit precoding which results in the effective channel L. Therefore, the first user does not see any interference from the other users and no further processing of its signal is required at the BS. However, each of the subsequent users sees interference from the preceding users and dirty paper coding is applied to eliminate this known interference.

Another technique called vector precoding jointly precedes the users' signals rather than applying dirty paper coding to the user signals individually. The vector precoding technique is shown in Figure 51. The desired signal vector d is offset by a vector l of integer values and this operation is followed by channel inversion, resulting in the transmitted signal x, given by

$$\mathbf{x} = \mathbf{H}^{-1}(\mathbf{d} + \tau \boldsymbol{l}) \tag{70}$$

where the vector l is chosen to minimize the power of x, i.e.

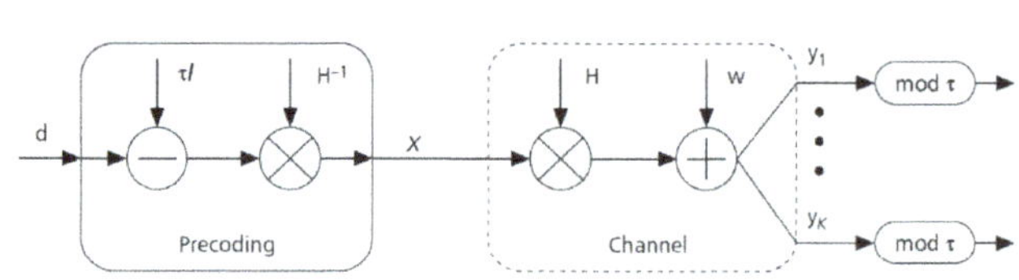

Figure 51: Vector precoding [48].

$$\boldsymbol{l} = \arg\min_{l'}\left\|\mathbf{H}^{-1}(\mathbf{d} + \tau \boldsymbol{l}')\right\| \tag{71}$$

The signal received at the k-th user is expressed as

$$y_k = d_k + \tau l_k + w_k \quad (72)$$

Where wk represents the Gaussian noise. A modulo operation is then applied to remove the offset k_l, as given by

$$\begin{aligned}(y_k)\bmod \tau &= (d_k + \tau l_k + w_k)\bmod \tau \\ &= (d_k + w_k)\bmod \tau\end{aligned} \quad (73)$$

Regularized vector precoding is a modification of vector precoding which uses regularized (MMSE) channel invers transmitter c ion in place of simple channel inversion. The d signal vector is then given by

$$\mathbf{x} = \mathbf{H}^H\left(\mathbf{H}\mathbf{H}^H + \alpha\mathbf{I}\right)^{-1}(\mathbf{d} + \tau \mathbf{l}) \quad (74)$$

where the vector l is chosen to minimize the norm of x and a = K/ P_T . Dirty paper coding techniques based on vector precoding approach the sum capacity of the MU-MIMO downlink channel, which is defined as the maximum system throughput achieved by maximizing the sum of the information rates of all the users [48].

Figure 52 shows the performance comparison of various channel inversion and dirty paper coding techniques for uncoded MU-MIMO downlink transmission with $M_T = 10$ transmit antennas and $M_R = 10$ single antenna UEs, using QPSK modulation. The vector pre coding techniques clearly outperform the others at high SNR. However, regularized channel inversion performs even better than regularized vector precoding in the low SNR region. A possible reason for the performance loss of regularized vector precoding at low SNR is the use of a finite cubical lattice in this algorithm [48]. Use of different lattice strategies may result in improved performance.

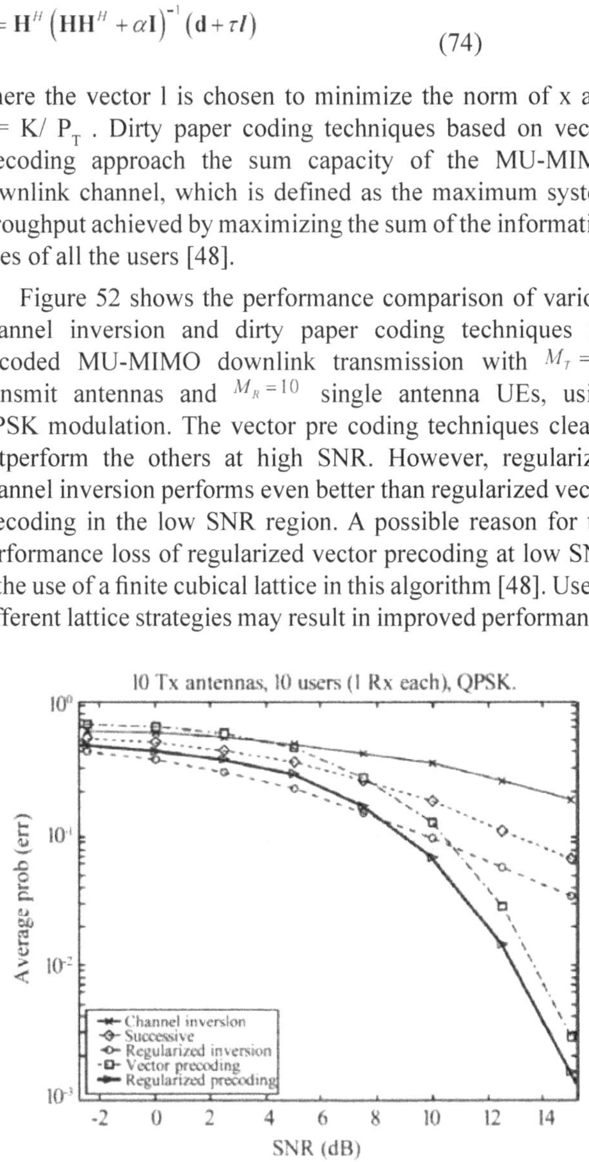

Figure 52: Performance comparison of various channel inversion and dirty paper coding techniques [48].

Tomlinson-Harashima Precoding

Tomlinson-Harashima precoding (THP) is a nonlinear precoding technique originally developed for SISO systems for temporal pre-equalization of ISI and is equivalent to moving the decision feedback part of the decision feedback equalizer (DFE) to the transmitter [59]. However, THP can also be applied to MU-MIMO downlink systems for MUI mitigation in the spatial domain. Two MU-MIMO downlink transmission schemes utilizing THP are described in [57]. The first one called SO THP combines SO and THP to improve performance by eliminating residual MUI. SO THP involves successive BD, reordering of the users and finally, HP. Figure 53 shows the block diagram of the SO THP system (taken from [57] with a slight notational change). Here P represents the combined precoding matrix for all users generated by SO, given by Equation 64. H is the channel matrix and D represents the combined demodulation (receive filtering) matrix. The lower triangular feedback matrix B is generated in the last step and is used for THP. In order to generate matrix B, the users are first arranged in the reverse order of precoding and then the lower diagonal equivalent combined channel matrix (which includes precoding and demodulation) is calculated, with singular values on the main diagonal. The elements in each row of this matrix are then divided by the corresponding singular values to obtain the feedback matrix B. The order in which THP precoding is applied to the users' data streams is opposite to the order in which their precoding matrices are generated. Therefore, THP precoding starts with the data stream of the first user whose precoding matrix P_1 was generated last.

Use of THP results in increased transmit power and for this reason, a modulo operator is introduced at the transmitter and the receiver so that the constellation points are kept within certain boundaries. At the receiver, each data stream is divided by the corresponding singular value before applying the modulo operator, which ensures that the constellation boundaries remain the same as at the transmitter. Detailed description of SO THP is provided in [60]

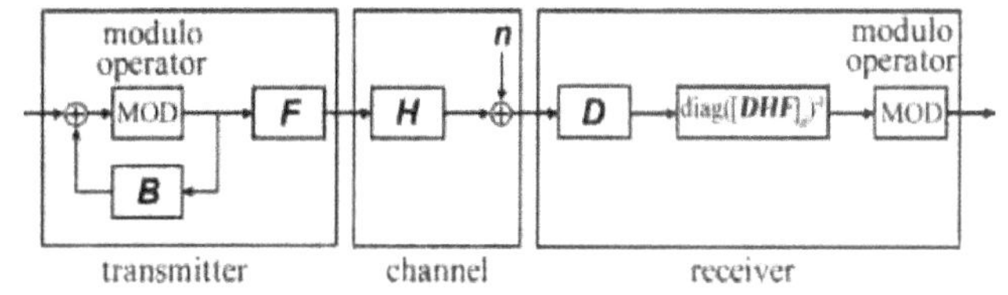

Figure 53: SO THP system block diagram [57].

The other scheme called MMSE THP [61] combines MMSE precoding and THP to eliminate the MUI below the main diagonal of the equivalent combined channel matrix. MMSE THP is an iterative precoding technique. The users are first arranged according to some optimal ordering criterion and the precoding matrix P is calculated column by column starting from the last user K. The i-th column of P corresponding to the i-th user is obtained using the corresponding i rows (first i rows for user K) of the channel matrix H according to the MMSE criterion, given by

$$\mathbf{P} = \beta\left(\mathbf{H}^H\mathbf{H} + \alpha\mathbf{I}_{M_T}\right)^{-1}\mathbf{H}^H \quad (75)$$

Where

$$\beta = \sqrt{\frac{P_T}{\mathrm{tr}\left(\mathbf{P}\mathbf{x}\mathbf{x}^H\mathbf{P}^H\right)}}, \quad \alpha = \frac{M_R\sigma_n^2}{P_T}$$

P_T represents the total transmit power, x is the datavector to be transmitted and σ_n^2 represents the variance of the zero-mean circularly symmetric complex Gaussian (ZMCSCG) noise. THP is then applied to eliminate the MUI seen by the i-th user from the previous i −1 users. Figure 54 compares the 10% outage capacity of SO THP and MMSE THP schemes for a MU-MIMO downlink system consisting of 4 single-antenna users and 4 BS transmit antennas, denoted as {1, 1, 1, 1} ´ 4 antenna configuration. Results for ZF channel inversion and a {2, 2} ´ 4 TDMA system are also provided.

Successive MMSE Precoding

The Successive MMSE (SMMSE) precoding scheme proposed in [57] addresses the problem of performance degradation associated with MMSE precoding when closely spaced receive antennas are used, like in case of multi-antenna UEs. SMMSE involves successively calculating the columns of the combined precoding matrix P where each column represents a beamforming vector corresponding to a particular receive antenna

Consider the system model of Subsection 7 where each of the K users is equipped with M_{R_i} receive antennas for $i = 1, \cdots, K$ and P_i represents the precoding matrix for the i-th user consisting of M_{R_i} columns, each corresponding to a receive antenna. For the j-th receive antenna of the i-th user, the matrix $\bar{\mathbf{H}}_i^{(j)}$ is defined as

$$\bar{\mathbf{H}}_i^{(j)} = \begin{bmatrix} \mathbf{h}_{i,j} & \mathbf{H}_1^T & \cdots & \mathbf{H}_{i-1}^T & \mathbf{H}_{i+1}^T & \cdots & \mathbf{H}_K^T \end{bmatrix}^T \tag{76}$$

where $\mathbf{h}_{i,j}^T$ represents the j-th row of the i-th user's channel matrix H_i. The corresponding column of P_i is then calculated using the MMSE criterion and is equal to the first column of the matrix

$$\mathbf{P}_{i,j} = \beta\left(\left(\bar{\mathbf{H}}_i^{(j)}\right)^H \bar{\mathbf{H}}_i^{(j)} + \alpha \mathbf{I}_{M_T}\right)^{-1} \left(\bar{\mathbf{H}}_i^{(j)}\right)^H \tag{77}$$

All columns of P_i are calculated in this manner and this process is repeated for all users to obtain the combined precoding matrix P. After precoding, the equivalent combined channel matrix is given by $\mathbf{HP} \in \mathbb{C}^{M_R \times M_R}$ which is block diagonal for high SNR values, resulting in a set of K SU-MIMO channels. Therefore, any SUMIMO technique e.g. eigen-beamform for capacity maximization or DET for maximum diversity and array gain can be the i-t user's equivalent channel matrix $\mathbf{H}_i\mathbf{P}_i$. SMMSE precoding has slightly higher complexity than BD [57]. A reduced complexity version called per-user SMMSE (PU-SMMSE) is proposed in [62].

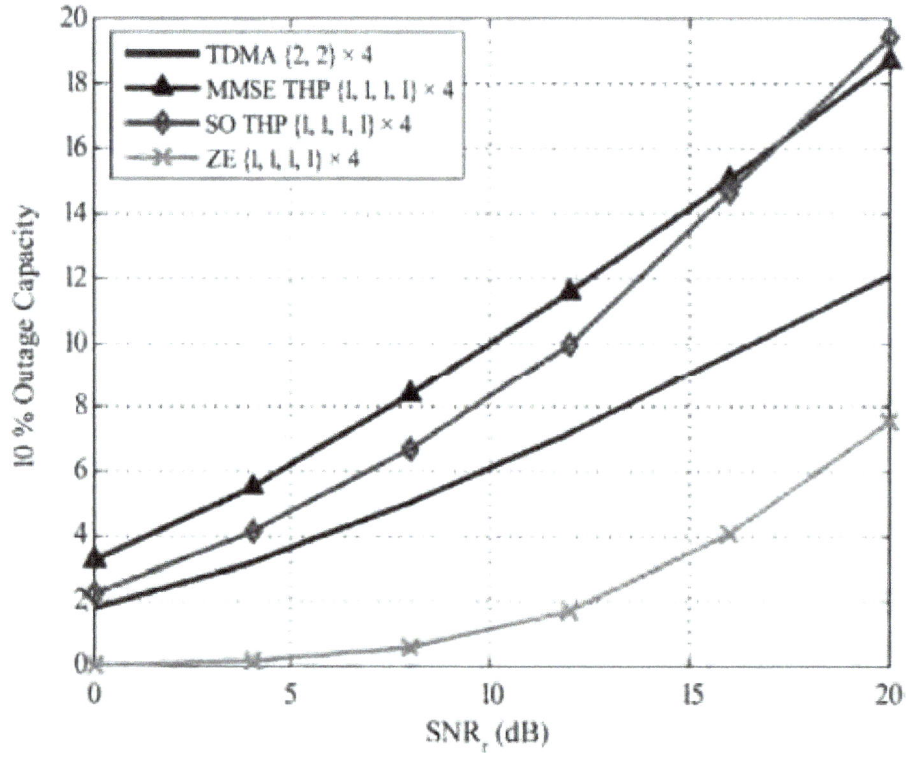

Figure 54: 10% outage capacity of SO THP and MMSE THP for {1, 1, 1, 1} ´ 4 configuration [57].

Figure 55 shows the BER performance comparison of SMMSE, SO THP and BD for {2, 2, 2} ´ 6 and MMSE THP for {1, 1, 1, 1, 1, 1} ´ 6 antenna configuration, in a spatially white flat fading channel. These results are based on diversity maximization for the individual users and water-filling is used for power allocation. SMMSE provided the best performance for the case of multi- antenna users while BD surpasses SO THP. Figure 56 shows the BER performance of SMMSE, SO THP and BD for {1, 1, 2, 2} ´ 6 and MMSE THP for {1, 1, 1, 1, 1, 1} ´ 6 configuration. Here SO THP outperforms the others. However, SMMSE performs better than MMSE THP and even SO THP at low SNR.

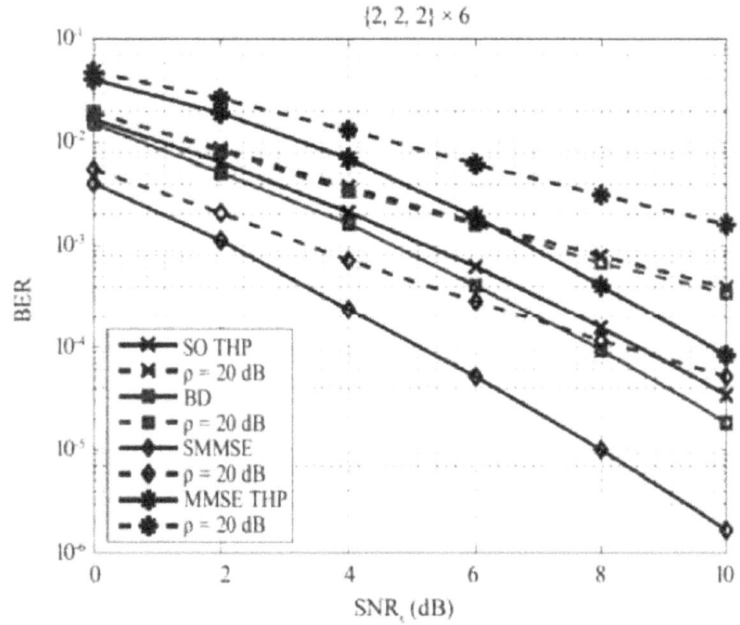

Figure 55: BER performance of SMMSE, SO THP, BD for {2,2,2} ′ 6 and MMSE THP for {1,1,1,1,1,1} ′ 6 configuration [57].

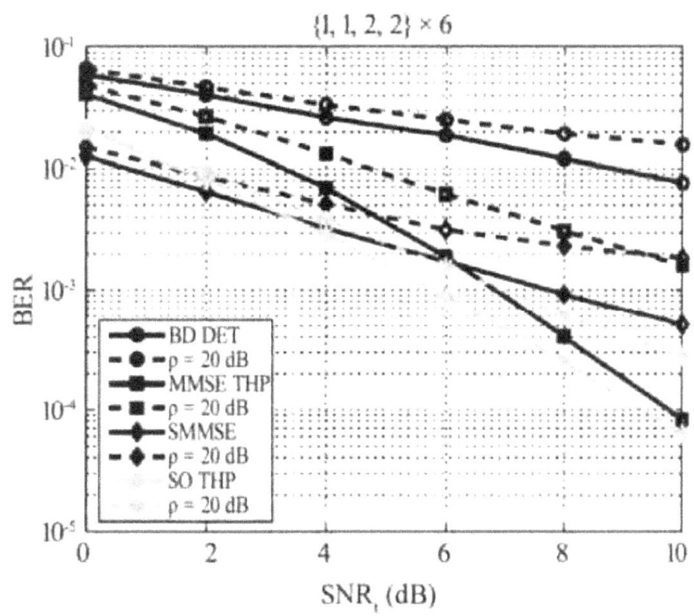

Figure 56: BER performance of SMMSE, SO THP, BD for {1,1,2,2} ′ 6 and MMSE THP for {1,1,1,1,1,1} ′ 6 configuration [57].

Iterative Linear MMSE Precoding

Two iterative linear MMSE precoding schemes are discussed in [55] for users with multiple antennas. Consider the system model of Figure 50 where P is the combined precoding matrix at the BS and V is the block-diagonal combined decoding matrix consisting of the decoding matrices V_i of all the users. In case of linear MMSE precoding which minimizes the MSE between $\hat{a}$ and a V_i is the linear MMSE receiver for user i and can be estimated locally at the corresponding UE. The first scheme called direct optimization, iteratively computes the MMSE solution using a numerical method. The SMMSE solution can be used as an initial guess for the free variables V_i with 1 = 1 for the free variable $\lambda \in \mathbb{R}$ An iterative process is then used which can lead to a true MMSE solution but not in all cases. The BD solution can also be used as an initial guess but that would result in slower convergence. The other scheme exploits the uplink/downlink duality [63] to obtain the true MMSE solution using an iterative algorithm. The resulting objective function is convex in this case. Detailed description as well as a practically implementable algorithm for this duality-based scheme is presented in [64].

The uncoded BER performance of BD, SMMSE, direct optimization and the duality-based scheme is compared in Figure 57 for {1, 2, 3} ′ 6 antenna configuration. The bit error rates are averaged over all users. DET is applied for BD and SMMSE while single stream (SS) transmission (consisting of a single data stream per user) is used for direct optimization and the duality-based scheme based on the algorithm from [64].

Independent Rayleigh fading channel perturbed by complex Gaussian noise is considered and QPSK modulation is used for transmission. Both iterative linear MMSE schemes outperform the other two by a large margin. The duality-based scheme shows slightly better performance than direct optimization

This difference is due to the nonconvex objective function used for direct optimization which occasionally causes the optimization routine to undesired minima. BD provides the worst performance because of the zero-forcing constraint.

Figure 58 shows the coded BER performance of these precoding schemes. A rate 1/2 turbo code is used for error correction. OFDM based transmission is considered where the precoding is applied on a per-subcarrier basis. The ITU Vehicular A channel model is used. Direct optimization and the duality-based scheme provide almost identical performance in this case, far better than BD and SMMSE.

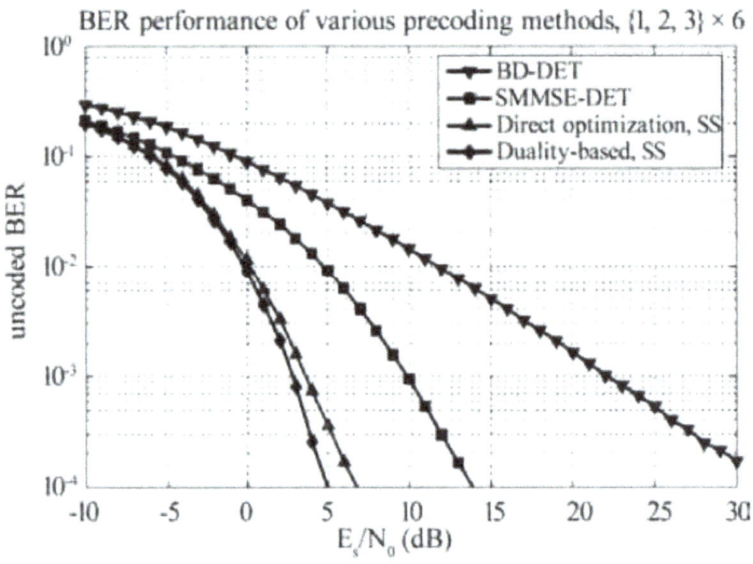

Figure 57: Uncoded BER performance of BD, SMMSE, direct optimization and duality-based scheme [55].

Partial CSI Feedback

Transmit precoding for downlink MU-MIMO transmission requires CSI feedback from the users. However, feedback information consisting entirely of the current state of the channel may not be accurate enough in case of rapidly varying channels. Downlink transmission schemes that utilize partial CSI consisting of long-term channel statistics along with some instantaneous channel information like SNR, SINR etc. provide a solution to this problem while reducing the feedback overhead. An interesting MU-MIMO downlink transmission scheme based solely on instantaneous channel norm feedback is proposed in [65]. MU-MIMO configuration with multiple base station (BS) antennas and a single antenna at each UE is considered. The proposed scheme can provide high multiuser diversity gain by optimizing resource allocation at the BS while simply utilizing the instantaneous channel norm feedback from the UEs.

Figure 59 shows the operation of the proposed system at the transmitter (BS). The BS initially transmits orthogonal pilot signals on all transmit antennas which are used by each UE to estimate the received signal energy i.e. the squared norm of the channel vector given by

Figure 58: Coded BER performance of BD, SMMSE, direct optimization and duality-based scheme [55].

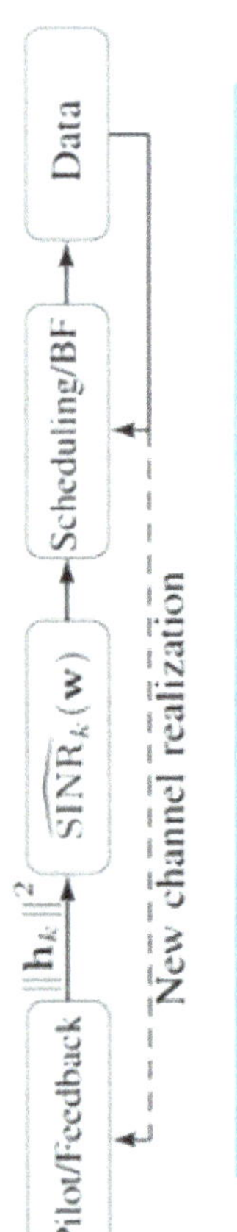

Figure 59: System operation at the transmitter [65].

$$\rho_k \triangleq \|\mathbf{h}_k\|^2 \tag{78}$$

where h_k represents the channel vector for the k-th UE from the I UEs scheduled for transmission. ρ_k is categorized as channel gain information (CGI) in [65]. This quantity is then fed back to the BS. The BS estimates the long-term channel statistics including the channel mean and the channel covariance matrix, defined as

$$\hat{\mathbf{h}}_k \triangleq \mathrm{E}\{\mathbf{h}_k \mid \rho_k\} \tag{79}$$

$$\hat{\mathbf{Q}}_k \triangleq \mathrm{E}\left\{\mathbf{h}_k \mathbf{h}_k^H \mid \rho_k\right\} \tag{80}$$

This slow varying statistical information is referred to as channel distribution information (CDI). The CGI feedback along with the CDI is used to estimate the SINR and the optimized beamforming weight vectors for each of the UEs scheduled for transmission. The SINR for user k is estimated as

$$\widehat{\mathrm{SINR}}_k = \frac{\mathbf{w}_k^H \hat{\mathbf{Q}}_k \mathbf{w}_k}{\sum_{i \in \mathcal{I} \setminus \{k\}} \mathbf{w}_i^H \hat{\mathbf{Q}}_k \mathbf{w}_i + \sigma_k^2} \tag{81}$$

Where w_k is the corresponding beamforming vector $\hat{\Gamma}_{ki} = \mathbf{w}_i^H \hat{\mathbf{Q}}_k \mathbf{w}_i$ is the MMSE estimate of the signal-to interferences power ratio (SIR) $\Gamma_{ki} = \mathbf{w}_i^H \mathbf{h}_k \mathbf{h}_k^H \mathbf{w}_i$ and σ_k is the AWGN power

The actual data transmission to the scheduled users then begins. Another set of users can later be scheduled to achieve better fairness and this process goes on until the CGI becomes outdated. At this point, BS transmits the pilot signals again and the whole process is repeated.

Multiuser Scheduling

The BS can only support a limited no. of simultaneous users for MU-MIMO downlink transmission with acceptable performance. The performance degrades in rank deficient scenarios and also when the users are spatially correlated. In case of multi-antenna users, the no. of users that can be supported simultaneously becomes even lesser In practical situations, the BS would generally server a larger no. of users than it can simultaneously support. Therefore, an efficient scheduling algorithm is required to select the group of users that will be spatially multiplexed by the BS at a certain time and frequency. The scheming algorithm should avoid grouping spatially correlated users and maximize system performance while maintaining fairness toward all users. Fairness ensures that all users are served including those with weak channels. Otherwise, the BS will transmit to the strong users only and the weaker ones will be ignored [59]. A fair scheduling scheme called the strongest weakest-normalized-sub channel-first (SWNSF) scheduling proposed in [66] enhances the coverage and capacity of MU-MIMO systems while requiring only a limited amount of feedback.

MU-MIMO Capacity

The maximum capacity of a MU-MIMO system is expressed in terms of the sum capacity (or sum-rate capacity) of the broadcast channel. As mentioned earlier the sum capacity represents the maximum achievable system throughput and is defined as the maximum sum of the downlink information rates of all the users [48]. Figure 60 illustrates the capacity region and the sum capacity of a MU-MIMO channel with two users. Clearly, achieving the sum capacity requires some tradeoff between the capacities of the individual users, depending on the shape of s the capacity region

It has been shown in [67] that the sum capacity of a Gaussian MIMO broadcast channel with an arbitrary no. of BS transmit antennas and multi-antenna users, is the saddle-point of a minimax problem and (assuming real-valued signals) is given by [49,67]

$$C_{BC} = \min_{\mathbf{R}_{nn} > 0, [\mathbf{R}_{nn}]_{k,k} = \sigma_n^2} \max_{\mathrm{Tr}(\mathbf{R}_{xx}) = P} \frac{1}{2} \log \frac{\det\left(\mathbf{H}\mathbf{R}_{xx}\mathbf{H}^H + \mathbf{R}_{nn}\right)}{\det\left(\mathbf{R}_{nn}\right)} \tag{82}$$

$$C_{MAC}\left(\mathbf{P}; \mathbf{H}^H\right) \triangleq \bigcup_{\{\mathbf{Q}_i \geq 0, \mathrm{Tr}(\mathbf{Q}_i) \leq P_i \forall i\}} \left\{ \begin{array}{l} (R_1, \cdots, R_K) \\ \sum_{i \in S} R_i \leq \frac{1}{2} \log\left|\mathbf{I} + \sum_{i \in S} \mathbf{H}_i^H \mathbf{Q}_i \mathbf{H}_i\right| \ \forall S \subseteq \{1, \cdots, K\} \end{array} \right\} \tag{83}$$

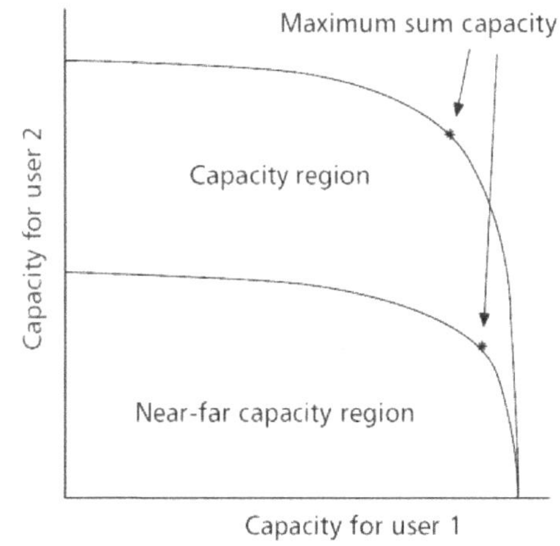

Figure 60: The capacity region and sum capacity of a twouser MU-MIMO channel [48].

where H is the channel matrix, R_{xx} is the transmit signal covariance matrix, R_{nn} is covariance matrix of the additive ZMCSCG noise with variance 2 s n and P represents the total transmit power. The sum capacity is obtained by iteratively computing the best transmit covariance matrix R_{xx} for a given noise covariance and then computing the least favorable noise covariance matrix R_{nn} for the given transmit covariance. It is also shown that decision-feedback precoding or dirty paper coding is the optimal precoding strategy capable of achieving the sum capacity. In [68], it has been proved that the capacity region of the Gaussian MIMO-BC with single-antenna users is equivalent to the dirty paper coding (DPC) rate region under a certain total transmit power constraint. The DPC achievable rate region for the case of multi-antenna users is formulated in [67] and i also discussed in [69]. Further research work is required on the capacity of fading MIMO broadcast channels.

The capacity of MIMO-MAC channels constitutes a relatively simple problem. The capacity of any MAC is given as the convex closure of the union of rate regions corresponding to every product input distribution $p(u_1)\cdots p(u_K)$ satisfying certain user-by-user power constraints, where $\mathbf{u}_k \in \mathbb{C}^{N \times 1}$ the k-th user's transmitted signal. However, the convex hull operation is not required for the Gaussian MIMO-MAC and only the Gaussian inputs need to be considered. This can be written as [69]

where $\mathbf{P} = (P_1, \cdots, P_K)$ represents the set of transmit powers corresponding to each of the K users, × denotes the determinant and R_i, Q_i and H_i represent the rate vector, spatial covariance matrix and the channel matrix respectively for the i-th user. A general expression for the capacity regions of fading MAC channels is also given in [69].

CONVEX OPTIMIZATION

Convex optimization methods provide a powerful set of tools for solving optimization problems expressed in convex form. However, most engineering problems are not convex when directly formulated and need to be reformulated in a convex form in order to apply convex optimization. Two methods are commonly used for this reformulation. The first method is to use a change of variables to obtain an equivalent convex form. The other is to remove some of the constraints so that the problem becomes convex, in such a way that the optimal solution also satisfies the removed constraints. Any problem, once expressed in convex form, can be optimally solved either in closed form using the optimality conditions derived from Lagrange duality theory e.g. Karush-KuhnTucker (KKT) conditions or numerically using iterative algorithms like the interior-point, cutting-plane and ellipsoid methods [70].

In recent years, convex optimization has gained significant importance in optimal joint transceiver design (transmit-receive beamforming) of MIMO systems based on linear precoding and equalization. Various design methods for linear multicarrier SU-MIMO transmitreceive beamformers, based on convex optimization are given in [71]. Two novel low-complexity multilevel water-filling solutions for the MAX-MSE and HARMSINR criteria are also proposed. The MAX-MSE method minimizes the maximum of the MSEs corresponding to the different substreams whereas HARM-SINR maximizes the harmonic mean of the SINRs. The ARITHBER method which minimizes the arithmetic m of the BERs, provides the best average BER performance and is considered as a benchmark in [71]. It is shown that when cooperating different subcarriers is allowed to improve performance, an exact optimal closed-form solution is obtained in terms of minimizing the rage BER which unifies all three optimization criteria mentioned above. In other words, MAX-MSE, HARM-SINR and ARITH-BER provide the same optimal solution for carrier-cooperative schemes

Figure 61 shows the average BER performance (at 5% outage probability) of ARITH-MSE, HARM-SINR, MAX-MSE and ARITH-BER for 2 ′ 2 MIMO configuration. The HIPERLAN/2 standard based on OFDM is used for the simulations with frequency selective fading in an indoor NLOS scenario. QPSK modulation is employed and perfect CSI is assumed at the transmitter and the receiver. In the absence of subcarrier cooperation ARITH-BER provides the best performance followed by MAX-MSE and HARM-SINR respectively. With subcarrier cooperation ARITH-BER, MAX-MSE and HARMSINR have identical performance which is optimal in the minimum average BER sense. A joint transceiver optimization scheme based on multiplicative Schur-convexity is proposed in [72] for THP precoded MIMO-OFDM systems. This scheme provides better BER performance than the aforementioned linear precoding schemes when the objective function is multiplicatively Schur-convex like in case of ARITH-BER, MAX-MSE or HARMSINR, and becomes equivalent to the optimal UCD scheme proposed in [37]. Convex optimization is also applicable to downlink beamforming in MU-MIMO systems as mentioned in [70] for the case of single-antenna users. The duality-based iterative MMSE precoding scheme proposed in [64] supports multi-antenna users and uses an iterative algorithm to solve the KKT optimality conditions.

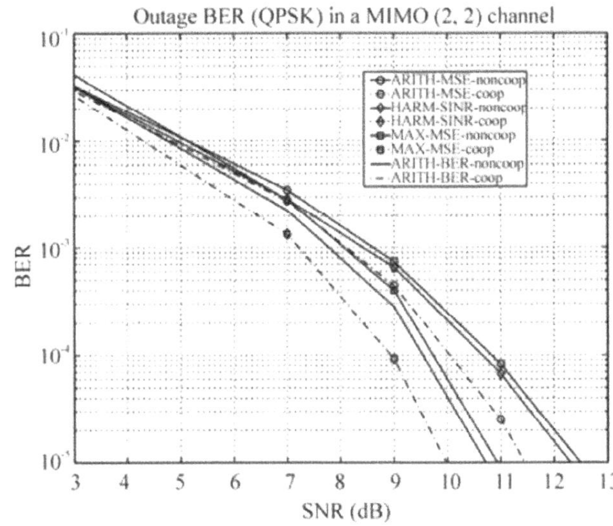

Figure 61: BER performance of ARITH-MSE, HARMSINR, MAX-MSE and ARITH-BER for 2 ′ 2 MIMO configuration using a single substream [71].

CONCLUSIONS AND FUTURE RESEARCH

Open-loop MIMO techniques provide a low-complexity solution for MIMO diversity and SM. STC or SFC e.g. STBC, orthogonal STBC (OSTBC), STTC, SFBC etc. can be used for diversity maximization. These schemes are also well-suited for transmission over high-speed mobile channels where link reliability is the primary concern rather than throughput maximization. Open-loop SM can be implemented by means of a simple V-BLAST system but at the cost of low BER performance. JDM MIMO systems like the CDA-SM-OFDM system which combines CDD and SM, provide better performance. Further research may be carried out to optimize the achievable diversity and multiplexing gain for enhancing system throughput while considering the impact of transmit and receive antenna correlations. The iterative turboMIMO systems are also capable of achieving high capacity but are more complex to implement. Turbo-MIMO systems may be improved further by using improved turbo codes and signal constellation shaping [73], improved code interleavers, and stratified processing [74].

Closed-loop MIMO systems like the SVD-based linear transceivers, are capable of achieving the SU-MIMO capacity by transmitting over the channel Eigen modes with optimal water-filling power allocation, provided perfect CSI is available at the transmitter and the receiver. Alternatively, DET can be employed to achieve the maximum diversity and array gain. Closed-loop STC like the closed-loop STBC schemes proposed in [75] which support more than two transmit antennas, also provide diversity maximization. The GMD-MIMO scheme decomposes the MIMO channel into identical subchannels and attempts to combine MIMO diversity and SM in an optimal manner. MAX-MSE, HARM-SINR and ARITH-BER methods based on convex optimization provide optimal average BER performance for multistream transmission when subcarrier cooperation is allowed. However, the GMD and convex optimization approaches assume the availability of full CSI at the transmitter and the receiver which is not always the case e.g. in rapidly changing mobile channels. Therefore, the performance of these schemes with imperfect feedback needs to be evaluated for practical implementation. Use of convex optimization methods for designing optimal linear transceivers utilizing partial CSI also need to be investigated.

For the MU-MIMO uplink, the LAST-MUD scheme based on V-BLAST detection provides good perform- ance for CDMA systems. The SMMSE MUD provides acceptable performance for MIMO-OFDM system The higher complexity Turbo-MUD scheme achieves better performance and can jointly detect the transmit antennas of each multi-antenna user. Use of different turbo detection strategies, joint detection of all users and extension to multicarrier systems may be investigated in future. The GA-assisted MUDs, particularly the TTCM-assisted IGA-MUD provide near-ML detection performance for SDMA-OFDM systems and also perform reasonably well in certain rank-deficient scenarios. The complexity is high but increases slowly with the no. of users as compared to the ML-MUD. GA-assisted MUDs can also incorporate joint channel estimation and symbol detection. Future research work may include extending the GA-assisted MUD schemes to support multi-antenna users and development of more efficient GAs to reduce system complexity. Use of artificial intelligence (AI) techniques like the radial basis function (RBF) based artificial neural networks (ANNs) should also be explored for multiuser detection.

Coming to the MU-MIMO downlink, SMMSE precoding performs reasonably well with manageable complexity. The nonlinear dirty paper coding techniques are capable of achieving the sum capacity of Gaussian multiuser channels with single-antenna users. The iterative linear MMSE techniques like direct optimization and the duality-based scheme which uses convex optimization, provide excellent uncoded and coded BER performance for single stream transmission. Further research is required to develop MU-MIMO downlink transmission schemes capable of supporting SM for individual multi-antenna mobile users. Transmission schemes based on partial CSI which require minimal CSI feedback from the users represent the most suitable choice for practical implementation. Convex optimization tools might be useful in designing joint transmit-receive beamforming systems less prone to errors resulting from imperfect CSI. Research efforts are also needed for developing efficient multiuser scheduling schemes that maintain fairness to the users while minimizing the loss of system capacity. Determining the sum capacity and capacity regions for fading MU-MIMO downlink channels with multi-antenna users is also an area for future research.

Accurate channel estimation is of prime importance in MIMO communications. Channel estimation errors may result in severe performance degradation. Therefore, research efforts are continuing for further improvements in this domain. A broadband wireless transmission technique called orthogonal frequency- and code-division multiplexing (OFCDM) [76] has recently gained prominence as a better alternative to OFDM. OFCDM readily supports MIMO techniques and extensive research would be required to realize the full potential of MIMOOFCDM systems.

REFERENCES

1. D. Gesbert, M. Shafi, D.-S. Shiu, P. J. Smith, and A. Naguib, "From theory to practice: An overview of MIMO space-time coded wireless systems," IEEE Journal on Selected Areas in Communications, Vol. 21, No. 3, pp. 281–302, April 2003.
2. Y. Jiang, J. Li, and W. W. Hager, "Joint transceiver design for MIMO communications using geometric mean decomposition," IEEE Transactions on Signal Processing, Vol. 53, No. 10, pp. 3791–3803, October 2005.
3. M. Jiang and L. Hanzo, "Multiuser MIMO-OFDM for next-generation wireless systems," Proceedings of the IEEE, Vol. 95, No. 7, pp. 1430–1469, July 2007.
4. K. W. Park, E. S. Choi, K. H. Chang, and Y. S. Cho, "An MIMO-OFDM technique for high-speed mobile channels," in Proceedings of Vehicular Technology Conference, Vol. 2, pp. 980–983, April 2003.
5. M. S. Gast, "802.11 wireless networks: The definitive guide," 2nd edition, O'Reilly, April 2005.
6. Wi-Fi Alliance, "WiFi CERTIFIED? 802.11n draft 2.0: Longer-range, faster-throughput,multimedia grade WiFi? networks," 2007. http://wi-fi.org/whiteppaper_80211n_draft2_technical.php.
7. FierceBroadbandWireless, "IEEE approves EWC 802.11n as first draft," 24 January 2006. http://www.fiercebroadbandwireless.com/story/ieee-approves-ewc-802-11n-as- first-draft/2006-01-25.
8. Y. Sun, M. Karkooti, and J. Cavallaro, "High throughput, parallel, scalable LDPC encoder/decoder architecture for OFDM systems," in Proceedings of IEEE Workshop on Circuits and Systems, pp. 225–228, October 2006.
9. H. N. Niu and C. Ngo, "Diversity and multiplexing switching in 802.11n MIMO systems," in Proceedings of 40th Asilomar Conference on Signals, Systems and Computers (ACSSC '06), pp. 1644–1648, 29 October–1 November 2006.
10. P. Kim and K. M. Chugg, "Capacity for suboptimal receivers for coded multiple-input multiple-output systems," Vol. 6, No. 9, pp. 3306–3314, September 2007.
11. S. Abraham, A. Meylan, S. Nanda, "802.11n MAC design and system performance," in Proceedings of IEEE International Conference on Communications (ICC 2005), Vol. 5, pp. 2957–2961, 16–20 May 2005.
12. Cisco Aironet 1250 Series Access Point Q&A. http://www. cisco.com/en/US/prod/collateral/wireless/ps5678/ps6973/ps8382/prod_qas0900aecd806b7c82.html.
13. S. Kim, S.-J. Lee, and S. Choi, "The impact of IEEE 802.11 MAC strategies on multi-hop wireless mesh networks," Wireless Mesh Networks, in Proceedings of 2nd IEEE Workshop on Wireless Mesh Networks (WiMesh 2006), pp. 38–47, 25–28 September 2006.
14. IEEE Std 802.16e?-2005 and IEEE Std 802.16?- 2004/Cor 1-2005 (Amendment and Corrigendum to IEEE Std 802.16-2004), "IEEE standard for local and metropolitan

area networks Part 16: Air Interface for fixed and mobile broadband wireless access systems amendment 2: Physical and medium access control layers for combined fixed and mobile operation in licensed bands and Corrigendum 1," 28 February 2006.

15. IEEE Std 802.16?-2004, "IEEE standard for local and metropolitan area networks Part 16: Air interface for fixed broadband wireless access systems," 1 October 2004.

16. J. G. Andrews, A. Ghosh, and R. Muhamed, "Fundamentals of WiMAX: Understanding broad-band wireless networking," Prentice Hall, 2007.

17. Kambourov, "MIMO aspects in 802.16e WiMAX OFDMA," WiMAX Tutorial, Siemens PSE MCS RA 2, 22 November 2006.

18. Nokia Siemens Networks, "Advanced antenna systems for WiMAX." http://www.nokiasiemensnetworks.com.

19. S. Nanda, R. Walton, J. Ketchum, M. Wallace, and S. Howard, "A high-performance MIMO OFDM wireless LAN," IEEE Communications Magazine, Vol. 43, No. 2, pp. 101–109, February 2005.

20. Agilent Technologies, "Mobile WiMAX 802.16 Wave 2 Features." http://www.home.agilent.com/agilent/editorial. jspx?action=download&cc=US&lc=eng&ckey=1213544&nid=-536902344.536910932.00&id=1213544.

21. IEEE C802.16m-07/069, "Draft IEEE 802.16m evaluation methodology document," IEEE 802.16 Broadband Wireless Access Working Group, 2007.

22. WiMAX Forum?, "Deployment of Mobile WiMAX? Networks by Operators with Existing 2G & 3G Networks," 2008. http://www.wimaxforum.org/technology/ downloads/deployment_of_mobile_wimax.pdf.

23. C. Ribeiro, "Bringing wireless access to the automobile: A comparison of Wi-Fi, WiMAX, MBWA, and 3G," 21st Computer Science Seminar, 2005.

24. B. M. Bakmaz, Z. S. Bojkovi?, D. A. Milovanovi?, and M. R. Bakmaz, "Mobile broadband networking based on IEEE 802.20 standard," in Proceedings of 8th International Conference Telecommunications in Modern Satellite, Cable and Broadcasting Services (TELSIKS 2007), pp. 243–246, 26–28 September 2007.

25. IEEE Std 802.20?-2008, "IEEE standard for local and metropolitan area networks Part 20: Air interface for mobile broadband wireless access systems supporting vehicular mobility–physical and media access control layer specification," 29 August 2008.

26. 3GPP TS 36.201 V8.1.0 (2007-11), "LTE physical layer – general description (Release 8)," 3GPP TSG RAN,2007.

27. J. Zyren, "Overview of the 3GPP long term evolution physical layer," White Paper, Freescale Semiconductor, Inc., 2007.

28. 3GPP TR 25.913 V7.3.0 (2006-03), "Requirements for evolved UTRA (E-UTRA) and evolved UTRAN (E-UTRAN) (Release 7)," 3GPP TSG RAN, 2006.

29. 3GPP TS 36.300 V8.4.0 (2008-03), "Evolved universal terrestrial radio access (E-UTRA) and evolved universal terrestrial radio access network (E-UTRAN); Overall description; Stage 2 (Release 8)," 3GPP TSG RAN, 2008.

30. 3GPP TS 36.211 V8.1.0 (2007-11), "Evolved universal terrestrial radio access (E-UTRA); Physical channels and modulation (Release 8)," 3GPP TSG RAN, 2007.

31. P. W. Wolniansky, G. J. Foschini, G. D. Golden, and R. A. Valenzuela, "V-BLAST: An architecture for realizing very high data rates over the rich-scattering wireless channel," in Proceedings of 1998 URSI International Symposium Signals, Systems, and Electronics, pp. 295–300, 29 September–2 October 1998.

32. S. T. Chung, A. Lozano, and H. C. Huang, "Approaching eigenmode BLAST channel capacity using V-BLAST with rate and power feedback," in Proceedings of IEEE VTC 2001 Fall, Vol. 2, pp. 915–919, 7–11 October 2001.

33. Q. P. Cai, A. Wilzeck, C. Schindler, S. Paul, and T. Kaiser, "An exemplary comparison of per antenna rate control based MIMO-HSDPA receivers," in Proceedings of 13th European Signal Processing Conference (EUSIPCO 2005), 4–8 September, 2005.

34. R. Gowrishankar, M. F. Demirkol, and Z. Q. Yun, "Adaptive modulation for MIMO systems and throughput evaluation with realistic channel model," in Proceedings of 2005 International Conference on Wireless Networks, Communications and Mobile Computing, Vol. 2, pp. 851–856, 13–16 June 2005.

35. M. I. Rahman, S. S. Das, E. de Carvalho, and R. Prasad, "Spatial multiplexing in OFDM systems with cyclic delay diversity," in Proceedings of IEEE Vehicular Technology Conference, pp. 1491–1495, 22–25 April 2007.

36. H. Busche, A. Vanaev, and H. Rohling, "SVD based MIMO precoding and equalization schemes for realistic channel estimation procedures," Frequenz Journal of RF-Engineering and Telecommunications, Vol. 61, No. 7–8, pp. 146–151, July–August 2007.

37. Y. Jiang and J. Li, "Uniform channel decomposition for MIMO communications," in Proceedings of 38th Asilomar Conference on Signals, System, and Computers, 7–10 November 2004.

38. S. Haykin, M. Sellathurai, Y. de Jong, and T. Willink, "Turbo-MIMO for wireless communications," IEEE Communications Magazine, Vol. 42, No. 10, pp. 48–53, October 2004.

39. M. Sellathurai and S. Haykin, "Turbo-BLAST for wireless communications: Theory and experiments," IEEE Transactions on Signal Processing, Vol. 50, No. 10, pp. 2538–2546, October 2002.

40. D. J. Love, R. W. Heath Jr., W. Santipach, and M. L. Honig, "What is the value of limited feedback for MIMO channels," IEEE Communications Magazine, Vol. 42, No. 10, pp. 54–59, October 2004.

41. Y. Yuda, K. Hiramatsu, M. Hoshino, and K. Homma, "A study on link adaptation scheme with multiple code words for spectral efficiency improvement on OFDM-MIMO systems," IEICE Transactions on Fundamentals, Vol. E90A, No. 11, pp. 2413–2422, November 2007.

42. Z. G. Zhou, H. Y. Yi, H. Y. Guo, and J. T. Zhou, "A partial feedback scheme for MIMO systems," in Proceedings of WiCom 2007, pp. 361–364, 21–25 September 2007.

43. Heidari, F. Lahouti, and A. K. Khandani, "Enhancing closed-loop wireless systems through efficient feedback reconstruction," IEEE Transactions on Vehicular Technology, Vol. 56, No. 5, pp. 2941–2953, September 2007.

44. H. R. Bahrami and T. Le-Ngoc, "MIMO precoding structures for frequency-flat and frequency-selective fading channels," in Proceedings of 1st International Conference on Communications and Electronics (ICCE '06), pp. 193–197, 10–11 October 2006.

45. M. Tsutsui and H. Seki, "Throughput performance of downlink MIMO transmission with multi-beam selection using a novel codebook," in Proceedings of IEEE VTC 2007-Spring, pp. 476–480, 22–25 April 2007.

46. K. W. Park and Y. S Cho, "An MIMO-OFDM technique for high-speed mobile channels," IEEE Communications Letters, Vol. 9, No. 7, pp. 604–606, July 2005.

47. A. Hutter, S. Mekrazi, B. N. Getu, and F. Platbrood, "Alamouti-based space-frequency coding for OFDM," Wireless Personal Communications: An International Journal, Vol. 35, No. 1–2, pp. 173–185, October 2005.

48. Q. H. Spencer, C. B. Peel, A. L. Swindlehurst, and M. Haardt, "An introduction to the multi-user MIMO downlink," IEEE Communications Magazine, Vol. 42, No. 10, pp. 60–67, October 2004.

49. Paulraj, R. Nabar, and D. Gore, "Introduction to space-time wireless communications," Cambridge, UK, Cambridge University Press, 2003.

50. S. Sfar, R. D. Murch, and K. B. Letaief, "Layered space-time multiuser detection over wireless uplink systems," IEEE Transactions on Wireless Communications, Vol. 2, No. 4, July 2003.

51. V. Stankovic and M. Haardt, "Improved diversity on the uplink of multi-user MIMO systems," in Proceedings of European Conference on Wireless Technology '05, pp. 113–116, 3–4 October 2005.

52. Santamaria, V. Elvira, J. Via, D. Ramirez, J. Perez, J. Ibanez, R. Eickoff, and F. Ellinger, "Optimal MIMO transmission schemes with adaptive antenna combining in the RF path," in Proceedings of 16th European Signal Processing Conference (EUSIPCO 2008), 25–29 August 2008.

53. N. Veselinovic, T. Matsumoto, and M. Juntti, "Iterative MIMO turbo multiuser detection and equalization for STTrC-coded systems with unknown interference," EURASIP Journal on Wireless Communications and Networking, Vol. 2004, No. 2, pp. 309–321, 2004.

54. Q. Spencer and M. Haardt, "Capacity and downlink transmission algorithms for a multi-user MIMO channel," in Proceedings of 36th Asilomar Conference on Signals, Systems, and Computers, pp. 1384–1388, November 2002.

55. B. Bandemer, M. Haardt, and S. Visuri, "Linear MMSE multi-user MIMO downlink precoding for users with multiple antennas," in Proceedings of IEEE 17th International Symposium on Personal, Indoor and Mobile Radio Communications (PIMRC'06), pp. 1–5, 11–14 September 2006.

56. Q. H. Spencer, A. L. Swindlehurst, and M. Haardt, "Zero-forcing methods for downlink spatial multiplexing in multiuser MIMO channels," IEEE Transactions on Signal Processing, Vol. 52, No. 2, pp. 461–471, February 2004.

57. V. Stankovic and M. Haardt, "Multi-user MIMO downlink precoding for users with multiple antennas," in Proceedings of 12th Wireless World Research Forum (WWRF), Toronto, ON, Canada, November 2004.

58. M. Costa, "Writing on dirty paper," IEEE Transactions on Information Theory, Vol. 29, pp. 439–441, May 1983.

59. V. Stankovic, M. Haardt, and G. D. Galdo, "Efficient multi-user MIMO downlink precoding and scheduling," in Proceedings of 1st IEEE International Workshop on Computational Advances in Multi-Sensor Adaptive Processing, pp. 237–240, 13–15 December 2005.

60. V. Stankovic and M. Haardt, "Successive optimization Tomlinson-Harashima precoding (SO THP) for multi- user MIMO systems," in Proceedings of IEEE International Conference on Acoustics, Speech, and Signal Processing (ICASSP), Philadelphia, PA, USA, March 2005.

61. M. Joham, J. Brehmer, and W. Utschick, "MMSE approaches to multiuser spatio-temporal Tomlinson-Harashima precoding," in Proceedings of 5th International ITG Conference on Source and Channel Coding (ITG SCC'04), pp. 387–394, January 2004.

62. M. Lee and S. K. Oh, "A per-user successive MMSE precoding technique in multiuser MIMO systems," in Proceedings of IEEE VTC 2007-Spring, pp. 2374–2378, 22–25 April 2007.

63. M. Schubert, S. Shi, E. A. Jorswieck, and H. Boche, "Downlink sum-MSE transceiver optimization for linear multi-user MIMO systems," in Proceedings of 39th Asilomar Conference on Signals, Systems, and Computers, pp. 1424–1428, October 2005.

64. Mezghani, M. Joham, R. Hunger, and W. Utschick, "Transceiver design for multi-user MIMO systems," in Proceedings of ITG/IEEE Workshop on Smart Antennas (WSA 2006), March 2006.

65. D. Hammarwall, M. Bengtsson, and B. Ottersten, "Acquiring partial CSI for spatially selective transmission by instantaneous channel norm feedback," IEEE Transactions on Signal Processing, Vol. 56, No. 3, March 2008.

66. C.-J. Chen and L.-C. Wang, "Enhancing coverage and capacity for multiuser MIMO systems by utilizing scheduling," IEEE Transactions on Wireless Communications, Vol. 5, No. 5, May 2006.

67. W. Yu and J. M. Cioffi, "Sum capacity of Gaussian vector broadcast channels," IEEE Transactions on Information Theory, Vol. 50, No. 9, September 2004.

68. H. Weingarten, Y. Steinberg, and S. Shamai, "The capacity region of the Gaussian MIMO broadcast channel," in Proceedings of ISIT 2004, 27 June–2 July

2004.

69. Goldsmith, S. A. Jafar, N. Jindal, and S. Vishwanath, "Capacity limits of MIMO channels," IEEE Journal on Selected Areas in Communications, Vol. 21, No. 5, pp. 684–702, June 2003.

70. D. P. Palomar, A. Pascual-Iserte, J. M. Cioffi, and M. A. Lagunas, "Convex optimization theory applied to joint transmitter-receiver design in MIMO channels," Space-Time Processing for MIMO Communications, edited by A. B. Gershman and N. D. Sidiropoulos, Chichester, England, Wiley, 2005.

71. D. P. Palomar, J. M. Cioffi, and M. A. Lagunas, "Joint tx-rx beamforming design for multicarrier MIMO channels: A unified framework for convex optimization," IEEE Transactions on Signal Processing, Vol. 51, No. 9, September 2003.

72. A. D'Amico, "Tomlinson-Harashima precoding in MIMO systems: A unified approach to transceiver optimization based on multiplicative Schur-convexity," IEEE Transactions on Signal Processing, Vol. 56, No. 8, August 2008.

73. M. Hochwald and S. ten Brink, "Achieving near-capacity on a multiple-antenna channel," IEEE Transactions on Communications, Vol. 51, No. 3, pp. 389–399, March 2003.

74. M. Sellathurai and G. Foschini, "A stratified diagonal layered space-time architecture: Information theoretic and signal processing aspects," IEEE Transactions on Signal Processing, Vol. 51, No. 11, pp. 2943–2954, November 2003.

75. S. Lambotharan and C. Toker, "Closed-loop space time block coding techniques for OFDM broadband wireless access systems," IEEE Transactions on Consumer Electronics, Vol. 51, No. 3, pp. 765–769, August 2005.

76. Y. Zhou, T.-S. Ng, J. Wang, K. Higuchi, and M. Sawahashi, "OFCDM: A promising broadband wireless access technique," IEEE Communications Magazine, Vol. 46, No. 3, March 2008.

CHAPTER 4

5G: CONVERGENCE OF WIRELESS COMMUNICATIONS

Raúl Chávez-Santiago, Michał Szydełko, Adrian Kliks, Fotis Foukalas, Yoram Haddad, Keith E. Nolan, Mark Y. Kelly, Moshe T. Masonta, and Ilangko Balasingham

Keywords: 5G Radio spectrum, Traffic offloading, Small-cells, Software defined radio, Software defined networking

INTRODUCTION

The ongoing deployment of the fourth generation (4G) of wireless mobile systems has prompted some telecommunication companies to consider further development towards fifth generation (5G) technologies and services. Since the appearance of the first generation (1G) system in 1981, new generations have emerged approximately every 10 years. Wireless communication generations typically refer to non-backwards-compatible standards following requirements specified by the International Telecommunication Union-Radiocommunication Sector (ITU-R). Examples of these specifications are International Mobile Telecommunications 2000 (IMT-2000) for 3G and IMT-Advanced for 4G.

Other standardization bodies like IEEE also develop wireless communication technologies, often for higher data rates albeit shorter transmission ranges in most cases. With these additional standards, it is common that predecessor systems occur on the market a few years before new cellular mobile generations. For instance, commercial mobile Worldwide Interoperability for Microwave Access (WiMAX) networks deployed since 2006 are considered predecessors to 4G. Later, first-release Long-Term Evolution (LTE) systems were marketed as 4G in Scandinavia and the United States in 2009 and 2010, respectively. However, these networks were not strictly compliant with the original IMT-Advanced technical requirements. In contrast, the latest 4G LTE-Advanced (LTE-A) systems will fully comply with the ITU-R specification and support peak downlink data rates of 100 Mbps and 1 Gbps for vehicular and pedestrian mobility, respectively. Future 5G systems are expected to provide significant gains over 4G like higher data rates, much better levels of connectivity, and improved coverage. An overview of the technical solutions considered for the "beyond-4G" cellular system is provided in, e.g., [3]. Various aspects of the next generation networks have been addressed in recent papers, e.g., [1, 14, 40, 43, 58]. Finally, comprehensive vision of the 5G networks and even beyond is presented in [60].

Nevertheless, there is still some debate on the technical characteristics that 5G must possess and the services it will provide. 5G is not a term officially used in any particular specification or in any official document made public by ITU-R or any standardization body. Moreover, some industry representatives have expressed skepticism towards 5G. Others, however, envisage the rollout of 5G for the early 2020s [52]. Enabling technologies for the next generation have started to be researched by consortia comprising key international mobile operators, network infrastructure manufacturers, and academic institutions. For instance, in October 2012, the University of Surrey in the United Kingdom secured £35 million in funding for the 5G Innovation Centre, 5GIC (http://www.surrey.ac.uk/ccsr/business/5GIC/), which offers testing facilities to mobile operators developing more energy and spectrum-efficient technologies for data rates beyond 4G capabilities.

In November 2012, the EU Project "Mobile and wireless communication Enablers for the Twenty-twenty Information Society," METIS 2020 (please see Table 1), started activities toward an international definition of 5G. In February 2013, the European Commission announced €50 million for research to deliver 5G mobile technology by 2020. Some recent EU-funded research projects that address the architecture and functionality for networks beyond 4G are listed in Table 1.

Citation: Raúl Chávez-Santiago, Michał Szydełko, Adrian Kliks, Fotis Foukalas, Yoram Haddad, Keith E. Nolan, Mark Y. Kelly, Moshe T. Masonta, Ilangko Balasingham 5G: The Convergence of Wireless Communications, DOI: 10.1007/s11277-015-2467-2.

Table 1: Examples of EU Projects researching technology beyond 4G

Project name	**Research area**	**Website**
METIS 2020	Laying the foundation for the future global mobile and wireless communication systems	www.metis2020.com
5GNOW	Development of new PHY and MAC layer concepts better suited for heterogeneous transmissions in 5G	www.5gnow.eu
SOLDER	Design and development of new spectrum overlay technology for efficient aggregation of heterogeneous bands (HetBands)	www.ict-solder.eu
iJOIN	Joint design/optimization of access network and backhaul, integrating small cells, heterogeneous backhaul, and centralized processing	www.ict-ijoin.eu
TROPIC	Distributed computing, storage, and radio resource allocation over cooperative femtocells	www.ict-tropic.eu
MCN	Cloud computing for future mobile network deployment and operation	www.mobile-cloud-networking.eu/site
COMBO	New integrated approaches for fixed/mobile converged broadband access/aggregation networks	www.ict-combo.eu
MOTO	Mobile Internet with terminal-to-terminal offloading technologies	www.fp7-moto.eu
PHYLAWS	Privacy concepts for wireless communications exploiting propagation properties	www.phylaws-ict.org

On the regulatory side, the ITU-R Working Party 5D (WP 5D) started two study items in February 2013, aimed at obtaining a better understanding of future technical aspects for mobile communications towards the definition of the next generation: (1) a study on IMT vision for 2020 and beyond, and (2) a study on future technology trends for terrestrial IMT systems. In light of these developments, the European COST Action[1] IC0905 "Techno-Economic Regulatory framework for Radio spectrum Access for cognitive radio/software defined radio," TERRA (http://www.cost-terra.org/), commenced activities to identify the role and impact that emerging radio technologies will have on the development of the next generation of mobile communications. This article provides an overview of these selected key technologies and the way they will facilitate the realization of 5G, along with some of the research challenges that must be addressed in the coming years. It has to be noted that the technologies described in the paper do not cover all solutions that could be considered for future wireless systems, and one can find numerous excellent proposals in the vast literature, e.g., [3]. In our work, we address the solutions that are in line with the vision of 5G network we present in the following chapter. The remainder of this paper is organized as follows; as mentioned before—our vision for 5G systems is described in Sect. 2, enhanced modulation schemes for 5G are the topic of Sect. 3, and traffic offloading techniques are presented in Sect. 4 as the alternatives for complementing future cellular networks. The roles of cognitive radio, software-defined radio, and software-defined networking as 5G-enabling technologies are outlined in Sects. 5, 6, and 7, respectively. A discussion on 5G infrastructure is presented in Sect. 8 and we conclude in Sect. 9.

VISION FOR 5G SYSTEMS

One central topic of debate is what the architecture of 5G systems will look like. The evolution from 1G to 4G was characterized mainly by a shift in the channel access method, i.e., FDMA →→ TDMA →→ CDMA →→ OFDMA, in conjunction with improved modulation and coding schemes. However, it is likely that 5G will not be a single air interface based on a single radio access technology (RAT) on the model of the previous generations. As modern communication systems approach the Shannon limit [19] and a wider range of devices demand wireless connectivity, some industry representatives envisage 5G as a "network of networks," i.e., a heterogeneous system comprising a variety of air interfaces, protocols, frequency bands, access node classes, and network types [20]. In that vein, 5G is not really about the deployment of a completely new single-technology solution (something that will replace LTE, for example), but how existing technologies and spectrum-usage schemes can be better combined. If 5G reflects this prognosis, one of the major challenges will be the seamless integration of evolved versions of currently existing wireless technologies and complementary new communication networks.

Expected Characteristics

The proliferation of mobile broadband services is expected

to increase tremendously in the coming years. According to some forecasts [15, 20], telecommunication networks in 2020 will have to support more than one thousand times today's mobile traffic volume. Currently, more than six billion wireless mobile terminals operate worldwide. Beyond 2020, the number of devices demanding wireless connectivity may reach fifty billion, comprising wireless sensors and machine communications [20]. Future mobile broadband systems are expected to provide users with the experience of radio access with "unlimited" performance, i.e., instantaneous delivery of large volumes of multimedia content over a highly stable wireless connection. Therefore, in addition to increased throughput, 5G systems must provide lower latencies, lower outage probability (i.e., better coverage), high bit rates over larger coverage areas, higher system spectral efficiency (data volume per area unit), higher versatility and scalability, lower infrastructure deployment costs, and higher reliability of communication links. Due to new types of connected devices (e.g., wireless sensors) in addition to current 2–4G mobile terminals, the requirements for data rate, latency, reliability, data payload, and power consumption will vary because of myriad different applications ranging from Internet of Things (IoT) to novel broadband services like ultra high resolution video and augmented reality [5]. Some applications may require peak data rates of several Gigabits per second and end-to-end latencies of a few milliseconds. For instance, wireless transmission of uncompressed lossless high definition (HD) video requires data rates of approximately 4 Gbps for 1080p resolution [32], which exceeds the capabilities of current wireless technologies. These top requirements must be set as goals for new research projects on 5G technologies; the European Commission has listed specific targets in the framework of the new program for research and innovation, Horizon 2020 (http://ec.europa.eu/research/horizon2020/), which commenced in 2014 and will run until 2020 [17]:

Network architecture and protocols capable of at least a tenfold increase in frequency reuse.

Technologies to avoid the spectrum crunch caused by explosive traffic growth.

Efficient versatile radio access infrastructures allowing the seamless convergence of mobile and fixed networks.

Real-time and flexible radio resource allocation (RRA) as a function of traffic distribution with quality of service (QoS) differentiation for multiple applications.

Reduction of the network energy consumption by a factor of 10.

End-to-end latency in the order of 1 ms.

Radio Spectrum for 5G

Typically, each generation of mobile networks has been assigned new frequency bands and a wider spectral bandwidth per radio channel. The above vision for 5G systems will require a much more aggregated spectrum to enable flexible bandwidth scaling and expansion. To this end, additional harmonized frequency bands must be allocated. In order to increase frequency reuse, the spectrum shall be used on a RAT-neutral basis, ideally by applying cognitive radio principles to small and large cells [4]. The additional spectrum for 5G may include 100 MHz of bandwidth below 1 GHz to improve rural wireless broadband access and 500 MHz of the band between 1 and 5 GHz for enhanced high data rate capacity [41]. The core 3rd Generation Partnership Project (3GPP) frequency bands of 900, 1800, 2100, and 2600 MHz will be used for capacity upgrades in LTE-A and High Speed Packet Access (HSPA) networks. LTE-A will also rely on the 700 MHz band across the ITU Region 1 after the World Radio Communication Conference (WRC) in 2015 [73]. The long-term vision is the convergence of broadcast and broadband services in joint multimedia networks covering the ultra-high frequency (UHF) band below 700 MHz. Small cell deployments will play a vital role in high-capacity hotspots, and the spectrum for that could come from the 3500 MHz band, where as much as 400 MHz of bandwidth is used for fixed broadband wireless access and satellite services. The result is that up to 1.5 GHz of the spectrum can be made available by 2020; at least 1 GHz will be the traditional exclusive spectrum and the rest can be unlocked through the use of new spectrum-sharing techniques like the exploitation of TV White Spaces (TVWS) through the cognitive radio. TVWS are large portions of the VHF/UHF spectrum that are now available in geographic regions where the switchover from analog to digital TV has been completed (http://www.whitespacealliance.org/). Moreover, the LTE time-division duplexing (TDD) service in the 3.5 GHz band, called LTE-Hi, is also considered for hot-spot areas where coverage is not a demand, and the highest possible throughput is required. In addition, unlicensed bands such as 2.4, 5, and 60 GHz offer inexpensive spectrum options for traffic offloading. Especially the incorporation of 60 GHz with 7 GHz of available bandwidth is an opportunity for small cell backhaul solutions. For mobile wireless communications beyond 4G, i.e., beyond 3GPP Release 12 LTE-A, multiple spectrum opportunities can be found in the frequency, time, code, and angle domains [33, 75, 77]. Table 2 summarizes the frequency-domain opportunities foreseen for the cognitive radio that can be used for 5G.

Ultra-Dense Radio Access Networks

A new concept foreseen in the context of 5G are Ultra-Dense Radio Access Networks (UDRANETs) [20]. UDRANETs are envisaged as low-power access nodes a few meters apart for indoor areas. The main goal of UDRANETs will be to provide an extremely high traffic capacity over highly-reliable short-range links. UDRANETs will likely operate in the 10–100 GHz frequency range, which has remained practically unused for commercial mobile wireless systems despite its potential to provide bandwidths of several hundreds of megahertz. New transmission and access technologies have to be developed and standardized for this type of networks, requiring spectrum allocation studies in millimeter waves (mmW) [8].

Table 2: Spectrum opportunities for 5G cognitive radios

Spectrum opportunity	Objective	Cognitive radio application
54–698 MHz	TV bands	Deployed under the carrier aggregation concept, i.e., using them as component carriers
2.7–2.9 GHz	Bandwidth scaling from 2.7 to 3.4 GHz for enhanced flexible spectrum usage	Not specified yet
3.4–3.6 GHz Band 43	A smooth exploitation of the Band 43 from WiMAX to LTE	Co-deployment on top of the macro cell layer
3.6–3.8 GHz	Contiguous carrier aggregation provision with the maximum 100 MHz bandwidth	Carrier aggregation application
3.8–4.2 GHz	Macro cell and small cell layers' deployment, i.e., heterogeneous networks (HetNets)	Carrier aggregation within HetNets
60 GHz unlicensed band	Small cell backhaul deployment	Not specified yet

Evolution toward 5G

In our view, the evolution toward 5G will occur in two simultaneous phases: (1) the enhancement of current cellular networks compliant with 3GPP standards, and (2) the integration of evolved cellular networks with complementary wireless communication systems based on standards released by the IEEE and other standardization bodies. For the first evolution phase, enhanced modulation and coding schemes are the most obvious way to achieve a higher spectral efficiency at the physical (PHY) layer, as it has been done in the previous generations of mobile networks. Nevertheless, as the existing wireless technologies operate very close to the Shannon limit, little improvement is expected from this approach for point-to-point radio links [19]. Therefore, the carrier aggregation (CA) technology in LTE-A [44] will contribute significantly to higher data rates through the dynamic utilization of multiple continuous or non-continuous spectrum bands. For the second phase, we envisage that the integration of 3GPP-compliant cellular networks with other wireless communication systems will materialize through traffic offloading. Nevertheless, the feature that will differentiate 5G from previous mobile communication generations is the anticipated extensive use of cognitive radio, software-defined radio, and software-defined networking, i.e., 5G-enabling technologies, in combination with evolution technologies, i.e., new modulation schemes, CA, and traffic offloading. The following two sections explain the evolution technologies whereas the 5G-enabling technologies are discussed thereafter.

MODULATION TECHNIQUES FOR HIGHER SPECTRAL EFFICIENCY

Solid spectrum access methods and RRA algorithms will be crucial for efficient spectrum management in evolved RATs for 5G. The requirements regarding the permitted power levels transmitted in and outside the nominal band in the form of spectrum masks and acceptable interference power induced into neighboring systems have an impact on radio resource management (RRM). Such issues are of particular importance in the context of carrier-aggregation-like approaches (considered for LTE/LTE-A systems [44]), or in scenarios where the coexistence of various wireless systems in the same geographical area is allowed. The latter case has been widely discussed in the context of cognitive radio networks [39, 55, 65]. Based on the obvious observation that better spectrum utilization can be achieved when signals are closer to each other in the frequency domain, future 5G systems must attempt to minimize the allowed level of out-of-band (OOB) and spurious emissions. Moreover, in non-contiguous multicarrier transmission schemes where a small portion of subcarriers can be switched-off, e.g., non-contiguous orthogonal frequency-division multiplexing (NC-OFDM) [49, 72], the need for efficient and significant reduction of the power emitted in narrow spectrum gaps is particularly important. Such a scheme can be seen as a special case of CA for narrower sub-bands. For instance, let us consider the requirements defined by the Office of Communications in the United Kingdom [57] for the so-called White Space Devices (WSD). Although the values provided therein are strictly devoted to cognitive-radio-based solutions, it is wise to analyze these data and project them into the guidelines for 5G systems, where the reduction of the interference observed by neighboring users must be as high as possible. In [57] it is stated that the out-of-band equivalent isotropically radiated power (EIRP) spectral density, P_{OOB}, expressed in terms of dBm per 100 kHz, must satisfy the following relation:

$$P_{\mathrm{OOB(dBm/100\,kHz)}} \leq \max\left(P_{\mathrm{IB(dBm/8\,MHz)}} - A, 84\right) \tag{1}$$

where P_{IB} is the in-block EIRP spectral density and A represents the adjacent frequency leakage ratio (AFLR), which varies from −74 to −54 dB, depending on the WSD class. These values have been adopted by the European Conference of Postal and Telecommunications Administrations-Electronic Communication Committee (CEPT ECC) and released in a dedicated report for WSD requirements, ECC Report 185 [18], and are proposed in the draft standard for WSD by the

European Telecommunications Standards Institute (ETSI) [22]. Similar values have also been identified by the Federal Communications Commission (FCC) [26], where the fixed adjacent channel emission limit is −72.8 dB in a 100 kHz bandwidth, measured relative to the total in-band power over a 6 MHz bandwidth. One can notice that the requirements put on WSD are higher than the corresponding parameters in technical standards for current cellular systems [21]. Therefore, it can be stated that effective solutions that will guarantee a strong reduction of the power radiated outside the nominal frequency band will be applied not only for cognitive radio systems, but can—or rather should—be considered for the future 5G networks. Such an approach will lead to a lower interference inducement into adjacent channels and, in consequence, to a better utilization of the radio resources.

However, establishing such challenging restrictions on the interference induced to the adjacent band will, on the one hand, improve the spectral efficiency, but on the other, will entail the application of spectrally efficient modulation schemes and sophisticated transceiver front-ends. This is due to the fact that there are two main sources of unwanted power emission outside the nominal band: (1) the modulation process resulting in high sidelobes in OFDM systems, and (2) nonlinear phenomena appearing in the terminal front-end, e.g., the influence of high power amplifiers that possess non-linear processing characteristics; other electronic components should also be considered [31]. The latter strongly depends on the physical modules implemented in the device front-end. It has been shown that quite effective hardware solutions are already available [37], i.e., the level of OOB power, measured very close to the nominal band, can be reduced to the level of −100 dBm per 100 kHz, even for signals occupying the frequency band of 30 MHz. Therefore, it is more critical to focus on the first source of unwanted OOB emission and strive to improve the modulation processing.

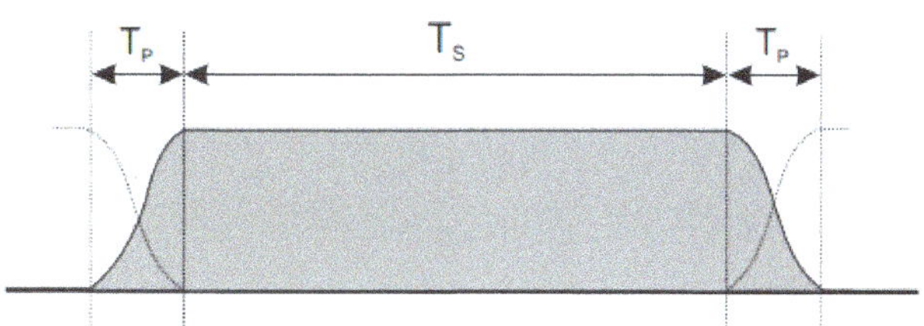

Figure 1: Illustration of OFDM symbol windowing; T_S is the duration of the OFDM symbol (including cyclic prefix), and T_P is the duration of the symbol pre- and post-fix.

It is evident that the application of sophisticated PHY techniques can result in a better protection of the transmitted data, as well as the reduction of OOB emission levels, thereby ensuring a better exploitation of the spectrum and creating new degrees of freedom for its management. In order to guarantee a high spectral efficiency together with a relatively low complexity of the receiver structure, OFDM has been applied to many wireless systems, including 4G LTE. Nevertheless, one of the main drawbacks of OFDM is a high OOB power emission due to high sidelobes of the sinc-like transfer function of the transmit pulse.

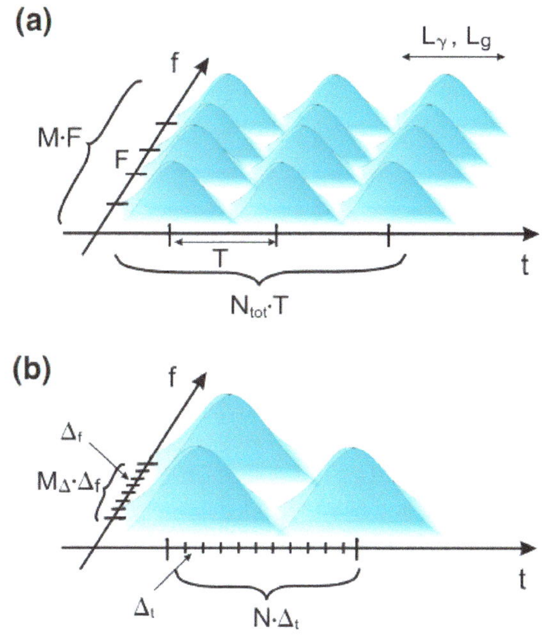

Figure 2: Time-frequency representation of the generic multicarrier frame

This undesirable phenomenon can be mitigated by extensive filtering and the use of sidelobe suppression algorithms, but even after applying such techniques, the amount of power radiated outside the nominal frequency band may remain high. To help address this issue, one technique involves extending the pulse duration to allow for smoother transitions between consecutive symbols. This is illustrated in Fig. 1 where the rectangular pulse is replaced by a square root raised cosine pulse with the time support of Ts+2TpTs+2Tp [67]. However, the cost associated with the extension of the duration of the time-domain symbol (the presence of a symbol prefix and postfix), is the reduction of the overall system efficiency (e.g., the rate expressed in bit/s/Hz/J). Another approach is to perform advanced filtering. However, it has to be taken into account that such signal processing improves the OOB reduction at the expense of some filtering distortions; an interesting analysis of the trade-off between interference and filtering distortions in OFDM systems can be found in [53]. Furthermore, the efficiency and the cost of various approaches are discussed in [24, 71]. These observations lead to the conclusion that new modulation processing technique should be applied; such techniques should guarantee a similar efficiency and robustness as OFDM, while outperforming the existing solutions in terms of OOB reduction. In this context, new modulation schemes have been promoted for mobile systems beyond 4G [30, 51], and cognitive radio solutions, e.g., the IEEE DySPAN-SC P1900.7 standard. Since rectangular (or almost rectangular) pulses, though simple in implementation and analysis, do not guarantee a low OOB emission, alternative solutions involve shaping the transmit and received pulses to result in a better energy concentration in the time-frequency plane. Let us first generalize multicarrier transmission by providing a generic signal description, and then indicate some of the most promising solutions that might be considered for the next generation of cellular networks. We denote $g(t)$ and $\gamma(t)$

as the pair of transmit and receive pulses. Thus, the generic transmitted multicarrier signal $s(t)$ and the estimation $\hat{c}_{m,n}$ of the user data, $c_{m,n}$ (e.g., Quadrature Amplitude Modulation (QAM) symbols), can be defined as:

$$s(t) = \sum_{m=0}^{M-1} \sum_{n \in \mathbb{Z}} c_{m,n} g_{m,n}(t) = \sum_{m=0}^{M-1} \sum_{n \in \mathbb{Z}} c_{m,n} g(t - nT) \exp(2\pi j m F(t - nT)) \quad (2)$$

and

$$\hat{c}_{m,n} = \int_{-\infty}^{\infty} s(t) \gamma_{m,n}^{*} \,\mathrm{d}t = \int_{-\infty}^{\infty} \sum_{m'=0}^{M-1} \sum_{n' \in \mathbb{Z}} c_{m',n'} g_{m',n'}(t) \gamma_{m,n}^{*} \,\mathrm{d}t \quad (3)$$

where M is the number of subcarriers, T is the time distance between consecutive pulses, F represents the distance between two neighboring pulses in the frequency domain, and $g_{m,n}(t)$ is the translated and modulated version of the original pulse, g(t)g(t). Thus, the generic time-frequency representation of the transmit frame consisting of $M \cdot N_{tot}$ pulses can be illustrated as in Fig. 2. In the OFDM case, the shape of both the transmit and received pulses is rectangular, the distance between the symbols in the time domain is set to T_S, and the distance between the adjacent subcarriers is reciprocal to the orthogonality time of the OFDM symbol. Relaxing these constraints, one can conclude that in the design process of the transmit-received pulse, various parameters have to be considered, such as the time and frequency support of the pulse, overlapping of pulses, and distance between pulses in both dimensions. Beside those parameters, also other criteria can be considered, such as the minimization of the minimum mean-square error (MMSE) of the assumed channel model [42], and others [9, 27, 46]. In the context of the reduction of OOB emission, the steepest possible power attenuation in the frequency domain is crucial, while maintaining the orthogonality or biorthogonality between pulses [27]. Intensive research has been done over the last decade in the area of efficient pulse shaping, and many newly proposed schemes offer clear advantages over purely rectangular shapes. Currently, among the various proposals that have been investigated, a class of pulses strictly devoted to filter bank-based multi-carrier (FBMC) systems seems quite promising [62]. Various pulse shapes have been considered for this modulation scheme [7, 23, 62, 64]. It has been shown that FBMC systems with isotropic orthogonal transfer algorithm (IOTA) filters can achieve an OOB attenuation at the level of −60 dB relative to the transmit power level in the nominal band. Moreover, good time-frequency localization of the pulses allows for removing the cyclic prefix typically applied in OFDM in order to combat the inter-symbol interference (ISI) problem. As a consequence, a higher spectral efficiency with a very low interference power induced to neighboring systems can be achieved at the expense of a slightly higher transceiver structure complexity. The FBMC transceiver could be implemented very effectively by means of inverse fast Fourier transform (IFFT) blocks followed by polyphase filters. However, despite its benefits, FBMC suffers from various other problems that still require attention, e.g., sensitivity to phase-noise or so-called intrinsic interference and its influence on signal detection [76]. A detailed analysis of the advantages and disadvantages of this solution has been presented in [62], and in extensive reports published within the framework of the PHYDYAS (http://www.ict-phydyas.org/) and EMPhAtiC (http://www.ict-emphatic.eu/) projects. In Table 3, we briefly summarize the main features of OFDM and FBMC signals in terms of their potential application in 5G systems.

Table 3: Comparison of OFDM and FBMC

Feature or parameter	OFDM	FBMC
Pulse shape	Pulse shape: rectangular (or almost rectangular in practical realizations due to various analogue filters)	Various shapes (e.g., IOTA, Gaussian, enhanced Gaussian pulse, Bellanger, etc.)
Cyclic prefix	Yes (up to 25 % of OFDM symbol duration)	No
Overlapping of pulses in time and frequency directions	Pulses do not overlap in time domain, orthogonality is preserved in frequency domain	Adjacent pulses overlap in both domains
AFLR	Poor (−13 dB for sinc-like shapes in frequency domain)	It can be kept very small (e.g., −60 dB for IOTA pulse)
Spectral efficiency	Limited by cyclic prefix	Limited by roll-off factor of the pulse
Complexity	It can be kept low; possible implementation with filter banks	It can be kept small, but always slightly higher than in OFDM; effective implementation with filter banks
Peak-to-average power ratio (PAPR)	High	High (it can be almost at the same level as for OFDM)

Another promising modulation scheme is NC-OFDM. In contrast to traditional OFDM signals, NC-OFDM enables a subset(s) of subcarriers to be switched off, allowing other signals to be transmitted in the vacated frequency bands. A low interference between neighboring signals can be achieved very efficiently by applying cancellation carriers (CCs) or windowing algorithms [49]. In such a case, the OOB emission can be minimized up to −60 dB in very narrow frequency gaps, where the power spectral density (PSD) of NC-OFDM signals with 320 subcarriers is presented. If implemented, it is conceivable that the above modulation schemes would serve as a supporting technology for RRA algorithms in 5G networks.

We will now outline the research activities focusing on generalized multicarrier schemes, where the constraints assumed in OFDM and even in FBMC schemes, are relaxed. The pair of the transmit and receive pulses, as well as the shape of the transmit frame, are designed in such a way that some specific predefined criteria are fulfilled. In [28, 38, 47, 59, 61, 63], the advantages of advanced waveforms are highlighted. In the context of 5G, which envisages the coexistence of various systems in close vicinity, together with the need for a better utilization of the frequency resources, the necessity of applying advanced waveforms is crucial. Although various proposals have been presented in the vast literature, FBMC and NC-OFDM schemes seem to be the most promising solutions at the current stage of development. The application of these waveforms will guarantee the fulfillment of the challenging OOB requirements (mentioned at the beginning of this section) that should be defined for 5G mobile terminals, as it has been done for WSD. At the same time, the spectral efficiency of these new modulation schemes will be kept at a very high level usually, outperforming the efficiency of today's OFDM solutions. Such an observation is confirmed not only by the research community, but also by the recent standardization activities. For two-tier heterogeneous networks (HetNets), like macrocell/femtocell layouts, the recently proposed Vandermonde-subspace frequency division multiplexing (VFDM) [10] overlay spectrum sharing technique has the potential to provide a spectral efficiency increase of up to 1 bit/s/Hz in HetNets equipped with cognitive radio capabilities. Thus, new sophisticated modulation schemes are the key evolution technologies toward 5G networks.

MOBILE TRAFFIC OFFLOADING

Smartphones, tablets, and other mobile broadband devices generate unprecedentedly large amounts of traffic. Mobile operators are facing great challenges to serve such a tremendous traffic growth with the current cellular infrastructure. It is evident that mobile networks beyond 4G will need to implement sophisticated traffic offloading strategies [2] beside the traditional network scaling of deploying more base stations per area. Traffic offloading consists in using complementary radio access networks to deliver data originally intended for mobile cellular systems, thereby decreasing the congestion on each individual radio link and respective backbone connection. Traffic offloading encompasses multiple solutions, which can be classified as overlay (sharing the same spectrum with the cellular access network) and non-overlay solutions; some of them are discussed below.

Cognitive Femtocells

Femtocell traffic offloading is based on the deployment of small, low-power cellular base stations at home or in other indoor areas, backhauled to the cellular core network by a conventional wired network [35]. Among several benefits of this solution is the fact that femtocells can handle both voice and data traffic with QoS assurance. However, the use of the same spectrum as overlaying macrocells complicates the search for available channels for femtocells in highly congested areas. Hence, intelligent interference management using cognitive radio and RRA coordination must be applied, especially in deployment scenarios where end users decide about the locations of the femtocells, e.g., uncoordinated home eNB. The highly-unpredictable interference with macrocells requires an opportunistic spectrum access within a hierarchical overlay system as follows: first, a sensing procedure at the macrocell level provides the knowledge about spectrum opportunities, which are subsequently exploited by the femtocells [36, 41, 66, 69].

Wi-Fi and White-Fi

Non-overlay traffic offloading via Wi-Fi networks has already been implemented by some cellular operators. Basically, when a mobile terminal is in the vicinity of a Wi-Fi hotspot, data traffic routing is switched to exploit its radio interface. This solution is attractive as it allows access to a free, unlicensed spectrum, thereby reducing the unnecessary congestion in expensive, licensed frequency bands. Strategic alliances between cellular operators and Internet service providers (ISPs) need to be established. Nevertheless, the Wi-Fi medium access control (MAC) protocol is not well-suited for a heavy traffic load and does not provide QoS differentiation. Consequently, this solution is currently implemented for best-effort traffic only, whereas voice services are still delivered via the cellular core network. One major challenge is to increase the system spectral efficiency by allowing more concurrent users on Wi-Fi networks, ideally for both best-effort and voice traffic. A possible way of achieving this is to implement Wi-Fi frequency reuse plans with slightly overlapping channels (Fig. 3). In a case study [12], computer simulations of an indoor IEEE 802.11b basic service set (BSS) subject to interference from surrounding cells demonstrated that transmissions of voice over Internet protocol (VoIP) are not significantly degraded in the 2.4 GHz band by the resulting co-channel interference (Fig. 4). For full VoIP offloading capabilities, however, the QoS mechanism already available in the IEEE 802.11 standards should be implemented. A more recent variant of this offloading approach is White-Fi, which aims at using the TVWS spectrum through cognitive radio for a transmission with the conventional Wi-Fi technology. In 2013, this solution was standardized in IEEE 802.11af.

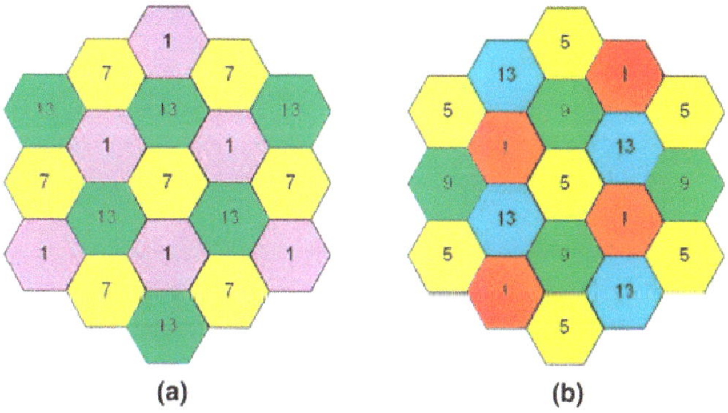

Figure 3: Indoor multicell Wi-Fi layout in which a BSS transmits VoIP traffic in the central cell with a radius equal to 30 m. **a** Configuration with 3 non-overlapping channels (13 overlapping channels are allocated in Europe on the 2.4 GHz band), and **b** configuration with 4 slightly overlapped channels

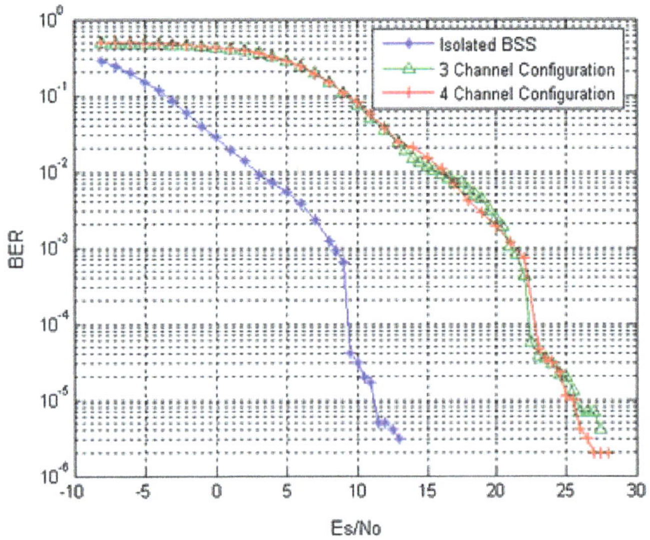

Figure 4: Bit-error-rate (BER) performance of the BSS VoIP transmission subjected to interference from surrounding cells in the layouts depicted in Fig. 3. In the isolated BSS case no interference from surrounding cells was present

Other Offloading Solutions

WiMAX can be considered as an offloading alternative, but it is more suitable as the backhaul for large-scale Wi-Fi networks. Furthermore, 3GPP compliant cellular networks have not considered interoperation with WiMAX so far; therefore, additional standardization would be required.

When communicating users are geographically close to each other, device-to-device (D2D) communication [16] can also help diminish traffic congestion on the cellular core network. D2D communication is an underlay to cellular networks in licensed frequency bands, contrasting with mobile ad-hoc networks (MANETs), which operate similarly but in the unlicensed spectrum.

The next generation of cellular systems could leverage offloading opportunities created by the integration of the aforementioned solutions and others that may be developed in the future. These solutions can be further extended through the use of cognitive radio, an approach we refer to as cognitive mobile-traffic offloading (CMTO). For instance, an outdoor mesh network of Wi-Fi and White-Fi hotspots backhauled to the cellular network through WiMAX links and/or wired broadband access can serve as traffic offloading alternatives in combination with femtocells and D2D links (Fig. 5).

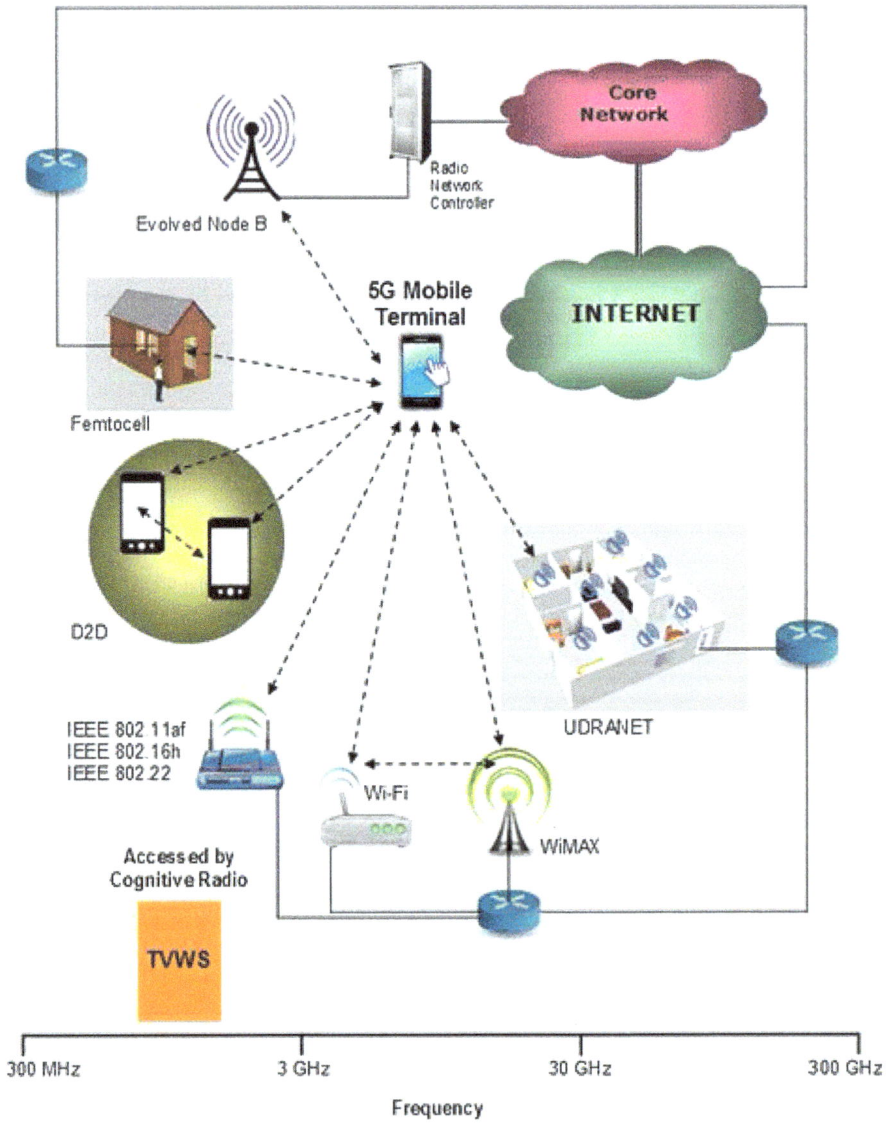

Figure 5: Traffic offloading approaches for 5G systems

Research Challenges

Traffic offloading solutions will allow the virtual expansion of the spectrum allocated to cellular networks, thereby increasing their capacity. However, important issues, like user authentication, network security, and service pricing [2] must be addressed first. The emerging software-defined networking technology has the potential to solve some of these problems, since this approach provides a radio network controller (RNC) with a real-time updated view of the state of each link and device in the network, as it will be explained later. From the user perspective, service continuity and same level of connectivity must be ensured independently of the access network [add: (cf. Sect. 6)].

UDRANETs will perform the role of traffic offloading alternatives for extremely high data rate applications, but the main challenge here is to produce low-cost mobile terminals that can operate in super-high frequency (SHF) and extremely-high frequency (EHF) bands. On the other hand, the use of mmW enables packing more antennas into a terminal, thereby facilitating the implementation of the massive multiple-input multiple-output (M-MIMO) for an improved throughput and a lower latency [50]. In combination with multiuser MIMO (MU-MIMO), cellular networks can benefit from a significant capacity increase. These topics represent many opportunities for research and innovation.

COGNITIVE RADIO

Cognitive radio (CR) has long been considered an enabling technology for the next generation of mobile communications [6]. The CR paradigm proposes an opportunistic utilization of the underused parts of licensed frequency bands, i.e., spectrum holes [68], by unlicensed (secondary) users and/or the efficient sharing of the licensed-exempt spectrum. For this sake, cognitive mobile terminals must acquire accurate real-time knowledge on transmission opportunities through RF spectrum scanning to identify the unoccupied radio channels/bands within the time-frequency resources table. Standards like IEEE 802.11af, IEEE 802.16h (cognitive WiMAX), and IEEE 802.22 are aimed at using CR techniques to enable sharing of the TVWS spectrum on a non-interfering basis with improved coexistence mechanisms.

The CR technology is now at an advanced stage and real-world trials and evaluations are being carried out. As an example, an experimental trial of CR and the dynamic spectrum access technology using the TVWS spectrum is being carried out in Ireland. This Irish research testbed is assessing the feasibility of taking an opportunistic spectrum usage approach in the 700 MHz band to establish an 18.3 km TVWS link connecting the Trinity College Dublin in Dublin city center to Intel Labs Europe (Ireland) based in Leixlip, Co. Kildare. Demonstrating the spectrum agility and the ability to avoid interference to incumbents in order to provide a robust link suitable for the support of new services is the key objective of this trial [56]. In future 5G systems, CR will be a key component to ensure interference coordination in 3GPP-compliant overlay networks as well as in different traffic offloading techniques. The 3GPP LTE standard has taken provisions to facilitate interference mitigation in overlay networks. Enhanced Intercell Interference Coordination (eICIC) was introduced in 3GPP Release 10 LTE-A to tackle interference issues in HetNets. In this approach, Almost-Blank-Subframes (ABSs) are introduced, which are mostly control channel frames with a very low power. A macrocell can configure ABSs, so that mobile terminals in small cells can send their data within ABSs to avoid interference with overlay macrocells. The use of CR can improve this approach resulting in a more efficient utilization of the radio resources with an even better interference coordination for two-tier HetNets, which ultimately leads to an enhanced system capacity and coverage [13]. CA for HetNets developed in compliance with 3GPP Release 12 LTE-A will introduce new CR-wise techniques for efficient spectrum allocation in heterogeneous wideband fading environments.

Research Challenges

In addition to CA in LTE-A systems, the CA of multiple and heterogeneous RATs (h-RATs) will also be a key feature of 5G communications, which will require CR application at all layers. Some of the associated challenges will be addressed by the recently established EU Project "Spectrum Overlay through Aggregation of heterogeneous Dispersed bands," SOLDER (see Table 1). The objectives of the project are: a non-continuous aggregation of heterogeneous bands, flexible multicarrier modulation, efficient MAC layer design and the aggregation of heterogeneous resources. Furthermore, although a number of CR solutions for underlay femtocells have been recently proposed, fewer CR approaches for underlay D2D communication are available in the literature. Research in this area is therefore required.

SOFTWARE DEFINED RADIO

The future 5G network infrastructure must have a sufficiently wide flexibility in order to cover different radio frequency (RF) parameter settings in a dynamic and adaptive way to allow efficient spectrum management. Reconfigurable platforms based on software-defined radio (SDR) will facilitate the dynamic air interface reconfiguration of the network nodes by software modifications, reflecting current traffic demands. The flexibility of the RF chains must be further reflected in the baseband processing capabilities, where the down-converted RF signals are to be processed. As 5G systems will need to exploit underutilized frequency bands to avoid the foreseen spectrum crunch, CR implementation on SDR platforms should consider the cooperation and interoperation of multiple radio technologies, e.g., through common radio resource management (CRRM). This implies that a reconfigurable platform must be capable to operate at different power levels, channel bandwidths and frequencies, modulation and coding schemes, modifying transmission parameters and characteristics according to the particular constraints of the radio technology standards in use; these constraints include unwanted emission in the operating band, AFLR, or intermodulation products. Promising SDR development initiatives include the GNU Radio (http://gnuradio.org/), an open-source software development toolkit for SDR implementation on various programmable platforms like the Universal Software Radio Peripheral (USRP) boards (http://www.ettus.com/), and the Open Base Transceiver Station (OpenBTS), which was recently used to demonstrate the implementation of a software-based Global System for Mobile Communications (GSM) base station on the Raspberry PI hardware platform (http://openbts.org/).

Looking at recent developments for LTE-A, the CA feature [70] with its maximum configuration of 5×205×20 MHz radio channels is a good starting point for future research on the incorporation of cognition, to further improve the flexibility of radio resources utilization. Nevertheless, the RF chains bandwidth and their tuning are considered here as a bottleneck, especially when a variety of different frequency bands are used for cellular networks in different markets. The multi-standard radio base station (MSR-BS) addresses these issues in the recent releases of 3GPP networks, enabling a single radio unit in the base station to simultaneously operate with multiple RATs at the same time without harming each other with unwanted RF emissions.

Research Challenges

Despite a significant progress in this field, further research is required for the practical implementation of SDR-based reconfigurable systems for 5G, such as an SDR-based MSR-BS. To fulfill the vision for 5G systems, the user equipment (UE) also needs to be fully reconfigurable [45] and able to tune the RF chain(s) to different aggregated radio channels and h-RATs, as dictated by the scheduler on the network side. Cognition aspects will have to be incorporated into self-organizing network (SON) features in future 3GPP standards releases, extending the dynamic network selection procedures. The major challenge in these cases is the design of the radio front-end to properly and efficiently handle the large operation bandwidth, from 700 MHz to 3.5 GHz. Furthermore, following the market trends, mobile terminals must be limited in size, power consumption, and cost.

SOFTWARE DEFINED NETWORKING

At the current stage of development, both CR and SDR technologies do not involve the control of the cellular core network. Until now, no coordination of traffic flow is possible at the core network, e.g., a UE cannot receive multiple different traffic flows from different eNBs simultaneously to increase the data rate. The revolutionary concept of software-defined networking (SDN) aims at providing a coordinator that has a global view of the network infrastructure, thereby facilitating a number of networking functionalities. In essence, the SDN concept consists in decoupling the forwarding plane and the control plane [48]. In this approach, the core devices, e.g., routers, do not make decisions on how and where to forward a given data packet; instead, the decision is taken by a central coordinator referred to as the controller. One of the protocols in charge of the communication between the controller and user devices is OpenFlow [54]. Data packets passing through interconnection devices (switches, routers, etc.) are classified into flows to make per-flow forwarding decisions. A flow is defined by a set of matching rules in 12 distinctive fields of a typical Ethernet/IP/UDP header (L2 and L3 addresses, virtual LAN (VLAN) information, ports, etc.). Each time a data packet affiliated to a specific flow enters a device, a counter is updated at the controller. This enables the controller to have a global view of the status of each network component. Therefore, the controller can make the decision to forward the traffic through a less congested route, or use a radio channel that momentarily experiences a good condition to forward the data packet to the end user [11]. SDN aims to support a much better integration of all the existing wireless networks (Wi-Fi, 2–4G, etc.) [74]. It would be possible to perform a seamless handover, not only within the same technology as already exists, but also across h-RATs. This feature will enable the offloading approaches described in Sect. 4. Moreover, SDN will greatly facilitate the management of chaotic deployments of a large number of small cells in LTE systems [34]. The ultimate goal of SDN is to completely abstract the forwarding plane to enable "network programmability" where the infrastructure is seen as a service that can be used by different applications above an intermediate layer referred to as the Network Operating System (NOS). This is the basis for the concept of cloud networks. In the context of 5G, we can refer to this approach as cloud-RAN or RAN as a Service (RANaaS).

Research Challenges

SDN entails some interesting drawbacks and challenges. For instance, security threats could be more severe in SDN. An attack on a controller could compromise the connection of a large coverage area. In the case of an inter-ISP handover, the collaboration between multiple controllers affiliated to different ISPs is required. However, to what extent and how to promote this kind of collaboration requires careful analysis, as it implies the exchange of network status information that most operators prefer to keep confidential for business purposes. More generally, the optimal balance between centralized and decentralized management needs to be established. It is clear that a fully distributed scheme does not allow the optimal utilization of the spectrum, since this cannot be achieved with partial or local knowledge; but on the other hand, it is not realistic to imagine a controller in charge of a very large area, since this could introduce too much latency in terms of answer-to-flow matching rules and device feedback processing.

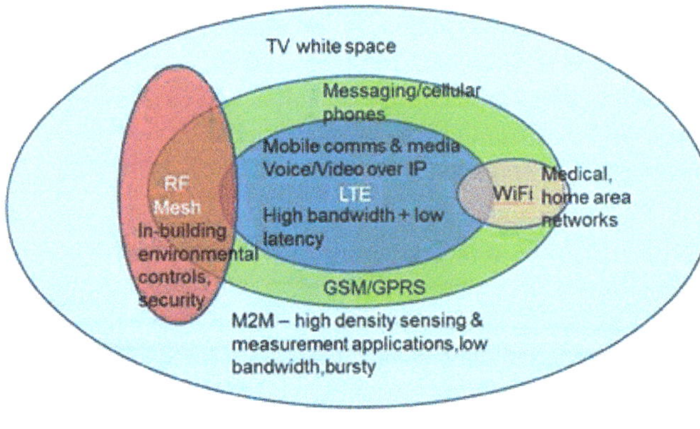

Figure 6: 5G as a blended infrastructure encompassing the range of cellular, local and personal area network, and short range technologies and their associated regulatory approaches, e.g., exclusive, light, licence-exempt, and shared spectrum models

BLENDED INFRASTRUCTURE

5G systems will leverage the availability and variety of the deployed infrastructure to better address the need for capacity, coverage, and high quality of service instead of relying on a single technology, topology, or regulatory approach [29]. In other words, unlike 4G, which involves exclusively-licensed

LTE in a predominantly cellular-based topology, 5G will focus on knitting together cellular, local and personal area networks, short range device technologies and topologies. This concept also extends to the regulatory domain; 5G communications will encompass a range of exclusive, light-licensing, license-exempt, and shared spectrum licensing schemes. Through this approach, the flow of data can be more efficiently abstracted from the physical infrastructure and regulatory approach. This approach is likely to increase in dominance over the coming years due to two main reasons: (1) it is unlikely that a single technology/standard will be able to provide a sufficient capacity and coverage to accommodate all market sectors; (2) networks seeking to maximize the utility of existing infrastructure in a bid to increase the return on capital investments will continue to pursue new avenues for offloading, data aggregation, active sharing, and licensed shared access options. As a result, they will position themselves as parts of a blended infrastructure, instead of a standalone network, as in the majority of operator models today. An example of how this blended infrastructure may be implemented is illustrated in Fig. 6. In this example, RF mesh and shared spectrum technologies, e.g., TVWS, complement a core cellular network technology, e.g., LTE, for capacity and high data-rate delivery, and 2–3G cellular technologies for coverage. Wi-Fi will continue to play an important part in home area networks and as part of operator data-offloading strategies. Short range devices, e.g., ZigBee and Z-Wave will dominate the in-building control sector. 5G will better leverage the advantages stemming from a blend of technology and spectrum-usage options. This could lead to usage scenarios in which the communications flow could better suit customers' priorities, e.g., the requirement for a high QoS and willingness to pay extra for the service, versus low cost communication requirement and acceptance of a lower QoS or higher latency services.

Societal Impact of 5G

From the societal perspective, 5G networks have the potential to improve the mobile broadband access in rural areas. The capital expenditure for deploying a large number of base stations and the low average revenue per user (ARPU) [25] has delayed the complete coverage of rural environments. By using TVWS and traffic offloading solutions, the deployment of 5G networks in rural areas will be possible at a lower cost thanks to more favorable propagation conditions in the VHF/UHF spectrum that directly translate into fewer base stations.

CONCLUSIONS

In the quest for higher data rates and lower latencies beyond 4G network capabilities, the next generation of wireless mobile communications has to adopt revolutionary ways of using the radio spectrum. In addition to better ways of exploiting the already allocated frequency bands through enhanced modulation schemes and traditional network scaling, an opportunistic access to the underused spectrum through cognitive radio will further increase the systems' capacity. Moreover, different offloading techniques will contribute to an "unlimited" access to large amounts of multimedia data anywhere and anytime. The software-defined radio technology will enable reconfigurable platforms for the implementation of the cognitive radio paradigm in underlay cellular networks and traffic offloading solutions, whereas software-defined networking will facilitate the programmable operation of the core network. The evolution towards 5G, as outlined in this paper, opens new research problems. Although we have briefly surveyed the key research projects recently launched in Europe, efforts to develop 5G technologies are being carried out worldwide. Further standardization for cooperation and interoperability between different radio access technologies will be needed, too. With this paper, we intend to motivate the search for innovative solutions to all these challenges.

ACKNOWLEDGMENTS

The work by Adrian Kliks has been funded by the Polish Ministry of Science and Higher Education within the status activity task Cognitive radio systems in 2015. The work by Keith Nolan was part funded by the European Unions Seventh Framework Programme for research, technological development and demonstration under Grant Agreement No. 600063 [the COOPERATE Project].

REFERENCES

1. Agyapong, P., Iwamura, M., Staehle, D., Kiess, W., & Benjebbour, A. (2014). Design considerations for a 5G network architecture. IEEE Communications Magazine, 52(11), 65–75. doi:10.1109/MCOM.2014.6957145.
2. Aijaz, A., Aghvami, H., & Amani, M. (2013). A survey on mobile data offloading: Technical and business perspectives. IEEE Wireless Communications, 20(2), 104–112.
3. Akyildiz, I. F., Gutierrez-Estevez, D. M., Balakrishnan, R., & Chavarria-Reyes, E. (2014). LTE-advanced and the evolution to beyond 4G (B4G) systems. Physical Communication,10, 31–60. doi:10.1016/j.phycom.2013.11.009.http://www.sciencedirect.com/science/article/pii/S1874490713000864.
4. Akyildiz, I. F., Lee, W. Y., Vuran, M. C., & Mohanty, S. (2006). NeXt generation/dynamic spectrum access/cognitive radio wireless networks: A survey. Computer Networks, 50(13), 2127–2159.
5. Akyildiz, I. F., Melodia, T., & Chowdury, K. R. (2007). Wireless multimedia sensor networks: A survey. IEEE Wireless Communications, 14(6), 32–39.
6. Badoi, C. I., Prasad, N., Croitoru, V., & Prasad, R. (2011). 5G based on cognitive radio.Wireless Personal Communications, 57(3), 441–464.
7. Bellanger, M. (2001). Specification and design of a prototype filter for filter bank based multicarrier transmission. In Proceedings of the IEEE international conferences acoustics, speech, and signal processing (ICASSP), Salt Lake City, UT (pp. 2417–2420).
8. Bleicher, A. (2013). Millimeter waves may be the future of 5G phones.http://spectrum.ieee.org/telecom/wireless/millimeter-waves-may-be-the-future-of-5g-phones.
9. Bolcskei, H., Duhamel, P., & Hleiss, R. (1999). Design of pulse shaping OFDM/OQAM systems for high data-rate transmission over wireless channels. In Proceedings of the IEEE international conference on communications (ICC), Vancouver, BC (pp. 559–564).

10. Cardoso, L. S., Kobayashi, M., Cavalcanti, F. R. P., & Debbah, M. (2013). Vandermonde-subspace frequency division multiplexing for two-tiered cognitive radio networks. IEEE Transactions on Communications, 61(6), 2212–2220.
11. Chaudet, C., & Haddad, Y. (2013). Wireless software defined networks: Challenges and opportunities. In Proceedings of the international IEEE conference on microwaves, communications, antennas and electronics (COMCAS), Tel-Aviv, Israel (pp. 21–23).
12. Chavez-Santiago, R., Haddad, Y., Lyandres, V., & Balasingham, I. (2015). Voip transmission in Wi-Fi networks with partially-overlapped channels. In Proceedings of IEEE wireless communications and networking conference (WCNC), New Orleans, USA.
13. Cheng, S. M., Lien, S. Y., Chu, F. S., & Chen, K. C. (2011). On exploiting cognitive radio to mitigate interference in macro/femto heterogeneous networks. IEEE Wireless Communications, 8(6), 40–47.
14. Chih-Lin, I., Rowell, C., Han, S., Xu, Z., Li, G., & Pan, Z. (2014). Toward green and soft: A 5G perspective. IEEE Communications Magazine, 52(2), 66–73. doi:10.1109/MCOM.2014.6736745.
15. Cisco. (2013). Cisco visual networking index: Global mobile data traffic forecast update. White paper by Cisco.http://www.cisco.com/en/US/solutions/collateral/ns341/ns525/ns537/ns705/ns827/white_paper_c11-520862.pdf.
16. Doppler, K., Rinne, M., Wijting, C., Ribeiro, C. B., & Hugl, K. (2009). Device-to-device communication as an underlay to LTE-Advanced networks. IEEE Communications Magazine, 47(12), 42–49.
17. European Commission. (2013). Horizon 2020-Work Program 2014–2015. Leadership in enabling and industrial technologies information and communication technologies.https://ec.europa.eu/digital-agenda/sites/digital-agenda/files/h2020 LEIT-ICT WP.
18. Electronic Communication Committee. (2013). Complementary report to ECC report.http://www.erodocdb.dk/Docs/doc98/official/pdf/ECCREP185.pdf.
19. Edwards, C. (2013). 5G mobile model challenged by 'spectrum crunch'. E&T Magazine, 8(8). http://eandt.theiet.org/magazine/2013/08/5g-searches-for-formula.cfm.
20. Ericsson. (2013). 5G radio access. White paper.http://www.ericsson.com/res/docs/whitepapers/wp-5g.pdf.
21. ETSI. (2012). LTE; evolved universal terrestrial radio access (E-UTRA); user equipment (UE) radio transmission and reception. ETSI Technical Specification ETSI TS 136 101 V9.12.0.
22. ETSI. (2013). White space devices (WSD); wireless access systems operating in the 470 MHz to 790 MHz frequency band; harmonized EN covering the essential requirements of article 3.2 of the R&TTE Directive. Standard Draft ETSI EN 301 598 V1.0.0.
23. Farhang-Boroujeny, B. (2011). OFDM versus filter bank multicarrier. Signal Processing Magazine, 28(3), 92–112.
24. Farhang-Boroujeny, B., & Kempter, R. (2008). Multicarrier communication techniques for spectrum sensing and communication in cognitive radios. IEEE Communications Magazine, 46(4), 80–85.
25. Fehske, A., Fettweis, G., Malmodin, J., & Biczok, G. (2011). The global footprint of mobile communications: The ecological and economic perspective. IEEE Communications Magazine, 49(8), 55–62.
26. Federal Communications Commission. (2012). Third memorandum opinion and order.http://transition.fcc.gov/Daily_Releases/Daily_Business/2012/db0405/FCC-12-36A1.pdf.
27. Feichtinger, H. G., & Strohmer, T. (1998). Gabor analysis and algorithms: Theory and applications (Applied and Numerical Harmonic Analysis). Boston, MA: Birkhuser.
28. Fettweis, G., Krondorf, M., & Bittner, S. (2009). GFDM-generalized frequency division multiplexing. In Proceedings of the IEEE 69th vehicular technology conference (VTC 2009-Spring), Barcelona, Spain (pp. 1–4).
29. Forum, W. I. (2013). Wireless innovation forums comments to ITU WP-5A regarding the working document towards a preliminary draft new report ITU-R [LMS CRS2] cognitive radio systems [(CRS) applications] in the land mobile service.http://groups.winnforum.org/d/do/6929.
30. Gameiro, A., (Ed.) (2010). Radio environment models. Deliverable of the QoSMOS project.http://www.ict-qosmos.eu/fileadmin/documents/Dissemination/Deliverables/files/DelPub/QoSMOS_WP3_D31.pdf.
31. Gharaibeh, K. M. (2012). Nonlinear distortion in wireless systems: Modeling and simulation with MATLAB. West Sussex: Wiley.
32. Gilbert, J. M., Doan, C. H., Emami, S., & Shung, C. B. (2008). A 4-Gbps uncompressed wireless HD A/v transceiver chipset. IEEE Micro, 28(2), 56–64.
33. GmbH, N. R. (2013). GPP LTE—A standardisation in Release 12 and beyond. White paper.http://www.nomor.de/uploads/fd/24/fd24709a64bc490a083a8eba6d3cc2cb/NoMoR_LTE-A_Rel12_and_Beyond_2013-01.pdf.
34. Gudipati, A., Perry, D., Li, L. E., & Katti, S. (2013). SoftRAN: Software defined radio access network. In Proceedings of the 2nd ACM SIGCOMM workshop on hot topics in software defined networking (HotSDN), Hong Kong, China (pp. 25–30).
35. Haddad, Y., & Porrat, D. (2009). Femtocell: Opportunities and challenges of the home cellular base station for the 3G. In IADIS international conference wireless applications and computing (WAC).
36. Haddad, Y., & Porrat, D. (2010). A two-tier frequency reuse scheme. In Personal, indoor and mobile radio communications workshops (PIMRC Workshops), 2010 IEEE 21st international symposium on (pp. 203–207). doi:10.1109/PIMRCW.2010.5670362.
37. Hammi, O., Kwan, A., & Ghannouchi, F. M. (2013). Bandwidth and power scalable digital predistorter for compensating dynamic distortions in RF power amplifiers. IEEE Transactions on Broadcasting, 59(3), 520–527.

38. Han, F. M., & Zhang, X. D. (2009). Wireless multicarrier digital transmission via Weyl–Heisenberg frames over time-frequency dispersive channels. IEEE Transactions on Communications, 57(6), 1721–1733.

39. Haykin, S. (2005). Cognitive radio: Brain empowered wireless communications. IEEE Journal on Selected Areas in Communications, 2(23), 201–220.

40. Hossain, E., Rasti, M., Tabassum, H., & Abdelnasser, A. (2014). Evolution toward 5G multi-tier cellular wireless networks: An interference management perspective. IEEE Wireless Communications, 21(3), 118–127. doi:10.1109/MWC.2014.6845056.

41. Huang, L., Zhu, G., & Du, X. (2013). Cognitive femtocell networks: An opportunistic spectrum access for future indoor wireless coverage. IEEE Wireless Communincations,20(2), 44–51.

42. Hunziker, T., Ju, Z., & Dahlhaus, D. (2007). Efficient channel description in time-frequency domain with application to flexible radio. In Proceedings of the15th European signal processing conference (EUSIPCO), Poznan, Poland (pp. 866–870).

43. Imran, A., & Zoha, A. (2014). Challenges in 5G: How to empower son with big data for enabling 5G. IEEE Network, 28(6), 27–33. doi:10.1109/MNET.2014.6963801.

44. Iwamura, M., Etemad, K., Mo-Han, F., Nory, R., & Love, R. (2010). Carrier aggregation framework in 3GPP LTE-advanced [WiMAX/lte Update]. IEEE Communications Magazine, 48(8), 60–67.

45. Janevski, T. (2009). 5G Mobile phone concept. In Proceedings of the 6th IEEE consumer communications and networking conference, Las Vegas, NV (pp. 1–2).

46. Jung, P., & Wunder, G. (2007). The WSSUS pulse design problem in multicarrier transmission. IEEE Transactions on Communications, 55(10), 1918–1928.

47. Kozek, W., & Molisch, A. F. (1998). Nonorthogonal pulseshapes for multicarrier communications in doubly dispersive channels. IEEE Journal on Selected Areas in Communications, 16(8), 1579–1589.

48. Kreutz, D., Ramos, F. M. V., Veríssimo, P., Rothenberg, C. E., Azodolmolky, S., & Uhlig, S. (2014). Software-defined networking: A comprehensive survey. CoRR abs/1406.0440.

49. Kryszkiewicz, P., & Bogucka, H. (2013). Out-of-band power reduction in NC-OFDM with optimized cancellation carriers selection. IEEE Communications Letters, 17(10), 1901–1904.

50. Larsson, E., Edfors, O., Tufvesson, F., & Marzetta, T. (2014). Massive MIMO for next generation wireless systems. IEEE Communications Magazine, 52(2), 186–195.

51. Lhetkangas, E., Lin, H., (Eds.) (2013). Deliverable D2.1. Requirement analysis and design approaches for 5G air interface. METIS Deliverable [online].https://www.metis2020.com/wp-content/uploads/deliverables/METIS_D2.1_v1.pdf.

52. Li, X., Gani, A., Salle, R., & Zakaria, O. (2009). The future of mobile wireless communication networks. In Proceedings of the international conference on Communication software and networks, Chengdu, China (pp. 554–557).

53. Maliatsos, K., Adamis, A., & Kanatas, A. G. (2013). Interference versus filtering distortion trade-offs in OFDM-based cognitive radios. Transactions on Emerging Telecommunications Technologies, 24, 692–708.

54. McKeown, N., Anderson, T., Balakrishnan, H., Parulkar, G., Peterson, L., Rexford, J., et al. (2008). OpenFlow: Enabling innovation in campus networks. ACM SIGCOMM Computer Communication Review, 38(2), 69–74.

55. Mitola, J. (2000). Cognitive radio—An integrated agent architecture for software defined radio. Ph.D. thesis. Royal Institute of Technology (KTH): Sweden.

56. Nolan, K. E., Kelly, M. Y., Forde, T., & Doyle, L. E. (2015). Development of a white space spectrum wireless interconnector. In Proceedings of the 10th international conference on cognitive radio oriented wireless networks (CROWNCOM), Doha, Qatar. April 21–23, 2015.

57. Ofcom. (2012). TV white spaces—A consultation on white space device requirements.http://stakeholders.ofcom.org.uk/binaries/consultations/whitespaces/summary/condoc.pdf.

58. Osseiran, A., Boccardi, F., Braun, V., Kusume, K., Marsch, P., Maternia, M., et al. (2014). Scenarios for 5G mobile and wireless communications: The vision of the metis project.IEEE Communications Magazine, 52(5), 26–35. doi:10.1109/MCOM.2014.6815890.

59. Pagadarai, S., Kliks, A., Bogucka, H., & Wyglinski, A. M. (2011). Non-contiguous multicarrier waveforms in practical opportunistic wireless systems. IET Radar, Sonar & Navigation, 5(6), 674–680.

60. Prasad, R. (2014). 5G: 2020 and beyond. River Publishers: Aalborg, Denmark.

61. Rui, Y., Hu, H., Li, M., Zhang, X., Yi, H., & Yang, Y. (2009). Comparing effects of carrier frequency offset on generalized multi-carrier and OFDM systems. In Proceedings of the IEEE international conference on communications (ICC), Dresden, Germany (pp. 1–6).

62. Sahin, A., Guvenc, I., & Arslan, H. (2013). A survey on prototype filter design for filter bank based multicarrier communications. http://arxiv.org/pdf/1212.3374.pdf.

63. Schafhuber, D., Matz, G., & Hlawatsch, F. (2002). Pulse-shaping OFDM/BFDM systems for time-varying channels: ISI/ICI analysis, optimal pulse design, and efficient implementation. In Proceedings of the 13th IEEE international symposium personal, indoor mobile radio communications (PIMRC), Lisbon, Portugal (Vol. 3, pp. 1012–1016).

64. Siohan, P., Siclet, C., & Lacaille, N. (2002). Analysis and design of OFDM/OQAM systems based on filterbank theory. IEEE Transactions on Signal Processing, 50(5), 1170–1183.

65. Stuber, G. L., Almalfouh, S. M., & Sale, D. (2009). Interference analysis of TV-band whitespace. Proceedings of the IEEE, 97(4), 741–754.

66. Sun, S., Ju, Y., & Yamao, Y. (2013). Overlay cognitive radio OFDM system for 4G cellular networks. IEEE Wireless Communications, 20(2), 68–73.

67. Sutton, P., Ozgul, B., Macaluso, I., & Doyle, L. (2010). OFDM pulse-shaped waveforms for dynamic spectrum access networks. In Proceedings of the IEEE symposium on new Frontiers in dynamic spectrum (DySPAN), Singapore (pp. 1–2).

68. Tandra, R., Mishra, S. M., & Sahai, A. (2009). What is a spectrum hole and what does it take to recognize one? Proceedings of the IEEE, 97(5), 824–848.

69. Wang, W., Yu, G., & Huang, A. (2013). Cognitive radio enhanced interference coordination for femtocell networks. IEEE Communications Magazine, 51(6), 37–43.

70. Wannstrom, J. (2013). Carrier aggregation explained. http://www.3gpp.org/Carrier-Aggregation-explained.

71. Weiss, T., Hillenbrand, J., Krohn, A., & Jondral, F. (2004). Mutual interference in OFDM-based spectrum pooling systems. In Proceedings of the IEEE 59th vehicular technology conference (VTC 2004-Spring), Milan, Italy (pp. 1873–1877).

72. Wyglinski, A. M. (2006). Effects of bit allocation on non-contiguous multicarrier-based cognitive radio transceivers. In Proceedings of the IEEE 64th vehicular technology conference (VTC 2006-Fall), Montreal, QC.

73. Xiao, J., Hu, R. Q., Qian, Y., Gong, L., & Wang, B. (2013). Expanding LTE network spectrum with cognitive radios: From concept to implementation. IEEE Wireless Communincations, 20(2), 12–19.

74. Yap, K. K., Sherwood, R., Kobayashi, M., Huang, T. Y., Chan, M., Handigol, N., McKeown, N., & Parulkar, G. (2010). Blueprint for introducing innovation into wireless mobile networks. In Proceedings of the 2nd ACM SIGCOMM workshop on virtualized infrastructure systems and architectures (VISA), New Delhi, India (Vol. 3, pp. 25–32).

75. Yucek, T., & Arslan, H. (2009). A survey of spectrum sensing algorithms for cognitive radio applications. IEEE Communincations on Surveys and Tutorials, 11(1), 116–130.

76. Zakaria, R., & Ruyet, D. L. (2012). A novel filter-bank multicarrier scheme to mitigate the intrinsic interference: Application to MIMO systems. IEEE Transactions on Wireless Communications, 11(3), 1112–1123.

77. Zhang, X., & Zhou, X. (2012). LTE-advanced air interface technology. Boca Raton, FL: CRC Press.

CHAPTER 5

ANALYSIS OF PRIVACY IN MOBILE TELEPHONY SYSTEMS

Myrto Arapinis, Loretta Ilaria Mancini, Eike Ritter, and Mark Dermot Ryan

Keywords: Privacy, Automatic verification, ProVerif, Mobile telephony, Pseudonym

INTRODUCTION

Mobile phones of the latest generation are relatively small but powerful devices that we keep always on and carry wherever we go. We rely more and more on services and apps running on them not only to make calls but to organize and enhance our daily life as well. For this reason, mobile phones are usually equipped with a variety of sensors and communication devices able to collect and/or communicate context-aware and personal data. This data collection and exchange activity take place even when the user is not aware of it. Hence, the user is often oblivious of potential security and privacy threats deriving from the use of these technologies and has little or no control over the generation, sharing and use of the data itself. Therefore, along with the benefits, the ubiquitous presence of smart, context-aware sensing and computing devices brings with it various concerns, mainly related to the security and privacy properties of such systems.

It is difficult to ensure that a system satisfies the desired security properties. The properties have to be formally defined. Every possible interaction of the adversary with the system has to be considered. The information the attacker gains from the execution of the system and the knowledge he can infer from it should be taken into account. This involves both careful scrutiny of the cryptographic primitives and of the protocol message flow. Formal methods have proved to be effective in highlighting weaknesses of protocols and hence enabling designers and programmers to patch them and strengthen their security guarantees, as well as in giving strong assurance on the properties a protocol achieves by proving the absence of attacks undermining the stated security properties. For example, conference management [1], electronic voting [2–4], single-sign-on [5, 6], cloud storage [7], RFID [8], TPM [9, 10] and mobile telephony protocols [11] have been scrutinized using manual and automatic verification techniques. Manual proof methods are lengthy and error prone, while automatic verification tools have to compromise in order to achieve decidability (at least for some subclass of processes) by bounding the number of agents and sessions and/or restricting the considered class of cryptographic functions.

Many of the security properties of mobile telephony systems can be modelled and analysed in terms of the classical concepts of confidentiality, authentication and integrity. However, new formal definitions of properties are required to reason about privacy and its different aspects and shades. Some of the most recurring privacy- related properties can be described in terms of location privacy, anonymity and unlinkability.

Modelling and verifying these properties is a difficult task since the sources of possible attacks are often hidden in implementation details or even in the protocol logic rather than in the cryptographic guarantees of the employed encryption algorithms. Moreover, the modelling of the privacy-related properties often relies on non-standard definitions of equivalence, and as a result, some of the properties are not supported by the currently available automatic verification tools. Hence, modelling and verifying these novel properties is one of the challenges requiring further development of the currently available automatic tools and, of course, of the theory supporting them.

Contributions

In this work, we present a thorough although not complete (i.e. we look only at some of the 2G/3G protocols) analysis of privacy in mobile telephony systems. We experimentally analyse these systems and point out flaws in the deployed

Citation: Myrto Arapinis, Loretta Ilaria Mancini, Eike Ritter, Mark Dermot Ryan, Analysis of privacy in mobile telephony systems, DOI: 10.1007/s10207-016-0338-9

policies. We formally analyse some of the protocols and identify vulnerabilities leading to user's privacy breaches which we show are feasible to implement on real networks. We also show that there are realistic solutions for ensuring privacy of mobile telephony subscribers and propose some privacy-friendly fixes of the flawed procedures. Moreover, we contribute to the formal analysis of privacy-related properties. In particular, we show how some definitions of unlinkability and anonymity as given in [12] can be adapted to be used for the automatic verification using the ProVerif tool. We use this technique to formally analyse the fixes we propose to patch the flawed 2G/3G procedures. In Sects. 3, 4, and 5, we will investigate the privacy provided by 2G and 3G system. In particular, in Sect. 3, we recall a well-known privacy weakness of mobile telephony which makes it possible to break user anonymity, and we reveal possible privacy leakages produced by signalling procedures of the 2G/3G protocol stack. Further, we conduct experiments to test the effectiveness of the changing pseudonym policies adopted by mobile telephony systems and present a replay attack affecting the TMSI reallocation procedure. In Sect. 4, we show an attack that affects only 3G systems and takes advantage of the information leakage resulting from the error messages produced by the execution of the AKA protocol. In Sect. 5, we demonstrate that it is possible to mount the attacks on real 3G networks. In Sect. 6, we provide privacy-enhancing versions of the procedures analysed in Sects. 3 and 4. Finally, in Sect. 7, we use formal verification to give guarantees on the privacy properties (anonymity and unlinkability) of the fixed procedures.

Privacy

As D'Introna points out in [13], privacy is a very important characteristic for the definition of social relationships and more crucially a fundamental property to enable autonomy and individuality. Knowing to be observed influences a person's behaviour; indeed, we adopt different kinds of behaviours depending on who is the observer. We chose to share different things with different people according to, for example, whether the relationship is intimate, social or work-related. Hence, privacy seems to be fundamental for the definition of oneself.

Privacy is a relatively novel concept that appears for the first time in a 1890 analysis in the Harvard Law Review in relation to a case of privacy invasion by the press [14], and there is no universally accepted definition of privacy. In the context of mobile telephony systems, privacy is a major concern [15–18] because, as previously mentioned, the quality, quantity and accuracy of collected personal data are of great magnitude. Furthermore, the user is often unaware of generating the data and has little or no control over its generation, exchange, transfer and use.

As in the real world, in the electronic world as well, privacy is a difficult property to define and we rather speak of privacy-related properties instead, which can be general ones or application specific. Moreover, privacy is a complex property to enforce and verify since it is multi-level and multi-protocol, i.e. it crosses more than one layer, if not all layers of the protocol stack, and typically has to be enforced by all protocols in order to be satisfied [19]. While some aspects of privacy are more suited to be enforced by policies as, for example, the sale of private information to third parties, others should be part of the verifiable properties embedded in the design of a system. However, privacy seems to be often in conflict with other system requirements as for example accountability, or in contrast with the nature of the system itself as in the case of location-based services. For these reasons, user's privacy is often overlooked or enforced by complex mechanisms which are difficult to verify by hand.

The most obvious privacy-related property is data confidentiality. However, it is generally not sufficient to ensure confidentiality in order to achieve privacy. For example, a user identifier could not be public and his identity could be disclosed by correlating his activity [20–22]. Moreover, multiple accesses to a system from different locations by the same identifier could directly or indirectly reveal a user's position or his movement patterns. Hence, anonymity and unlinkability of a user's activity and access to a system are important aspects of a user's privacy. Location privacy is one of the privacy aspects which has been the subject of many studies [23–25]. Several techniques have been proposed to protect a user's location, for example obfuscation, pseudonymity and mix zones [23]. Another interesting privacy-related property which can be for example desirable for private information stored on portable devices is forward privacy which requires that the secrecy of certain information holds even after the device has been corrupted. For example, it may require that session keys are not retrievable even in the event that the master keys were disclosed after the device was corrupted. Formal definitions of anonymity and unlinkability are given in [12], definitions of untraceability and forward privacy are given in [26], and a definition of voter's privacy is given in [27]. These definitions are all equivalence-based and often pose challenges for the currently available automatic verification tools because of the definition of the property itself, because of the structure of the protocol or because of the algebraic properties of the involved cryptographic primitives.

In this work, we are concerned about mobile telephony systems' privacy. We carry out both experimental and formal analysis in order to test the level of privacy provided to mobile telephony users over the radio link. In particular, we experimentally analyse the use of pseudonyms and expose some critical scenarios which makes it possible to violate a user's privacy. Moreover, we expose some protocol's flaws resulting in breaches of the anonymity and/or user unlinkability (as defined in [12]). We show that it is possible to implement the attack we found on a real network using an hacked femtocell. We propose some lightweight fixes of the flawed 2G/3G procedures. Finally, we provide the theoretical framework for the automatic verification of the unlinkability and anonymity of the fixed 2G/3G procedures.

Security Protocols and Formal Methods

Security protocols aim to protect sensitive data in particular when communicating over an unprotected connection, where there is little or no control over the flow of information, and the attacker can intercept, manipulate, replay, inject, replace, substitute and compare messages. This model is sometimes called the Dolev-Yao attacker model [28]. This makes the design of security protocols a notoriously difficult task since it does not only involve the protection of data through some sort of smart cryptographic algorithm but also the ability of foreseeing how the flow of messages could be used to retrieve, deduce and get access to sensitive information. Details hidden in the protocol logic often allow to break a protocol security without even breaking the underlying cryptography. Examples are the attacks recently found on TLS [29], SAML [6, 30] and OAuth [5]. These attacks do not take advantage of weak cryptography but rely on the logic and interaction between concurrent protocols at the same and/or different layers of the stack. Another example is the traceability attack on the French e-passport which allows one to trace an e-passport holder by performing a replay attack [8]. Hence, to declare a protocol secure, one should check it against any possible adversary and any possible interaction. Moreover, one would need to state what being secure means for the protocol, i.e. should specify the security properties the protocol aims to achieve, and this would depend on the purpose of the protocol and on the application requirements. Hence, to evaluate the security of a protocol, it is very important to rigorously define both the protocol itself and the desired security properties.

Formal methods help establishing the security of protocols in three ways:

- Rigorously modelling security protocols and the attacker model
- Rigorously defining security properties
- Evaluating the protocols against the desired properties

There are two different formal approaches to the problem of protocol security. The computational approach is closer to the actual protocol implementation, represents messages as bitstrings, cryptographic functions as polynomial time algorithms and the adversary as any probabilistic polynomial time algorithm and can give strong guarantees on the security of the cryptographic algorithms. However, proofs in this setting can be very long, difficult, error prone and not very accessible to careful scrutiny. Moreover, they are not very well suited for automation.

We will use symbolic methods instead. Symbolic methods abstract from the details of the cryptographic algorithms. Messages are represented by terms and cryptographic primitives are represented by function symbols which can be applied to terms. The properties of cryptographic primitives are abstracted by algebraic properties. In general, cryptographic primitives are assumed to be perfect. Hence, the adversary can have any interaction with the protocol, injecting, modifying, replaying and intercepting messages but cannot break cryptography.

As in the case of computational methods, in the symbolic world the security of protocols can be established by means of long, tedious and error-prone manual proofs or by manually exhibiting a counterexample, i.e. an attack. However, lots of work focus on automating the proofs and/or search for counterexamples. In this area, there are still plenty of challenges and room for improvements and researchers who continuously aim to expand automatic verification to larger classes of protocols and equational theories for both reachability and equivalence-based properties.

Some of the most famous formalisms in the field of symbolic methods are process calculi, constraint systems and strand spaces. In this work, we will use the applied pi-calculus [31] which is a calculus for modelling cryptographic protocols offering flexibility in the range of cryptographic primitives represented by function symbols and equational theories and making the adversary knowledge explicit thanks to the frame construct. Many results obtained using different formalisms such as spicalculus and constraint systems can be transposed into the applied pi-calculus. We describe some of the most relevant.

The general problem of security is shown to be undecidable even for a bounded number of sessions for both reachability-based properties [32] and for equivalence-based properties [33]. Restricting the calculus and or to the equational theory can yield decidability results.

One of the most common restrictions is to consider only a finite number of sessions. Automatic verifiers for bounded processes can be very useful to find protocol flaws; however, when they do not find any attack, no guarantees of the absence of attack can be derived for a number of sessions greater than the one used to run the verifier on the protocol. To obtain general security results, an unbounded number of sessions should be considered. Another common restriction to the calculus consists in forbidding else branching, and the equational theory is often required to be convergent.

A decision algorithm for trace equivalence of bounded processes with no else branching and a pre-defined signature consisting of pairing, symmetric and asymmetric encryption is given in [34] along with a prototype implementation for checking deducibility and static equivalence. However, for deciding, trace equivalence is required to implement a further procedure that generates a pair of constraints for each possible interleaving. The AKiss tool implements two procedures for bounded processes with no else branches and for convergent rewrite systems with the finite variant property. One procedure under-approximates trace equivalence and can be used to prove protocols correct. The other over-approximates trace equivalence and can be used to discard incorrect protocols. The tool can be used to check observational equivalence of determinate processes since in this case observational equivalence coincides with the under-approximation of trace equivalence.

The decidability of observational equivalence of "*simple*" processes with no replication and with no else branches for subterm convergent equational theories is shown in [35]. The decision procedure relies on the decidability of the equivalence of constraint systems given in [36]. This procedure has not been implemented so far. To the best of our knowledge, the only tool able to automatically verify both reachability-based [37] and equivalence-based [38] properties for unbounded processes with else branches and convergent theories is the ProVerif tool [39]. Blanchet, Abadi and Fournet [38] introduce the concept of bi-process, which is a pair of processes that differ only in the choice of some term. ProVerif can prove observational equivalence of bi-processes. However, the procedure is sound but not complete, meaning that the tool can prove that a property holds on the given protocol model, but when it outputs an attack trace, this could be a false attack. Moreover, the tool might not terminate.

BACKGROUND: MOBILE TELEPHONY SYSTEMS

Global System for Mobile Communication (GSM) and

Universal Mobile Telecommunications System (UMTS) are the most widely used mobile telephony standards with billions of users worldwide. GSM was developed by European Telecommunications Standards Institute (ETSI) in order to promote a common standard for the European cellular telephony replacing the multitude of first-generation standards. UMTS is specified and maintained by the Third Generation Partnership Project (3GPP); it was introduced in 1999 to offer a better support for mobile data applications by increasing the data rate and lowering the costs of mobile data communications. Furthermore, UMTS offers an improved security architecture with respect to previous mobile communication systems such as GSM. Both GSM and UMTS have been improved and extended several times. We will use the terms 2G and 3G to indicate the wider set of standards including GSM and UMTS, respectively. Most of the protocols and issues presented in the following sections are common to both 2G and 3G systems, when this will not be the case, it will be explicitly pointed out. In the rest of this paper, we will adopt a unified terminology to address mobile telephony systems components. However, 2G and 3G standards may not use the same terminology even when addressing components having the same functions/purposes.

In the next section, we will introduce the 2G/3G network architecture and we will describe in more details their security features. We will then summarize some of the well-known weaknesses of mobile telephony systems, along with relevant work on 2G/3G security and privacy.

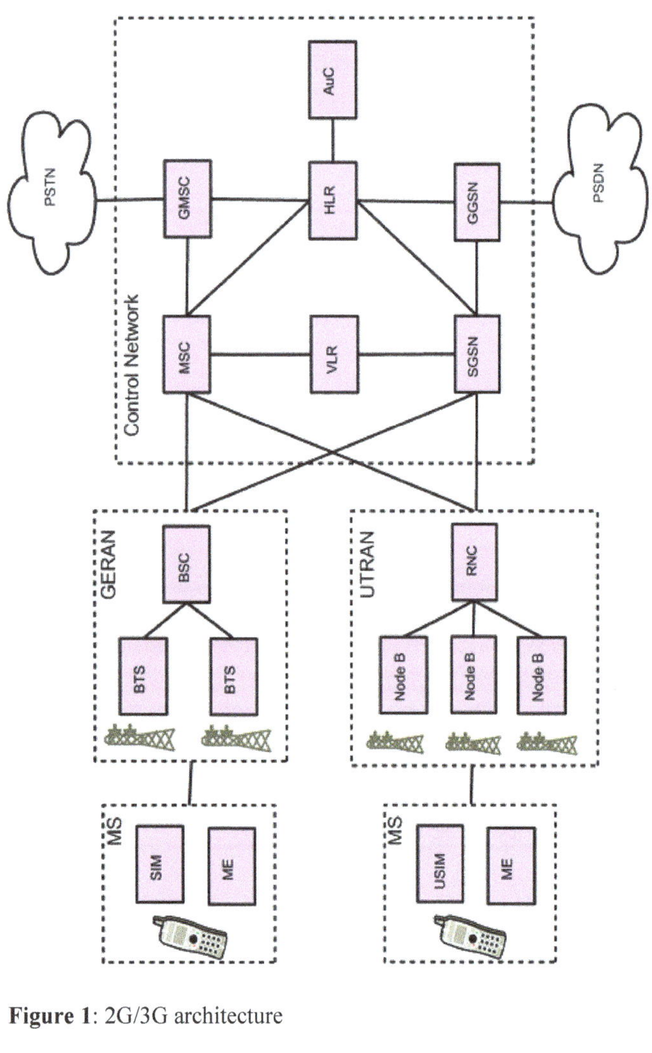

Figure 1: 2G/3G architecture

GSM/UMTS ARCHITECTURE

The 2G/3G network architecture, depicted in Fig. 1, integrates both GSM and UMTS components. The user side of the network consists of mobile stations (MS). This term is used in mobile telephony systems to indicate both the mobile equipments (ME), such as mobile phones, and the so-called SIM or USIM card (Universal Subscriber Identity Module) in 2G/3G systems, respectively. The (U)SIM card identifies the user as a legitimate subscriber within a mobile telephony operator network. To access the services offered by a mobile operator, a MS connects through radio communication technology to the UTRAN (UMTS Terrestrial Radio Access Network) or GERAN (GSM/EDGE[1] Radio Access Network) network that is a GSM access network. A mobile station directly communicates with a BTS (base transceiver station) or Node B which covers the area the MS is located in. One or more Nodes B are connected to a Radio Network Controller (RNC), and one or more BTS are connected to a Base Station Controller (BSC) defining a cell. A group of cells forms a Location Area. RNCs and BSCs manage the radio resources and inter-cell handover. They are the interface between the mobile station and the core network. The core network offers circuit-switch and packet-switch services. The mobile switching centre (MSC) and gateway mobile switching centre (GMSC) offer inter- and intra-network circuit-switching domain services, respectively, and interface the 2G/3G systems with the traditional fixed telephony network. The Serving GPRS[2] Support Node (SGSN) and the Gateway Serving GPRS Support Node (GGSN) offer, respectively, inter- and intra-network packet-switching domain services as well as connecting 2G/3G networks with the Internet. Within the core network, the Home Location Register (HLR) stores permanent sensitive information of subscribers such as identities, service profiles and activity statuses. These informations are linked to the SIM and recorded when stipulating a contract with the mobile network operator (MNO). 2G/3G systems offer roaming capabilities between different mobile network operators and between the different technologies (provided that the mobile equipment supports them), meaning that a mobile station can be connected to a visited network, the serving network (SN), which might be different from the subscriber's home network (HN) and which could be using a different standard. Each subscriber has a long-term identifier IMSI (International Mobile Subscriber Identity) that is stored in the (U)SIM and a temporary identifier TMSI (Temporary Mobile Subscriber Identity), allocated by the serving network to protect the subscriber's identity privacy. The Visitor Location Register (VLR) stores temporary informations about subscribers visiting a given location area of the serving network and maintains TMSI/IMSI associations. The network operator and each subscriber share a different unique long-term secret key used for authentication purposes. This key is stored in the (U)SIM. The authentication centre (AuC) is a protected database storing associations between subscriber identities (IMSI) and long-term keys.

In the rest of this work, we only consider a simplified network architecture, since we are interested in third-party attackers having access only to the radio path and not to the wired network infrastructure. This architecture involves simply the mobile stations and the network. The network models both the UTRAN/GERAN Base Station that the MS is directly communicating with and the complex structure of databases and servers connected with it and forming the UMTS/GSM control network. In particular, we do not distinguish between serving network and home network. Hence we abstract away from any communication within the network and model only communication between mobile stations and the network. This abstraction allows us to hide details which are uninteresting for the purposes of our analysis and keep the models used for verification small, but at the same time precisely model the interactions over-the-air between MS and network, which are the subject of our analysis.

GSM/UMTS security features

The intended security features of 2G and 3G systems are slightly different, as 3G systems aim to overcome the weaknesses of 2G systems such as the lack of mutual authentication and the use of weak ciphering algorithms.

2G systems aim to provide the following security features [40]:

- Subscriber identity confidentiality The IMSI is not made available or disclosed to unauthorized individuals, entities or processes.
- Subscriber identity authentication The corroboration by the mobile network communicating with a mobile station that the subscriber identity (IMSI or TMSI), sent by the mobile subscriber to identify itself, is indeed the one claimed.
- User data confidentiality The user information exchanged on traffic channels is not made available or disclosed to unauthorized individuals, entities or processes.

The 3G communication system aims to improve the security features offered to mobile telephony subscribers, requiring for example mutual authentication between the user and the network and specifying more precisely the provided user's privacy properties in terms of identity confidentiality or anonymity, location confidentiality and user untraceability or unlinkability. The security properties stated by the 3G standard are as follows [41]:

User identity confidentiality

- *User identity confidentiality* The permanent user identity (IMSI) of a user to whom a service is delivered cannot be eavesdropped on the radio access link;
- *User location confidentiality* The presence or the arrival of a user in a certain area cannot be determined by eavesdropping on the radio access link;
- *User untraceability* An intruder cannot deduce whether different services are delivered to the same user by eavesdropping on the radio access link.

Entity authentication

- *User authentication* The serving network corroborates the user identity of the user;
- *Network authentication* The user corroborates that he is connected to a serving network that is authorized by the user's HN to provide him services; this includes the guarantee that this authorization is recent.

Confidentiality

- *Cipher algorithm agreement* The MS and the SN can securely negotiate the algorithm that they shall use subsequently;
- *Cipher key agreement* The MS and the SN agree on a cipher key that they may use subsequently;
- *Confidentiality of user/signalling data* User/signalling data cannot be overheard on the radio access interface;

Both 2G and 3G systems rely on an authentication and key agreement (AKA) protocol, on the use of encryption of the confidential data transmitted on the radio channel and on the use of periodically changing temporary identities in order to achieve the above- mentioned security properties. Moreover, 3G systems use integrity protection of the sensitive data. Unlike the 2G authentication protocol, the 3G AKA protocol allows MS and SN to establish not only a ciphering key but also an integrity key and achieves mutual authentication of MS and HN.

Previous work on 2G/3G security and privacy

Most of the work on security of mobile telephony systems concern properties such as secrecy, integrity and authentication [42–44]. There are only few formal and experimental studies concerning the level of privacy provided to users by mobile telephony systems [45–47].

2G vulnerabilities The identification procedure, consisting in the request of the user identity by the network followed by a cleartext reply containing the user identity, is acknowledged in the standard as a breach of the user identity confidentiality [41, p. 19, s. 6.2]. This procedure is exploited by the well-known "IMSI catcher" attack, which is the best known attack to mobile telephony users' privacy. It consists in forcing a mobile phone to reveal its identity (IMSI) [48,49] by triggering the identification procedure from a fake operator base station (configured with the corresponding mobile network and country code settings). Moreover, the 2G fake base station can then force the attached mobile station to use no encryption for data and signalling communication, and hence, it can capture ingoing and outgoing plaintext traffic. Until fairly recently, implementing an IMSI catcher required specialized software and equipment such as base stations. However, such devices have become more and more affordable thanks to software emulation [50].

Foo Kune et al. [51] present a study on the use of the paging procedure to locate mobile telephony users. They perform a tracking attack relying on passive sniffing of paging response messages triggered by placing silent phone calls (obtained by hanging up before the receiving phone rings) to the victim phone. This technique allows to reveal the presence of the victim in an area monitored by the attacker. Munaut and Nohl [43] previously outlined a similar technique. They performed a GSM sniffing attack, which allows to eavesdrop a GSM phone call by using a modification of the osmocom-BB [52] open-source implementation of the GSM protocol stack and an old Motorola mobile phone. Differently from Foo Kune et al., they used a silent SMS to trigger the paging responses needed to locate the victim. Although these works take advantage of the fact that a TMSI is allocated for a long time window, they do not analyse the security and privacy provided by the TMSI reallocation procedure. Moreover, in order to perform the attack, the adversary needs to know the mobile number of the victim.

Engel showed at the 25C3 conference [53] how network signalling messages, triggered when sending/receiving SMS messages, can be used to localize mobile telephony users. He suggests that network operators should use home routing, i.e. forwarding through the home network, as a countermeasure to this SMS tracking attack. This attack requires access to the intra-network communication infrastructure, which although possible may require subscription to a pay per query service. In Sects. 3 and 4, we will analyse the privacy of the more exposed over-the-air communication which is available to any attacker with a radio-enabled device. We will not assume to have control over the less easily accessible intra-network communication channel.

The *gsmmap* project [45, 46] assesses and visually renders on a map the level of security and privacy provided by 2G and 3G network operators across the world. The data about the security of 2G systems are gathered using a variant of the open-source GSM protocol stack developed within the osmocom-BB project, while data about the security of 3G networks can be collected using some specific model of mobile phones and downloading the gsmmap specific app. The project relies on volunteers to contribute the needed data; hence, the map is not complete and some of the results are estimations based on the available data. In particular, gsmmap aims to check whether network operators are protecting the users from well-known attacks by adopting countermeasures such as the use of A5/3 encryption, padding randomization and full authentication for outgoing calls and SMS to prevent impersonation and interception, and the use of regular TMSI updates and home routing to prevent Engel's SMS tracking attack. However, the list of security protection mechanisms gsmmap is looking for in order to evaluate the security offered by mobile telephony systems is continuously updated according to the most recent weaknesses found and the suggested countermeasures.

Golde, Redon and Seifert [54] showed that it is possible to hijack a mobile terminating service such as the reception of a phone call or an SMS message and even perform a targeted or large- scale denial of service by answering paging requests faster than the intended recipient. They implemented their attacks on GSM networks, but speculate they would work on UMTS and LTE as well since the attacked procedure, i.e. the paging procedure is common to the three systems.

3G weaknesses resulting from 3G/2G-interoperability 3G mobile stations are as well vulnerable to 2G IMSI catcher; in fact 3G systems are fully interoperable with 2G systems, and 3G mobile stations can roam in 2G networks. A 3G mobile stations can be induced to attach to a 2G base station by broadcasting a stronger signal with respect to the 3G one. To avoid this, 3G mobile stations can be configured to use only 3G networks. To the best of our knowledge, the only implementation of a pure 3G IMSI catcher is the one presented in [55] and is realized using a modified femtocell. Previously proposed attacks on 3G security exploit the vulnerabilities which are propagated from GSM to 3G when providing interoperability between the two systems. Most of the reported attacks of this kind take advantage of well-known weaknesses of the GSM authentication and key agreement protocol, such as the lack of mutual authentication and the use of weak encryption. These attacks allow an active attacker to violate the user identity confidentiality, to eavesdrop on outbound communications [56] and to masquerade as a legitimate subscriber obtaining services which will be billed on the victim's account [44]. However, these attacks cannot be carried out on pure 3G networks, because they rely on the lack of mutual authentication in GSM and on the possibility of downgrading the communication from 3G to GSM.

3G specific To the best of our knowledge, the only attack that does not rely on GSM/3G interoperability has been presented by Zhang and Fang in [57]. This attack is a variant of the false base station attack and takes advantage of the fact that the mobile station does not authenticate the serving network. It allows the redirection of the victim's outgoing traffic to a different network, for example, a network which uses a weaker encryption algorithm or one which charges higher rates than the victim's one. Zhang and Fang's attack concerns impersonation, service theft and data confidentiality.

LTE privacy The LTE authentication and key agreement protocol suffers from a linkability attack [58] similar to the one we present in Sect. 4.2.

Previous formal analysis of 3G protocols The 3G AKA protocol in its pure form (i.e. with no GSM support) has been formally proved to meet some of the specified security requirements [59], such as authentication and confidentiality of data and voice communication. However, privacy- related properties such as unlinkability and anonymity, which are the focus of our work, are not analysed in [59]. The framework applied in [59] cannot be used to specify unlinkability and anonymity properties, let alone reason about them. The formal framework we used allows us to precisely define and verify privacy-related properties. Hence, we can discover privacy attacks on the modelled protocols and propose solutions which we then formally prove to satisfy the desired privacy properties.

Other work on 3G privacy enhancement A new framework for authentication has been proposed to provide subscriber privacy with respect to the network [47]. In particular, the authors aim to achieve MSs anonymity with respect to the serving network and location privacy of mobile stations with respect to the home network. To achieve this purposes, they propose a new mechanism for the location update and a three-way handshake protocol to be used instead of the currently used 3G AKA protocol. However, this work is not supported by a formal model of the AKA protocol, nor does it provide a formal verification of the properties enforced by the proposed protocols. Moreover, their attacker model considers the network as not fully trusted, while in this work we are only concerned with third-party attackers controlling the radio link communications.

The feasibility of implementing the solutions we propose to fix the identification procedure, IMSI paging procedure and AKA protocol in the current 2G/3G infrastructure is discussed in [60].

ANALYSIS OF MOBILE SYSTEMS' PRIVACY: 2G/3G WEAKNESSES

In this section, we describe some of the mobile systems protocols that are dealing with the identity of mobile phone users and point out breaches of the user's privacy, which expose a subscriber's identity and allow an attacker to identify the presence of a target mobile phone (MS) in a monitored area or even track its movements across a set of monitored areas. In Sect. 4, we illustrate a privacy breach that exploits a procedure specific of the 3G system. The attacker considered in this section is capable of sending, receiving and sniffing[3] messages over-the-air. Further, we assume that the attacker has unlimited access to the radio link between the mobile station and the base station but no access to the communication within the core network or between the base station and the core network. As we will see, the attacker does not need to know any keys, nor perform any cryptographic operation. In fact, the vulnerabilities exposed in this section take advantage of the poor management of the user identities, due to the fact that at the time mobile telephony systems were designed, active attacks were considered too costly and hence unlikely to be performed. In fact, mounting an attack required very expensive equipment such as a base station. However as we briefly discussed in Sect. 2.3, nowadays the development of open- source devices and software has lowered significantly these costs.

In the rest of this section, we present two procedures of the mobile telephony protocol stack that involve the mobile station identity, namely the identification procedure and the paging procedure, and we show how these procedures can be used to violate a user's privacy. Furthermore, we introduce the TMSI reallocation procedure that is the procedure implementing the pseudonym changing mechanism adopted by mobile telephony systems in order to provide anonymity and unlinkability. We experimentally evaluate the effectiveness of the pseudonym changing mechanism on real networks. We finally show a replay attack on the TMSI reallocation procedure which is enabled by the fact that the reuse of previously established keys is allowed by the standard and is indeed a policy adopted by real network operators (see Sect. 3.8).

Identification procedure

The identification procedure [61] allows the network to request the credentials of a mobile station. In particular, the network can request a mobile station's IMSI that identifies the user's SIM card or better identifies the user as a legitimate one having a contract with the MNO, or a mobile station's IMEI, which identifies the specific handset. The identification procedure is always initiated by the network. It is typically used when a mobile station connects to a network for the first time. Once the network identifies the MS, it can initiate a procedure to establish a shared short-term encryption key and then usually assigns a temporary identity, TMSI, to the MS. The temporary identity is transmitted to the MS in an encrypted fashion and is used instead of the IMSI in the following communications.

The identification procedure consists of two messages: the identity request message and the identity response message. The parameter IMSI_REQ specifies that the identity requested by the network is the IMSI. The identity response message is sent in clear along with the requested identity (see Fig. 2).

Figure 2: Identification procedure. The identity of the mobile phone is sent in clear on the radio path

Identification procedure attack

The identification procedure is acknowledged in the standard as breaching the user identity confidentiality. It allows a passive attacker to overhear an MS's IMSI. Moreover, any attacker controlling a tool able to act as a base station can trigger at any moment the identification procedure and capture the identity response and thus can trace the presence of a particular subscriber in the range of his device. Moreover, he can track the movements of the subscriber by installing tracking devices in different areas. This attack to the subscribers' privacy is known as IMSI catcher attack [48, 49, 62]. Typically, the attacker would control a fake BS that just sends an identity request and waits for the response.

Paging procedure

The paging procedure defined in [63] is used to locate a mobile station in order to deliver a Mobile Terminated (MT) service to it, for example an incoming call or SMS message. The network sends paging requests in all the most recently visited location areas in order to locate the mobile station. In fact, most MS are idle most of the time to save battery so the network does not know which base station is currently providing the best reception level to the MS. The paging request message is sent on a Common Control Channel (CCCH) and contains the identity of one or more mobile stations. The paging procedure is typically run after a TMSI is assigned to the mobile station by the network, and hence, the TMSI should be used to identify a MS in the paging request. However, the IMSI can be used when the TMSI is not known by the network.

A mobile station receiving a paging request message checks whether one of the identities included in the message is its own. If the check succeeds, the MS establishes a dedicated channel to allow the delivery of the service. On this dedicated channel, the MS sends a paging response message containing its IMSI in GSM networks (see Fig. 3), the most recently assigned TMSI in UMTS networks (see Fig. 4).

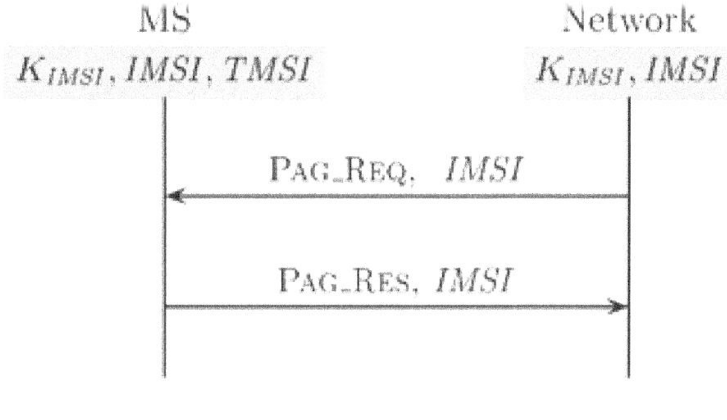

Figure 3: IMSI paging procedure in GSM: the paging request mes-

sage is sent on a common radio channel and contains the long-term identity of the paged mobile station IMSI in cleartext. The paging response message also contains the IMSI and is sent in clear on a dedicated channel

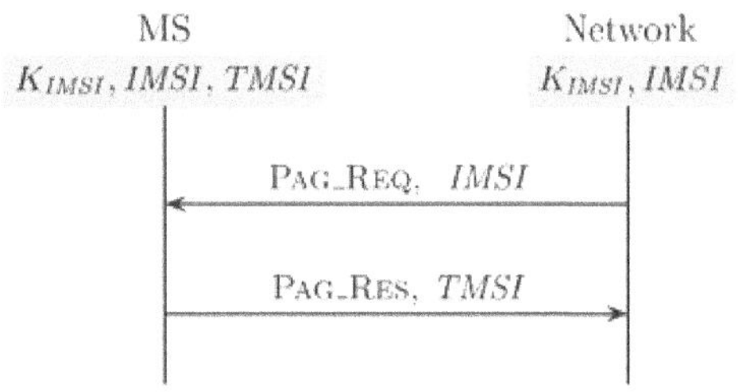

Figure 4: IMSI paging procedure in UMTS: the paging request message is sent on a common radio channel and contains the identity of the paged mobile station (in this case the IMSI) in cleartext. The paging response message contains the mobile phone's temporary identity TMSI and is sent in clear on a dedicated channel

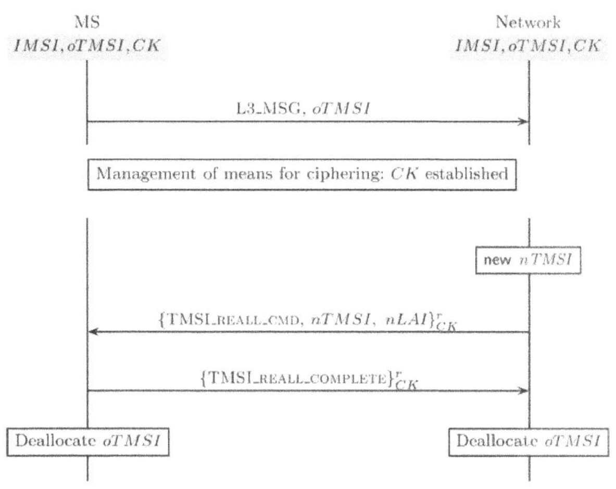

Figure 5: TMSI reallocation procedure: the new TMSI along with the current Location Area Identifier (LAI) is sent in an encrypted message to the mobile station in order to avoid users' linkability

IMSI paging attack

The possibility of triggering a paging request for a specific IMSI allows an attacker to check a specific area for the presence of mobile stations whose identity is known to the attacker and in 3G networks to correlate IMSI and TMSI. The observation of the related paging response allows the correlation of the victim's IMSI with his current TMSI. In practice, an attacker would need to confirm the link between the paged IMSI and the related TMSI by replaying the attack several times.

TMSI reallocation procedure

A mobile station (MS) is uniquely identified by means of its IMSI. If a third party that eavesdrops on the radio link was able to identify wireless messages as coming from a particular mobile phone, he would be able to track the location of the mobile phone user in real time. This could lead to stalking and other forms of harassment, as well as more mundane invasions of privacy. To avoid over-the-air attackers from identifying and linking a user's transactions, a temporary identity called TMSI is assigned by the network and is used to identify the mobile station in protocol messages sent in clear over-the-air. The mobile station identity (its TMSI, if available, or its IMSI) is always included in the first message sent from the MS to the network after the establishment of a dedicated channel. This allows the network to identify the MS before delivering a service to it. For example, the identity is carried in location update requests, CM (Call Management) requests and paging responses. The use of TMSIs avoids the exposure of the long-term unique identity (IMSI) and hence aims to provide third-party anonymity and unlinkability to mobile telephony subscribers. Temporary identities are periodically updated by the network by means of the *TMSI reallocation procedure*.

The 3GPP standard specifies that a new TMSI should be assigned at least at each change of location area. Besides this constraint, the choice of how often a new assignment is performed within a location area is left to the network operators [61]. In order to prevent an adversary linking the old TMSI with the new one, the assignment of a new TMSI is performed in ciphered mode. The session key used to encrypt the new TMSI is established by executing the AKA protocol.

The TMSI reallocation assigns a new pseudonym (TMSI) to a mobile station. The new TMSI is sent to the mobile station in an encrypted fashion. Figure 5 depicts the TMSI reallocation procedure as defined in the 3GPP standard [41, 61]:

The mobile station sends a first message on a dedicated channel. This message contains the current MS's temporary identity *oTMSI*;

Upon receipt of this message, the network can identify the MS and establish means for ciphering of the subsequent communication on the dedicated channel;

The rest of the communication is then encrypted and consists of a TMSI reallocation command message containing a new pseudonym *nTMSI* randomly chosen by the network and the current location area *nLAI* (the area within which *nTMSI* is meaningful);

This message is followed by a TMSI reallocation complete message which is sent by the MS to acknowledge the completion of the reallocation procedure.

If the network does not receive the expected acknowledgement from the MS, it maintains both*oTMSI* and *nTMSI* as valid pseudonyms for the IMSI. The network can perform a TMSI reallocation at any time, while a dedicated channel is established. The standard does not fully specify how often this procedure should be performed. However, it mandates that it should at least be performed at each change of location [61]. The standard defines two options for the management of the means for ciphering (i.e. to establish the ciphering key CK):

(1) Either a fresh ciphering key is established by executing the authentication procedure;

(2) Or a previously established ciphering key can be restored by means of the security mode setup procedure, which allows the MS and the network to agree on a ciphering algorithm.

Subscriber privacy experimental analysis

In this subsection, we will present the results of our experimental analysis of the use of TMSIs in mobile networks and highlight critical scenarios from a privacy point of view. Our aim is to analyse whether the changing pseudonyms policy adopted by mobile telephony systems guarantees user privacy as intended. In particular, two aspects appear to be important:

1. TMSI reallocation will protect user privacy only if TMSIs are re-allocated often enough and at the right times (e.g. when users move between locations). The 3GPP standard does not rigorously define the conditions under which TMSI reallocation takes place, leaving the choice to the network operators. As our experiments show, this leaves users open to privacy abuses.
2. The success of TMSI reallocation requires that a wireless eavesdropper cannot link the new TMSI to the old one. Encrypting the TMSI in the allocation message is necessary (but may not be sufficient) to ensure that. It turns out that other factors, in particular the use of fresh session keys for each TMSI reallocation, are also necessary to guarantee unlinkability of old and new TMSIs. The 3GPP standard does not mandate this, again leaving user privacy subject to choices made by network operators.

We monitored over-the-air communications of idle and active MSs in order to understand how real networks implement user identity confidentiality through the use of TMSIs, both in terms of frequency of reallocation and ciphering keys used. Our experiments confirm that the reuse of previously established keys is a commonly adopted policy. However, we show that in case the reuse of encryption keys is adopted for the execution of the TMSI reallocation procedure, this enables a linkability attack which makes it possible to link old and new TMSIs.

Our experiments were carried out using an old GSM Motorola C115 mobile phone in France, UK, Greece and Italy and using SIM cards from all the major UK, Greek and Italian network operators.[4]

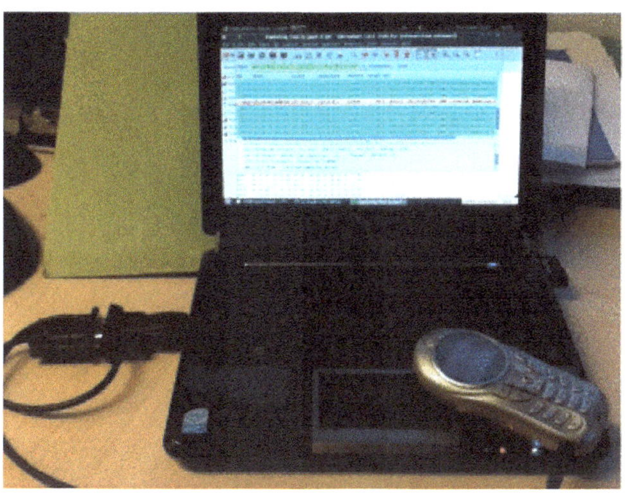

Figure 6: Experimental tools

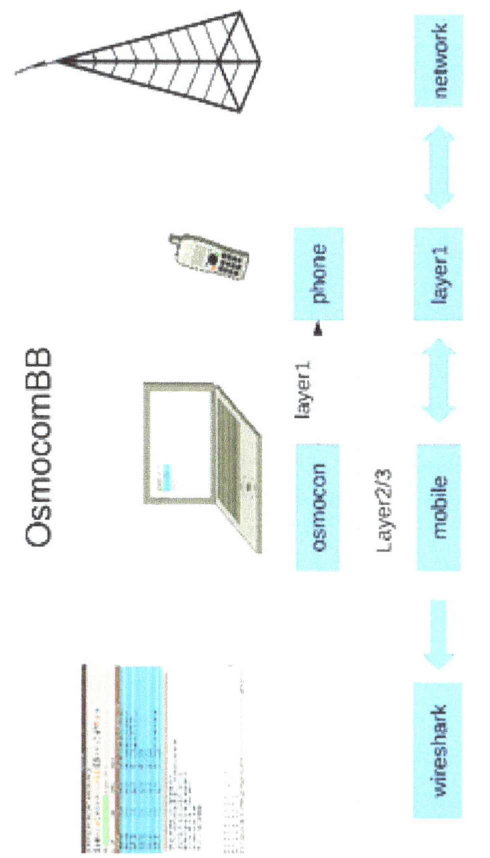

Figure. 7: Osmocom-BB architecture

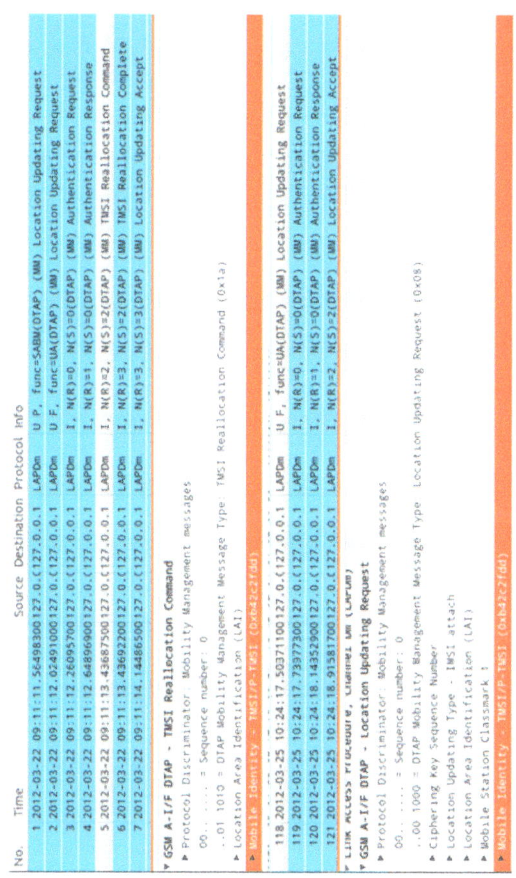

Figure. 8: Trace of a UK Vodafone SIM card obtaining a new TMSI (0xb42c2fdd) on 22/03/12. The same TMSI is still in use on 25/03/12 after 3 days from its allocation

Experimental settings and scenarios

The Motorola C115 has a TI Calypso baseband chipset which is supported by the Osmocom-BB project [52]. The Osmocom-BB project includes an open-source implementation of the GSM baseband and various other applications aiming to implement a GSM mobile station. The radio communication functions are implemented in the firmware which is flashed from a laptop into the mobile phone through the Osmocon software, by means of a T191 unlock cable (Fig. 6). The firmware implements layer 1 of the GSM protocol stack, while layers 2 and 3 are implemented in specialized applications running on the laptop and communicating with the mobile phone through the T191 cable (Fig. 7). In particular, we used the "mobile" application which implements layer 2 and 3 of the GSM protocol stack to provide all the basic functions of a mobile phone (network registration, location update, making and receiving calls, and sending and receiving SMSs). The mobile phone activities are logged on a shell terminal, and the radio communication is encapsulated in UDP packets sent to a configurable IP address. This traffic can be captured through the Wireshark network traffic analyser [64]. Interactions with the mobile phone are enabled by a telnet command interface. This allows one to manually select a network, start phone calls, send SMSs and service requests, etc.

We captured over-the-air messages using the "mobile" application in different settings:

1. Mobile station in idle state and not moving;
2. Mobile station in idle state and moving across two urban areas;
3. Mobile station involved in activities such as receiving or starting phone calls, receiving or sending SMSs, and requesting services as for example call diversions.

Since the 3GPP standard merely gives guidelines, real networks differ in the implementation details of the TMSI reallocation. To understand whether the different implementations achieve the privacy guarantees they were intended for, we analysed the traffic captured with the mobile application. In particular, we are interested in finding out whether the frequency of TMSI reallocation execution is high enough to defeat passive and active tracking attacks, whether the policy of changing TMSI at least at each change of location is actually implemented so to obtain at least location-dependent privacy and whether the frequency of execution of the TMSI reallocation procedure is related to the amount of activity of the MS (i.e. to how often the TMSI is exposed to overhearing).

Findings/results

The 3GPP standard relies on the use and frequent reallocation of TMSIs in order to provide user's untraceability. In particular, it mandates that the TMSI reallocation should be performed whenever the MS moves between "location areas" (identified by location area identifiers, LAIs). However, it is known that location areas often span over several square kilometres, and a subscriber's movements are typically confined within one or two location areas [65, 66]. So location areas may be too large to trigger TMSI reallocations in practice. We report on three different scenarios showing that some of the actual implementations of the strategy for changing pseudonyms to avoid tracking are not offering enough privacy guarantees to the mobile telephony subscribers. Our observation and their consequences on users' privacy are discussed in this section.[5]

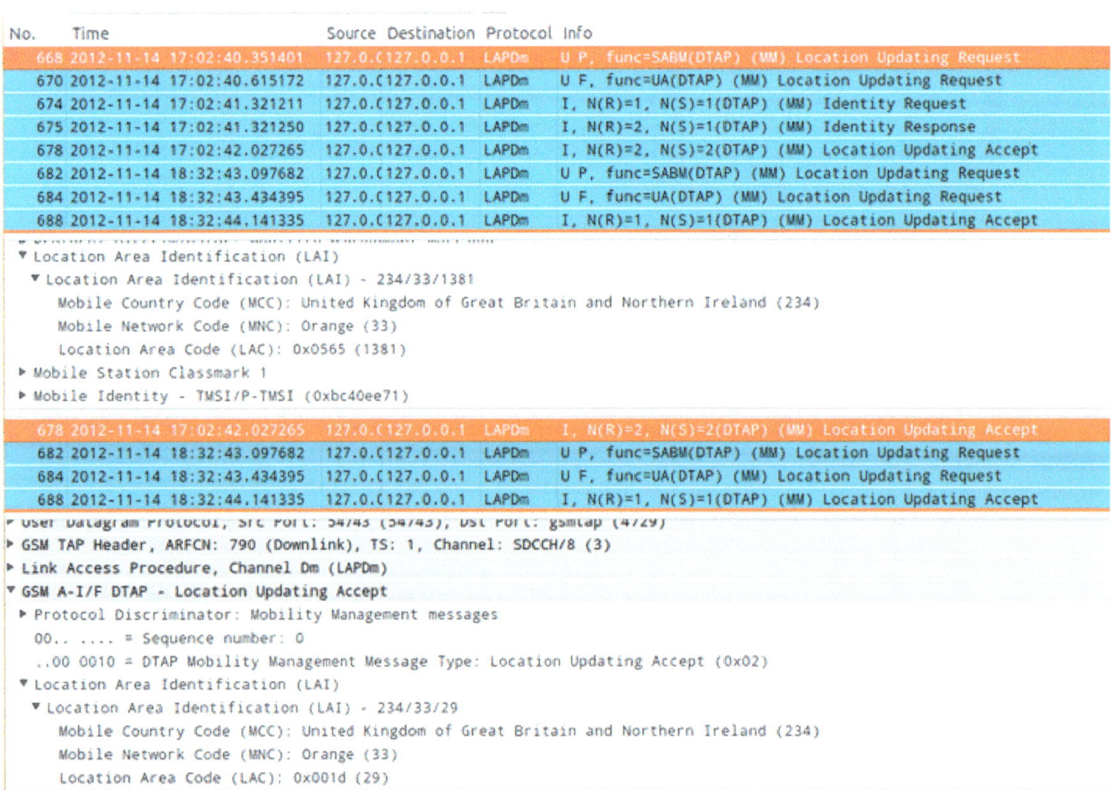

Figure 9: Trace of a UK Orange SIM card. The TMSI used at location 234/33/1381 (packet no. 668) is accepted at location 234/33/29 (packet no.678), while the 3GPP standard mandates a TMSI reallocation at each change of location

The TMSI reallocation procedure is rarely executed Although in the standard the privacy offered to mobile phone bearers is based on frequent updates of TMSIs, our experiments show that the same TMSI can be allocated for several hours and even days. Moreover, turning on and off the MS does not usually result in a new TMSI being allocated. As an example, Fig. 8 shows that a TMSI allocated on 22/03/2012 had not been updated by 25/03/2012, making the phone trackable for a period of 3 days. This behaviour can be observed for the major UK, Greek, French and Italian network operators. An attacker could take advantage of the long life of a TMSI and monitor a few sub-areas using short-range devices in order to obtain a fine-grained tracking of his victim within a same LAI.

We observed that the major UK network operators and the Vodafone and TIM Italian operators rarely execute the TMSI reallocation even in presence of MS activity, but the first message sent by a MS when requesting or receiving a service contains its TMSI and hence exposes it to eavesdropping third parties. As mentioned in Sect. 2.3, TMSI liveness makes it possible to locate mobile telephony users without alerting them. The attack consists in paging the victim and hence provoking a paging response. To reduce the set of answering TMSIs to the victim's one, the attacker must repeat the process several times because more than one MS could be sending a paging response at the same time and it is possible only if the TMSI is not reallocated even in case of activity exposing the TMSI (e.g. receiving calls). The attack in [51] thus relies on the low frequency of TMSI reallocations and demonstrates that changing pseudonyms, as mechanism to provide location privacy, is not effective without a policy for changing of pseudonyms which takes into account the actual exposure of the pseudonym caused by the mobile station activity.

A change of location area does not imply a change of TMSI Although such a change is mandated by the 3GPP standard. We observed this behaviour when capturing the signalling messages of a mobile station moving by coach between different cities in the UK, using the Orange and the O2 networks where we observed the same pseudonym being accepted in different location areas with no further execution of the TMSI reallocation procedure. Assuming an average speed of 70 km/h, we observed that a new TMSI was assigned after about 45 min (about 53 km) and a second one after about 60 min (about 70 km), while we observed a change of LAI every 5 min on average, and hence, a new TMSI should have been allocated, on average, about every 3 km. Figure 9 shows an example trace where a TMSI used at location 234/33/1381 (packet no. 668) is accepted at a different location 234/33/29 (packet no. 678).

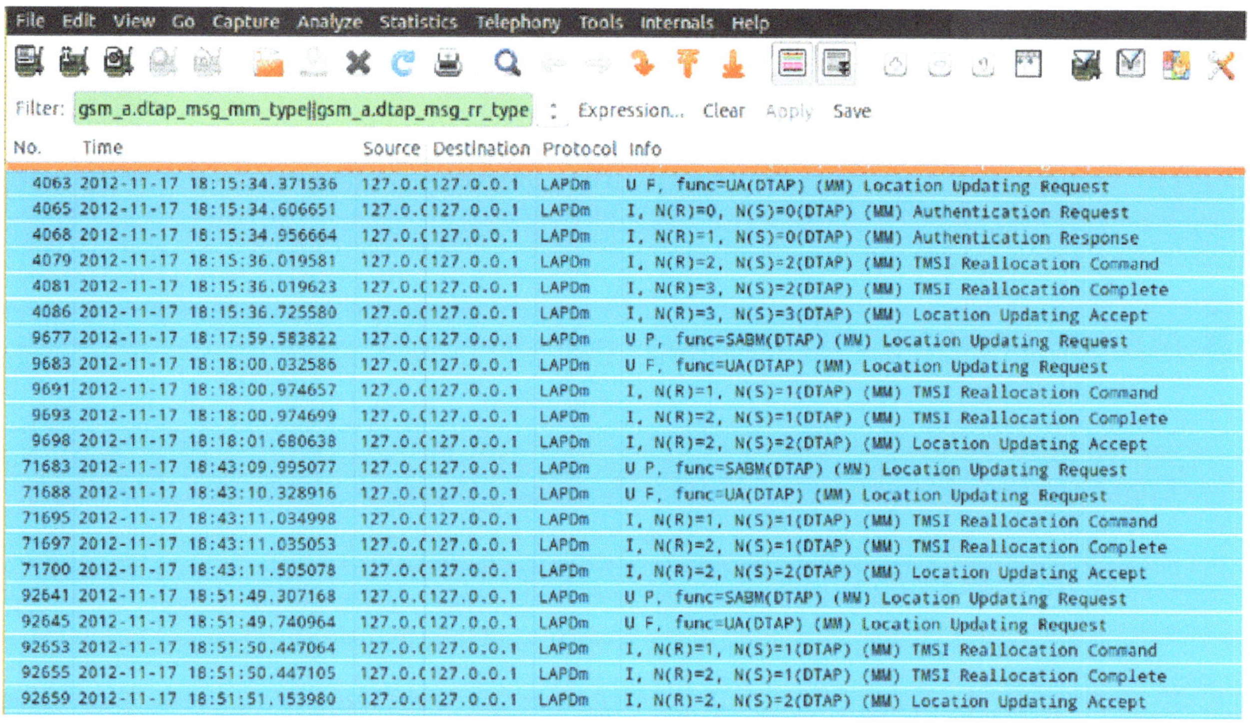

Figure 10: Trace of a UK Lebara SIM card attached to the Vodafone network while travelling on a train. The TMSI reallocation procedure is executed by reusing a previously established key. The MS first performs a location update (packet no. 4063), then the authentication procedure to establish a ciphering key (packets 4065, 4068), followed by the TMSI reallocation procedure (packets 4079, 4081). The following three TMSI reallocations (packets 9691, 9693, 71,695, 71,697, 92,653, 92,655) are executed without first performing the authentication procedure and hence reusing the previously established ciphering key

The fact that a TMSI was accepted in two neighbouring LAIs contradicts the specification that a TMSI reallocation should be performed at least at each change of location. However, changing pseudonym when changing location area would provide location-dependent privacy to the user since it would prevent passive tracking across different LAIs. The combination of the two behaviours reported so far (i.e. keeping the same TMSI for a long period of time and not changing it when changing location area) enables the attacker to both track his victim within an area and follow him across different areas without doing any extra effort other than passively sniffing.

Previously established keys are restored and used to encrypt the TMSI reallocation procedure Our captures confirm that the reuse of previously established keys is a policy adopted by real networks. In particular, we show that this policy is adopted for the execution of the TMSI reallocation procedure. The experiments we performed show that major UK and Italian network operators[6] reuse previously established keys instead of performing the authentication procedure before each execution of the TMSI reallocation procedure. Figure 10shows a trace from a UK Lebara SIM card attached to the Vodafone network performing a location update (packet no. 4063). Then the execution of the authentication procedure establishes a new ciphering key (packets 4065, 4068), and consecutively, the TMSI reallocation procedure (packets 4079, 4081) is executed. The subsequent TMSI reallocations (packets 9691, 9693, 71695, 71697, 92653, 92655) are executed without previously performing the authentication procedure and hence reusing the previously established ciphering key.

The reuse of a previously established ciphering key enables a replay attack. We describe the TMSI reallocation replay attack in Sect. 3.9.

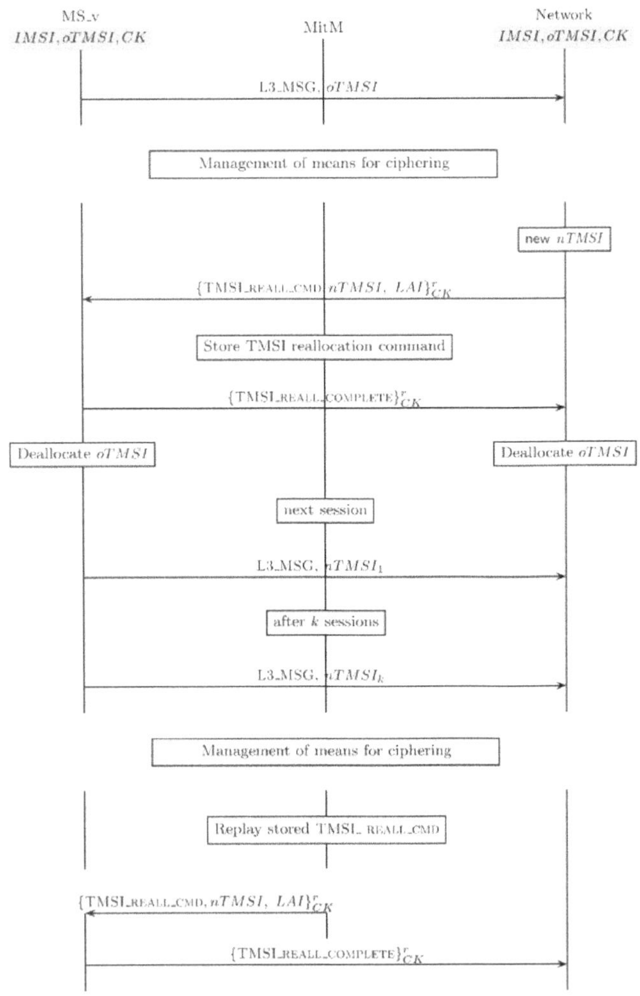

Figure 11: TMSI reallocation procedure attack

TMSI reallocation replay attack

So far in Sect. 3, we analysed the TMSI reallocation procedure from an *experimental* point of view. In this section, we demonstrate a replay attack on this procedure which allows a third party to violate a user's privacy in spite of the reallocation protocol. This attack is enabled by the 3GPP standard policy allowing to restore previously established keys. The experiments reported in Sect. 3 confirm that this policy is commonly adopted across real mobile telephony operators. We now show that a linkability attack is enabled by the reuse of session keys making it possible to link old and new TMSIs on real networks. In particular, a replay attack such as the one depicted in Fig. 11 could be mounted.

An attacker, controlling a radio device able to sniff and inject messages over-the-air, first captures a TMSI reallocation command (the second message in Fig. 11). Later on, when the MS has possibly already changed its pseudonym but not yet established new keys, the attacker can replay the captured TMSI reallocation command (one message before last in Fig. 11). Since reuse of the session key *CK* is allowed, the victim's MS successfully decrypts the reallocation message and sends the TMSI reallocation complete message. This allows the attacker to distinguish the victim's MS from any other that would not successfully decrypt the message and thus would not send any reply, even though in the meantime a different TMSI ($nTMSI_k$ in Fig. 11) was assigned to the victim's MS.

Note that although the TMSI reallocation suffers from a replay attack, from a theoretical point of view, we have not investigated the feasibility of the attack in practice. In particular, it could be difficult to capture a legitimate TMSI reallocation command message to be replayed during the attack, since this message is encrypted. Moreover, the actual replay of the message is not trivial because the correct frame number should be in use by the MS when decoding it, in order to obtain the correct decryption. Further work would be needed to assess the practicality of this attack on both 2G and 3G networks.

Discussion

The experiments we conducted show how the adoption of pseudonyms is not a sufficient condition to ensure the privacy of mobile telephony users and that real network implementations leave plenty of room to tracking attacks. We suggest network operators should adopt activity- related policies in order to prevent active tracking attacks. In general, the execution of the TMSI reallocation procedure should be more frequent even when the MS is in idle state, so to prevent mere passive tracking. Moreover, we suggest that a MS should be able to trigger the TMSI reallocation procedure in order to tackle active attacks exposing the TMSI without involving the network as for example directly injecting paging messages on the paging channel.

In particular, mobile network operators should enforce the policy of changing pseudonyms at each change of location area. This would reduce the probability of tracking across different LAIs. Furthermore, considering that Foo Kune et al. [51] are able to determine the presence of a target MS in an area by triggering the paging procedure on average 8 times, we suggest that this threshold should be used as maximum number of TMSI exposing messages a MS can send in clear over-the-air before a reallocation procedure is initiated by the network. This would lower the probability of successful tracking attacks. Moreover, as a countermeasure to simple passive tracking, a reallocation procedure should periodically be triggered even in case the TMSI exposing activity has been under the suggested threshold. This would avoid the possibility of determining for example that a MS has not moved in a given time frame. We suggest to keep the time threshold fairly low. A sensible time threshold could be devised by estimating the time window for 8 TMSI exposing activity to occur when the MS is subject to average or low usage load. Using pseudonyms is a good mechanism to ensure the user's privacy, provided that there is enough possibility of mixing within the network, which is usually the case in mobile telecommunication networks. However, the efficiency of the pseudonym change strategy depends on many factors which the 3GPP standard leaves as implementation choices.

We showed that the implementation choices made by real network operators do not provide a satisfactory level of privacy and leave space for different kinds of tracking attacks. Moreover, we showed that the loose standard specification can produce implementations of the TMSI reallocation procedure which are subject to a linkability attack.

The TMSI reallocation procedure is adopted in 3G+ networks as well. However, our experiments were conducted on the GSM network, and further work and specialized equipment would be needed so to investigate the usage scenario of TMSIs and TMSI reallocation in real 3G++ networks.

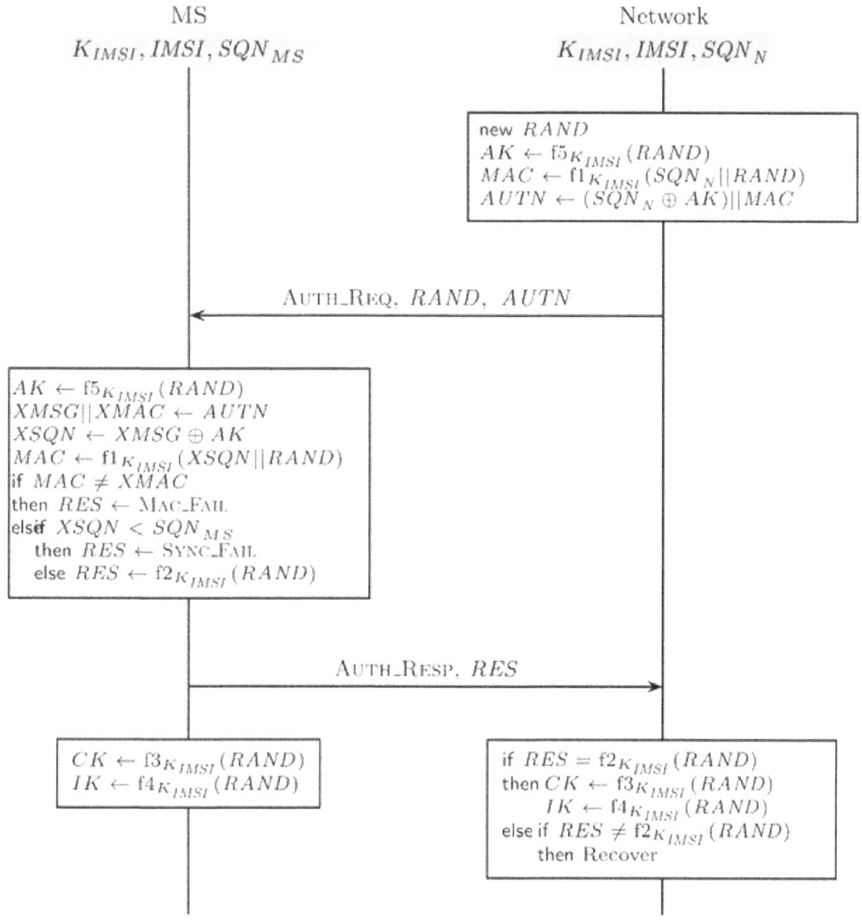

Figure 12: 3G authentication and key agreement

ANALYSIS OF MOBILE SYSTEMS' PRIVACY: PURE 3G LINKABILITY ATTACK

In the previous section, we showed how the paging procedure can be exploited to perform a simple linkability attack and we presented an experimental analysis of the identity management thanks to which we exposed the presence, in real mobile networks, of critical behaviour from a privacy point of view. Most of these privacy critical scenarios lead straightforwardly to breaches of the mobile telephony users' privacy. In this section, we consider a "Dolev Yao" attacker [28] who has full access only to the radio link, where he can sniff, inject, replay and modify messages, but cannot break the cryptography involved in the protocol, i.e. cannot encrypt/decrypt without knowing the required encryption/decryption keys. In this section, we show a replay attack that allows to track the presence of a user in a monitored area or across a set of monitored areas. This attack concerns the 3G authentication and key agreement protocol.

3G authentication and key agreement protocol

The Authentication and Key Agreement (AKA) protocol achieves mutual authentication between a MS and the network and establishes shared keys to be used to secure the subsequent communications.

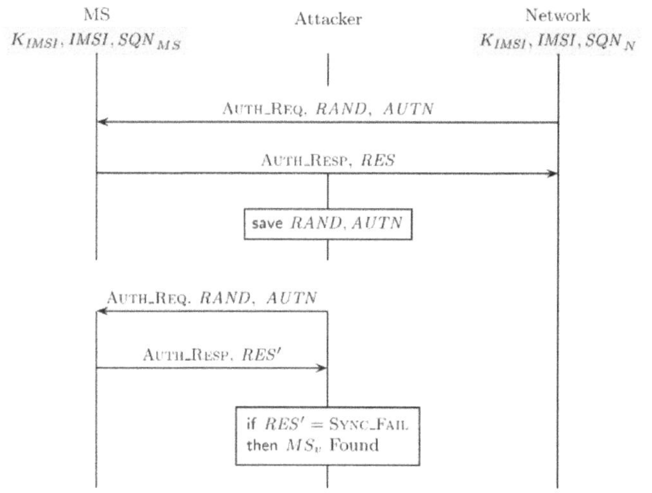

Figure 13: AKA protocol linkability attack

The keys are not exchanged during the protocol but computed locally by the MS and the SN. According to [41], the authentication procedure is always initiated by the SN for the purpose of:

- Checking whether the identity provided by the MS is acceptable or not.
- Providing parameters enabling the MS to calculate a new UMTS ciphering key.
- Providing parameters enabling the MS to calculate a new UMTS integrity key.
- Allowing the MS to authenticate the network.

Each MS with identity IMSI shares with the network a different secret long-term key, K_{IMSI}assigned to the subscriber by the mobile operator and stored in the USIM. The secret key allows the MS and the network to compute shared ciphering and integrity session keys to be used for encryption and integrity check of communications.

A successful authentication procedure establishes a so-called *security context* which identifies the set of keys to be used for secrecy and integrity purposes; a Key Set Identifier (KSI) is used to retrieve the established security context through different sessions. Once a security context is established, it is considered valid until the authentication procedure is next executed or the deletion of KSI is requested by the network.

The 3G AKA protocol [41], depicted in Fig. 12, consists in the exchange of two messages: the authentication request and the authentication response. Before sending an authentication request to the MS, the network computes the authentication data: a fresh random challenge RAND, the authentication token AUTN, the expected authentication response $f2_{KIMSI}$(RAND), the integrity key *IK* and the encryption key *CK* (see Fig. 12). The functions f1, f2, f3, f4 and f5, used to compute the authentication parameters, are keyed cryptographic functions computed using the shared key K_{IMSI}, see [68] for more details. The authentication function f1 is used to calculate the message authentication code MAC; f2 is used to produce the authentication response parameter RES; the key generation functions, f3, f4 and f5, are used to generate the ciphering key *CK*, the integrity key *IK* and the anonymity key*AK*, respectively.

The network always initiates the protocol by sending the authentication challenge RAND and the authentication token AUTN to the mobile station. AUTN contains a MAC of the concatenation of the random number with a sequence number SQN_N generated by the network using an individual counter for each subscriber. A new sequence number is generated either by increment of the counter or through time-based algorithms as defined in [41]. The sequence number SQN_N allows the mobile station to verify the freshness of the authentication request to defend against replay attacks (see Fig. 12).

The MS receives the authentication request, retrieves the sequence number SQNN and then verifies the MAC (condition MAC = XMAC in Fig. 12). This step ensures that the MAC was generated by the network using the shared key K_{IMSI} and thus that the authentication request was intended for the mobile station with identity IMSI. The mobile station stores the greatest sequence number used for authentication, so far SQN_{MS}. This value is used to check the freshness of the authentication request (condition XSQN < SQN M S in Fig. 12) to avoid replay attacks

The mobile station computes the ciphering key CK, the integrity key IK and the authentication response RES and sends this response to the network. The network authenticates the MS by verifying whether the received response is equal to the expected one (RES = $f2_K$ (RAND)). The authentication procedure can fail on the MS side either because the MAC verification failed, or because the received sequence number XSQN is not in the correct range with respect to the sequence number SQN_{MS} stored in the mobile station. In the former case, the mobile station sends an authentication failure message indicating MAC failure (Mac_Fail) as the failure cause. In the latter case, the authentication failure message indicates synchronization failure (Sync_Fail) as the failure cause. When a MAC failure occurs, the network may initiate the identification procedure. When a synchronization failure occurs, the network performs re-synchronization

After successful authentication, the SN sends a security mode command message to the MS, indicating which one of the allowed algorithms to use for ciphering and integrity checking of the following communications.

3G AKA protocol linkability attack

To detect the presence of a victim mobile station MS_v, in one of his monitored areas, an active attacker just needs to have previously intercepted one legitimate authentication request message containing the pair (RAND, AUTN) sent by the network to MS_v. The captured authentication request can now be replayed by the adversary each time he wants to check the presence of MS_v in a particular area. In fact, thanks to the error messages, the adversary can distinguish any mobile station from the one the authentication request was originally sent to. On reception of the replayed authentication challenge and authentication token (RAND, AUTN), the victim mobile station MS_v successfully verifies the MAC and sends a synchronization failure message. However, the MAC verification fails when executed by any other mobile station, and as a result, a MAC failure message is sent. The implementation of few false BSs would then allow an attacker to trace the movements of a victim mobile station, resulting in a breach of the subscriber's untraceability. The proposed attack is shown in Fig. 13. Note that this attack affects only 3G mobile systems; in fact 2G systems adopt a different authentication protocol which does not provide mutual authentication (i.e. the mobile station does not authenticate the network) and does not involve error and recovery procedures in case the authentication of the network fails. Thus, the distinguishing attack on the error messages cannot be performed in 2G networks.

IMPLEMENTATION OF SOME 3G PROTOCOLS ATTACKS

In order to test the attacks presented in Sects. 3.4 and 4.2 in a deployed telecommunication network, we used a commercially available femtocell. Although the particular femtocell hardware is tied to the network operator SFR, the proposed attacks are not. Indeed, we tested the attacks using mobile phones registered to different operators, hence just using SFR as serving network. The authentication token *AUTN* is still provided by the victim's Home network. So by testing our attacks on T-Mobile, O2, SFR and Vodafone victim MSs, we establish that all these tested networks are vulnerable to the attacks described above. However, we want to stress here that our implementation has the only purpose of showing the feasibility of our attacks and confirm that real cellular networks follow the 3GPP standard specifications and thus are vulnerable to the proposed attacks. The same attacks could be mounted by appropriately programming a USRP [69], which is a hardware device able to emit and receive radio signals. In this case, one could obtain wider range attack devices in order to monitor larger areas.

Femtocell architecture

A femtocell is a device that acts as a small base station to enhance 3G coverage and connectivity, especially inside buildings with otherwise bad coverage. Its coverage radius ranges from 10 to 50 m. It connects mobile phones to the network of the corresponding MNO (mobile network operator) using an existing wired Internet connection provided by the femtocell user, not the operator. 3G femtocells, also called Home Node B (HNB), support most of the functionalities provided by a typical 3G base station (Node B), e.g. physical layer (radio signalling) functions. In addition, the HNB establishes an authenticated secure tunnel over the Internet with the network of the operator. Using this encrypted connection, the femtocell forwards all radio signalling and user-generated traffic to the GANC (GAN Controller), which is connected to the core network of the operator (refer to [70] for more details of the femtocell architecture).

The communication between the femtocell and the GANC is based on the Generic Access Network (GAN) protocol. The GAN protocol was originally designed to allow mobile communication over Wi-Fi access points. The protocol was standardized by MNOs in 2004 [71] and led to the GAN specification [72, 73] in 2005. This specification has been adopted and extended to be used in femtocell environments [74]. The femtocell uses this protocol to forward communication from a mobile station via the GANC to the network or vice versa. The MS does not need any special GAN support; it just connects to the femtocell in the same way as it connects to a standard base station. The femtocell maps all Layer-3 radio signalling to TCP/IP-based GAN messages and passes them to the GANC. Thus, it transparently encapsulates all traffic generated by the phone and the network.

Attack procedure

For the purpose of implementing our attacks (Sects. 3 and 4.2), we use a compromised femtocell like the one described in [55]. More specifically, we reproduce the hacking performed in [55] to gain root access of our femtocell and redirect the traffic to a Man-in-the-Middle (MitM) GAN proxy, positioned between the femtocell and the GANC. We use this MitM GAN proxy as entry point for message injection. In particular, using the MitM GAN proxy, we can inject messages into the connection between the MNO and the femtocell. The femtocell forwards these messages to the mobile phone, making them appear as legitimately delivered by the MNO. To perform the attacks, we intercept, modify and inject 3G Layer-3 messages into the communication from the base station to the mobile phone in both directions, GANC-to-femtocell and femtocell-to-GANC. We redirect all the traffic between the femtocell and the GANC to our GAN proxy.

The GAN traffic is cleartext travelling over an IP Sec tunnel for which we own the key material, thanks to the initial rooting/hacking of the femtocell. Additionally, we developed a set of applications which allow us to intercept, manipulate or insert selected messages and distinguish different types of GAN messages. This allows us, for example, to cache subscribers information used to perform the attacks. In particular, we store the random challenge RAND, the authentication token AUTN, the TMSI and the IMSI of our victim MS. This information is directly extracted from the traffic that is passed through the MitM GAN proxy.

IMSI Paging Procedure Attack To perform the IMSI paging attack, our software crafts a paging message encoding the necessary paging headers and parameters and a mobile station identity, i.e. one of the previously stored victim IMSIs. The crafted paging request is then sent by the GAN proxy to the femtocell. When the victim mobile phone receives the IMSI paging request, it readily answers with a paging response containing the victim's TMSI. Thus, by injecting a paging request, we can check whether a phone belonging to a designated victim is in the area covered by our device. In case of success, the phone generates the paging response, while a failed attempt generates no message. In general, it is possible that more than one phone replies to a paging request during the same time slot. However, one can repeat this procedure multiple times and correlate the timing and TMSI usage from the multiple replies as in [51].

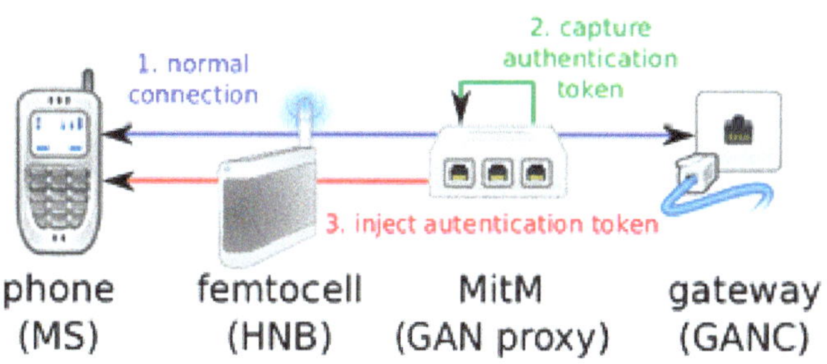

Figure 14: Experimental attack setup

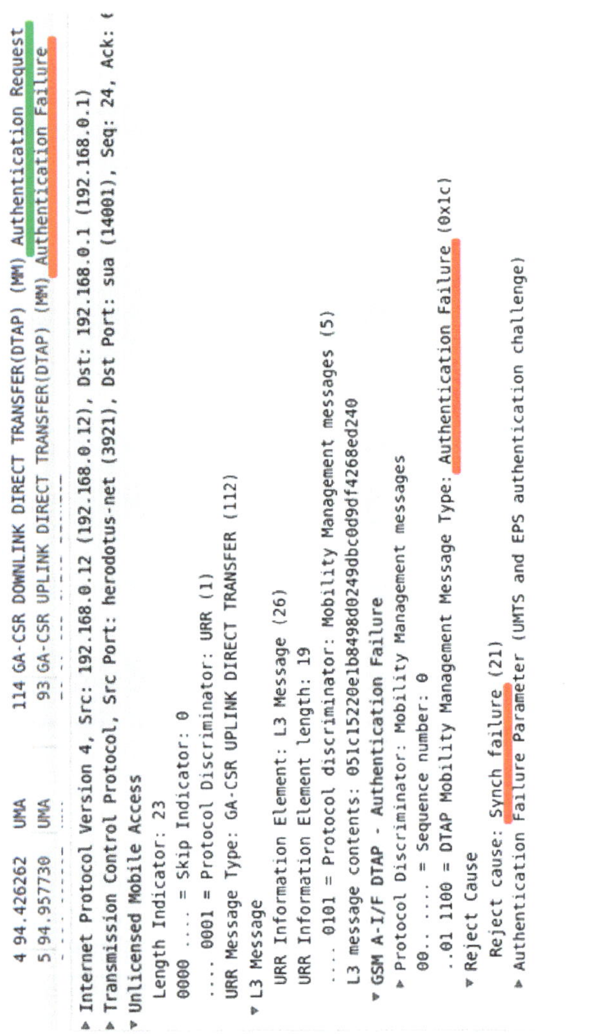

Figure 15: Linkability attack: victim found

AKA Protocol attack To perform the AKA attack, we replay a given authentication message for a specific target for which the GAN proxy cached the legitimate authentication data, i.e. RAND,AUTN. These data are sent unencrypted on the radio link and could be captured with any equipment capable of sniffing the radio link. As soon as a dedicated channel is allocated to the MS, e.g. after being paged or when initiating a phone call, our software crafts an authentication request AUTH_REQ using the previously cached RANDand AUTN, i.e. replays a previous request. This request is encapsulated into a GAN message and sent to the femtocell. The femtocell takes care of delivering the authentication request message on the dedicated channel assigned to the MS, as illustrated in Fig. 14. The phone performs a validation of the authentication request and answers with the authentication response. If the response to the replayed authentication is a Synchronization Failure (Fig. 15), then the MS on this dedicated channel is the victim's phone, and the victim is indeed in the femtocell area. Otherwise, if the response to the replayed authentication is a MAC failure (Fig. 16), the attacker needs to inject the same message to the other mobile stations in his area in order to find out whether the victim MS is present or not.

The 3G AKA protocol is performed at each new session in the femtocell setting. This makes the caching of the authentication parameters very easy. Though we do not have the tools to test whether this applies when connecting to a typical Node B, we tested the 3G/GSM interoperability scenario by using the Osmocom-BB software and we observed that in this setting the execution of the AKA protocol can be triggered by calling for example the victim mobile phone a given number of times (by hanging up within a short time window, this activity can be made non-detectable by the victim [51]). For instance, our experiments showed that the execution of the AKA protocol on the UK Vodafone network can be triggered by calling six times the victim mobile phone and hanging up before it even rings.

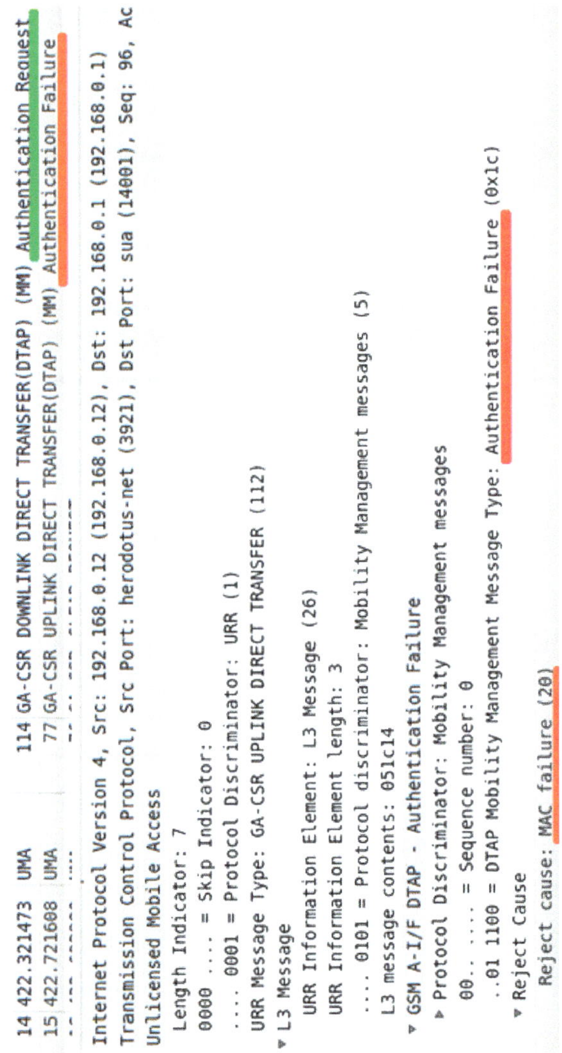

Figure 16: Linkability attack: victim not found

To illustrate the use of our attacks, consider an employer interested in tracking one of his employee's accesses to a building. He would first use the femtocell to sniff a valid authentication request. This could happen in a different area than the monitored one. Then the employer would position the device near the entrance of the building. Movements inside the building could be tracked as well by placing additional devices to cover different areas of the building. Similarly, these attacks could be used to collect large amount of data on users' movements in defined areas for profiling purposes, as an example of how mobile systems have already been exploited in this direction is available in [75]. If devices with wider area coverage than a femtocell are used, the adversary should use triangulation to obtain finer position data.

PRIVACY-FRIENDLY FIXES

Despite the fact that mobile telephony systems adopt temporary identities to avoid linkability and to ensure anonymity of mobile telephony subscribers, active attackers can exploit the identification and the paging procedure to break both anonymity and unlinkability. Moreover, the TMSI reallocation procedure and the AKA protocol provide a way to trace mobile telephony users without the need to identify them in any way. These attacks on user privacy are not only theoretical but can be implemented in practice, as described in Sects. 3 and 5. Hence, the analysed procedures are a real threat for users' privacy, and countermeasures should be taken to provide an effectively privacy-friendly mobile telephony system. In this section, we propose some privacy-friendly solutions tackling the issues exposed in Sects. 3, and 4. In Sect. 7, we show how to model and verify anonymity and unlinkability of the proposed fixed procedures using an automatic protocol verifier called ProVerif . We propose a set of countermeasures involving symmetric and public key-based cryptography. The public key infrastructure we propose is easy to deploy because we only require one public/private key pair per mobile network operator and none for the mobile stations. More generally, the solutions we present require only small changes to the current security architecture and to the cryptographic functions currently used in 3G. Hence, we believe our solutions may be implemented in a cost-effective way and thus could realistically be adopted by the telecommunication operators in future generation mobile telephony systems. However, our solutions do not aim to be full protocol specifications. Their role is to show that realistic privacy-friendly solutions are possible without dramatic changes to the current infrastructure. In this section, we propose solutions to fix the identification procedure, the IMSI paging procedure, the TMSI reallocation procedure and the AKA protocol. Indeed, the problem of privacy is a multilayer/multiprotocol problem [19] which requires all protocols at all layers to satisfy the desired properties.

Public key infrastructure

We propose the adoption of a public key infrastructure (PKI) providing each MNO with a private/public key pair. The public key of a user's network provider can be stored in the USIM. This public key makes it possible for a mobile station to encrypt privacy-related information such as the IMSI and deliver them to the network in a confidential manner. We do not require a public/private key pair to be assigned to the mobile stations. The adoption of such a PKI can also solve the problem exposed by Zhang and Fang in [57] concerning the lack of serving network authentication in the current 3G infrastructure.

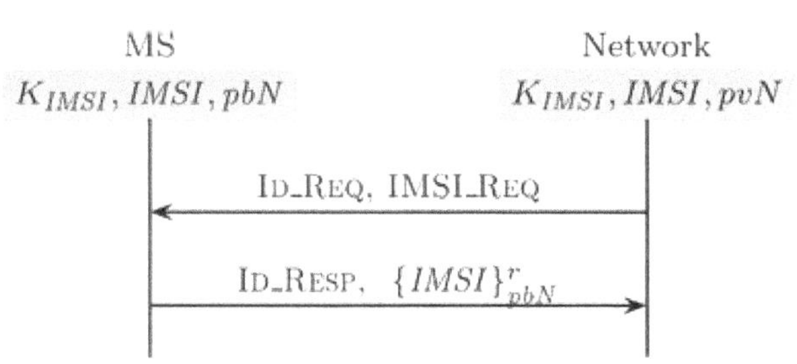

Figure 17: Identification procedure fix. The identity response is encrypted with the public key of the network. The r denotes randomized encryption

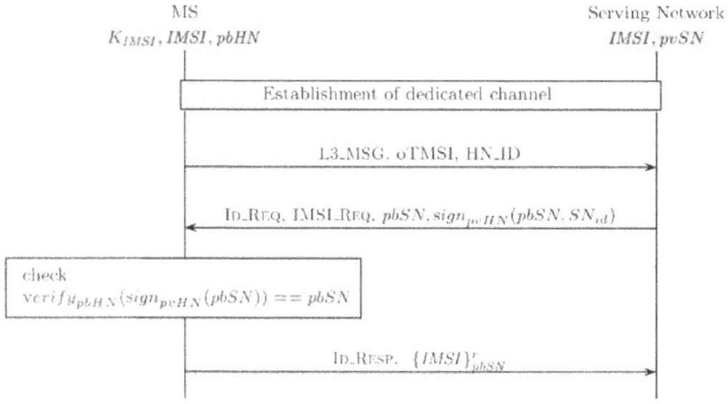

Figure 18: Identification procedure fix when roaming. The identity response is encrypted with the public key of the serving network. The r denotes randomized encryption

Fixes of the analysed procedures

Protecting the Identification Procedure The identification procedure exposes the IMSI of a MS (the IMSI is sent in cleartext upon request by the network). Hence, it breaches both anonymity and unlinkability. According to the standard, the use of the identification procedure should be limited as much as possible, to avoid a passive attacker overhearing the IMSI. However, as discussed in Sect. 2, the cost of devices allowing active attacks is constantly decreasing. An active attacker can simply trigger the identification procedure to retrieve the identity of all users in a monitored area [50]. We propose a solution to tackle this well-known attack (IMSI catcher attack). The fixed version of the identification procedure (Fig. 17) involves two messages: the first is sent by the network to ask for the IMSI, and the second, the identity response, is the randomized encryption of the IMSI of the mobile station using the public counterpart (pbN) of the private key of the network operator (pvN). The identification procedure always happens on a dedicated channel and after the MS sent a first Layer 3 message containing its temporary identity TMSI. We require an MS attached to a network different from its HN to include the HN identity in the first L3 message sent when attaching to the SN. The SN can then send its public key and certificate signed by the mobile station's HN along with the identity request message.

For example, an MS that has just been switched on would attach, by asking for a dedicated channel, to the (allowed) network offering the best signal and would then send an (L3 message) IMSI attach message (which is actually a type of location update message) containing its current TMSI. In our solution, this message would carry the HN identity. Assuming that the SN does not recognize the TMSI, then it would send an identity request message carrying its public key pbSN and certificate signed by the MS's home network sign(pvHN,(pbSN,SNid)) The MS would then reply sending the identity response message containing the randomized encryption of the IMSI of the mobile station using the public counterpart (pbSN) of the private key of the network operator (pvSN). This solution is illustrated in Fig. 18

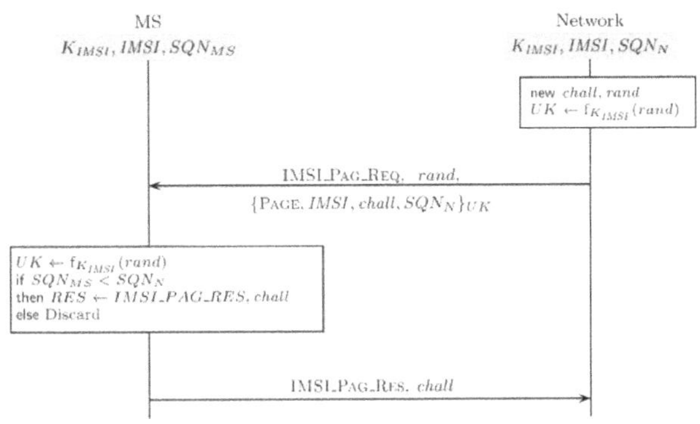

Figure 19: Paging procedure fix. The paging request is encrypted with the unlinkability key *UK*

Protecting the IMSI paging procedure To protect the paging procedure, we propose to encrypt the paging request using a shared session key *UK*, which we call unlinkability key. This key is generated by applying a new one-way keyed function f to the long-term key K_{IMSI} and a random number rand contained in the paging request. This key should be used for privacy-preserving purposes only. Furthermore, we require the encrypted request message to include a random challenge challchall and a sequence number *SQN*. The network stores the random challenge and checks it against the one sent by the MS in the paging response (Fig. 19). The aim of the *SQN* is to ensure freshness of the paging request and avoid replay attacks. The *SQN* should be handled in the same way as in the AKA protocol. A MS receiving a legitimate IMSI paging request should discard it if the *SQN* is not in the correct range. The use of this procedure should still be kept minimal (preferring the paging with TMSI whenever possible) to avoid burdening the signalling communication with cryptographic operations. In fact, each MS has to decrypt and check all the received encrypted IMSI paging to determine whether it is the recipient (note that TMSI paging is still sent in cleartext). To enable the IMSI paging by a serving network, the encrypted paging request may be precomputed by the home network and IMSI paging vectors could be sent in bulk to the serving network the MS is attached to. An IMSI paging vector (chall,rand, $\{PAGE,IMSI,chall,SQN_N\}_{UK)}$ would contain the challenge, the random and the encrypted paging; in this way, the unlinkability key *UK* and the sequence number SQN_N only have to be shared between the MS and the home network.

As an alternative solution, one could slightly relax the privacy requirements and allow a passive attacker to discover that the victim recently visited a certain location area. In this case, the IMSI paging request could be sent in clear over-the-air as it already is in the current systems. However, the paging response would be encrypted using the network's public key. This would eliminate the burden for the MS of decrypting all the received IMSI paging requests.

Fixing the TMSI reallocation procedure The solution we propose to fix the TMSI reallocation procedure does not require any change in the security architecture of mobile telephony systems. We only require the standard to specify that the reuse of the encryption key, *CK*, is not permitted when the key is used to execute the TMSI reallocation procedure, i.e. the establishment of a new symmetric encryption key before the execution of the TMSI reallocation procedure should be mandatory. This would avoid the possibility of replay attacks to be mounted. However, frequent executions of the authentication procedure could burden the radio communication and slow down the delivery of mobile telephony services. Alternative solutions are possible, as for example the introduction of a sequence number in the TMSI reallocation command, similarly to the one used to avoid replay attacks against the authentication and key agreement protocol [41]. We illustrate this solution in Fig. 20. The network sends a sequence number $SQNS_N$ along with the TMSI reallocation command. The MS checks whether the received sequence number is in the expected range ($SQN_{MS} \leq SQN_{SN}$). If so, it carries on with the reallocation of the TMSI. Otherwise, the MS aborts the TMSI reallocation execution, hence avoiding replay attacks.

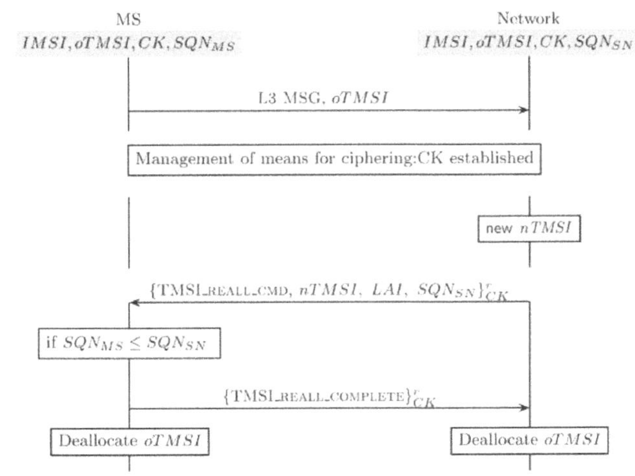

Figure 20: TMSI reallocation procedure fix: this fix uses the SQN to ensure the freshness of the reallocation command

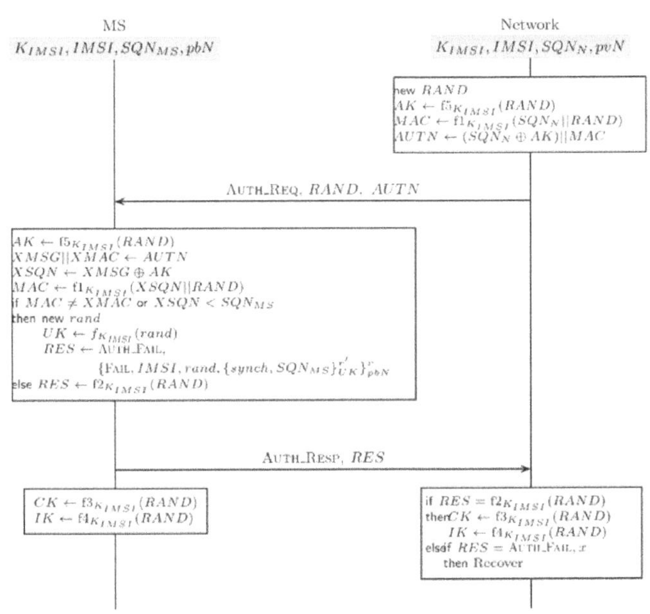

Figure 21: The fixed AKA protocol. The error messages are encrypted using the network public key

Fixing the AKA Protocol The AKA protocol is a threat for the unlinkability of 3G subscribers because the error messages sent in case of authentication failure leak information about the identity of the subscriber. To avoid this information leakage, the error messages sent in case of any type of failure should look indistinguishable from an attacker's point of view.

Moreover, the 3G standard stipulates [41] different procedures to recover from each of the two kinds of failure, but this is a source of additional information flow that can be used to launch our privacy attack. In the solution we propose, we solve this problem since error recovery can be performed within the network without the need to trigger further procedures over-the-air. Indeed, all the parameters needed for error recovery are sent in the error message allowing the recovery procedure to be carried within the network.

The fixed version of the AKA protocol (Fig. 21) carries on as specified by the standard. The network sends RAND,AUTN and waits for a response. The response is $RES=f_2K_{IMSI}(RAND)$, as in the standard, in case the checks of MAC and sequence number are successful. If either of these checks fails, an error message is sent to the network. The failure message is now encrypted with the public key of the network pbNpbN and contains a constant FAIL, the IMSI and the current sequence number SQN_{MS} of the MS. The IMSI sent encrypted in the error message allows the network to check the identity of the MS without triggering the identification procedure. The current sequence number of the mobile station enables the network to perform resynchronization with the Authentication Centre (AuC, the server storing subscribers authentication data) of the operator of the mobile station, if needed. SQN_{MS} is sent encrypted with the unlinkability key (as defined in the fixed paging procedure) in order to authenticate the error message to the network as coming from the MS with permanent identity IMSI. The network can deduce the cause of the failure from the IMSI and SQN_{MS} contained in the error message. Upon receipt of this authentication failure message, the action performed for error recovery purposes should be the same regardless of the type of failure occurred. Indeed any difference in behaviour would be a source of additional information flows. Note that, as suggested in [60], simpler solutions such as not sending any error message or sending a constant error message are also possible. However, these solutions do not allow the network to perform neither resynchronization nor any other sort of error recovery procedure since no information on the cause of the failure is given to it.

Discussion of the proposed fixes

While the fix we propose for the TMSI reallocation procedure and for the identification procedure is intuitive and straightforward, this is not the case for the other two procedures. In particular, we take care of maintaining the style of mobile telecommunication protocols and at the same time ensuring privacy. We introduce the unlinkability key, a new session key generated for privacy purposes, instead of using the long-term key K_{IMSI} (as in the 3G AKA) and make use of the sequence number SQN for freshness purposes (this is needed to avoid user linkability caused by replay attacks). We maintain the authentication flow of the AKA and modify only the way error messages are dealt with by including error recovery information inside the error message (this avoids the triggering by the network of specific procedures in order to perform error recovery depending on the occurred error).

Our proposed fixes use public key cryptography; intuitively, there is no way to avoid that, since if a mobile station's TMSI is unknown to the serving network (hence the need to perform the identification procedure), then there is no shared key by which they can communicate privately. The additional costs associated with deploying and using public key cryptography are in fact small for the two following reasons.

Firstly, only mobile telephony *operators* are required to have a public/private key pair. Neither subscribers nor mobile phone equipments or USIMs need to have their own public/private key pair. The operator's public key could be stored in the USIM of the mobile station, as it is already the case for the IMSI and the long-term key K_{IMSI}. The home network can act as a certifying authority for the public key of the different serving networks (see below). Thus, the public key infrastructure is similar to that used on the web, where corporations (not users) have certified keys.

Secondly, the computationally expensive public key encryption and decryption are required only for the identification protocol and when the AKA protocol fails. The execution of the identification and the IMSI paging procedures should anyway be kept minimal according to the currently deployed standard. Moreover, failures during the execution of the AKA protocol rarely occur according to our experiments. Hence, the computational overhead of the public key cryptography is not significant. Moreover, it is possible to delegate the encryption and decryption to the mobile equipment, instead of executing them on the USIM. This would not weaken the security properties of the 3G procedure, since the mobile equipment in the current architecture has already access to the IMSI, while the network public key is publicly available information.

For roaming purposes, each home network (HN) can act as certifying authority of the serving network (SN) for its own subscribers. The public key *pbHN* of the HN could be stored in the USIM. At registration time with a SN, the MS would declare its HN, and the SN would provide the MS with its public key *pbSN*, together with a certificate from the mobile station's HN (signskHN(pbSN) Hence, a mobile station would only need to obtain a certified version of the SN's public key and verify it using its own network provider public key. This would provide, in a efficient way, the MS with the necessary public keys to execute our fixed versions of the protocols.

The introduction of cryptographic operations on the mobile equipment side could be a source of denial-of-service (DoS) attacks aiming to consume the battery load of victim MSs. To mitigate the effect of such attacks, the mobile phone's software could rate limit the phone's willingness to respond to authentication, IMSI paging and identity request messages, so to guarantee a minimum battery life time even in case of attempted DoS attacks. We have calculated that responding to such requests on average once per minute would consume an additional one-tenth of battery life.

AUTOMATIC VERIFICATION OF THE FIXED PROTOCOLS

Many deployed protocols have subsequently been found to be flawed [6, 76–78]. In this perspective and in order to increase the confidence one can have in the solutions proposed in Sect. 6, we show how to formally analyse our proposed fixes *w.r.t.* privacy. We use the formalization of privacy-related properties as given by Arapinis et al. in [12], namely strong unlinkability and strong anonymity, and adapt them to obtain definitions suitable for automatic verification using the ProVerif tool [79].

ProVerif Encoding

We use the automatic verification tool ProVerif [79] which takes in input processes written in a syntax similar to the applied pi-calculus [31] one. In the rest of this section, we use the ProVerif syntax (see Fig. 22) to describe how we modelled the fixed protocols. However, we omit the full formal details for which we refer the reader to [37, 80].

Cryptographic primitives are modelled as functions and messages are represented by *terms* built over an infinite set of names a,b,c, an infinite set of variables x,y,z,... and a finite set of function symbols $f_1,\dots,f_n$. The effect of applying function symbols to terms is described by a set of reduction rules.

Example 1: Using functions and reduction rules, we can define cryptographic functions.

```
fun senc/3, pub/1, aenc/3,
    f/2, f1/2, f2/2, f3/2, f4/2, f5/2.
reduc sdec(k, senc(k, m, r))= m.
reduc adec(k, aenc(\texttt{pub}(k), m, r))= m.
```

The functions senc and model randomized symmetric and asymmetric encryption, respectively. In particular, the reduction rules allow to retrieve the plaintext m from its encryption given the knowledge of the decryption key k. The functions f/2, f1/2, f2/2, f3/2, f4/2, f5/2 model the one- way functions used in mobile telephony protocols.

```
P, Q, R ::=                     plain processes
0                               null process
P | Q                           parallel composition
!P                              replication
new n; P                        name restriction
if M = N then P else Q          conditional
let M = D in P else Q           destructor application
in(M, x); P                     message input
out(M, N); P                    message output
```

Figure 22: ProVerif syntax

We briefly describe the informal semantics of the calculus (formal details can be found in [37,80]. The null process does nothing. P | Q represents the parallel execution of P and Q. The replication !P of a process P acts like the parallel execution of an unbounded number of copies of P. The name restriction new n;P creates a new name n whose scope is restricted to the process P and then runs P. The message input in(M, x); P represents a process ready to input from the channel M. The message output out(M, N); P describes a process that sends a term N on the channel M and then behaves like P. The let construct tries to rewrite D and matches the result with M; if this succeeds, then the variables in M are instantiated accordingly and P is executed; otherwise, Q is executed. The conditional checks the equality of two terms M and N and then behaves as P or Q accordingly. We will omit the else branch of a let or a conditional when the process Q is 0.

Example 2: A system consisting of multiple mobile stations MS, with identity imsi, and long-term private key sk running along with the serving network, SN, can be modelled by the process:

```
S = new pvN; let pbN = pub(pvN) in
      out(c, pbN);
      !new sk; new imsi; !new sqn; (SN|MS).
```

The privacy properties we verify are expressed in terms of *observational equivalence*. Intuitively, two processes P and Q are observationally equivalent denoted by P≈Q, if any interaction of P with the adversary can be matched with an interaction of Q (and vice versa, i.e. all interactions of Q can be matched by P) and the same input/output behaviour is observed.

The ProVerif tool can prove diff-equivalence of biprocesses, which implies observational equivalence. Biprocesses are pairs of processes which differ by some choice of terms; this choice is written choice[M, M']. For example, to test whether the processes out(c, a) and out(c, b) are equivalent, one would check the following biprocess using ProVerif : out(c, choice[a, b]). Diff-equivalence is stronger than observational equivalence; this means that when ProVerif terminates producing an attack trace, the attack could be a false attack. However, when ProVerif terminates with a proof, this means that, under the abstractions/assumptions made when modelling the protocols, there are no attacks to the verified property.

The encoding of the equivalence representing unlinkability is challenging since the processes to be tested do not have the same structure, and because of this, it is not straightforward to see how to build the biprocess representing them. In the next two sections, we show how we built the biprocess to test strong unlikability and the biprocess to test strong anonymity, respectively.

Table 1: ProVerif results on the on fixed procedures

Properties	Identifi-cation	IMSI paging	TMSI re-allocation	3G AKA
Unlinkabil-ity	√√	√√	√√	√√
Anonymity	√√	√√	√√	√√

NA not applicable √√ proved to hold ×× attack found

Strong unlinkability

In our mobile phone scenario, the strong unlinkability property holds when the situation where mobile stations access services multiple times looks the same as the ideal situation where each mobile station accesses the services at most once, i.e. where by construction unlinkability holds. Formally, we want the process S, defined in Example 2, to be observationally equivalent to the process S_{UNLINK} defined as follows:

```
S_UNLINK = new pvN; let pbN = pub(pvN) in
             out(c, pbN);
             !new sk; new imsi; new sqn; (SN|MS).
```

The absence of the replication before the new sqn construct means that in S_{UNLINK} each MS executes the protocol at most once. The above-mentioned observational equivalence can be verified with ProVerif , defining S and S_{UNLINK} as the following biprocess $P_{VUNLINK}$, where sk1, sk2 are long- term keys and imsi1, imsi2 are long-term identities:

```
PVUNLINK = new pvN; letpbN = pub(pvN)in
 out(c. pbN);
 !new sk1; new imsi1;
 !new sk2; new imsi2; new sqn;
 let(sk. imsi) = choice[(sk1. imsi1). (sk2. imsi2)]
 in(SN|MS).
```

We have that the left side of the choice represents a system where a mobile station (with identity imsi1 and key sk1) may execute the protocol many times, while the right side represents a system where mobile stations execute the protocol at most once (the identity imsi2 and the key sk2 are always different and can be used at most once for the execution of the protocol). Hence, we reduce the problem of testing strong unlinkability to the diff-equivalence of a biprocess. ProVerif proves that the strong unlinkability property is satisfied by our models of the fixed identification, paging, TMSI reallocation and AKA protocols as described in Sect. 6 (see Table 1).

Strong anonymity

In our mobile phone scenario, strong anonymity requires a system in which a mobile station M_{SV} with publicly known identity $imsi_V$ executes the protocol to be indistinguishable from a system in which the MSV is not present at all. Such a system obviously preserves $imsi_V$'s anonymity. Formally, we want the system S, defined as in Example 2, to be observationally equivalent to the system SV defined as follows:

```
S_V = new pvN; letpbN = pub(pvN)in
      out(c. pbN);
      !new sk; new imsi; (!new sqn; (SN|MS))
      |new sk; !new sqn; (SN|MS_V).
```

In the system S_V, the mobile station M_{SV} with publicly known identity $imsi_V$ can run the protocol. The mentioned observational equivalence can be translated in the following ProVerif biprocess PVANON, where $imsi_V$ and $imsi_{ms}$ are permanent mobile station identities:

```
free imsi_v.

let PV_ANON =  new pvN; let pbN = pub(pvN) in
    out(c, pbN);
    (!new sk; new imsi;
        (!new sqn;  (MS | SN)))
    |(new sk; new imsi_ms;
        let  imsi=choice[imsi_V, imsi_ms ] in
        !new sqn;  (SN | MS)).
```

The left side of the choice represents a system where the mobile station with public identity $imsi_V$ can run the protocol. Our fixes of the identification procedure, paging procedure, TMSI reallocation and 3G AKA protocol as described in Sect. 6 are proved by ProVerif to satisfy anonymity (see Table 1).

Automatic verification results and remarks

We run the ProVerif tool on the 2G/3G identification procedure, on the 2G/3G IMSI paging procedure, on the 2G/3G TMSI reallocation procedure and on the 3G AKA protocol, in order to confirm that the tool would have detected the breaches of the privacy properties present in the 3GPP standard procedures. Even though the coding of the protocols in ProVerif is straightforward, the coding of the observational equivalences defining the privacy properties in term of bi-processes is not. In fact the biprocess structure is symmetrical, while the definitions of anonymity and unlinkability are not. We showed in the previous section how we obtained a symmetrical definition in terms of bi-processes. Moreover, we had to take particular care in avoiding false attacks that could be reported by the tool due to its abstractions. Indeed, we formally define privacy properties through observational equivalence; however, ProVerifadopts a stronger equivalence relation called diff-equivalence (≈diff). In particular, diff-equivalence can distinguish between the execution of different branches of a conditional statement even in the following case:

```
1 reduc geterr(err(x,z,y,y))=macFail;
2       geterr(err(x,x,y,z))=synchFail.
3
4 let AKA_MS = new r_ms; in(c, x);
5   let (xrand, xautn) = x in (
6   let (msg, xmac) = xautn in (
7   let ak = f5(k, xrand) in (
8   let xsqn = sdec(ak, msg) in (
9   let mac = f1(k, (xrand, xsqn)) in (
10   if (xmac, xsqn) = (mac,sqn) then (
11       let res = f2(k, xrand) in (
12       let ck = f3(k, xrand) in (
13       let ik = f4(k, xrand) in (
14       out(c, res); in(c, xmsg)))))
15   else (let err_msg =
16           geterr(err(mac, xmac, sqn, xsqn)) in
17           out(c, err_msg)))))).
```

Figure 23: Code of the MS side of the original AKA protocol

Table 2: ProVerif results on original 3GPP Procedures

Properties	Identification	IMSI Paging	TMSI reallocation	3G AKA
Unlinkability	××	××	××	××
Anonymity	××	××	FA	√√

NA not applicable √ proved to hold × attack found fa false attack

```
if a = a then P else P
```

$$\approx_{diff}$$

```
if a = b then P else P
```

and hence, although the above processes are observationally equivalent (PP is executed regardless the result of the if statement evaluation), they do not satisfy diff-equivalence. We are dealing with this issue in our code for the verification at lines 1–2, 10 and 15–16 of the code of the original AKA protocol in Table 7.4. We check the MAC and the sequence number (line 10) in the same conditional statement, so to avoid false attacks due to the evaluation of the conditional. For the same reason, we introduce the functions err and geterr (lines 1–2) to determine the error message (lines 15–16) and avoid the use of an if statement (Fig. 23).

As expected, the verification with the ProVerif tool fails to prove the anonymity of the IMSI paging procedure and gives a false attack when verifying the anonymity of the TMSI reallocation procedure. Unlinkability is not satisfied by the original IMSI paging, TMSI reallocation and 3G AKA protocols (see Table 2). In case of the IMSI paging procedure, ProVerif exhibits actual attack traces. In the case of the original 3G AKA protocol, the anonymity property is proved to hold, while the verification of the unlinkability property fails. Although the trace provided by ProVerif is a false attack, it does give a hint of the real attack by highlighting the test of the MAC received from the network as the source of the problem. The modelling of unlinkability and anonymity into diff-equivalences we showed in the previous section can in general be adopted for protocols which do not require an initialization phase preceding the main protocol procedure. Hence, our method is not specific for the analysed protocols and shows how to automatically verify unlinkability and anonymity on a wide class of protocols. The ProVerif code used for the automatic verification is available online [81] and in part in "Appendix".

Table 3: Results of the automatic verification of the fixed procedures

Properties	Identification	Paging	TMSI reallocation	AKA
Secrecy				
IMSII	√√	√√	√√	√√
KIMSI	NA	√√	√√	√√
CK, *IK*	NA	NA	NA	√√
Confidential information	NA	NA	NA	√√
Authentica-tion	NA	NA	NA	√√
Integrity	NA	NA	NA	√√

NA not applicable √√ proved to hold ×× attack found

Note that for verification purposes in our models of MS and SN, we use randomized symmetric encryption to conceal the sequence number SQN instead of using the exclusive-OR. Indeed, even if the theory allows to write a set of reduction rules to model the xor function, the ProVerif tool cannot deal with its algebraic properties. The use of randomized encryption anyway would achieve stronger properties with respect to the secrecy of the sequence number, and we hence recommend the adoption of this modification in the standard protocol.

The model of the TMSI reallocation procedure requires the use of a state (memory cell) where to store the newly assigned TMSI to be used in the following session. The state is initialized during an initialization phase. We use a private channel to model the state. An auxiliary process (MEM) makes the content of the state available to read and write from the channel. In particular, the process used for the verification of the unlinkability property of the fixed procedures in the presence of the memory cell is coded as follows.

```
let MEM = in(mem,x);out(mem,x).

process new pvlI;
  let pblI = pub(pvlI) in out(c, pblI);
  (! (new sk1; new imsi1;new otmsi1; new mem;

     (* memory initialisation *)
     out(mem,otmsi1);

     (! (new sk2; new imsi2; new osqn;
     new otmsi2; new sqn_p;
     let imsi = choice[imsi1, imsi2] in (
     let k = choice[sk1, sk2] in (
     let otmsi = choice[otmsi1,otmsi2] in (
        (MS) | (SN) | (MEM)))))))
```

This abstraction may produce false attacks and non-termination problems (see [82]). For this reason, the encoding of the TMSI reallocation procedure is not the one you would obtain with a straightforward encoding of the description of the protocol in applied pi-calculus. In particular, the TMSI of a single-session mobile stations is created during the *i*th session of the multi-session mobile station and used in the i+1th session (see lines 105, 108, 156 of the code in "Appendix". Intuitively, this gives the TMSI used by the single-session mobile station and the one used by the multi-session one the same amount of freshness in the ProVerif model. Hence, the position of the new construct at lines 105 and 156 in the code in "Appendix" is critical to the success of the verification process.

Authentication, secrecy and integrity The main purpose of the 3G AKA protocol is to provide mutual authentication and establish session keys to be used for integrity protection and secrecy. Hence, our analysis would not be complete without ensuring that our privacy-preserving version of the 3G AKA protocol still achieves the goals it was originally designed for. We verify mutual authentication and integrity properties as injective correspondence properties. We prove using ProVerif that the original properties of the 3G AKA protocol are preserved by our fixes; the verification results are shown in Table 3.

The full code used for the verification of the security and privacy of the 3GPP procedures analysed in this paper when run in isolation and when run in parallel is available online [81]. In the "Appendix", we report the code used to verify the unlinkability of all the fixed procedures when running in parallel.

CONCLUSIONS

In this work, we presented a thorough (although not complete) experimental and formal analysis of users' privacy in mobile telephony systems. In particular, we experimentally analysed the use of pseudonyms and pointed out weaknesses in the deployed policies. We showed how these weak policies make it possible to violate a user's privacy and proposed some solutions to strengthen the pseudonyms management in mobile telephony systems. We also exposed some protocol's vulnerabilities resulting in breaches of the anonymity and/or user unlinkability. We showed with a prototype implementation that the breaches we found translate into actual attacks on real networks. To countermeasure these attacks, we proposed solutions which are realistic and lightweight but require some changes to the existing infrastructure. The fixes we proposed show that privacy-friendly solutions could be adopted by future generation of mobile telephony systems. However, these are not meant to be full protocol specifications, and implementation details are left for the mobile telephony operators to specify. Finally, we provided a theoretical framework for the automatic verification of the unlinkability and anonymity using the ProVerif tool. We used this framework to automatically verify the fixed 2G/3G procedures. Further investigation of mobile telephony users' privacy should be undertaken in order to obtain a complete picture of the privacy offered by the interactions of the many protocols at different levels of the stack. Currently, our demonstration of the attacks presented in Sect. 5 relies on particular hardware/software using closed-source implementation of the 3G protocol stack and radio signalling functions. It would be interesting and beneficial for further research in the area of mobile telephony systems to investigate the possibility of implementing open-source testing equipment, such as a 3G base station and mobile phone, using low-cost hardware, e.g. USRP and the GNU radio software. Further work is needed in order to confirm experimentally the replay attack presented in Sect. 3.9. This would allow to check whether there are or not mechanisms in place (not stated in the standard) to thwart this attack by preventing replayed messages from being accepted by the Mobile Station. Also, a thorough and methodical analysis of the level of privacy achieved by different privacy policies would be of great interest. However, this would possibly require collecting further data about user mobility, aggregation areas, population density, network coverage and user base per geographical area. The impact of the adoption of the proposed TMSI reallocation policies on the network performances should be studied and related to the level of achieved user's privacy in order to carefully balance these equally important aspects of mobile telephony systems. The overall privacy of mobile telephony systems and at each layer of the protocol stack requires further investigation and would possibly offer interesting challenges for the development of formal methods as well. In particular, the proliferation of location-based services makes the analysis of application layer privacy highly desirable from a user's point of view.

ACKNOWLEDGMENTS

We would like to thank Kevin Redon, Nico Golde and Ravi Borgaonkar for the proof of concept implementation of the attacks presented in this paper. We are also grateful to Chris J. Mitchell for lively discussion about our ideas on mobile telephony protocols.

Appendix: ProVerif code

ProVerif code used to verify the unlinkability of the fixed procedures when running in parallel. Note that the fixed TMSI reallocation procedure executes the AKA protocol prior to the reallocation so to obtain a fresh encryption key.

```
(* public communication channel *)
free c.

(* constant values *)
fun Fail/0.
fun reject/0.
fun page/0.
fun pagingReq/0.
fun pagingResp/0.
fun imsiReq/0.

(* UMTS AKA protocol specific mac and
   key generation functions *)
fun f1/2.
fun f2/2.
fun f3/2.
fun f4/2.
fun f5/2.

(* New key generation function *)
fun f/2.

(* symmetric key encryption function *)
fun senc/3.
fun sdec/2.
equation sdec(k, senc(k, r, m)) = m.

(* public key generation function *)
fun pub/1.

(* public key encryption function *)
fun aenc/3.
reduc adec(k, aenc(pub(k), r, m))= m.

let AKA_MS =
   new r_ms;
   in(c, x);
   let (xrand, xautn) = x in (
   let (msg, xmac) = xautn in (
   let ak = f5(k, xrand) in (
   let xsqn = sdec(ak, msg) in (
   let mac = f1(k, (xrand, xsqn)) in (
   if (xmac, xsqn) = (mac, osqn) then (
      let res = f2(k, xrand) in (
      let ck = f3(k, xrand) in (
      let ik = f4(k, xrand) in (
      out(c, res);
      in(c, xmsg)))))
   else (
      out(c, aenc(pbN, r_ms, (Fail, imsi, osqn)))
   )))))).

```

```
let AKA_SN =
   new rand;
   new r_sn;
   new s;
   new r;
   let mac = f1(k, (rand, osqn)) in (
   let res = f2(k, rand) in (
   let ck = f3(k, rand) in (
   let ik = f4(k, rand) in (
   let ak = f5(k, rand) in (
   let autn = (senc(ak, r_sn, osqn), mac) in (
   let av = (rand, res, ck, ik, ak) in (
   out(c, (rand, autn));
   in(c, xres);
   if xres = res then (
      out(c, senc(ck, r, s))
   )
   else (
      out(c, reject))))))))).

let ID_MS =
   new r;
   in(c, req);
   if req = imsiReq then (
      out(c, aenc(pbN, r, imsi))).

let ID_SN =
   out(c, imsiReq);
   in(c, ximsi).

let PAGING_MS =
   in(c, x);
   let (msgtype, xrand, xblob) = x in (
   if msgtype = pagingReq then (
      let (xpage, ximsi, =sqn_p, xchall) =
            sdec(f(k, xrand), xblob) in (
      if xpage = page then (
            if imsi = ximsi then (
       out(c, (pagingResp, xchall))))))).

let PAGING_SN =
   new rand;
   new chall;
   new r_sn1;
   let UK = f(k, rand) in (
   out(c, (pagingReq, rand, senc(UK,
        r_sn1, (page, imsi, sqn_p, chall))));
   in(c, pres)).

let TMSI_MS =
   in(mem, xmem);
   new mr;
   out(c, xmem);
   in(c, y);
   let (=TMSI_REALL, yid) = sdec(ck, y) in
         (out(c, senc(ck, mr, COMPLETE));
     out(mem,choice[yid, otmsi2])).

 let TMSI_SN =
    new nid;
    new sr;
    in(c, z);
    out(c, senc(ck, sr,
                (TMSI_REALL, nid)));
    in(c, w).

let FixedTMSI_MS =
   new r_ms;
   in(c, x);
   let (xrand, xautn) = x in (
   let (msg, xmac) = xautn in (
   let ak = f5(k, xrand) in (
   let xsqn = sdec(ak, msg) in (
   let mac = f1(k, (xrand, xsqn)) in (
   if (xmac, xsqn) = (mac, osqn) then (
      let res = f2(k, xrand) in (
      let ck = f3(k, xrand) in (
      let ik = f4(k, xrand) in (
      out(c, res);
      TMSI_MS))))
   else (
      out(c, aenc(pbN, r_ms, (Fail, imsi,
           osqn))))))))).

let FixedTMSI_SN =
   new rand;
   new r_sn;
   new s;
   new r;
   let mac = f1(k, (rand, osqn)) in (
   let res = f2(k, rand) in (
   let ck = f3(k, rand) in (
   let ik = f4(k, rand) in (
   let ak = f5(k, rand) in (
   let autn = (senc(ak, r_sn, osqn),
                        mac) in (
   let av = (rand, res, ck, ik, ak) in (
   out(c, (rand, autn));
   in(c, xres);
   if xres = res then (
      TMSI_SN)
   else (
      out(c, reject))))))))).

let MS = (AKA_MS|ID_MS|PAGING_MS|FixedTMSI_MS).
let SN = (AKA_SN|ID_SN|PAGING_SN|FixedTMSI_SN).
let MEM = in(mem,x);out(mem,x).

process new pvN; let pbN = pub(pvN) in
                    out(c, pbN);
(! (new sk1; new imsi1;new otmsi1; new mem;
  out(mem,otmsi1);
(! (new sk2; new imsi2; new osqn;
new otmsi2; new sqn_p;
let imsi = choice[imsi1, imsi2] in (
let k = choice[sk1, sk2] in (
let otmsi = choice[otmsi1,otmsi2] in (
    (MS) | (SN) | (MEM)))))))
```

REFERENCES

1. Arapinis, M., Bursuc, S., Ryan, M.: Privacy supporting cloud computing: ConfiChair, a case study. In: Degano, P., Guttman, J. (eds.) Principles of Security and Trust, Lecture Notes in Computer Science, vol. 7215, pp. 89–108. Springer, Berlin Heidelberg, Tallinn, Estonia (2012)
2. Cortier, V., Wiedling, C.: A formal analysis of the Norwegian E-voting protocol. In: Proceedings of the 1st International Conference on Principles of Security and Trust (POST'12), Lecture Notes in Computer Science, vol. 7215, pp. 109–128. Springer, Tallinn, Estonia (2012)
3. Grewal, G.S., Ryan, M.D., Bursuc, S., Ryan, P.Y.A.: Caveat Coercitor: Coercion-Evidence in Electronic Voting. In: Proceedings of the 2013 IEEE Symposium on Security and Privacy, SP'13, pp. 367–381. IEEE Computer Society, Washington, DC, USA (2013)

4. Cortier, V., Smyth, B.: Attacking and fixing helios: an analysis of ballot secrecy. J. Comput. Secur. 21(1), 89 (2013)CrossRef
5. Bansal, C., Bhargavan, K., Delignat-Lavaud, A., Maffeis, S.: Discovering concrete attacks on website authorization by formal analysis. J. Comput. Secur. 22(4), 601 (2014) CrossRefMATH
6. Armando, A., Carbone, R., Compagna, L., Cuellar, J., Tobarra, L.: Formal analysis of SAML 2.0 web browser single sign-on: breaking the SAML-based single sign-on for Google Apps. In: Proceedings of the 6th ACM Workshop on Formal Methods in Security Engineering, FMSE'08, pp. 1–10. ACM, New York, NY, USA (2008)
7. Bansal, C., Bhargavan, K., Delignat-Lavaud, A., Maffeis, S.: Keys to the cloud: formal analysis and concrete attacks on encrypted web storage. In: Principles of Security and Trust, pp. 126–146. Springer, Berlin, Heidelberg, Rome, Italy (2013)
8. Chothia, T., Smirnov, V.: A traceability attack against e-passports. In: Proceedings of the 14th International Conference on Financial Cryptography and Data Security, FC'10, pp. 20–34. Springer, Berlin, Heidelberg (2010)
9. Delaune, S., Kremer, S., Ryan, M.D., Steel, G.: A formal analysis of authentication in the TPM. In: Degano, P., Etalle, S., Guttman, J. (eds.) Formal Aspects of Security and Trust, Lecture Notes in Computer Science, vol. 6561, pp. 111–125. Springer, Berlin, Heidelberg (2011)
10. Batten, I., Xu, S., Ryan, M.: Dynamic measurement and protected execution: model and analysis. In: Abadi, M., Lluch Lafuente, A. (eds.) Trustworthy Global Computing, Lecture Notes in Computer Science, vol. 8358, pp. 45–63. Springer International Publishing, Buenos Aires, Argentina (2014). doi:10.1007/978-3-319-05119-2_4
11. Arapinis, M., Mancini, L., Ritter, E., Ryan, M., Golde, N., Redon, K., Borgaonkar, R.: New privacy issues in mobile telephony: fix and verification. In: Proceedings of the 2012 ACM Conference on Computer and Communications Security, CCS'12, pp. 205–216. ACM, New York, NY, USA (2012)
12. Arapinis, M., Chothia, T., Ritter, E. Ryan, M.: Analysing unlinkability and anonymity using the applied pi calculus. In: Computer Security Foundations Symposium (CSF), 2010 23rd IEEE, CSF, pp. 107–121. IEEE, Edinburgh (2010)
13. Introna, L.D.: Privacy and the computer: why we need privacy in the information society. Metaphilosophy 28(3), 259 (1997)CrossRef
14. Warren, S.D., Brandeis, L.D.: The right to privacy. In: Harvard Law Review, vol. 4, issue 5, pp. 193–220. The Harvard Law Review Association (1890)
15. Cas, J.: Privacy in pervasive computing environments—a contradiction in terms? IEEE Technol. Soc. Mag. 24(1), 24 (2005)CrossRef
16. Ryan, M.D.: Cloud computing privacy concerns on our doorstep. Commun. ACM 54(1), 36 (2011)CrossRef
17. Shilton, K.: Four billion little brothers? Privacy, mobile phones, and ubiquitous data collection. Commun. ACM 52(11), 48 (2009)CrossRef
18. Krontiris, I., Freiling, F., Dimitriou, T.: Location privacy in urban sensing networks: research challenges and directions [Security and Privacy in Emerging Wireless Networks]. IEEE Wirel. Commun. 17(5), 30 (2010) CrossRef
19. Avoine, G., Oechslin, P.: RFID traceability: a multilayer problem. In: Patrick, A., Yung, M. (eds.) Financial Cryptography and Data Security, Lecture Notes in Computer Science, vol. 3570, pp. 125–140. Springer Berlin, Heidelberg, Roseau, The Commonwealth Of Dominica (2005). doi:10.1007/11507840_14
20. Narayanan, A., Shmatikov, V.: De-anonymizing social networks. In: Proceedings of the 2009 30th IEEE Symposium on Security and Privacy, SP'09, pp. 173–187. IEEE Computer Society, Washington, DC, USA (2009)
21. Narayanan, A., Shmatikov, V.: How To Break Anonymity of the Netflix Prize Dataset. arXiv preprint arXiv:cs/0610105 (2006)
22. Barbaro, M., Zeller Jr., T.: A face is exposed for aol searcher no. 4417749. The New York Times (2006). http://www.nytimes.com/2006/08/09/technology/09aol.html?_r=0
23. Beresford, A.R., Stajano, F.: Location privacy in pervasive computing. IEEE Pervasive Comput. 2(1), 46 (2003)CrossRef
24. Krumm, J.: A survey of computational location privacy. Pers. Ubiquitous Comput. 13(6), 391 (2009)CrossRef
25. Görlach, A., Heinemann, A., Terpstra, W.: Survey on location privacy in pervasive computing. In: Robinson, P., Vogt, H., Wagealla, W.(eds.) Privacy, Security and Trust within the Context of Pervasive Computing, The International Series in Engineering and Computer Science, vol. 780, pp. 23–34. Springer, USA (2005)
26. Brusó, M., Chatzikokolakis, K., Hartog, J.D.: Formal verification of privacy for RFID systems. In: Proceedings of the 2010 23rd IEEE Computer Security Foundations Symposium, CSF'10, pp. 75–88. IEEE Computer Society, Washington, DC, USA (2010)
27. Kremer, S., Ryan, M.: Analysis of an electronic voting protocol in the applied pi calculus. In: Proceedings of the 14th European Conference on Programming Languages and Systems, ESOP'05, pp. 186–200. Springer, Berlin, Heidelberg (2005)
28. Dolev, D., Yao, A.C. (1981) On the security of public key protocols. In: Foundations of Computer Science, Annual IEEE Symposium on 0, 350
29. Bhargavan, K., Lavaud, A., Fournet, C., Pironti, A., Strub, P.: Triple handshakes and cookie cutters: breaking and fixing authentication over TLS. In: Security and Privacy (SP), 2014 IEEE Symposium on, pp. 98–113. IEEE, USA (2014)
30. Armando, A., Carbone, R., Compagna, L., Cuellar, J., Pellegrino, G., Sorniotti, A.: From multiple credentials to browser-based single sign-on: are we more secure? In: Camenisch, J., Fischer-Hbner, S., Murayama, Y.,

Portmann, A., Rieder, C. (eds.) Future Challenges in Security and Privacy for Academia and Industry, IFIP Advances in Information and Communication Technology, vol. 354, pp. 68–79. Springer, Berlin (2011)

31. Abadi, M., Fournet, C.: Mobile values, new names, and secure communication. In: Proceedings of the 28th ACM SIGPLAN-SIGACT Symposium on Principles of Programming Languages, POPL'01, pp. 104–115. ACM, New York, NY, USA (2001)
32. Durgin, N.A., Lincoln, P., Mitchell, J.C., Scedrov, A.: Undecidability of bounded security protocols. In: Workshop on Formal Methods and Security Protocols, FMSP'99, Trento, Italy (1999)
33. Hüttel, H.: Deciding framed bisimilarity. Electron. Notes Theor. Comput. Sci. 68, 1–18 (2002)CrossRefMATH
34. Cheval, V., Comon-Lundh, H., Delaune, S.: Automating security analysis: symbolic equivalence of constraint systems. In: Proceedings of the 5th International Conference on Automated Reasoning, IJCAR'10, pp. 412–426. Springer, Berlin, Heidelberg (2010)
35. Cortier, V., Delaune, S.: A method for proving observational equivalence. In: Proceedings of the 2009 22nd IEEE Computer Security Foundations Symposium, CSF'09, pp. 266–276. IEEE Computer Society, Washington, DC, USA (2009)
36. Baudet, M.: Deciding security of protocols against off-line guessing attacks. In: Proceedings of the 12th ACM Conference on Computer and Communications Security, CCS'05, pp. 16–25. ACM, New York, NY, USA (2005). doi:10.1145/1102120.1102125
37. Blanchet, B.: Automatic verification of correspondences for security protocols. J. Comput. Secur. 17(4), 363 (2009)CrossRef
38. Blanchet, B., Fournet, C.: Automated verification of selected equivalences for security protocols. In: Proceedings of the 20th Annual IEEE Symposium on Logic in Computer Science, LICS'05, pp. 331–340. IEEE Computer Society, Washington, DC, USA (2005)
39. Blanchet, B.: An efficient cryptographic protocol verifier based on prolog rules. In: Proceedings of the 14th IEEE Workshop on Computer Security Foundations, CSFW'01, pp. 82–96. IEEE Computer Society, Washington, DC, USA (2001)
40. 3GPP, Technical specification group services and system aspects; security aspects (release 4), TS 42.009. Technical Specification TS 42.009 V4.3.0, 3rd Generation Partnership Project (2010)
41. 3GPP, Technical specification group services and system aspects; 3G security; security architecture (release 9), TS 33.102. Technical Specification TS 33.102 V9.3.0, 3rd Generation Partnership Project (2010)
42. Biryukov, A., Shamir, A., Wagner, D.: Real time cryptanalysis of A5/1 on a PC. In: Proceedings of the 7th International Workshop on Fast Software Encryption, FSE'00, pp. 1–18. Springer, London, UK (2001)
43. Nohl, K., Munaut, S.: Wideband gsm sniffing. http://events.ccc.de/congress/2010/Fahrplan/attachments/1783_101228.27C3.GSM-Sniffing.Nohl_Munaut.pdf (2010)
44. Ahmadian, Z., Salimi, S., Salahi, A.: New attacks on UMTS network access. In: Conference on Wireless Telecommunications Symposium, WTS'09, pp. 1–6. IEEE, Prague (2009)
45. Nohl, K., Melette, L.: Defending mobile phones. http://events.ccc.de/congress/2011/Fahrplan/attachments/1994_111217.SRLabs-28C3-Defending_mobile_phones.pdf (2011). http://www.gsmmap.org. 28th Chaos Communication Congress (28C3)
46. May, S., Melette, L.: Distributed gsm security analysis. https://program.sigint.ccc.de/fahrplan/system/attachments/21/original/120518.GSMMAP-SIGINT.pdf (2012). http://www.gsmmap.org. SIGINT 2012
47. Koien, G., Oleshchuk, V.: Location privacy for cellular systems; analysis and solution. In: Privacy Enhancing Technologies Symposium, vol. 3856 (2006)
48. Fox, D.: IMSI-Catcher. Datenschutz und Datensicherheit (DuD) 21, 539 (1997)
49. Strobel, D.: IMSI Catcher. In: Seminar Work, Ruhr-Universitat Bochum (2007)
50. Paget, C.: Practical cellphone spying. Def Con 18 Hacking Conference (2010).http://www.defcon.org/html/defcon-18/dc-18-speakers.html
51. Kune, D.F., Koelndorfer, J., Hopper, N., Kim, Y.: Location leaks over the GSM air interface. In: Annual Network & Distributed System Security Symposium. NDSS (2012)
52. Welte, H., Munaut, S.: A. Eversberg, other contributors. OsmocomBB.http://bb.osmocom.org (2011)
53. Engel, T.: Locating mobile phones using signalling system 7.http://events.ccc.de/congress/2008/Fahrplan/attachments/1262_25c3-locating-mobile-phones.pdf (2008). 25th Chaos Communication Congress (25C3)
54. Golde, N., Redon, K., Seifert, J.P.: Let me answer that for you: exploiting broadcast information in cellular networks. In: Presented as part of the 22nd USENIX Security Symposium (USENIX Security 13), pp. 33–48. USENIX, Washington, DC (2013).https://www.usenix.org/conference/usenixsecurity13/technical-sessions/presentation/golde
55. Golde, N., Redon, K., Borgaonkar, R.: Weaponizing femtocells: the effect of rogue devices on mobile telecommunications. In: Annual Network & Distributed System Security Symposium. NDSS (2012)
56. Meyer, U., Wetzel, S.: A man-in-the-middle attack on UMTS. In: ACM Workshop on Wireless Security. WiSe (2004)
57. Zhang, M., Fang, Y.: Security analysis and enhancements of 3GPP authentication and key agreement protocol. IEEE Trans. Wirel. Commun. 4(2), 734 (2005)CrossRef
58. Hahn, C., Kwon, H., Kim, D., Kang, K., Hur, J.: A privacy threat in 4th generation mobile telephony and its countermeasure. In: Wireless Algorithms, Systems, and Applications, pp. 624–635. Springer (2014)
59. 3GPP, Technical specification group services and system aspects; 3G security; formal analysis of the 3G authentication protocol (release 4), TS 33.902. Technical

Specification TR 33.902, V4.0.0, 3rd Generation Partnership Project (2001)

60. Khan, M.S.A., Mitchell, C.J.: Another look at privacy threats in 3G mobile telephony. In: Information Security and Privacy, pp. 386–396. Springer (2014)
61. 3GPP, Technical specification group core network and terminals; mobile radio interface layer 3 specification; core network protocols; stage 3 (release 9), TS 24.008. Tech. Rep. TS 24.008 V9.4.0, 3rd Generation Partnership Project (2010)
62. Wehrle, D.: Open Source IMSI-Catcher für GSM. Master Thesis (2009)
63. 3GPP, Technical specification group radio access network; radio resource control (RRC); protocol specification (release 10), TS 25.331. Tech. Rep. TS 25.331, V10.5.0, 3rd Generation Partnership Project (2011)
64. Combs, G.: Wireshark (1998). http://www.wireshark.org
65. Gonzalez, M.C., Hidalgo, C.A., Barabasi, A.L.: Understanding individual human mobility patterns. Nature 453, 779 (2008).http://www.nature.com/nature/journal/v453/n7196/suppinfo/nature06958_S1.html
66. Bayir, M., Demirbas, M., Eagle, N.: Discovering spatiotemporal mobility profiles of cellphone users. In: World of Wireless, Mobile and Multimedia Networks Workshops, 2009. WoWMoM 2009, pp. 1–9. IEEE, Kos, Greece (2009)
67. Arapinis, M., Mancini, L.I., Ritter, E., Ryan M. (2013). http://www.markryan.eu/research/mobile/
68. 3GPP, Technical specification group services and system aspects; 3G security; cryptographic algorithm requirements (release 10), TS 33.105. Technical Specification TS 33.105 V10.0.0, 3rd Generation Partnership Project (2011)
69. Ettus. USRP (2009). http://www.ettus.com
70. 3GPP, Security of Home Node B (HNB) / Home evolved Node B (HeNB), TS 33.302. Technical Specification TS 33.302 v11.2.0, 3rd Generation Partnership Project (2011)
71. Kineto Wireless Inc. official Unlicensed Mobile Access presentation website.http://www.smart-wi-fi.com/ (2010)
72. 3GPP, Generic Access Network (GAN); Mobile GAN interface layer 3 specification, TS 44.318. Technical Specification TS 44.318 v9.2.0, 3rd Generation Partnership Project (2010)
73. 3GPP, Generic Access Network (GAN); Stage 2, TS 43.318. Technical Specification TS 43.318 v9.0.0, 3rd Generation Partnership Project (2010)
74. Zhang, J., de la Roche, G.: Femtocells: Technologies and Deployment. Wiley, New York (2009)
75. P.I. Ltd.: Footpath. http://www.pathintelligence.com (2004)
76. Lowe, G.: Breaking and fixing the Needham–Schroeder public-key protocol using FDR. In: Proceedings of the Second International Workshop on Tools and Algorithms for Construction and Analysis of Systems, TACAS, pp. 147–166. Springer, London, UK (1996)
77. Cervesato, I., Jaggard, A.D., Scedrov, A., Tsay, J.K., Walstad, C.: Breaking and fixing public-key Kerberos. Inf. Comput. 206, 402 (2008)MathSciNetCrossRefMATH
78. Bortolozzo, M., Centenaro, M., Focardi, R., Steel, G.: Attacking and fixing PKCS#11 security tokens. In: Proceedings of the 17th ACM Conference on Computer and Communications Security, CCS'10, pp. 260–269. ACM, New York, NY, USA (2010)
79. Blanchet, B.: Proverif: Cryptographic protocol verifier in the formal model.http://www.proverif.ens.fr/ (2001)
80. Blanchet, B., Abadi, M., Fournet, C.: Automated verification of selected equivalences for security protocols. J. Log. Algebraic Program. 75(1), 3 (2008) MathSciNetCrossRefMATH
81. Arapinis, M., Mancini, L.I., Ritter, E., Ryan, M.:http://www.markryan.eu/research/UMTS/ (2012)
82. Arapinis, M., Mancini, L.I., Ritter, E., Ryan, M.: Privacy through pseudonymity in mobile telephony systems. In: 21st Annual Network and Distributed System Security Symposium (NDSS14), p. 14. Internet Society, San Diego, CA, USA (2014)

CHAPTER 6 PERFORMANCE IMPROVEMENT OF WIRELESS COMMUNICATIONS USING FREQUENCY HOPPING SPREAD SPECTRUM

Naser Hossein Motlagh

Keywords: Frequency Hopping Spread Spectrum, Adaptive Frequency Hopping, Performance Evaluation

INTRODUCTION

In this paper the focus is to improve wireless communication performance by adaptive frequency hopping which is implemented by selecting sets of communication channels and adaptively hopping sender's and receiver's frequency channels and determining the channel numbers with less interference. Also the work investigates whether the selected channels are congested or clear then a list of good channels can be generated and in practice to use detected frequency channels as hopping sequence to improve the performance of communication and finally the quality of service. The Fourier transform mathematical modules are used to convert signals from time domain to frequency domain and vice versa. The mathematical modules are applied to represent the frequency and simulate them in MATLAB and as result the simulated spectrums are analysed. Then a simple two-state Gilbert-Elliot Channel Model [1,2] in which a two-state Markov chain with states named "Good" and "Bad" is used to check if the channels are congested or clear in case of interference. Finally, a solution to improve the performance of wireless communications by choosing and using "Good" channels as the next frequency hopping sequence channel is proposed.

REVIEW OF RELATED THEORIES

Spread Spectrum

Spread spectrum is a digital modulation technology and a technique based on principle of spreading a signal among many frequencies to prevent interference and signal interception [3]. As the name shows it is a technique to spread the transmitted spectrum over a wide range of frequencies. It started to be employed by military applications because of its Low Probability of Intercept or demodulation, interference and anti-jamming from enemy side. The idea of spreading spectrum is to spread a signal over a large frequency band to use greater bandwidth than the data bandwidth while the power remains the same. And as far as the spread signal looks like the noise signal in the same frequency band it is difficult to recognize the signal which this feature of spreading provides security to the transmission. Compared to a narrowband signal, spread spectrum spreads the signal power over a wideband and the overall SNR is improved because only a small part of spread spectrum signal is affected by interference. In sender and receiver sides of a communication system one spreading generator is located based on the spreading technique they synchronize the received modulated spectrum. Shannon capacity Equation is the basis for spread spectrum systems, which typically operate at a very low SNR, but use a very large bandwidth in order to provide an acceptable data rate per user. Applying spread spectrum principles to the multiple access environments is a development occurring over the last decade [4].

Frequency Hopping Spread Spectrum (FHSS)

Frequency hopping spread spectrum is a transmission technology used in wireless networks and a technique to generate spread spectrum by hopping the carrier frequency FHSS uses narrow band signal which is less than 1 MHz, In this method data signal is modulated with a narrowband carrier signal that hops in random and hopping happens in pseudo-random predictable sequence in a regular time from frequency to frequency which is synchronized at both ends. FHSS improves privacy. It is a powerful solution to avoid interference and multi-path fading (distortion). It decreases narrowband interference, increases signal capacity, and improves the signal to noise ratio. The efficiency of bandwidth is high and it is difficult to intercept.

Citation: N. Motlagh, "Performance Improvement of Wireless Communications Using Frequency Hopping Spread Spectrum" International Journal of Communications, Network and System Sciences, Vol. 3 No. 10, 2010, pp. 805-810. doi: 10.4236/ijcns.2010.310108.

Also this transmission can share a frequency band with many types of conventional transmissions with minimal interference. For frequency hopping a mechanism must be defined to transmit data in a clear channel and to avoid the congested channels. Frequency hopping is the periodic change of transmission frequency and hopping happens over a frequency bandwidth which consists of numbers of channels. Channel which is used as a hopped channel is instantaneous bandwidth while the hopping spectrum is total hopping bandwidth. Frequency hopping categorized into slow hopping and fast hopping which by slow hopping more than one data symbol is transmitted in same channel and by fast hopping frequency changes several times during one symbol. Hopping sequence means which next channel to hop. There are two types of hopping sequence: random hopping sequence and deterministic hopping sequence. The focus of this work is on slow and deterministic frequency hopping sequence. In a frequency hopping network, there can be different number of receivers which one sender is designed as base which is responsible to transmit the synchronization data to the receivers.

Adaptive Frequency Hopping

Adaptive frequency hopping (AFH) is a system in which devices constantly change their operating frequency to avoid interference from other devices to improve commutation performance. AFH classifies channels as "Good" or "Bad" and adaptively selects from the pool of Good channels. Bad channels are the channels with interference. The idea of using AFH is to hop only over Good channels, which means to choose the frequency channels that have less interference. For using AFH there must be a mechanism to choose "Good" and "Bad" channels. Received signal strength indication (RSSI) tells each channel quality to generate a list for "bad channels". The system and principle of a proposed AFH scheme are illustrated in [5], assuming that there is a duplex transceiver system. The system is an ordinary frequency hopping system which uses a number of narrowband channels.

Channel and Interference

Compared to the other kinds of wireless communications, high frequency communication is selectively fading because of the multi-path propagation and abundance of interference from the others. Interference always exists in any wireless system. Bit error rate is highly important for the performance improvement of the communication systems. Every frequency channel due to interferences and fading shows different signal to noise ratio. In some of the frequency channels there are stronger SNR and these channels are more suitable for the transmission. Adaptive frequency hopping is a powerful solution and a technique that deals with different kind of interference, noise and fading. For the simplicity of the work the focus is only on the interference as the main disturbance in achieving a desired and suitable transmission quality and neglects all the other disturbance resources such as other noises and fading

GILBERT-ELLIOT CHANNEL MODEL

Markov Chain

Bit error models generate a sequence of noise bits (where 0's represent good bits and 1's represent bit errors) to produce output bits, and modulo 2 to the input bits must be added. Models are grouped into two classes: memoryless models and models with memories [6]. In memoryless models the noise bits are produced by a sequence of independent trials that each trial has the same probability P(0) of producing a correct bit and probability P(1) = 1 – P(0) of producing a bit error

The actual measurement from the communication channels indicates that these channels are with memories, for example the probability of 100th bit error is dependent on the 99th bit. For modelling of such kind of probabilistic situation a commonly technique is used which is Markov chain. This technique helps to make the bit error probability depend on the states. The use of Markov chain in bit error models has been introduced by Gilbert-Elliot for the first time. Gilbert-Elliot channel model based on Markov chain has two states G (Good) and B (Bad). In state G, transmission is error-free and in sate B the link has probability h of transmitting a bit correctly. Figure 1 shows a transition diagram and bit error probabilities for Markov chain. The model has three independent parameters (p, P and h) to describe the error performance of wireless links. The situation of small p is where transition jumps from B to G and the capital P is where transition jumps from G to B. Also the states B and G tend to persist and the model simulates bursts of errors

The parameters p, P and h are not directly observable and therefore must be determined from statistic measurements of the error process. It's also important to note that runs of G alternates with runs of B. The run length has geometric distributions, with mean 1/P for the G-runs and 1/p for the B-runs

Geometric Distribution

The run lengths of Good and Bad states can be expressed by geometric distribution in which for the Good runs, mean value of 1/P and for the Bad runs, the mean value of 1/p is used. The time fraction in both of Good and Bad states based on persistence in each state can be calculated, for example the fraction of time spent in B state is:

$$P(B) = \frac{P}{P+p} \tag{1}$$

The sequence of states cannot be reconstructed from the sequence of bits in the error process, because both of 0's and 1's (the Good bits and Bad bits) are produced in the B state and since bit errors happen only in state B with probability of 1-h then the probability of error is:

$$P(1) = P(1,B) = P(B)P(1 \mid B) = (1-h)\frac{P}{P+p} \tag{2}$$

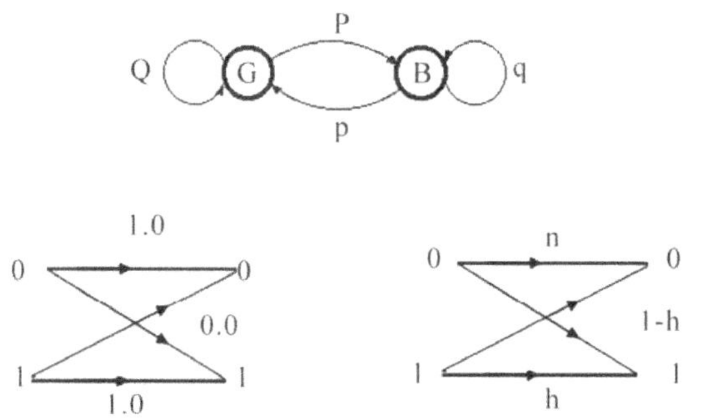

Figure 1: Transition diagram and bit error probabilities model.

The bits of the error process (runs of 0's and 1's) and the distribution of run lengths of 0's (error gaps) and 1's (error bursts) are observable to determine model parameters.

Parameter Estimation

The determination of the three parameters p, P, and h from measurements of the error process requires that parameters be expressed as functions of three other parameters that are directly observable. For Markov chain parameter estimation the functions have been proved formerly [6]:

$$\mu_{EB} = \frac{1}{1-q(1-h)} \quad (3)$$

$$\mu_{EG} = \frac{h(1-Q)+(1-q)}{(1-h)(1-Q)[1-q(1-h)]} \quad (4)$$

$$\sigma^2_{EB} = \frac{\sqrt{q(1-h)}}{1-q(1-h)} \quad (5)$$

$$\sigma^2_{EG} = \frac{(1-h)(qJ+p-Q)J(J+1)}{[1-q(1-h)](J-L)(1-J)^3} + [J \leftrightarrow L] - \mu^2_{EG} \quad (6)$$

In the Equations m $_{EB}$ is the mean error bust length and m $_{EG}$ is the mean error gap length, σ^2_{EB} is the variance of the error burst distribution and σ^2_{EB} is the variance of error gap distribution. J and L are defined as:

$$2J = Q + hq + \sqrt{(Q+hq)^2 + 4h(p-Q)} \quad (7)$$

$$2L = Q + hq - \sqrt{(Q+hq)^2 + 4h(p-Q)} \quad (8)$$

MATLAB MODELLING

Gilbert-Elliot Modelling

Gilbert-Elliot channel model is used for modelling a telecommunication channel. For obtaining the parameters of this model, first a sequence of data bit is given to the transmitter and then from the receiver side the transmitted data is received as output data. With the input sequence and output sequence, bit error sequence can be calculated easily. By having this bit error sequence and the method of parameter estimation in [6] the model parameters can be calculated

For this reason channel simulation is done with Simulink. To obtain the bit sequence of input and output, two variables with names "in" and "out" are used. With XORing the input and output bit sequences the bit error sequence is calculated. By setting bit error sequence at argument of function marcov, Markov parameters can be achieved from the output of function marcov. In function marcov by using the function coef, the sequence of error burst and error gap can be calculated. After calculation of statistical parameters of these two sequences, Markov parameters can be then calculated by function fsolve which solves nonlinear Equations.

Defining Markov Chain Parameters

To obtain Markov parameters in Matlab, a function of marcov is created as follow.

```
error_seq = xor(in,out);

z = marcov(error_seq);

z = fsolve(@solv,[.1 .1 .1],[],meb,meg,veg);
```

In this function the error sequence is first inputted to the function of coef then the output of sequence is obtained as 0's and 1's.

For example assume there is a sequence of:

```
error_seq = [0 1 0 0 0 1 1 1 0 1 1 0 1 0 1 1 1 1 0 0 0]
```

Then at the output of function coef will obtain:

```
error_burst_seq = [1 3 2 1 4]

error_gap_seq = [1 3 1 1 1 3]
```

Now from the output error_burst_seq and error_ gap_ seq which is the sequence of error runs it can be seen that the length of the run of the errors has come in order of their happenings. Next step is to calculate the mean value and the variance of the sequence.

Channel Performance Evaluation

100 communication channels are evaluated and channel performances are categorized based on Gilbert-Elliot channel model. Gilbert-Elliot model is used for modelling a real communication channel and evaluating the performance of the channels, in which first a bit sequence is sent through a channel and then its bit error sequence is computed. Using bit error sequence helps to find out the parameters of the model. Markov parameters can be used to find following two functions: Fraction of time spent in state B (Bad) from Equation (1) and probability of the error from Equation (2).

To evaluate the channel performance based on Gilbert-Elliot Markov chain model the information about bit error sequence is collected to simulate the channel model with Matlab. Additive white Gaussian noise (AWGN) channels with 100 random input powers are used in simulation.

First the percent of time is computed which each channel spends in state B or in the other word the probability of being in state B that multiplied by 100. Figure 2 shows the result of each channel being in Bad state.

The achieved result from Figure 2 helps to categorize the Channels based on three different groups as "Bad Channels",

"Good Channels" and "Very Good Channels" by identifying two threshold values and categorizing those decides to transmit data over "Very Good" and "Good" channels then by such transmission the performance of the communication system can be improved. Then the error probability in Bad state for each channel is computed. Figure 3 shows these probabilities for 100 different channels.

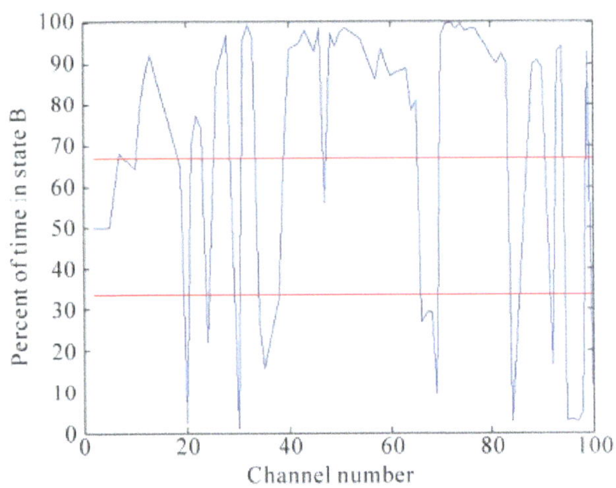

Figure 2: Percent of time that each channel spends in state B.

Testing

Gilbert-Elliot channel model is used to simulate the error process and correctly reproduce all of its statistical properties. To validate the model, the error process generated by the model must be compared to the measured error process. For testing, the program bit error sequence is generated using Markov chain model. Two programs are made as follow: marcov_gen is a bit error sequence generator for Markov parameters and marcov_test tests the bit error sequence and the output is displayed in workspace.

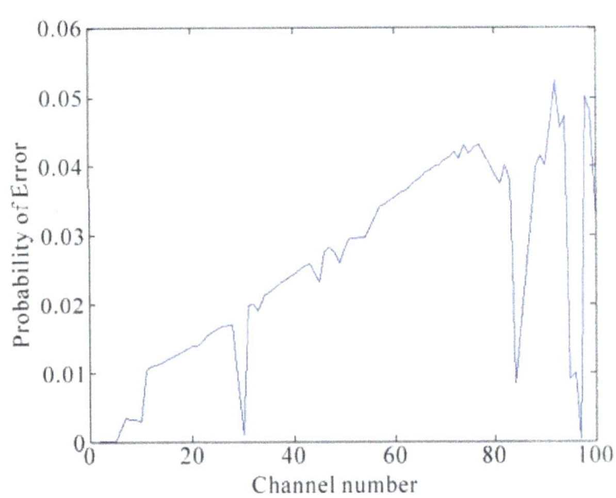

Figure 3: Error probability being in Bad state for each channel.

The objective of the parameter estimation is to choose values of the model parameters that generate error burst and error gap distributions that reassembles the corresponding measured distributions as close as possible. Therefore for testing the mean and variance of error burst and error gap of regenerated error sequence are calculated and compared by statistical parameters of channel bit error sequence, where the result is shown in Table 1

A: Statistical parameters channel error sequence.

B: Statistical parameters of regenerated error sequence.

For testing, first Markov model parameters of a channel error sequence are computed, then a sequence of the model is generated and statistical parameters are computed. The statistical parameters must be as equal as channel error sequence. It is important to mention that first state of the Markov model in function marcov_gen chooses the probability 0.5, so sometimes two different answers can be seen and that the nearest one to the error sequence statistic is the correct one.

EVALUATION OF FREQUENCY HOPPING

To design the frequency hopping (FH) model, MATLAB Simulink has been used. The spreader at transmitter section is an M-FSK modulation but the input of the modulator is hopping index. This section consists of a PN Sequence Generator, an Assemble Packets block and a Goto block as shown in Figure 4.

The design of frequency hopping spreader is shown in Figure 5. The spreader part consists of M-FSK modulator base (with M equal to 64), a From block (Hop index that is created in previous step), a To Frame block and a Multiplication block. The block parameter of FSK modulator is 64 in M-FSK number and it means that there are 64 hopping sections. These sub-bands are selected by the hop indexes.

The design of frequency hopping despreader, is the same as spreader section but the output of M-FSK modulator block is complex conjugated as shown Figure 6.

This frequency hopping model is used for evaluation of three different modulations: QAM, QPSK, GFSK, and compares the performance with the situation without frequency hopping. Performance evaluation is based on BER values under two situations (with and without FH) versus normalized signal-to-noise ratio (SNR) measured by E_b/N_0 values of the channel, as shown in Figures 7-9

From Figure 7 it can be seen that applying FH with QAM modulation does not lead to a sensible improvement in performance or significant reduction of BER. From Figure 8 it can be seen that applying FH with QPSK modulation gives a good result and reduces significantly BER compared to without FH at same level of SNR. From Figure 9 it can be seen that applying FH with GFSK modulation reduces dramatically BER compared to without FH at same level of SNR and lead to a much higher performance.

Table 1: Statistical parameters of channel error and regenerated error sequence.

	error burst mean	error gap mean	error gap variance	
SNR = 3dB Input power = 1	1.0568	18.1134	319.4516	A
	1.0492	18.4713	319.3184	B
SNR = 3dB Input power = 2	1.1456	7.7868	48.9981	A
	1.1500	8.0421	55.0026	B
SNR = 3dB Input power = 3	1.2201	5.6570	25.5754	A
	1.2271	5.5166	24.5044	B

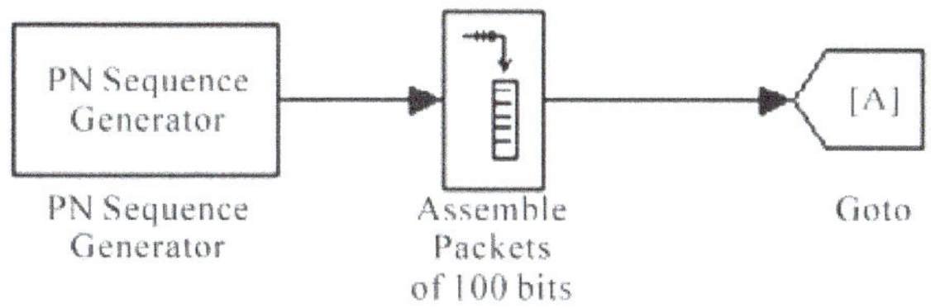

Figure 4: Model of frequency hopping in Simulink.

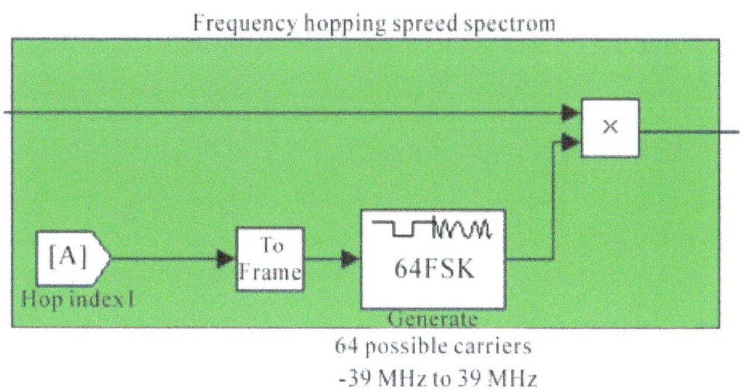

Figure 5: Design of frequency hopping spreader.

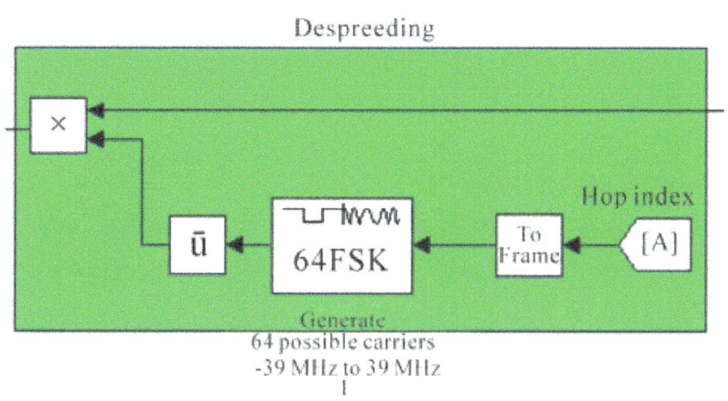

Figure 6: Design of frequency hopping despreader.

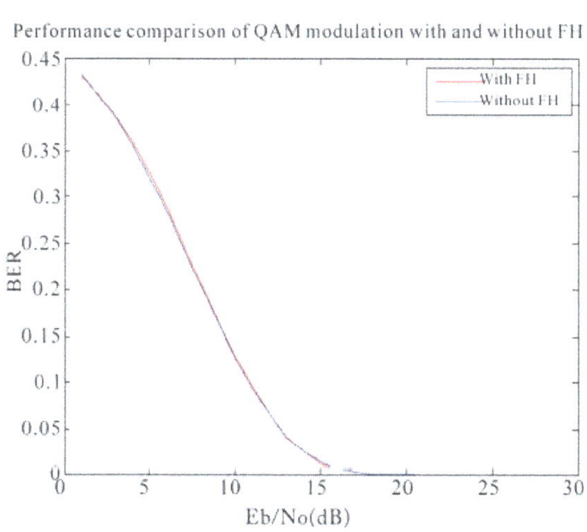

Figure 7: QAM modulation.

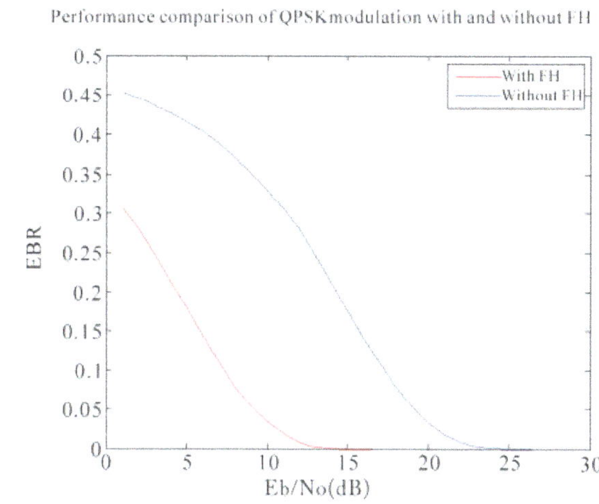

Figure 8: QPSK modulation.

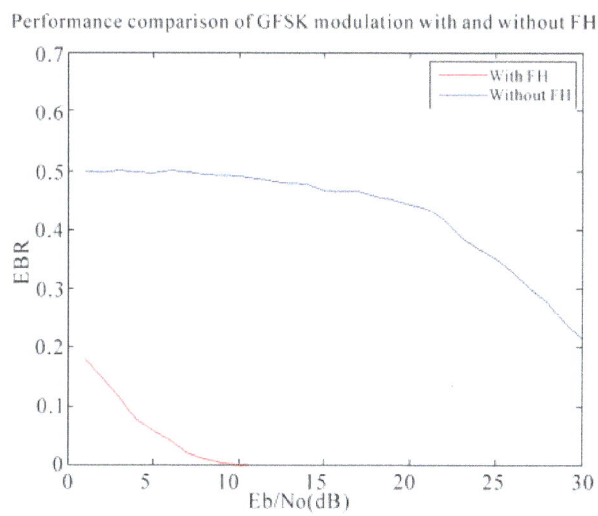

Figure 9: GFSK modulation.

In overall, based on the evaluation results it can be concluded that applying the designed FH schemes with certain modulations can improve their communication performances, especially at weak SNR levels as most cases of short range wireless communications have.

CONCLUSIONS

As a result it can be concluded that adaptive frequency hopping is a powerful technique to deal with interference and Gilbert-Elliot channel model is a good technique to analyze the situations of channels by categorizing the channel conditions based on their performance as Good or Bad, and then apply adaptive frequency hopping which hops frequencies adaptively by analyzing the state of the channel in case of environmental problems such as interferences and noises to improve the communication performance. Frequency hopping spread spectrum is modelled with MATLAB and three different modulations i.e. QAM, QPSK and GFSK are studied to investigate which of these modulations are good to apply with FHSS model. The simulation results show that applying FHSS with QAM modulation dose not lead to a remarkable reduction of BER, but with QPSK modulation gives a good result and reduces BER at lower SNR, while in GFSK modulation shows a significant reduction of BER and lead to a high performance.

ACKNOWLEDGEMENT

The author gratefully acknowledges Dr Yang Liu for his valuable support and guidance in this research work.

REFERENCES

1. E. N. Gilbert, "Capacity of Burst-Noise Channels," Bell System Technical Journal, Vol. 39, 1960, pp. 1253-1265.
2. E. O. Elliott, "Estimates of Error Rates for Codes on Burst-Noise Channels," Bell System Technical Journal, Vol. 42, 1963, pp. 1977-1997.
3. R. E. Ziemer, R. L. Peterson and D. E. Borth, "Introduction to Spread Spectrum Communications," Prentice Hall, Englewood Cliffs, New Jersey, 1995.
4. R. J. Bates and D. W. Gregory, "Voice & Data Communications Handbook," McGraw-Hill Osborne Media, Berkeley, CA, 2001.
5. J. Zander and G. Malmgren, "Adaptive Frequency Hopping in HF Communications," IEE Proceedings Communications, Vol. 142, No. 2, 1995, pp. 99-105.
6. J. J. Lemmon, "Wireless Link Statistical Bit Error Model," Institute for Telecommunication Sciences, 2002.

CHAPTER 7 WIRELESS COMMUNICATIONS ON LINEAR TOPOLOGIES

Daniele De Caneva, and Pier Luca Montessoro

Keywords: Wireless Sensor Networks, Linear Topology, Pervasive Devices, Ad-Hoc Network

INTRODUCTION

Many routing protocols have been designed for Wireless Sensor Networks (WSNs) considering nodes that operate in a mesh topology. For specific application scenarios, however, a mesh topology may not be appropriate or simply not corresponding to the natural node deployment. Bridge [1] or pipeline [2] monitoring applications are examples where the position of sensor nodes is predetermined by the physical structure and application requirements. In this applications, where it is clearly present a privileged dimension, it is quite natural to take advantage of it.

So far, little focus has been given to efficient MAC protocols for low-power, wireless communications over linear topologies. This paper presents WiWi (Wireless Wire): a contention-free MAC protocol based on synchronous multi-hop transmission along a chain of independent nodes. The original and main purpose of WiWi is to perform the virtualization of a wired link by means of an ad-hoc network, made of a chain of tiny short-ranged transceivers with limited power capabilities. Nonetheless, WiWi is not limited to end-to-end communication but it can be profitably used to collect data along the path.

In the WiWi architecture, devices are displaced in order to build up a linear (or curvilinear) strip. WiWi does not require any routing table or complex calculation for message delivery: this makes it feasible to apply the protocol even on very simple transceivers, with limited memory and processing capabilities. Moreover, it is ready for hardware implementations that could be realized on tiny pervasive devices. The issue regarding synchronization of nodes along the network is addressed by choosing fixed-size messages. WiWi takes advantage of linear topology and synchronous communication to provide deterministic and predictable latency and throughput in both directions. As will be later described, they can be configured by modifying protocol parameters in order to fulfill the nodes' capabilities and the application requirements. The paper is organized as follows. After this introduction, Section 2 presents interesting related work. The WiWi architecture, protocol, performance and applications are discussed in Section 3, whereas Section 4 shows the prototypal implementation. A fault-tolerant WiWi node architecture is presented in Section 5, and Section 6 draws some conclusions tracing some expectations for future work

RELATED WORK

Design issues and tradeoffs that need to be considered for power-constrained WSNs with low data rate links have been addressed and studied in noteworthy works [3-5].

A series of studies on routing in ad hoc networks and WSNs face the problem of optimization on behalf of higher layer parameters, such as efficient localization, propagation, resiliency, and so on, proposing a wide variety of algorithmic solutions. Some approaches also examine cross-layer issues that aim at minimizing energy consumption and computation [6]. It is likely to observe that topology in most studies is often in a second order matter, since nodes are expected to be mobile or to be deployed taking random positions in a field. More recently, deployment of wireless sensors has been studied and optimized to achieve coverage and connectivity [7]. Topology is important for any type of network because it has great impact on the performance of the system. Limited research has been conducted on the effect that well-defined topologies have on protocols for wireless networking [8]. The focus, however, has been on mobile networks rather than the ones with regular topologies or with a fixed node placement. The case of patterned WSNs is known in literature and well described in [9]. On the

---

Citation: D. Caneva and P. Montessoro, "Wireless Communications on Linear Topologies," International Journal of Communications, Network and System Sciences, Vol. 3 No. 12, 2010, pp. 925-933. doi: 10.4236/ijcns.2010.312126.

other hand, most works are interested in demonstrating how the topology of a WSN impacts on the performance of a given MAC or routing protocol. Our perspective, instead, aims at simplifying protocol requirements and device complexity starting from a very specific application, in order to propose a reliable and efficient solution for this and similar problems. Quite common methodologies in WSN-related protocol development [10] recommend that protocols should reduce the number of contentions to improve power saving, as well as using shorter packet lengths. The receiver usage time, however, tends to be higher for protocols that require the mobile nodes to sense the medium before attempting a transmission. In our system, the protocol has been optimized on behalf of these main goals. Moreover, devices' link layer is capable to perform and keep synchronization during the whole lifetime of the network

In literature we can find some examples of algorithms and protocols that are specifically aimed for linear topologies. MERR [11] is a routing protocol whose reference scenario is a network made by sensors deployed over a linear topology. MERR deals with the problem of finding the best route from every node to a common control center. With MERR, Zimmerling et al. propose a distributed protocol where each node independently chooses the best relay node among its neighbors.

In [12] is presented an algorithm whose aim is to minimize the routing path and, at the same time, balance the load. In particular that work covers the special case of a network where nodes are located in a narrow strip with a width at the most $\sqrt{3}/2$ times the communications range of each node.

Both [11, 12] are focused on routing problems without considering the underlying MAC protocol, which could become a real bottleneck for network performances.

In [13] is presented DiS-MAC (Directional Scheduled MAC). This protocol has been developed for wireless sensor networks that show a linear topology. It reaches the considerable channel utilization of 1/2, but requires every node could direct the radiation beam of its antenna and suffers from being unidirectional. WiWi recalls some DiS-MAC features; in particular both protocols avoid interferences between simultaneous transmissions by alternating transmissions between adjacent nodes. However, WiWi presents many important advantages: it does not require directional antennas, it provides bidirectional communication over a single RF channel (thus providing support for an end-to-end acknowledgement), and it can be configured in order to make the bandwidth and latency fit the application needs.

WIWI ARCHITECTURE AND COMMUNICATION PROTOCOL

The development of WiWi was originated by the need for a system able to emulate a wired link by means of an ad hoc network constituted by nodes distributed along a strip. The purpose of this emulation is to handle scenarios where a single hop wireless link is not feasible and a wired link is not practical. An example could be given by a speleologist going deep down into the bowels of the Earth, who can deploy the wireless network while it goes further with the exploration in order to maintain a communication channel with the outside world. Other examples can be found in all those situations where a multihop link is required, in particular those bounded to monitoring applications.The design of WiWi architecture is based on the observations about wireless communications presented by Min and Chandrakasan in [3]. In particular the authors of [3] showed that even if a power law often describes the radiated power necessary to transmit over a distance d and path loss n, this term alone fails to consider the energy overheads of the hardware. Min and Chandrakasan suggested an energy consumption model with an additional distance-independent term in order to take account of overheads caused by transmitter and receiver electronics (such as PLLs, VCOs, LNA, bias currents, etc.)

The proposed model is the following:

$$E(d) = \alpha + \beta d^n \qquad (1)$$

where α is the distance-independent term. Additionally, Min and Chandrakasan collected the estimated values of the model's term for a set of short-range radios having maximum output power up to +20 dBm

Starting from these data, authors of [3] noted that for short-range radios typically used in MANET research, the value of α dominates the value of the path loss term. This has severe consequences on multi-hop wireless communications that try to reduce energy consumption by adding intermediate relay nodes in order to reduce the path loss term. This strategy affects only the path loss term (βd^n) limiting the value of d, however this is useful only for long range radio links where that term is dominant. The energy model presented in [3] shows that a two hop link consumes less energy than one hop link when

$$\alpha < \beta d^n \left[1 - 2^{(1-n)}\right] \qquad (2)$$

Min and Chandrakasan noted that with the exception of the μAMPS-1 custom radio, this inequality never holds for typical path losses

Similar considerations where made in [14] where Bhardwaj et al. introduced the concept of characteristic distance of a transceiver. This distance is strictly bounded to the transceiver characteristics and minimizes power consumption in a multi-hop communication. In fact, the characteristic distance represents the optimal tradeoff between distance-independent and distance-dependent terms in power consumption relation

WiWi has been designed as a synchronous multi-hop communication scheme where nodes are intended to be deployed with mutual distances approximately close to the characteristic distance (Figure 1). No other assumption is made over node deployment: the chain of nodes is not required to strictly follow a straight path as it can bend. Moreover, in order to minimize the possible production costs, we assumed that pervasive devices could be absolutely identical one to each other, in the sense that no factory-set unique ID is required

A synchronous architecture is not very common in typical sensor networks described in literature, nevertheless it is our opinion that its deterministic behavior better suites the aim of WiWi .A detailed description of the synchronous communication pattern is presented in the following subsection.

Basic Communication Pattern

The communication between WiWi nodes is synchronous, based on fixed size packets, and follows a staggered pattern like the model presented in DMAC [15]. However, unlike the DMAC protocol, which is designed to handle tree topologies, in WiWi the synchronism is intrinsic in the communication model and bidirectional data flows are supported.

Figure 2 shows the communication model and the different handling of downstream and upstream data flows. In this and following diagrams the vertical axis is depicted downward, according to the direction of the downward flow that provides synchronization to all the nodes. The master node, source of synchronization, is at position n = 0. The downstream data flow is generated by the head of the chain which acts as master end point. This data flow (the gray one in Figure 2) proceeds downwards from the head of the chain to the tail following a strictly staggered pattern: a node sends a packet to the next one, which in turn immediately forwards the packet further down along the chain. This stream is responsible of maintaining overall network synchronization too: every node resynchronizes its clock upon the start of the incoming downstream packets. The upstream flow follows the same principle of passing messages along the chain, but between the reception of a packet and its forwarding, the node waits four time slots in order not to collide with the downstream one. In Figure 2, the upstream flow is depicted in white blocks, while the arrows of different patterns follow the advancement of different upstream packets.

This scheme combines very good performance with high regularity that makes its implementation easy. In fact, once a node is synchronized with the downstream flow, its activity pattern is receive-transmit-idle-transmit receive-idle (R-T-I-T-R-I) regardless its position in the chain. Moreover, in WiWi, nodes require no explicit addressing because within the range of transmission there is only one receiving node, i.e. the destination of the next hop for the packet

It is worth noting that the transmission scheme is designed to avoid interferences among different nodes. In fact, from the receiver point of view, when an adjacent node is transmitting, the closest other transmitting node is three hops away. So, considering the common assumption that the interference radius is twice the nominal transmission one [13] this three-hop safe distance grants a good resiliency from inter-node interference.

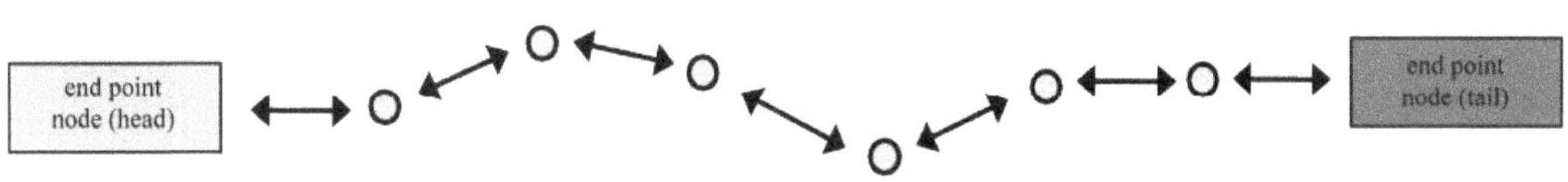

Figure 1: WiWi topology: The linear strip as a chain of nodes.

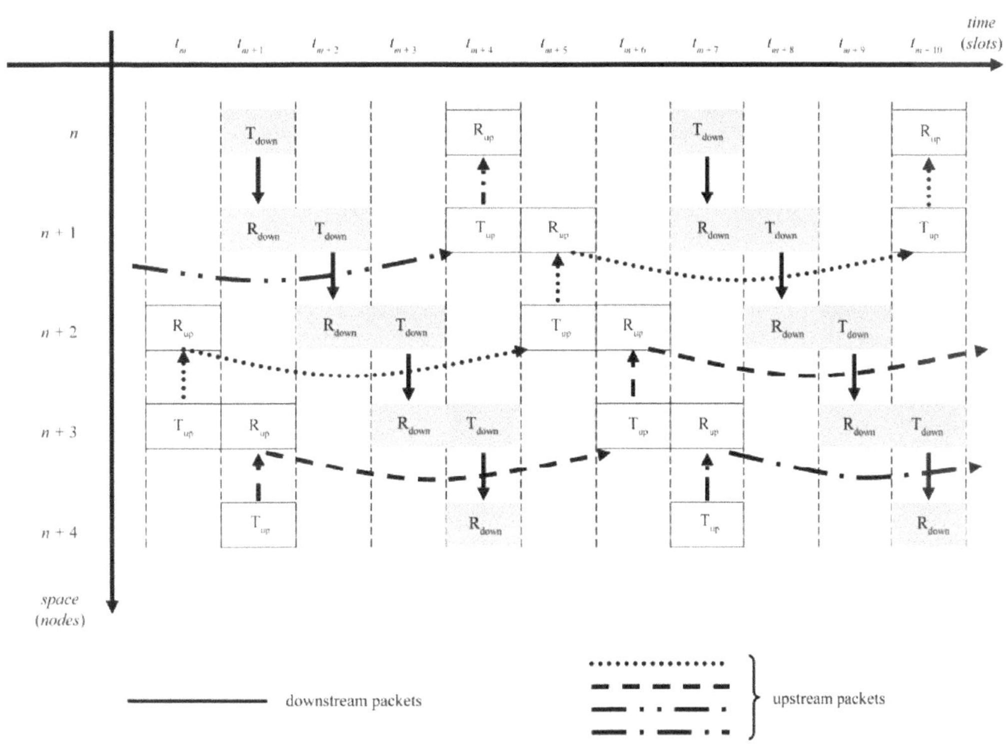

Figure 2: Bidirectional staggered transmission.

Performance Evaluation and Advanced Patterns

A performance evaluation of this scheme of the communication pattern is straightforward. Let N be the number of hops of the link, B the maximum raw bit rate and S the time slot associated with a single packet transmission (which will also include a communication safety time in order to handle clock drifts between consecutive nodes). Both downstream and upstream flows use one time slot every six, therefore each one provides a bit rate of nearly B/6 and the overall channel utilization is equal to 1/3 (1/6 for each stream). Actually, the bit rate will necessarily be lower than B/6 due to non-idealities of hardware (e.g. clock drifts). It is worth noting that a channel utilization equal to 1/3 is the optimum for a bidirectional communication, being the maximum allowed by the minimum three-hops safe distance chosen to avoid interference between simultaneous transmissions. Additionally, the throughput of both streams equal to 1/6 of the transceiver rate has to be considered a good tradeoff between performance and latency, since in [16] the maximum operational limit of a unidirectional wireless chain is called to be equal to 1/4.

The downstream flow has the minimum latency allowed by a store-and-forward pattern, being the delay equal to N·S. On the other hand the upstream flow delay is five times larger: since a packet can be sent every six time slots, there is, in average, an additional constant delay of three time slots per hop. However, WiWi architecture is flexible: the receivetransmit pattern can be adjusted to fit the desired tradeoff between latency, bandwidth and symmetry of upstream and downstream flows, at the cost of a lessthan-optimal channel utilization. Even the distance between simultaneously active transmitters could be changed depending on the RF layer requirements. Figure 3 shows a pattern similar to the previous one but providing asymmetrical throughput (B/5 downstream and B/10 upstream) and asymmetrical latency (N·S downstream and 10 N·S upstream, plus in average 2.5 time slots and 5 time slots respectively to wait for the transmission window).

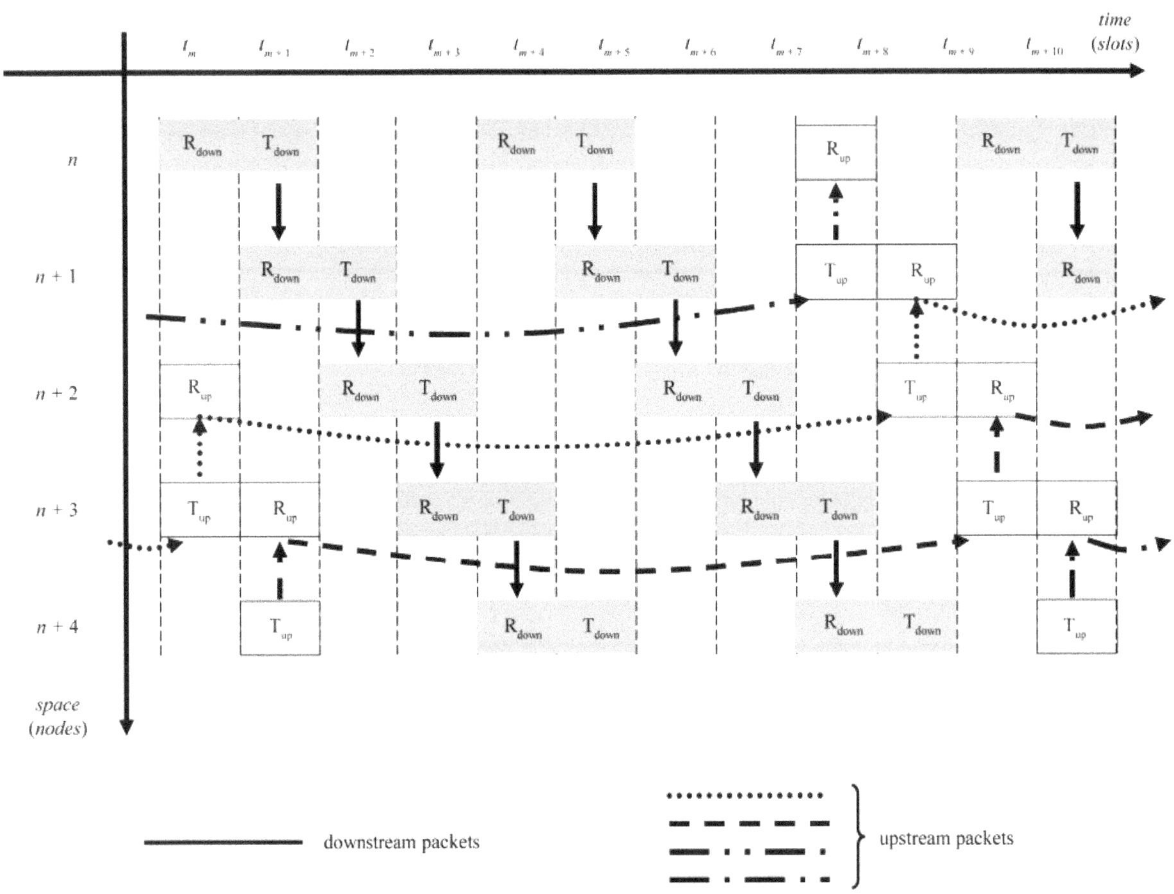

Figure 3: Bidirectional, staggered transmission with asymmetric throughput and latency over a WiWi link.

Figure 4 shows another pattern where both bandwidth and latency are symmetrical: a transmission occurs every eight time slots in each direction, providing an overall channel utilization of 25%, whereas the latency is 2N·S. The block arrows show the travel, in space and time, of downstream and upstream packets. Unfortunately, in this case the receive (R), transmit (T) and idle (I) pattern is no longer independent from the cluster position along the chain: even-order clusters run R-R-T-T-I-I-I-I, whereas odd-order clusters behave R-I-T-R-I-T-I-I, thus requiring a bit more complex chain setup. WiWi does not provide acknowledgment to guarantee the correct packet exchange between nodes: datagram transmission is much more flexible in the scenarios in which WiWi is supposed to operate. This choice implies that the nodes could be very simple because they do not have to store more than two packets at a time in their internal memory. Anyhow, error correction codes could be adopted to increase the reliability of communications along the WiWi chain. In addition, an acknowledgement system could be implemented by higher layer protocols between the head and the tail of the strip. In scenarios where the end points are many hops away from each other and the packet error rate is particularly severe, this solution could be responsible of poor latency performance due to frequent retransmissions. This problem could be mitigated by displacing intermediate endpoints, which corresponds to deploy many WiWi strips in sequence instead of one single strip.

Strip of Sensors: WiWi as a Conveyor Belt

So far, WiWi has been presented as a protocol for end-to-end communication on a strip of short-range wireless devices. But the WiWi synchronous nature makes it a powerful tool to collect data from a strip of sensors. Conventionally, in routing-oriented sensor networks, a sensor places its reading in a packet that finds its way through the network, fulfilling at each hop the chosen MAC rules. In WiWi each node can host one or more sensors. The data flow that continuously traverses the strip can be seen as a conveyor belt on which the sensors place their readings. Several application-level protocols can be adopted to handle possible conflicts: for example, a flag that marks each time slot when vacant, or a more sophisticated token-based protocol

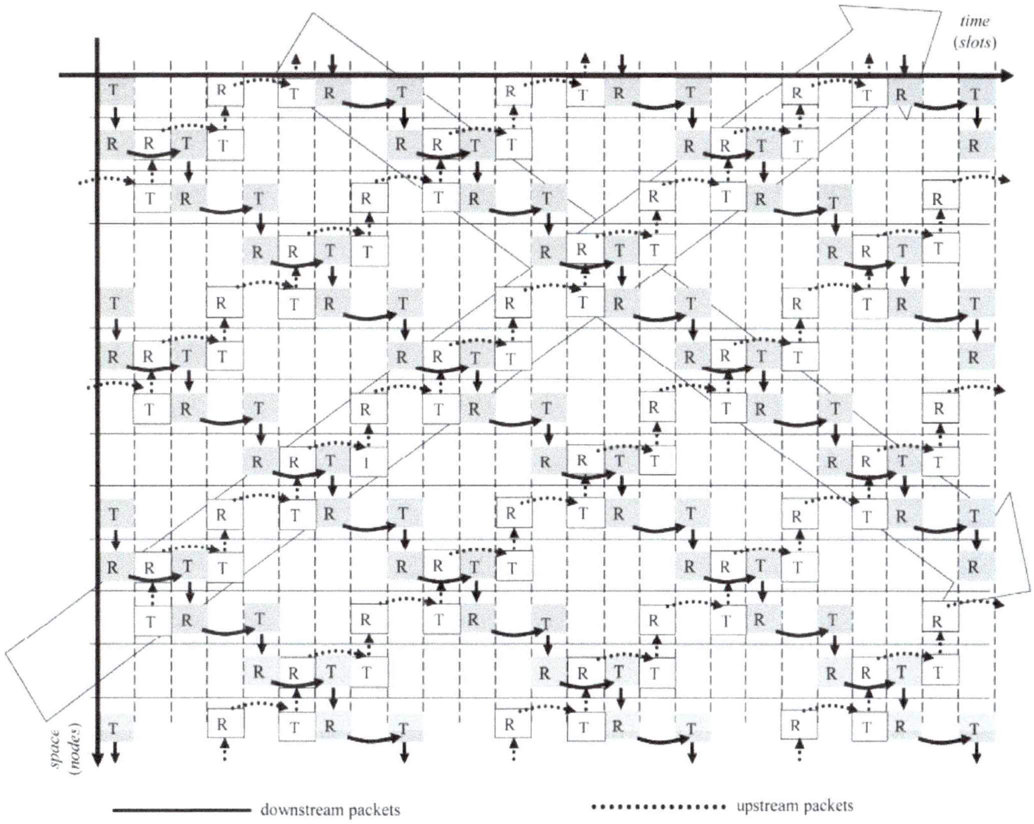

Figure 4: Bidirectional, staggered transmission with symmetric throughput and latency over a WiWi link.

The WiWi application as conveyor belt for a strip of sensor has been tested using accelerometers in our prototypal implementation, as described in the next section.

TESTING AND PROTOTYPING

WiWi has been validated in both fault-free and faulty scenarios (see Section 5) by simulation using the Omnet++ platform [17], but the obtained results are limited by the validity of the transmission model that, in the real world, should include RF hardware performance, battery levels, environment, interferences, etc. In order to verify the feasibility of the proposed approach and show the effectiveness of the system, we developed a prototype for two demo applications: end-to-end communications and strip of sensors.

The prototype (Figure 5) accommodates a Microchip microcontroller (PIC16F689), I/O hardware for debugging purposes, and a transceiver. Two different transceiver modules by Aurel [18] have been tested; their bit rates and transmission ranges are reported in Table 1.

Table 1: Transceivers performance.

Transceiver	Modulation	Bit rate	Transmission range
XTR-CYP-2 .4 GHz	GFSK, 78 channels 2.4 GHz band	up to 64 Kbit/s	45 m
XTR-VF-2.4 LP	GFSK, 98 channels 2.4 GHz band	up to 1 Mbit/s	10 m

Figure 5: Hardware prototype (5 cm × 6 cm).

The prototype is equipped with a switching step-up voltage regulator in order to reduce power consumption and make possible the use of two AA batteries as power supply. USB connectivity has been included in order to expedite firmware debugging and enable WiWi endpoints to exchange data with any common PC. The first application developed with this prototype emulated a chat application between two computers, using WiWi as channel to deliver messages from one user to the other (Figure 6). In the second application WiWi has been used to aggregate and transport data coming from 3-axis accelerometer sensors connected to WiWi nodes (Figure 7). Data coming from sensors were preprocesses by nodes in order to filter noise and obtain the maximum acceleration for each axis in 10 milliseconds time windows. Since this kind of application doesn't need an extremely high throughput, 500 ms time slots have been used, much longer that the 1 ms packet length. This way proper throughput and consumption performance have been achieved since each node had more time to spend in low power mode. At each step along the chain, data measurement pertaining to the traversed node was inserted into a specific field within the current packet. Since this data-collecting application is essentially unidirectional, the downstream flow was delegated to remote control and management of the nodes. This flow was much less intense, therefore downstream packets were shorter (one quarter of upstream packets), resulting in additional energy saving, since idle time has been further increased.

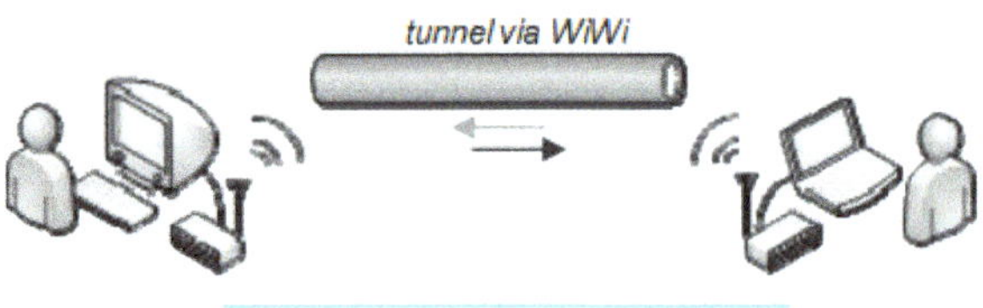

Figure 6: Chat demo application.

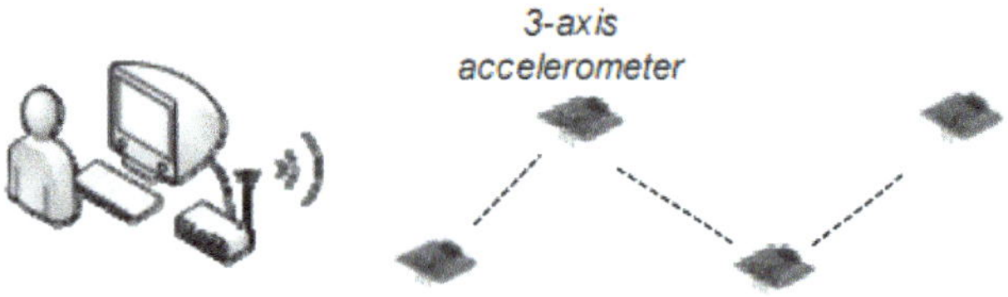

Figure 7: Sensoristic demo application.

DESIGNING A FAULT-TOLERANT WIWI NODE

An obvious weakness of chain topologies is that a single node failure can block the whole communication channel. Nevertheless, if the topology of the physical environment supposed to host the chain is strictly linear, neither backup paths nor meshed topologies can be deployed. Since multiple nodes within the same transmission range are not suitable for the WiWi synchronous protocol due to interferences and performance loss, a fault-tolerant WiWi node has been designed.

A fault tolerant WiWi node is made of multiple transceiver modules (Figure 8), each one able to handle the whole communication protocol. Their activity is scheduled on behalf of a round robin policy. Just one module at a time is effectively involved in packet forwarding, while the other ones act as backup modules. The modules are ordered, and the first one is the module on duty, responsible for packet forwarding. During the receive time slot every module receives and stores the packet coming from the previous node in the chain. In the subsequent transmission slot, the module on duty forwards the packet, while at the same time all backup modules sense the transmission to be sure it occurs. If the first backup module perceives the loss of the module on duty, after a short "sense time slot" it forwards the packet. The remaining backup modules keep listening to the channel in order to be sure that a backup module has reacted. The process is iterated until a backup module forwards the packet or there are no more backup modules left. Figure 9 shows this technique giving an example of redundancy and failure management over four modules: at the beginning of the transmit slot the module on duty (number 0) should forward the packet (time t_1). Modules 1, 2 and 3 start sensing the transmission for the duration of the s1 sensing window. If no transmission occurs, at time t_2 node 1 should backup module 0 transmitting the packet. Remaining modules 2 and 3 continue to sense, in order to backup module 1 (with module 2 at time t_3) and, if still needed, module 2 (with module 3 at time t_4). In the example the packet is forwarded by the last backup module.

After a module failure, active backup modules redefine their order. This way, the same technique can be used to provide fault tolerance and to schedule sleep-times for load balancing and energy saving

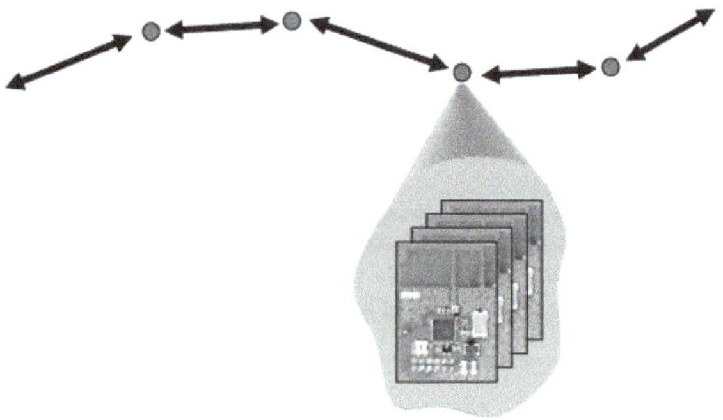

Figure 8: Hardware prototype.

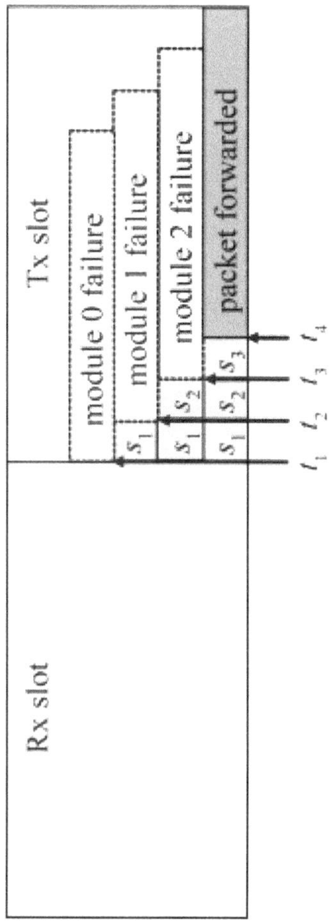

Figure 9: Redundancy mechanism.

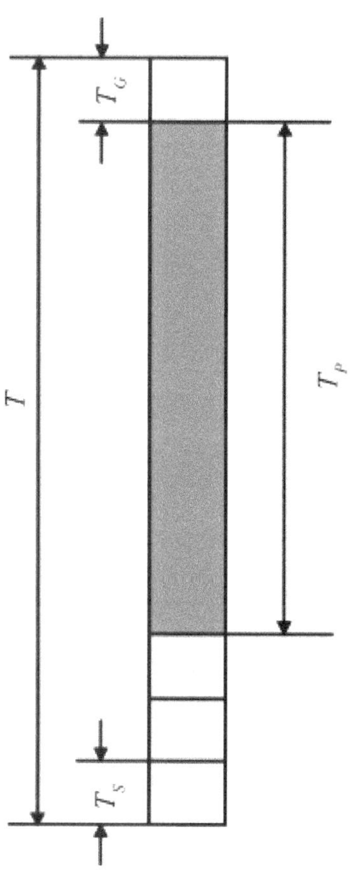

Figure 10: Time slot components.

This fault tolerance extension to the basic WiWi protocol grants a redundancy equal to the number of backup modules. The slot time upon which WiWi is based must be extended to make room to the sensing time windows, leading to a minor loss in throughput and latency performances.

The extended time slot can be modeled as in Figure 10. Being N_{max} the maximum number of nodes belonging to a pool, the duration of a slot is represented by three components:

- T^P: time needed to receive the packet, depending only on packet size and transceiver bit-rate;
- T^S: sensing time window: sum of the sensing time and the time needed to commute the radio module from sensing state to transmitting state;
- T^G: the inter-packet gap, a period comprehensive of the time needed to switch from the receiving state to the transmitting state, the time needed to store and elaborate the packet, and a safety margin to counteract clock drifts.

Hence, with N_{max} modules in each node, the time dedicated to packet transmission is equal to $T_S(N_{max}-1)+T_P$ the total duration T of a slot is:

Thanks to the decoupling of the tiers composing the architecture, the other performance relations remain valid. In fact throughput is equal to $T_{pck}/(6T_S)$ while latency is equal to NT_S for the downstream flow and $(N + 3)T_S$ for the upstream flow.

The backup technique requires modules within each pool to be dynamically enumerated and ordered. Optionally, the architecture could include a set of algorithms for modules deployment and management that do not require any manual configuration or the assignment of unique node ID at firmware/hardware level. This aspect is not discussed here because several algorithms for cluster organization can be found in literature and easily applied to WiWi.

CONCLUSIONS AND FUTURE WORK

A new MAC protocol for linear topologies, called WiWi, has been presented. It is based on a chain of short-range communication devices. Predictable latency and bandwidth are provided by the synchronous communication scheme and by specific transmit-receive patterns that have been shown in Section 3. Fault tolerant node design has been discussed as well. WiWi has been simulated and tested on hardware prototypes. Strategies to use WiWi in different scenarios are currently under study, in particular to use the strip as a fundamental brick for more complex topologies. Moreover, different channels can be used to handle overlapping strips, thus allowing a variety of topologies that combines TDMA and FDMA techniques. In tree topologies, an asymmetric configuration of the protocol can be exploited to benefit the links from the root to the leaves, making WiWi a challenging solution to control and deliver data to a large number of devices in widespread applications.

As a synchronous technique, WiWi is well suitable to include strategies, already known in literature, to schedule stand-by periods and wake-up events to save energy. Since each strategy may be optimal only for one or few kind of applications, another current research direction is to define some detailed application scenarios and select for them the optimal energy-saving schemes.

ACKNOWLEDGEMENTS

Authors wish to thank Giampietro Tecchiolli from Eurotech S.p.A. for his encouragements and his interest in WiWi application scenarios. We also wish to thank Francesco Cossettini and Manuel Nobile for their contribution in the simulations of the protocol and the development of the hardware prototype.

REFERENCES

1. S. Kim, S. Pakzad, D. Culler, J. Demmel, G. Fenves, Glaser and M. Turon, "Health Monitoring of Civil Infrastructures Using Wireless Sensor Networks," Proceedings of 6th International Symposium on Information Processing in Sensor Networks, Cambridge, 25-27 April 2007, pp. 254-263.
2. I. Jawhar, N. Mohamed and K. Shuaib, "A Framework for Pipeline Infrastructure Monitoring Using Wireless Sensor Networks," Proceedings of Wireless Telecommunications Symposium, Pomona, 26-28 April 2007, pp. 1-7.
3. R. Min and A. Chandrakasan, "Top Five Myths about the Energy Consumption of Wireless Communication," ACM SIGMOBILE Mobile Computing and Communications Review, Vol. 7, No. 1, 2003, pp. 65-67.
4. A. Youssef, J. Haslett and S. Magierowski, "Design Issues for Sensor Network RF Receivers," Proceedings of Canadian Conference on Electrical and Computer Engineering, Vancouver, 22-26 April 2007, pp. 1535-1538.
5. C. Schurgers, V. Tsiatsis, S. Ganeriwal and M. Srivastava, "Optimizing Sensor Networks in the Energy-LatencyDensity Design Space," IEEE Transactions on Mobile Computing, Vol. 1, No. 1, 2002, pp. 70-80.
6. Y. Zhang and L. Cheng, "Cross-Layer Optimization for Sensor Networks," Proceedings of New York Metro Area Networking Workshop, New York, September 2003, pp. 1-3.
7. X. Bai, S. Kuma, D. Xua, Z. Yun and T. H. La, "Deploying Wireless Sensors to Achieve both Coverage and Connectivity," Proceedings of 7th ACM International Symposium on Mobile Ad Hoc Networking and Computing, Florence, 22-25 May 2006, pp. 131-142.
8. A. Conti and D. Dardari, "The Effects of Nodes Spatial Distribution on the Performance of Wireless Sensor Networks," Proceedings of IEEE 59th Vehicular Technology Conference, Milan, Vol. 5, 17-19 May 2004, pp. 2724- 2728.
9. H. M. F. AboElFotoh, E. S. El Mallah and H. S. Hassanein, "On the Reliability of Wireless Sensor Networks," Proceedings of IEEE International Conference on Communications, Istanbul, Vol. 8, 11-15 June 2006, pp. 3455- 3460.
10. J. Chen, K. M. Sivalingam and P. Agrawal, "Performance Comparison of Battery Power Consumption in Wireless Multiple Access Protocols," Wireless Networks, Vol. 5, No. 6, 1999, pp. 445-460.
11. M. Zimmerling, W. Dargie and J. M. Reason, "EnergyEfficient Routing in Linear Wireless Sensor Networks," Proceedings of IEEE International Conference on Mobile Adhoc and Sensor Systems, Pisa, 8-11 October 2007, pp. 1-3.
12. L. Zhang and J. Gao, "Load Balanced Short Path Routing in Wireless Networks," IEEE Transactions on Parallel and Distributed Systems, Vol. 17, No. 4, April 2006, pp. 377-388.
13. T. Karveli, K. Voulgaris, M. Ghavami and A. H. Aghvami, "A Collision-Free Scheduling Scheme for Sensor Networks Arranged in Linear Topologies and Using Directional Antennas," Proceedings of 2nd International Conference on Sensor Technologies and Applications, Cap Esterel, 25-31 August 2008, pp. 18-22.
14. M. Bhardwaj, T. Garnett and A. Chandrakasan, "Upper Bounds on the Lifetime of Sensor Networks," Proceedings of IEEE International Conference on Communications, Helsinki, 11-14 June 2001, pp. 785-790.
15. G. Lu, B. Krishnamachari and C. S. Raghavendra, "An Adaptive Energy-Efficient and Low-Latency MAC for Data Gathering in Sensors Networks," Proceedings of 18th IEEE International Parallel and Distributed Processing Symposium, Santa Fe, 26-30 April 2004, pp. 224- 231.
16. J. Li, C. Blake, D. Couto, H. Lee and R. Morris, "Capacity of Ad Hoc Wireless Networks," Proceedings of the 7th ACM Annual International Conference on Mobile Computing and Networking, Rome, 16-21 July 2001, pp. 61-69.
17. Omnet++ Community, 2010. http://www.omnetpp.org
18. Aurel S.p.A., 2010. http://www.aurel.it

CHAPTER 8

A RANDOM ACCESS PROTOCOL WITH MULTI-PACKET RECEPTION FOR INFRASTRUCTURE-LESS WIRELESS AUTONOMIC NETWORKS

Jahangir H. Sarker and Hussein T. Mouftah

Keywords: Ad hoc networks, Autonomic, Cost, Multi-packet reception, Node transmitting probability, Sensor networks, Slotted ALOHA

INTRODUCTION

Selecting an efficient protocol for sharing a common broadcast wireless channel among a set of mobile nodes is a challenging task. The conventional and emerging wireless networks, such as satellite, cellular, wireless local area, ad hoc, sensor and wireless mesh networks, all face the same challenge. Among the protocols for random access through a common wireless channel, Slotted ALOHA is one of the most attractive medium access control (MAC) protocol due to its simplicity and low delay (under light load) for bursty traffic. The capacity of Slotted ALOHA system can be improved significantly by receiving multiple packets in a time slot [1]. Multiple packets can be received in a time slot using directional antennas and multiple receivers, with and without capture effect [2–4]. Direct sequence code division multiple access (DS-CDMA) is another random multiple access, where mobile nodes transmit packets in the same channels using different orthogonal (or quasi-orthogonal) spreading codes [5]. The multi-packet reception (MPR) capability can be achieved using direct sequence spread spectrum (DS-SS) Slotted ALOHA [6]

The channel state information (CSI) directly influences the quality of the physical (PHY) layer. The MPR capability can be enhanced in the PHY layer by the proper estimation of CSI. The performance of MAC layer can be improved by mutual interactive of MAC-PHY perspective [7,8]. The MPR capability in wireless network can be enhanced by collision resolution [9]. An infinite user population model for the analysis of random access protocols jointly assigned by MPR and retransmission diversity is proposed and analyzed in [10]. A centralized multi-queue service room MAC protocol (MQSR) can optimize the same protocol by avoiding unnecessary empty slots for light traffic and excessive collision for heavy traffic [11]. The optimized MQSR has superior throughput and delay performance as compared to, for example, the Slotted ALOHA with the optimal retransmission probability [11]. Another MPR, called the dynamic queue protocol, which offers a much simpler implementation and only marginal performance degradation [12]. Recently, optimum transmission policy using perfect decentralized CSI is introduced and analyzed [13]. Unfortunately, these MAC schemes [11–13] require a central controller to coordinate the transmissions of the client mobile nodes

In all the above mentioned MPR protocols [1–13], mobile users transmit their packets to a central entity (base station). Recently, there has been an increasing interest in data transmissions over infrastructure-less wireless networks. In those networks, the communications between two nodes may be direct or via one or more other mobile nodes without the aid of a fixed infrastructure or centralized management. The nodes in those networks should form the network autonomic, i.e., self-optimized [14–17]. Ad hoc, sensor and mesh networks belong to those networks. A random access protocol for infrastructure-less wireless autonomic networks is introduced and analyzed in this paper. In those networks each mobile node either will be in a transmitting mode or in a receiving mode or in an idle mode. If the mobile nodes can control the network traffic to obtain the maximum throughput, without requiring a central controller to coordinate the transmissions of the client mobile nodes, a network is autonomic, i.e., self-optimized

The paper is organized as follows. Section 2 describes the system model and assumptions. The capacity of a mobile node with MPR capacity is described in Sect. 3. Section 4 provides optimum throughput for autonomic networks. The optimum

Citation: Jahangir H. Sarker · Hussein T. Mouftah, A Random Access Protocol with Multi-Packet Reception for Infrastructure-Less Wireless Autonomic Networks, DOI 10.1007/s11277-010-9982-y.

number of MPR capability by considering the maximum gain is estimated in Sect. 5. Conclusions are provided in Sect. 6.

SYSTEM MODEL AND ASSUMPTIONS

Let us consider an area, where mobile nodes are uniformly distributed. Assuming that each node can cover an area of a. If γ is the average number of nodes in the area a, then the distribution of nodes is Poisson Point Process. In infrastructure-less wireless networks, the distribution of mobile nodes as a Poisson Point Process is a widely accepted phenomena and used in many research papers [18,19,25]. The mobile nodes can be divided into three parts: transmitting nodes, receiving nodes, and idle nodes (neither transmitting nor receiving). The whole radio frequency is divided into equal parts. The number of radio frequencies (channels) is the same as the number of receiving mode nodes. All transmitting mode nodes know their receiving nodes and their corresponding channel. Let us assume that the radio channels in a multi-hop infrastructure-less network is divided into time frames with equal duration called time slot. Each mobile node transmits its data packets by fitting those time slots. These data slots are used when few nodes are transmitting packets and few other nodes receive packets. Because of that, in infrastructure-less networks, mobile nodes cannot be in a transmission mode and in a receiving mode at the same time. Assume that each node is in the transmitting mode, receiving mode and idle mode with probability, b, c and d, respectively. The probability that n packets are ready to transmit to the direction of a given receiving node in a random access system without any central entity is [19]

$$Tr = \frac{(b/c)^n}{n!} e^{-(b/c)}$$

Equation (1) depends exclusively on the ratio of node transmitting probability b, and node receiving probability c. It is important to note that the probability equation derived in Eq. (1) is independent of the average number of transmitting nodes or receiving nodes within the range of operation (area a). The arrival model of aggregate traffic to the direction of any node or any receiving node for an infrastructure-less wireless network is given by Eq. (1), which is Poisson Point Process.

CAPACITY OF A MOBILE NODE WITH MPR CAPABILITY

The MPR allows receiving multiple packets from several transmitters in a transmission slot. Let us assume that each mobile node has M MPR capacity. A mobile node can receive a maximum of M packets in each time slots. If more than M packets are transmitted to it, then all packets are unsuccessful. Ideally the MPR capacity, M, should be the same as the maximum number of packets transmitting to any receiver per time slot. This could require a large MPR capacity that could be too costly. On the other hand, if the MPR capacity is very low, then the packet unsuccessful probability will be high. Therefore, the question arises: what is a reasonable MPR capacity that will provide sufficient performance in an infrastructure-less network?

According to Eq. (1), we can also reduce the packet arrival to any mobile node by reducing the ratio of transmission probability of each mobile node and the receiving probability of each mobile node. The network traffic can be controlled very precisely by each mobile node, without any central information. Let us assume that each node can receive a maximum of M packets per time slot. The average number of received packet by each mobile node using MPR capability of M is

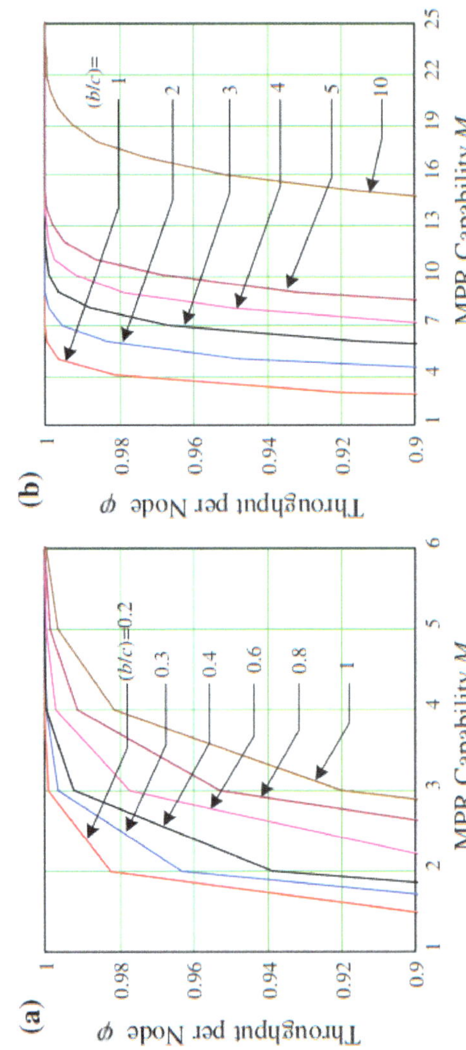

Figure 1: Normalized throughput per node. a (b/c) ≤ 1; b (b/c) ≥ 1

$$\Gamma = \sum_{l=0}^{M} l \frac{(b/c)^l}{l!} e^{-(b/c)} = e^{-(b/c)} (b/c) \sum_{k=0}^{M-1} \frac{(b/c)^k}{k!} \tag{2}$$

Equation (2) shows the average number of received packets by any mobile node using M MPR capacity. From Eq. (1), it can be shown that the average number of ready to send packets to a given receiving node is (b/c). Let us define the normalized throughput as the ratio of average received packets using MPR to the average number of packets transmitted to the direction of a given mobile node

$$\varphi = \frac{\Gamma}{(b/c)} = e^{-(b/c)} \sum_{k=0}^{M-1} \frac{(b/c)^k}{k!} \tag{3}$$

Figure 1 shows the normalized throughput with the variation of MPR capacity M using Eq. (3). The average number of unsuccessful packets by each mobile node per time slot is

$$\Delta = \sum_{l=M+1}^{\infty} l \frac{(b/c)^l}{l!} e^{-(b/c)} = e^{-(b/c)}(b/c)\left[e^{(b/c)} - \sum_{k=0}^{M-1} \frac{(b/c)^k}{k!}\right]$$
$$= (b/c)\left[1 - e^{-(b/c)} \sum_{k=0}^{M-1} \frac{(b/c)^k}{k!}\right] = (b/c)[1-\varphi] \quad (4)$$

Equation (4) shows the average number of unsuccessful packets by a given mobile node using M MPR capacity. Let us define the packet retransmission probability as the ratio of average unsuccessful packets using MPR to the average number of transmitted packets per time slot to the direction of a given mobile node

$$\delta = \frac{\Delta}{(b/c)} = 1 - e^{-(b/c)} \sum_{k=0}^{M-1} \frac{(b/c)^k}{k!} = 1 - \varphi \quad (5)$$

Figure 2 shows the packet retransmission probability per time slot per node using Eq. (5). Note that, Figs. 1 and 2 show that the average arrival traffic (b/c) plays a very important role for achieving higher normalized throughput and/or lower packet retransmission probability. In the proposed system model, this parameter [(b/c)] can be controlled very easily. Using the technology described in [20], each mobile node can measure the network traffic or the traffic transmitted to it. Each mobile node is also able to transmit less traffic by reducing its ratio of transmission probability and receiving probability (b/c), without any central information.

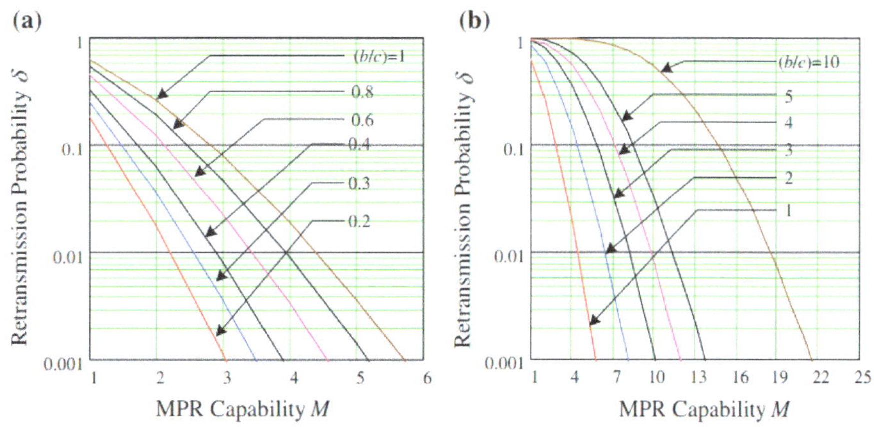

Figure 2: Retransmission probability. a (b/c) ≤ 1; b (b/c) ≥ 1

The direct communications between mobile nodes have also been studied in [21–24]. The optimum transmission probability of each mobile node to obtain the maximum throughput, without using of any central information, has not been studied in [21–24]. The authors believe that the study in this paper is the first attempt to combine the five factors together. These five factors are: autonomic (i.e., self-optimizing), half-duplex, Slotted ALOHA, MPR capability and infrastructure-less network.

THE OPTIMUM THROUGHPUT FOR AUTONOMIC NETWORKS

In an infrastructure-less wireless network, the mobile nodes in transmitting modes transmit their packets to the direction of mobile nodes in the receiving modes. Each mobile node in receiving mode can receive a maximum of M packets in a time slot [15]. According to Eq. (2), it can be said that for a given value of MPR capability M, the throughput, , increases initially, with the increase of arrival rate, (b/c). It reaches its maximum value in a certain value of arrival rate, (b/c). We will call this value as optimum arrival rate, $(b/c)_{opt}$. The throughput, , again starts to decrease and goes to finally zero, when the arrival rate, (b/c), increases further.

To obtain the optimum arrival rate, $(b/c)_{opt}$, differentiate Eq. (2), with respect to (b/c)

$$\frac{d\Gamma}{d(b/c)} = \begin{cases} e^{(b/c)} - (b/c)e^{-(b/c)} & \text{for } M = 1 \\ e^{-(b/c)}\left\{\sum_{k=0}^{M-1} \frac{(b/c)^k}{k!} - \frac{(b/c)^M}{(M-1)!}\right\} & \text{for } M > 1 \end{cases} \quad (6)$$

We can evaluate the optimum traffic load, $(b/c)_{opt}$, by setting $\frac{d\Gamma}{d(b/c)} = 0$, in Eq. (6). The optimum arrival rates, $(b/c)_{opt}$ and the corresponding maximum throughputs, opt, for different values of MPR capability, M, are evaluated numerically and provided in Table 1.

The optimum packet arrival rate, $(b/c)_{opt}$ can be considered as the ratio of optimum transmission probability of each mobile node, b_{opt} and the optimum receiving probability of each mobile node, c_{opt}. Therefore, we can write

Table 1: The maximum throughput and other parameters to obtain the maximum throughput

M	$(b/c)_{opt}$	$b_{opt}/(1-d)$	S_{opt}
1	1	0.5	0.368
2	1.618	0.618	0.84
3	2.27	0.694	1.371
4	2.945	0.747	1.942
5	3.64	0.784	2.544
6	4.349	0.813	3.168
7	5.071	0.835	3.812
8	5.804	0.853	4.472
9	6.546	0.867	5.145
10	7.297	0.879	5.831
20	15.116	0.938	13.131
30	23.285	0.959	20.907
40	31.661	0.969	28.956
50	40.18	0.976	37.191
60	48.805	0.98	45.561
70	57.515	0.983	54.038
80	66.293	0.985	62.602
90	75.129	0.987	71.237
100	84.014	0.988	79.934

$$\frac{b_{opt}}{c_{opt}} = (b/c)_{opt} \tag{7}$$

Now we will evaluate the optimum transmission probability of each mobile node, b_{opt}. Since $b_{opt} + c_{opt} + d = 1$, thus the Eq. (7) can be rewritten as

$$b_{opt} = (1-d)\frac{(b/c)_{opt}}{1+(b/c)_{opt}} \tag{8}$$

Since the maximum throughput can be obtained without any centralized feedback information, Eq. (8) can be considered as a basic equation for infrastructure-less wireless autonomic networks. The ratio of the transmission probability of each mobile node and the inverse of a mobile node is in an idle mode, $b_{opt}/(1-d)$, for a given MPR capability, M, is fixed and is given in Table 1.

Let us consider an optimum random access protocol for infrastructure-less wireless autonomic network design example. Each mobile node has 3 MPR capabilities. According to Table 1, $(b/c)_{opt}$ is equals to 2.27. We will consider two separate cases for this design model.

In the first case, each mobile node has sufficient traffic to transmit. Then the probability that a mobile node in an idle mode, d, can be set equals to zero. The optimum transmission probability of each mobile, b_{opt}, will be set as(2.7)/(1+2.7) = 0.73. The optimum receiving probability of each mobile node, c_{opt}, will be set as $c_{opt} = 1 - b_{opt} - d = 1 - 0.73 - 0 = 0.27$

In the second case, each mobile node has not sufficient traffic to transmit. The optimum transmission probability of each mobile, b_{opt}, is 0.6. According to the system model, the optimum arrival rate, $(b/c)_{opt}$ should be 2.7. According to Eq. (7), the optimum receiving probability of each mobile node, $c_{opt} = \frac{b_{opt}}{(b/c)_{opt}} = \frac{0.6}{2.7} = 0.22$. Finally, the probability that a mobile node is in an idle mode, d, will be set as $d = 1 - b_{opt} - c_{opt} = 1 - 0.6 - 0.22 = 0.18$.

From the above mentioned two cases, it can be said that although the optimum transmission probability of each mobile, b_{opt}, are different, the throughputs will be same. According to Table 1, the throughputs will be 1.371. The reason is that we are able to optimize the arrival rate, $(b/c)_{opt}$, is equals to 2.7.

MAXIMIZING THE GAIN CONSIDERING PERFORMANCE AND COST

It is already shown that the network and nodes performance increases with the increase of MPR capability. On the other hand, the cost (including research and manufacturing) of each mobile node increases with the increase of MPR capability. In general the optimal MPR capability of a mobile node is a function of the throughput performance and the network cost

Let us consider a throughput utility function, U(M), that corresponds to the level of satisfaction from the throughput and the network cost utility function, C(M), that corresponds to the level of dissatisfaction from the network cost. Then the optimal operating point would correspond to the MPR capability that gives maximum positive difference between the level of satisfaction and dissatisfaction. Since each node has MPR capability, it is convenient to consider the throughput utility function, U(M) as the optimum throughput per node. Therefore

$$U(M) = e^{-(b/c)_{opt}} \sum_{k=0}^{M-1} \frac{(b/c)_{opt}^k}{k!} \tag{9}$$

Let the node cost has a constant component equal to the cost of a node with a receiver of one MPR capability and a component proportional to the cost of a MPR capability that grows exponentially with receiver capacity due to increased complexity and power consumption. Then, assuming that the dissatisfaction from the cost is proportional to the cost we consider the following cost utility function form

$$C(M) = \lambda + \mu M^{\omega} \tag{10}$$

Figure 3a shows the throughput utility function that is compared with the cost utility function: $C_1(M)$ with $\lambda = 0.36$, $\mu = 0.003$, $\omega = 2$. Obviously the cost utility function parameters λ, μ, and ω depend on many factors and will change in time so the given examples are for illustration purposes only. According to Fig. 3a, the optimum MPR capability is 5, since it provides maximum positive gain over cost.

Figure 3b shows the throughput utility function that is compared with the cost utility function: $C_2(M)$ with $\lambda = 0.5$, $\mu = 0.0025$, $\omega = 2$. In the considered example, the optimal MPR capability, corresponding to the maximum net-satisfaction level is 6. Note that in the second example of the cost utility function, the receiver capacity equal to one is not acceptable since the net-satisfaction level is negative. This example illustrates that mobile node without MPR capability allows applying more expensive nodes due to the negative-gains in the network throughput.

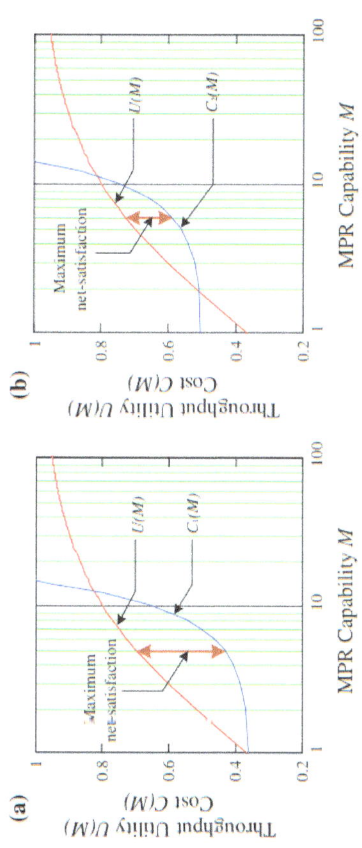

Figure 3: a Max. gain for M = 5; b Max. gain for M = 6

CONCLUSIONS

A new random access protocol with MPR capability for infrastructure-less wireless autonomic networks is introduced and analyzed. In an infrastructure-less wireless autonomic network, mobile nodes may communicate with each other directly without a central entity (base station), using the proposed random access protocol. The average traffic arrival rate to each mobile node is the ratio of the transmission probability of each mobile node and the receiving probability of each mobile node. For a given average traffic arrival rate, the normalized throughput per mobile node, ϕ, increases with the increase of the MPR capability, M. The reason is that the probability of success increases with higher number of the MPR capability. Therefore, the packet retransmission probability, δ, also decreases with the increase of the MPR capability.

The average traffic arrival rate to each mobile node plays an important role for the performance of infrastructure-less wireless networks. Each mobile node is able to calibrate its ratio of transmission probability and receiving probability to control the network traffic load very precisely to obtain the maximum throughput. Table 1 shows the parameters for the proposed random access protocol that can be used for infrastructure-less wireless autonomic networks. The throughput utility, U(M), and thus the satisfaction level increases with the increase of the MPR capability. On the contrary, the cost utility, C(M), increases and thus dissatisfaction level increases with the increase of the MPR capability. The optimum number of MPR capability can be considered by considering the highest satisfaction gain. The difference between the throughput and cost utilities defines the net-satisfaction for given MPR capability and the optimal MPR capability should maximize this metric. It is also shown that for some cases it may be not practical to implement an infrastructure-less network without the MPR capability

REFERENCES

1. Tong, L., Naware, V., & Venkitasubramaniam, P. (2004). Signal processing in random access. IEEE Signal Processing Magazine, 21(5), 29–39.
2. Lau, C. T., & Leung, C. (1990). A slotted ALOHA packet radio system with multiple antennas and receivers. IEEE Transactions on Vehicular Technology, 39(3), 218–226.
3. Lau, C., & Leung, C. (1988). Throughput of slotted ALOHA channel with multiple antennas and receivers. Electronics Letters, 24(25), 1540–1542.
4. Shinomiya, T., & Suzuki, H. (2000). Slotted ALOHA mobile packet communication systems with multiuser detection in a base station. IEEE Transactions on Vehicular Technology, 49(3), 948–955.
5. Abramson, N. (1994). Multiple access in wireless digital networks. Proceedings of the IEEE, 82(9), 1360– 1370.
6. Chen, C. C., Tjhung, T. T., Chew, Y. H., & Wang, E. (2005). Multiple delay capture probability and performance of DS-SS slotted ALOHA packet radio system. In IEEE international conference on communications 2005, ICC 2005 (Vol. 2, pp. 895–900).
7. Li, T., Wang, H., & Tong, L. (2006). Hybrid Aloha: a novel medium access control protocol. In IEEE ICASSP 2006 proceedings (Vol. 4, pp. IV 257–IV 260).
8. Wang, H., & Li, T. (2007). Hybrid ALOHA: a novel MAC protocol. IEEE Transactions on Signal Processing, 55(12), 5821–5832.

9. Zheng, P. X., Zhang, Y. J., & Liew, S. C. (2006). Analysis of exponential backoff with multipacket reception in wireless networks. In 31st IEEE conference on local computer networks, proceedings 2006 (pp. 855–862).
10. Samano-Robles, R., Ghogho, M., & McLernon, D. C. (2008). An infinite user model for random access protocols assisted by multipacket reception and retransmission diversity. In IEEE work. On signal proc. advances in wireless comm., 2008, SPAWC 2008 (pp. 111–115).
11. Zhao, Q., & Tong, L. (2003). A multiqueue service room MAC protocol for wireless networks with multipacket reception. IEEE/ACM Transactions on Networking, 11, 125–137.
12. Zhao, Q., & Tong, L. (2004). A dynamic queue protocol for multi-access wireless networks with multipacket reception. IEEE Transactions on Wireless Communications, 3(6), 2221–2231.
13. Ngo, M. H., Krishnamurthy, V., & Tong, L. (2008). Optimal channel-aware ALOHA protocol for random access in WLANs with multipacket reception and decentralized channel state information. IEEE Transactions on Signal Processing, 56(6), 2575–2588.
14. Al-Meshhadany, T., & Ajib, W. (2007). New CDMA-based MAC protocol for ad hoc networks. In IEEE vehicular technology conference, 2007, VTC-2007 fall (pp. 91–95).
15. Aljnidi, M., & Leneutre, J. (2008). Security solutions in mobile autonomic networks. In 3rd International conference on information and communication technologies, ICTTA 2008 (pp. 1–6).
16. Moursy, A., Ajbar, I., Perkins, D. & Bayoumi, M. (2008). Empirical model-based adaptive control of MANETs. In IEEE INFOCOM 2008 (pp. 1–6).
17. Bellavista, P., Corradi, A., & Magistretti, E. (2005). Lightweight autonomic dissemination of entertainment services in widescale wireless environments. IEEE Communications Magazine, 43, 94–101.
18. Bettstetter, C. (2005). Connectivity of wireless multihop networks in a shadow fading environment. Wireless Network, 11(5), 571–579.
19. Sarker, J. H., & Mouftah, H. T. (2008). Stability of multiple receiving nodes slotted ALOHA for wireless ad hoc networks. In IEEE GLOBECOM 2008 (pp. 1–5).
20. Tsatsanis, M. K., Zhang, R., & Banerjee, S. (2000). Network-assisted diversity for random access wireless networks. IEEE Transactions on Signal Processing, 48, 702–711.
21. Ghez, S., Verdu, S., & Schwartz, S. C. (1988). Stability properties of slotted Aloha with multipacket reception capability. IEEE Transactions on Automatic Control, 33(7), 640–649.
22. Chlamtac, I., & Farago, A. (1994). An optimal channel access protocol with multiple reception capacity. IEEE Transactions on Computers, 43(4), 480–484.
23. Bao, J. Q., & Tong, L. (1999). Performance analysis of slotted Aloha random access ad-hoc networks with multipacket reception. In IEEE MILCOM 1999 (Vol. 1, pp. 251–255).
24. Mergen, G., & Tong, L. (2002). Random scheduling medium access for wireless ad hoc networks. In IEEE MILCOM 2002 (Vol. 2, pp. 868–872).
25. Chen, Q., Kanhere, S. S., & Hassan, M. (2009). Analysis of per-node traffic load in multi-hop wireless sensor networks. IEEE Transactions on Wireless Communications, 8(2), 958–967.

CHAPTER 9

RECONFIGURABLE MICROSTRIP DOUBLE-DIPOLE ANTENNAS FOR PERSONAL WIRELESS COMMUNICATIONS

Abdelnasser Eldek, Abubaker Abdallah, and Mahmoud Manzoul

Keywords: Reconfigurable Antennas, Dipole, Wideband, Multiband

INTRODUCTION

As communication devices become smaller due to greater integration of electronics, the antenna becomes a significantly larger part of the overall package volume. This results in a demand for similar reductions in antenna size. However, reducing antenna size without significantly impacting gain and efficiency is a challenging task. As integration increases, a single antenna is often required to support two or more of the many wireless services across a broad frequency range. Multiband, wideband and reconfigurable antennas are being developed to meet this need [1]. Multi-band and wideband antennas require complicated filters with stringent requirements to improve their out-of-band noise-rejection performance. These filters, moreover, are generally bulky and expensive. In contrast, frequency-reconfigurable antennas have great potential for reducing production cost and offer better out-of-band noise rejection [2]. In addition, compared to multiband and broadband antennas, reconfigurable antennas offer the advantages of compact size, similar radiation pattern for all designed frequency bands, efficient use of electromagnetic spectrum and frequency selectivity useful for reducing the adverse effects of co-site interference and jamming [3]. Dual frequency reconfigurable microstrip antennas can offer additional advantages of frequency reuse for doubling the system capability and polarization diversity for good performance of reception and transmission or to integrate the receiving and transmitting functions into one antenna for reducing the antenna size [4,5]. Microstrip antennas are attractive candidates for reconfigurable antenna technology because of their inherent low profile, light weight, low cost, and ease of fabrication and installation [2,6-13]. Number of attractive reconfigurable antenna designs were presented in [2] with the objective of reduction of the number of MEMs utilizing antenna symmetry to decrease the cost. The concept of a frequency-reconfigurable rectangular ring slot antennas fed by slotlines or CPW is presented in [7]. A reconfigurable patch antenna is obtained in [8] by inserting slits at the non-radiating edges of the patch. A reconfigurable Yagi antenna is presented in [9] to operate at 2.4 and 5.78 GHz for wireless communications. A reconfigurable slot dipole antenna was presented in [10] in the X-band. Patch antenna with polarization diversity using switchable L-shaped slots is presented in [11]. A reconfigurable dipole antenna with harmonic trap was presented in [12] where the operating frequency is controlled by a reconfigurable dipole length, while the higher-order modes are eliminated using a reconfigurable l/4 open circuit stub. A compact multiband antenna has been redesigned to be reconfigurable with embedded switches to allow selective addressing of these bands [13].

In this paper, three microstrip-fed double-dipole antenna designs [14-16] to support multi-band, wideband, and single-band operation modes are presented for personal wireless communications. The reconfiguration for the microstrip antennas is carried out by connecting or disconnecting (switching on or off) appropriate rectangular strips to model the MEM switches. The shape and number of these strips along with the switching state and position is used to control the operating frequency of the antenna without changing the feeding mechanism. The antennas' geometries and their results will be explained in the next two sections. The full-wave electromagnetic simulations and analysis for the presented antenna are performed using the commercial computer software package Ansoft High Frequency Structure Simulator

Citation: A. Eldek, A. Abdallah and M. Manzoul, "Reconfigurable Microstrip Double-Dipole Antennas for Personal Wireless Communications," Wireless Engineering and Technology, Vol. 2 No. 2, 2011, pp. 60-69. doi: 10.4236/wet.2011.22009.

(HFSS) [17], which is based on the finite element method. Measurements of the return loss and radiation patterns are also conducted to verify the simulation results and demonstrate the feasibility of the proposed configurations.

ANTENNAS' GEOMETRIES AND PARAMETRIC STUDY

Three double-dipole antennas were designed for different frequency bands and depicted in **Figure 1**. The double-dipole antenna consists of two dipoles with similar or different lengths depending on the required operating frequencies. A simplified balun is used to feed the dipoles. In each dipole, the left half is printed on the top substrate layer and fed by a microstrip line, while the right half is printed on the bottom layer and connected to the truncated ground plane.

The length of the dipole is approximately half the wavelength at the operating frequency, while its distance from the truncated ground plane is quarter wavelength for the best performance. Thus, if the antenna is required to operate at a lower frequency (f_L) and a higher frequency (f_H), then the shorter dipole will be closer to the ground plane. In this case, the antenna will operate at multiple bands. A broadband design can be achieved by making the longer and the shorter dipoles to resonate at closer frequencies. This can be achieved by reversing the positions of the two dipoles. With the contribution of the coplanar strip (CPS) line length to the antenna length, the difference between their lengths become small, and the two operating bands will be close and can merge in one relatively wide frequency band.

The next step is to merge these two modes in one so that the length of the closer dipole is controlled by the lower operating frequency in the wideband mode, and the length of the second dipole is controlled by the higher operating frequency in the multiband mode. Next, a proper matching circuit between the two dipoles is designed to eliminate the mutual effect between the two modes. Therefore, the shape of the CPS connecting the two dipoles is modified as needed by the design to obtain the required modes. Finally, when adding the ground plane, it will have an effect on the performance; therefore, the last step is to shape the ground plane to reduce this effect as shown in **Figure 1**. A smooth transition from the microstrip line mode to the two CPS line mode is used for this purpose.

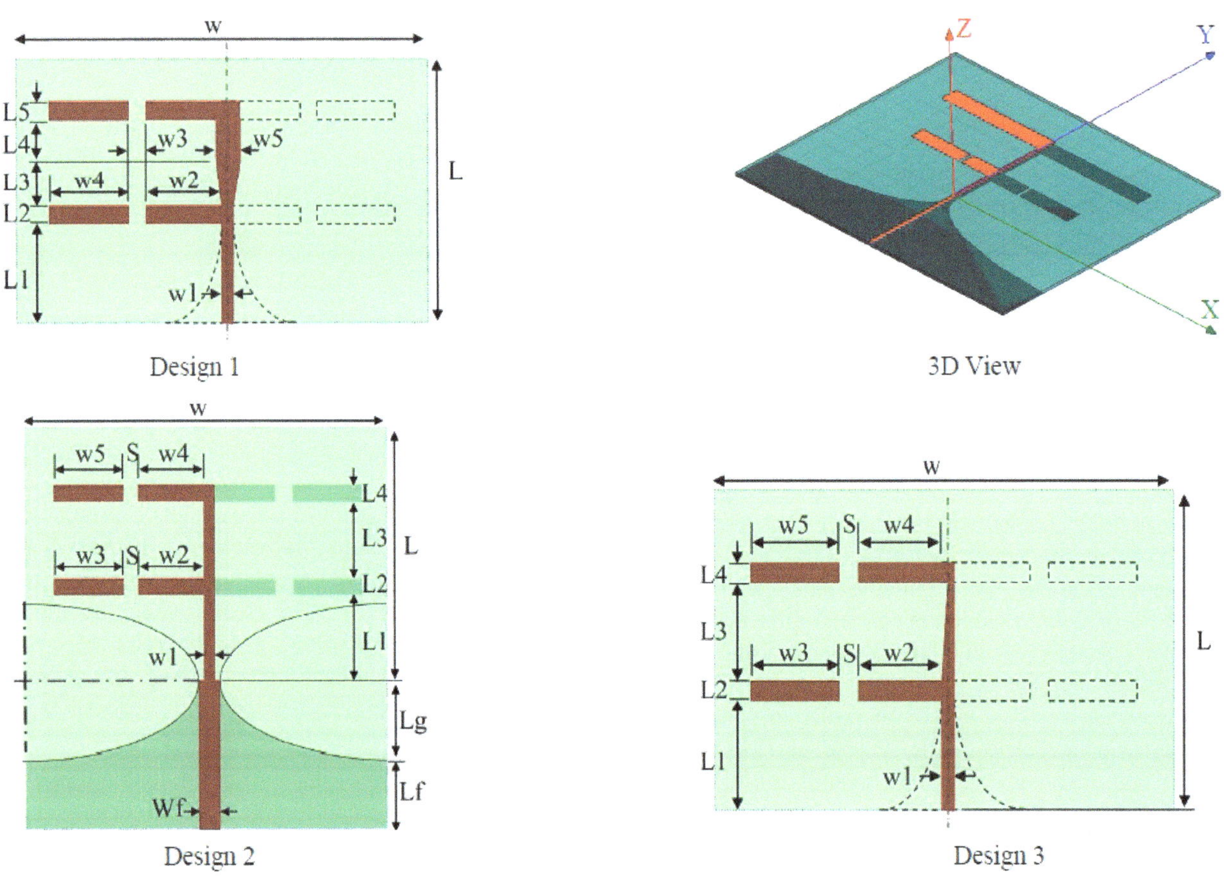

Figure 1: Antennas' geometries and paramters.

Parametric study is performed on each antenna and depicted in Figures 2 to 11 for Design 1. This study is done in the multiband mode to know the effect of each parameter, and select the best dimensions, and tune the antenna to obtain the best bandwidth in the broadband mode without affecting the multiband mode. The dimensions of the three designs resulted from the parametric study are shown in **Table 1**.

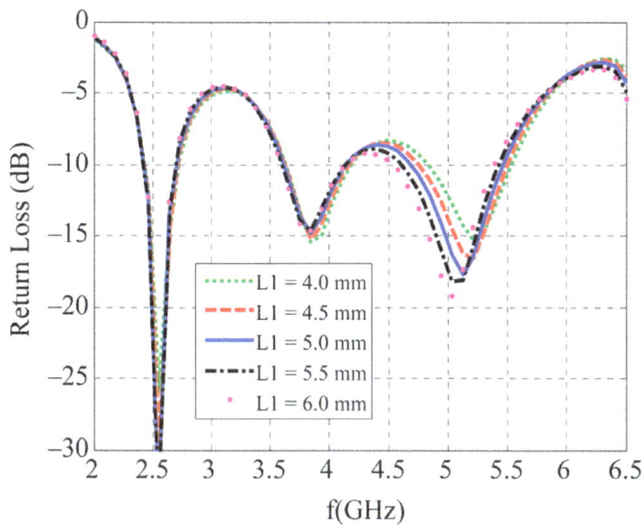

Figure 2: Effect of L1.

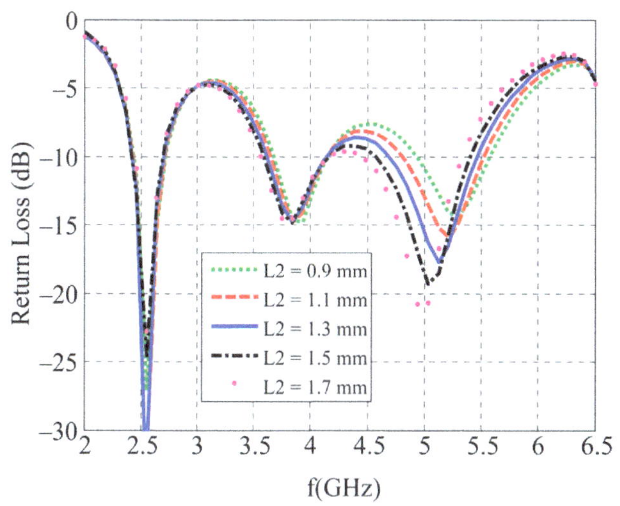

Figure 3: Effect of L2.

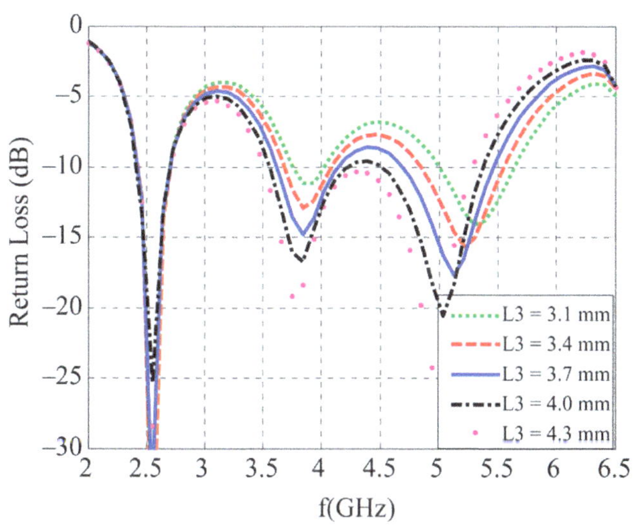

Figure 4: Effect of L3.

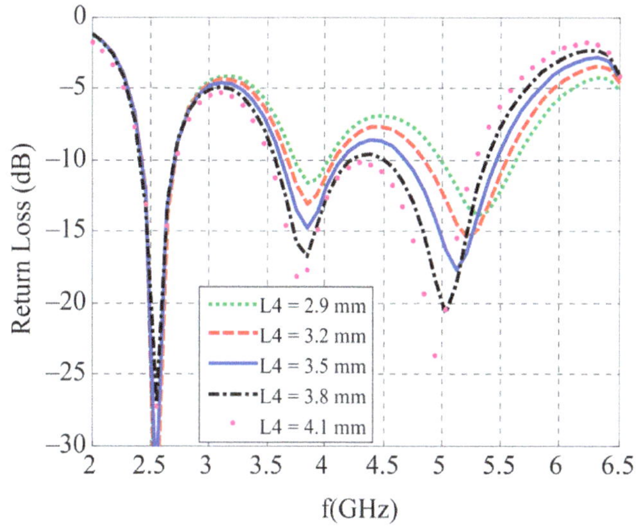

Figure 5: Effect of L4.

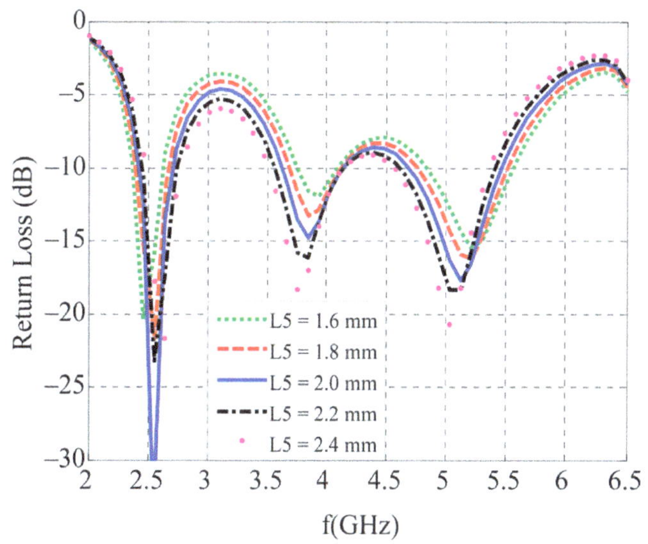

Figure 6: Effect of L5.

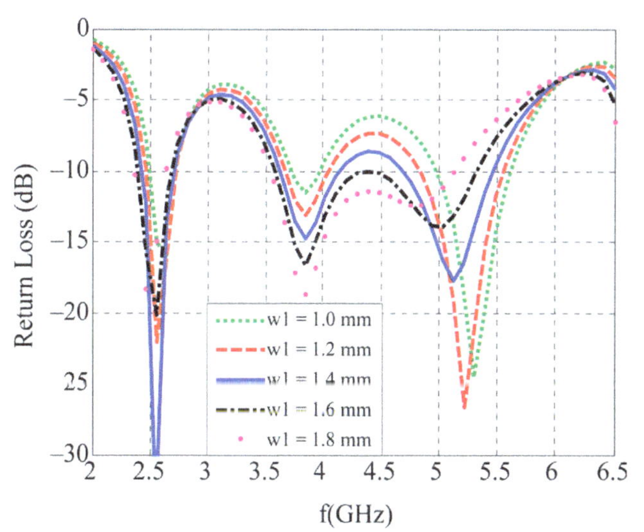

Figure 7: Effect of w1.

RESULTS

The return loss and radiation patterns are computed using Ansoft HFSS for Designs 1, 2 and 3. The reconfiguration is carried out by connecting or disconnecting appropriate rectangular strips to model the MEM switches. The rectangle width is "S" and its height is the same like the strip on which Pin diode or MEMs can be installed.

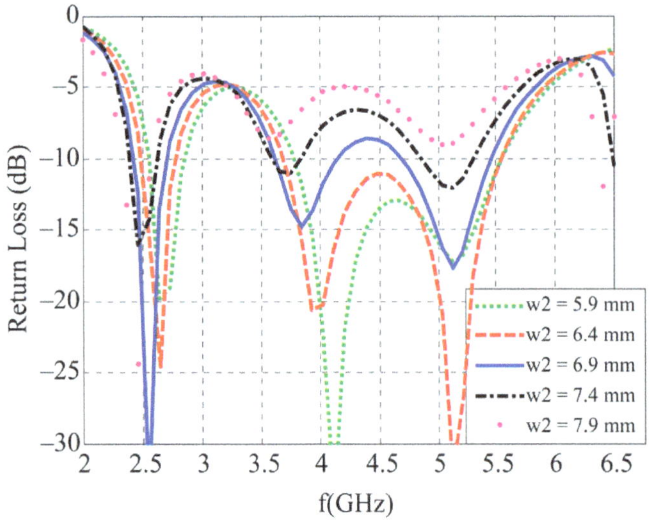

Figure 8: Effect of w2.

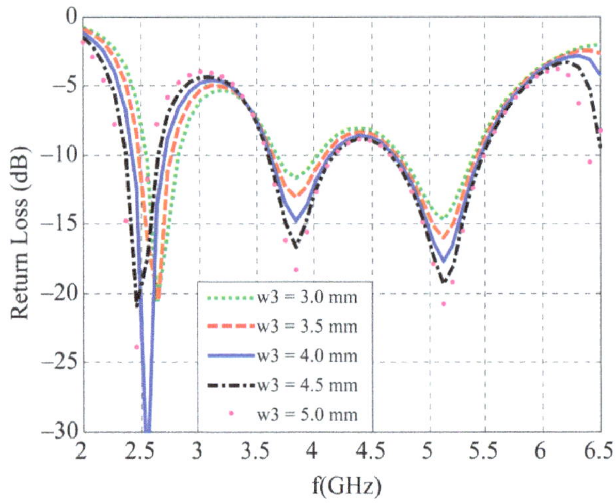

Figure 9: Effect of w3.

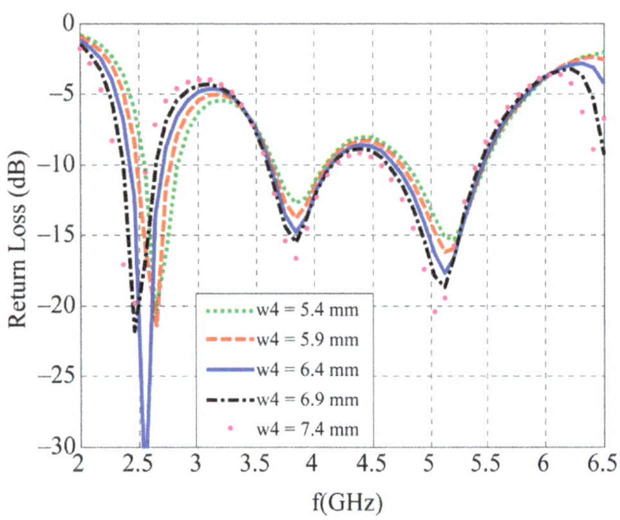

Figure 10: Effect of w4.

Since we have four switches, we can create 16 cases starting from Case 0000 when all switches are open, to Case 1111 when all switches are closed. However, only 10 different combinations are considered after excluding the similar cases such as 0010 and 0001.

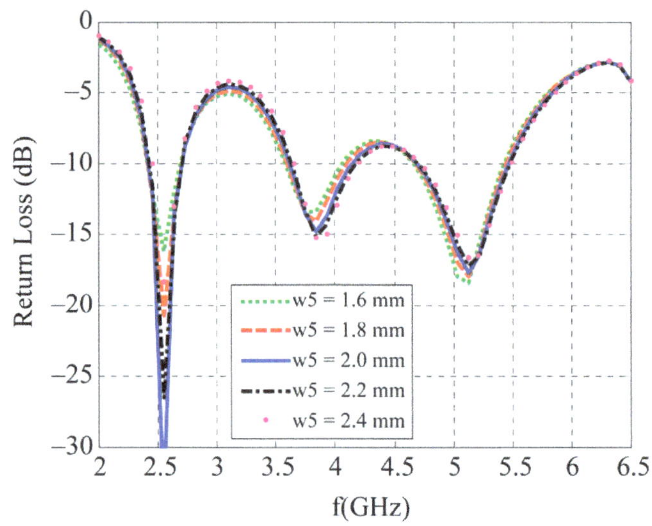

Figure 11: Effect of w5.

	Design 1	Design 2	Design 3
w	46.2	93.5	144.5
L	18.9	46.3	89
w1	0.6	1.3	3.5
w2	7	9	22
w3	1	16.8	31
w4	9.8	19.8	32
w5	1.2	12.8	30
L1	5	12	42
L2	1.3	4.3	3
L3	3.1	13	36
L4	2.9	5	3
L5	1.6	N/A	N/A
S	N/A	1	1

Table 1: Dimensions (mm) of the proposed designs.

Figure 12 shows the multiband (0011) and broadband (1100) cases in addition to the cases when all switches are ON/OFF (0000 and 1111). These cases are the symmetrical cases where S1 and S3 are the same as S2 and S4, respectively. The operating bands of the three designs in this figure are summarized in Tables 2, 3, and 4 for Designs 1, 2, and 3, respectively. With only four cases out of ten, Design 1 covers 90% of the band between 2.22 and 6.28 (95% bandwidth), Design 2 can operate in 90% of the band between 1.64 and 3.2 (64% bandwidth), and Design 3 also covers almost all the band between 0.83 and 2.34 (95% bandwidth). In the broadband Case, the three Designs provide from 22% to 56% bandwidth. **Figure 13** shows some unsymmetrical cases which show that we still can make these antennas operate at other frequencies.

These results show many advantages for these designs. First, we can switch between different frequency bands without changing the antenna. Second, we can also change between multi-band and broadband modes. In addition, the overall antenna operating bandwidth is at least 1.7 times the maximum bandwidth of all cases.

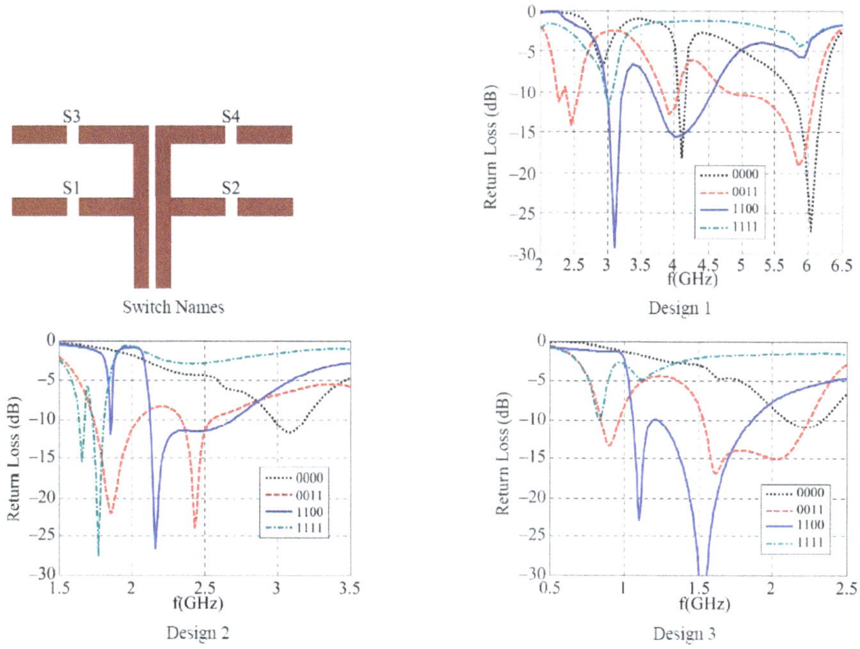

Figure 12: Return Loss (dB) for the symitrical cases.

	Case	*Band 1*	*Band 2*	*Band 3*
Design No. 1	0000	4.02 - 4.15 (3%)	5.77 - 6.28 (8.5%)	N/A
	0011	2.22 - 2.55 (14%)	3.82 - 4.05 (6%)	4.75 - 6.10 (25%)
	1100	2.96 - 3.25 (9%)	3.65 - 4.55 (22%)	N/A
	1111	2.95 - 3.10 (5%)	N/A	N/A

Table 2: Operating frequency bands and bandwidth for Design 1 in the symitrical cases.

	Case	*Band 1*	*Band 2*
Design No.2	0000	2.95 - 3.20 (8%)	N/A
	0011	1.73 - 2.05 (17%)	2.36 - 2.53 (7%)
	1100	1.85 - 1.86 (0.5%)	2.13 - 2.73 (24%)
	1111	1.64 - 1.68 (2.4%)	1.73 - 1.80 (4%)

Table 3: Operating frequency bands and bandwidth for Design 2 in the symitrical cases.

	Case	*Band 1*	*Band 2*
Design No.3	0000	2.1 - 2.34 (11%)	N/A
	0011	0.83 - 0.98 (17%)	1.52 - 2.25 (39%)
	1100	1.05 - 1.86 (56%)	N/A
	1111	0.83 - 0.84 (1%)	N/A

Table 4: Operating frequency bands and bandwidth for Design 3 in the symitrical cases.

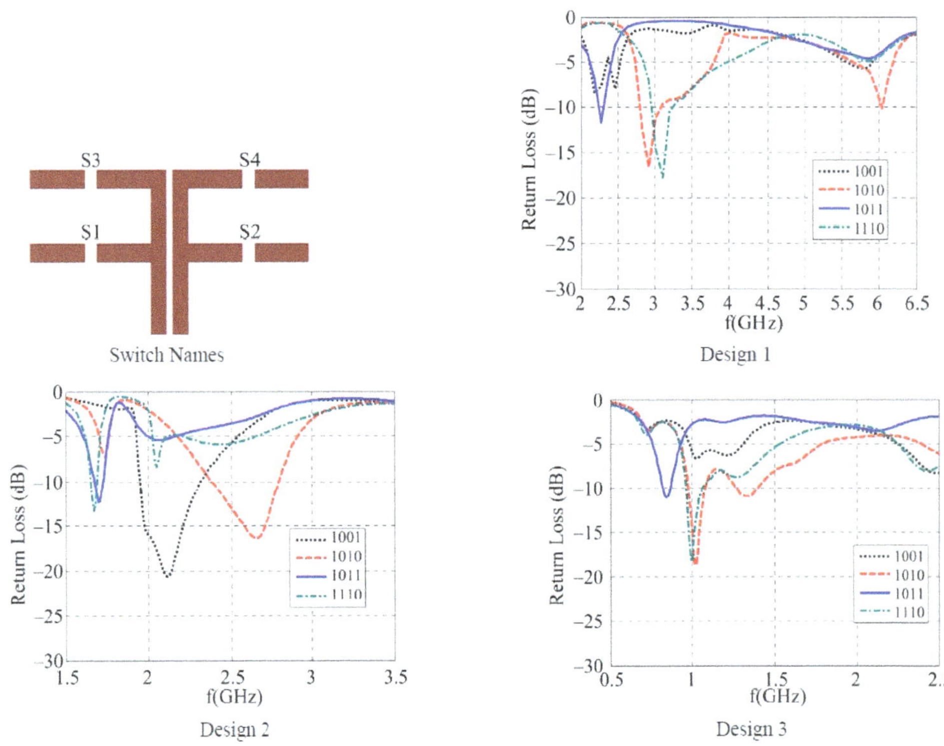

Figure 13: Return Loss (dB) for the unsymitrical cases.

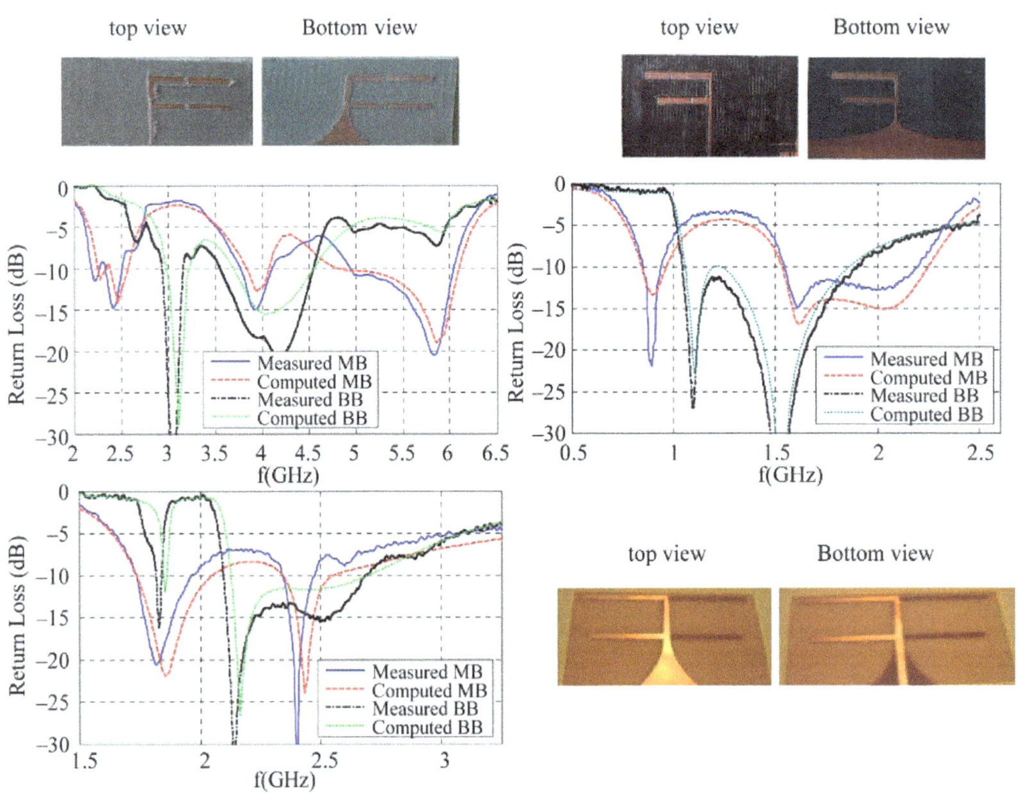

Figure 14: Measured and computed return loss and antenna prototypes.

Prototypes from the three designs are fabricated using LPKF milling machine. The return loss is measured using vector network analyzer for the different reconfigurable cases. **Figure 14** shows a good agreement between the measured and computed results in the multi-band (MB) and broadband (BB) modes. Finally, the radiation patterns are computed and measured using anechoic chamber. Figures 15 to 17 show the co-polarized fields at selected resonance frequencies for Designs 1, 2 and 3, respectively, for the MB and BB cases. The lower half is cropped because of the symmetry. An acceptable match is obtained between the measured and computed results. Currently, we are in the process of sending antenna prototypes to specialized for integrating reconfigurable enabling technology such as MEMs in these designs.

CONCLUSIONS

Three novel reconfigurable microstrip antenna designs are presented for wireless communications applications. The proposed antennas are combination of two dipole antennas of similar and different lengths. By changing the dipoles' lengths, we were able to change the antenna operating frequency as well as the operation mode.

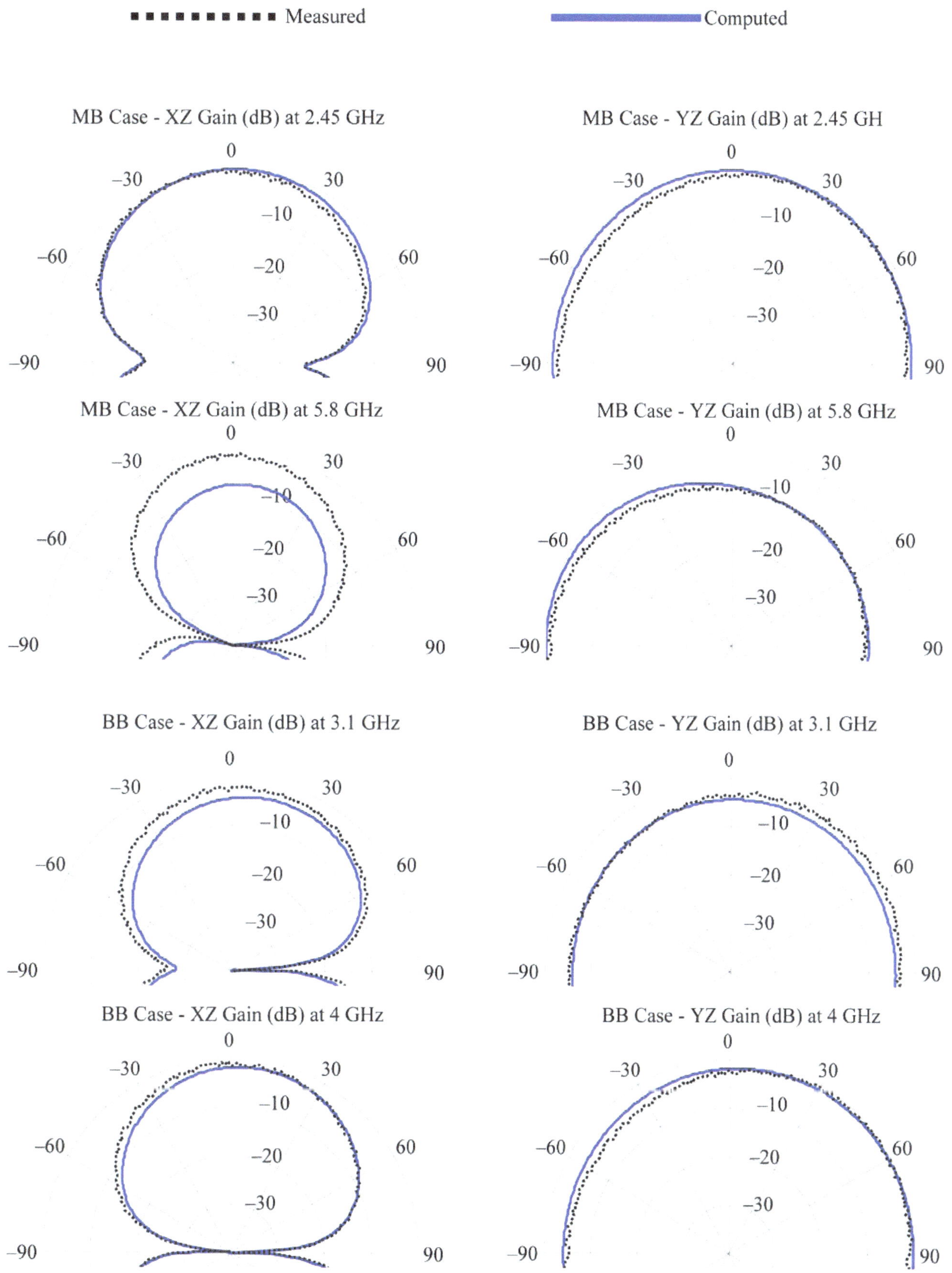

Figure 15: Radiation Patterns for Design 1 in multiband and broadband modes.

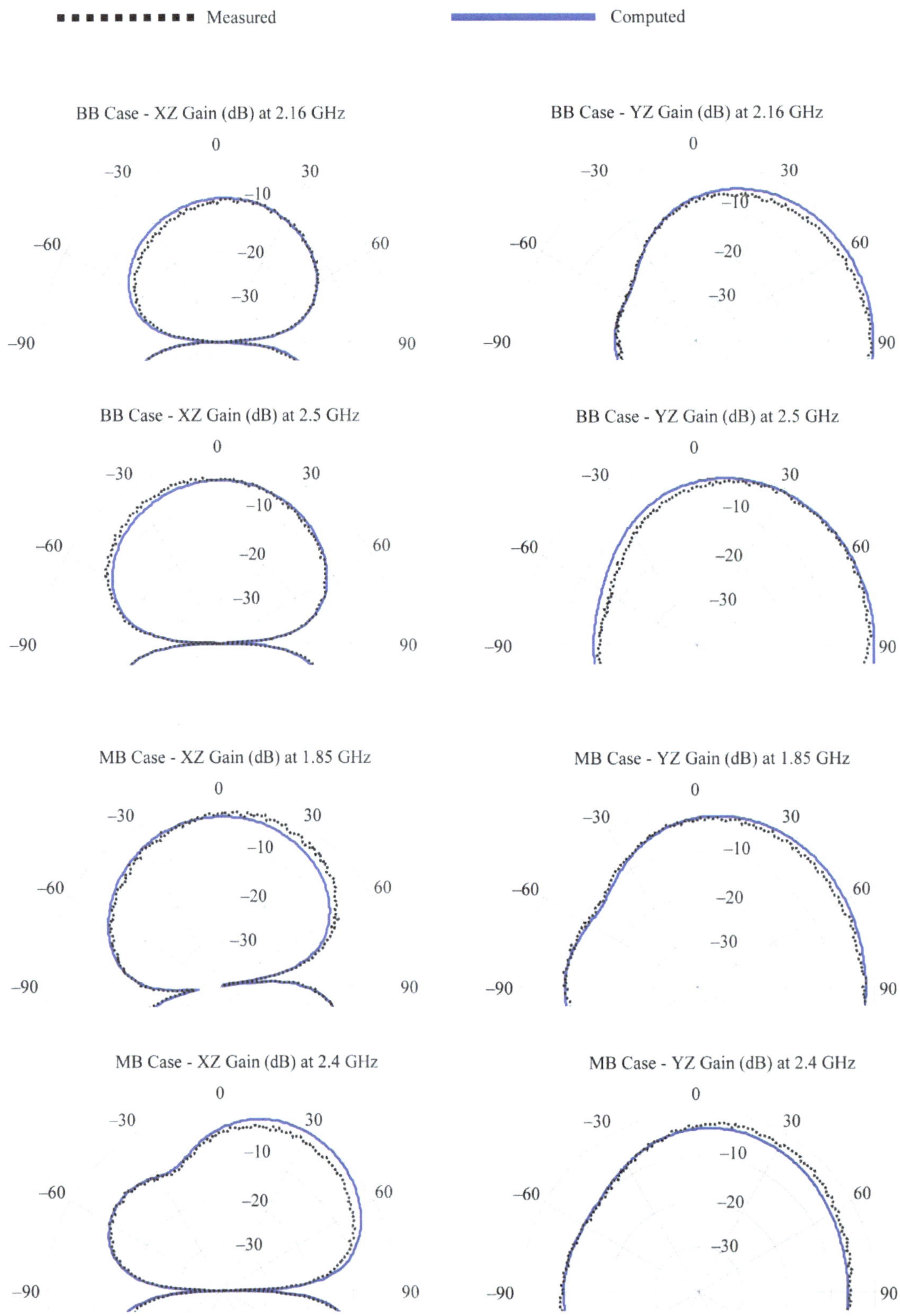

Figure 16: Radiation Patterns for Design 2 in multiband and broadband modes.

The antenna is merging two operating modes: Multi-band and Broadband in a single design. The presented designs cover different frequency bands: 0.83 - 2.34 GHz, 1.64 - 3.2 GHz, and 2.22 - 6.28 GHz. A wide overall bandwidth of 95% was obtained and a wide single case operating bandwidth of 56% was also provided.

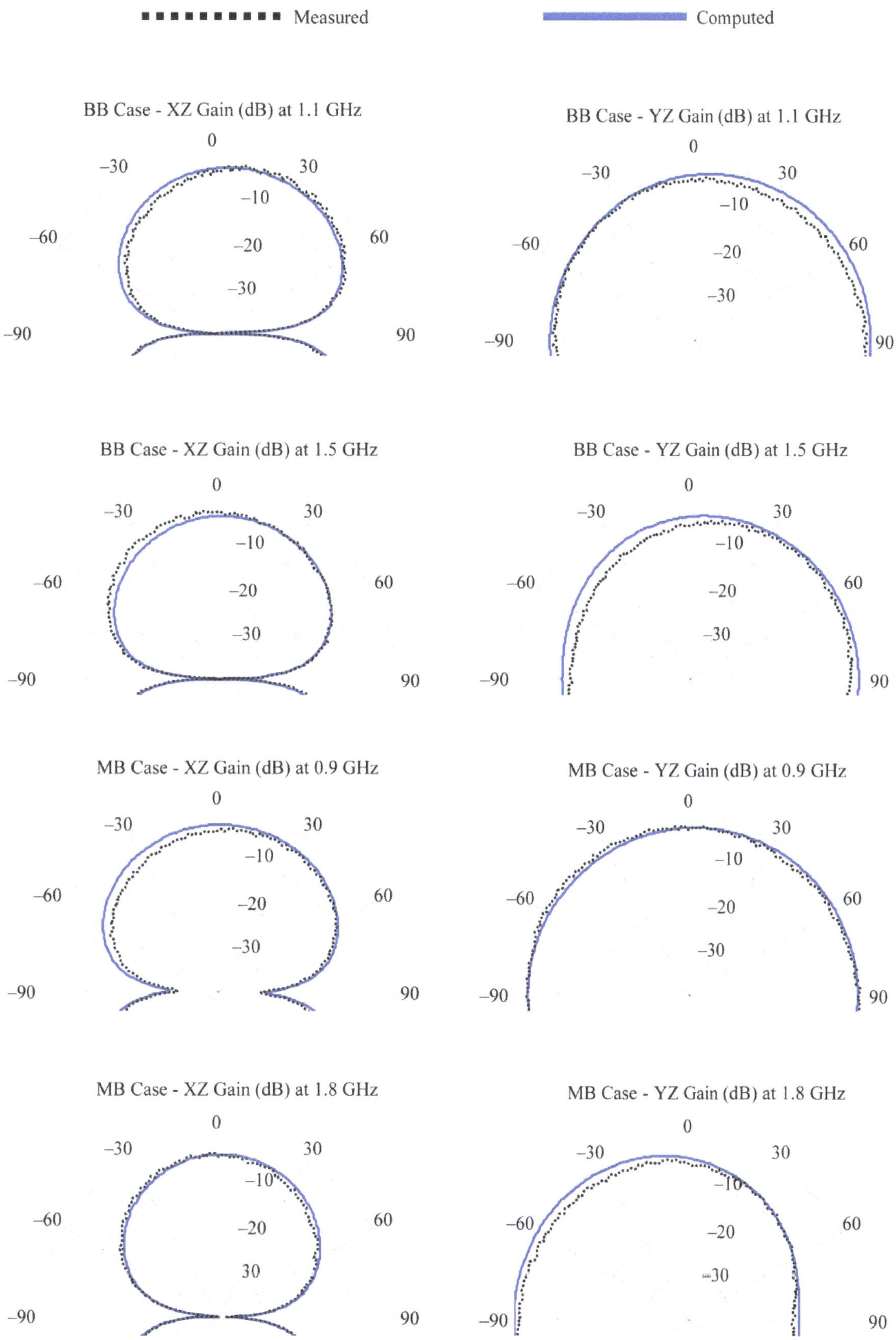

Figure 17: Radiation Patterns for Design 3 in multiband and broadband modes.

ACKNOWLEDGEMENTS

The authors would like thank the Air Force Research Laboratory (AFRL)—Minority Leaders Program (MLP), and Clarkson Aerospace Corp for funding this research through contract number FA8650-05-D-1912. Many thanks are also to Mr. William Munn and Mr. Christopher Munn from Jackson State University for their help in building the proposed antennas.**ACKNOWLEDGEMENTS**

REFERENCES

1. C. B. Dietrich, Jr., R. M. Barts, W. L. Stutzman and W. A. Davis, "Trends in Antennas for Wireless Communications," Microwave Journal, vol. 46, no. 1, January 2003, p. 22. [Paper reference:1]
2. S. N. Yang, C. N. Zhang, H. K. Pan, A. E. Fathy and V. K. Nair, "Frequency Reconfigurable Antennas for Multiradio Wireless Platforms," IEEE Microwave Magazine, vol. 10, no. 1, February 2009, pp. 66-83. doi:10.1109/MMM.2008.930677
3. D. Peroulis, K. Sarabandi and P. B. K. Katehi, "Design of Reconfigurable Slot Antennas," IEEE Transactions on Antennas and Propagation, Vol. 53, No. 2, 2005, pp. 645- 654.doi:10.1109/TAP.2004.841339
4. S. V. Shynu, G. Augustin, C. K. Aanandan, P. Mohanan, and K. Vasudevan, "A Reconfigurable Dual Frequency Slot Loaded Microstrip Antenna Controlled by Pin-Diodes," Microwave and Optical Technology Letters, vol. 44, No. 4, 2005, pp. 374-376.
5. S. V. Shynu, G. Augustin, C. K. Aanandan, P. Mohanan, and K. Vasudevan, "Design of Compact Reconfigurable Dual, Frequency Microstrip Antennas Using Varactor Diodes," Progress in Electromagnetics Research, Vol. 60, 2006, pp. 197-205.doi:10.2528/PIER05120101
6. W. H. Weedon, W. J. Payne and G. M. Rebeiz, "MEMsSwitched Reconfigurable Antennas," IEEE Antennas and Propagation Society International Symposium, Boston, vol. 3, 8-13 July 2001, pp. 654-657.
7. K. C. Gupta, J. Li, R. Ramadoss and C. J. Wang, "Design of Frequency-Reconfigurable Rectangular Slot Ring Antenna," IEEE Antennas and Propagation Society International Symposium, Salt Lake City, vol. 1, 16-21 July 2000, p. 326.
8. S. Xiao, B. Z. Wang and X. S. Yang, "A Novel Frequency-Reconfigurable Patch Antenna," Microwave and Optical Technology Letters, vol. 36, no. 4, February 2003, pp. 295-297.doi:10.1002/mop.10746
9. P. F. Wahid, M. A. Ali and B. C. Deloach, "A Reconfigurable Yagi Antenna for Wireless Communications," Microwave and Optical Technology Letters, vol. 140, no. 2, July 2003, pp. 140-141. doi:10.1002/mop.10997
10. A. Eldek, A. Z. Elsherbeni, C. E. Smith and K. F. Lee, "Wideband Slot Antennas for Radar Applications," Proceedings of IEEE Radar Conference, Huntsville, 5-8 May 2003, pp. 79-84.
11. S. Raghavan, T. Shanmuganantham and M. S. K. Kumar, "Reconfigurable Patch Antenna with Switchable L-Shaped Slots for Circular Polarization Diversity," Microwave and Optical Technology Letters, vol. 50, no. 9, September 2008, pp. 2348-2350.doi:10.1002/mop.23674
12. Mirkamali, P. S. Hall and M. Soleimani, "Reconfigurable Printed-Dipole Antenna with Harmonic Trap for Wideband Applications," Microwave and Optical Technology Letters, vol. 48, no. 5, May 2006, pp. 927-929. doi:10.1002/mop.21521
13. N. Zhang, S. N. Yang, H. K. Pan, A. E. Fathy, S. El-Ghazaly and V. K. Nair, "A Reconfigurable Multiband Patch Antenna for Wireless Applications Using MEMS Switches," Microwave and Optical Technology Letters, vol. 51, no. 8, August 2009, pp. 1892-1896.doi:10.1002/mop.24474
14. F. Tefiku and C. A. Grimes, "Design of Broad-Band and Dual-Band Antennas Comprised of Series-Fed PrintedStrip Dipole Pairs," IEEE Transactions on Antennas and Propagation, vol. 48, no. 6, 2000, pp. 895-900. doi:10.1109/8.865221
15. A. Eldek, "Pattern Stability Optimization for Wideband Microstrip Antennas for Phased Arrays and Power Combiners," Microwave and Optical Technology Letters, vol. 48, no. 8, August 2006, pp. 1492-1494. doi:10.1002/mop.21740
16. A. Eldek, "Design of Double Dipole Antenna with Enhanced Usable Bandwidth for Wideband Phased Array Applications," Progress in Electromagnetics Research, Vol. 59, 2006, pp. 1-15. doi:10.2528/PIER06012001
17. Ansoft Corporation, "HFSS: High Frequency Structure Simulator Based on the Finite Element Method," Version 12, 2008.

CHAPTER 10 A WIDEBAND CIRCULARLY POLARIZED PATCH ANTENNA FOR 60 GHZ WIRELESS COMMUNICATIONS

Rongguo Zhou, Duixian Liu, and Hao Xin

Keywords: Patch Antenna, Circular Polarization, Millimeter Wave, Wireless Communication

INTRODUCTION

Demands of modern communication and sensor systems for more bandwidth, higher resolution and compactness lead to operating frequency up to the millimeter wave (mmWave, f > 30 GHz) or even sub-mmWave regime. In particular, the 60 GHz ISM (the industrial, scientific and medical) band is available in many countries. For example, in the US, the ISM band spans from 57 to 64 GHz with a worldwide overlap of 5 GHz of bandwidth for unlicensed use. As a result of the large bandwidth available, the 60 GHz band is very attractive for a number of applications, including wireless personal-area networks (WPAN) and local-area networks (WLAN). As the battery lifetime is a major bottleneck for many portable devices, the 60 GHz band has a distinct benefit because the large bandwidth can be utilized to tradeoff bandwidth efficiency for low power consumption, while still maintaining high data rate [1]. In addition, due to the atmospheric absorption (mainly oxygen molecules) peak around 60 GHz, communication link at this band is inherently secure and has less interference among users, which are ideal for indoor applications. However, at mmWave frequencies, higher material losses, fabrication tolerances and packaging issues often hinder the performance of wireless front-ends. Therefore, as an important front-end component, mmWave antennas that are wideband, efficient and packaged to be compatible with integrated circuits are highly desirable. Moreover, previous wireless channel propagation studies have shown that circular polarization can effectively suppress multi-path fading and inter-symbol interference (ISI) [2-4].

Citation: R. Zhou, D. Liu and H. Xin, "A Wideband Circularly Polarized Patch Antenna for 60 GHz Wireless Communications," Wireless Engineering and Technology, Vol. 3 No. 3, 2012, pp. 97-105. doi: 10.4236/wet.2012.33016.

Various techniques to improve antenna performances such as bandwidth and efficiency have been reported in the literature. As shown in [5], cavity backing can isolate an antenna from its surroundings, suppress surface waves and increase antenna bandwidth. The detailed characteristics of various cavity backed antennas are summarized in [6]. Another approach to improve the bandwidth and efficiency of the conventional patch antenna is to minimize the substrate dielectric constant, as demonstrated in [7, 8]. It is also known that the antenna gain can be considerably increased by covering the antenna with a high permittivity superstrate [9, 10]. The fundamental effects of the substrate-superstrate structure on printed-circuit antennas are explored in [11], demonstrating that the antenna radiation efficiency can be optimized to approximately 100% by selecting the proper materials and dimensions of the substrate and superstrate.

All of the above mentioned techniques for improving the antenna performances can be applied in the mmWave antenna design. Several cavity backed wide-band antennas have been demonstrated in the V-band in [12-16]. In addition, metamaterial/EBG (electromagnetic band gap) based antennas with increased antenna gain have been reported in [17-21]. A good review of various mmWave integrated circuit antennas can be found in [22]. Furthermore, several topologies of circularly polarized antenna have been proposed for the 60 GHz wireless communications [23-28].

In this paper, we present the detailed design of a 60 GHz left-hand circular polarized antenna [27] that has a good overall performance by combining various techniques reported in [5-16,29-31]. This patch antenna incorporates a diagonal slot at the center and features a superstrate and an air cavity backing to achieve desired performances including wide bandwidth, high efficiency and low axial ratio. The metal frame underneath the antenna layer serves as the cavity backing, useful for antenna bandwidth enhancement, as well as mechanical support for the antenna, making the antenna much more stable than using a supporting pin (only feasible at low frequencies) [7]. The microstrip-fed patch antenna is packaged with a flip-chip CPW interface that is fully compatible to semiconductor integrated circuits (ICs). In [29-31], linear polarized folded dipole antennas for 60 GHz applications using this superstrate/substrate topology and packaging scheme have been demonstrated. In this work, by optimizing the slot length of the patch antenna, a 6 dB axial ratio bandwidth of 22.7% is achieved, which is much greater than the 1.1% axial ratio bandwidth in [32]. Moreover, this antenna achieves a wide bandwidth of more than 26%.

A prototype antenna is fabricated and characterized using a probe-based measurement setup as described in [29]. The demonstrated patch has a dimension of 1050 mm × 1050 mm, while the final packaged antenna including feed lines has a dimension of 4849 mm × 5555 mm. Beyond the initial report in [27], the detailed design procedure of the circularly polarized antenna, including the design of the microstrip-fed patch antenna and the comparison of the performances of the antenna with different feeding interfaces, is described in this paper. The measured antenna efficiency is also obtained, demonstrating greater than 75% efficiency over the entire measured frequency range. In addition, the measured co-polarized and cross-polarized radiation patterns of the antenna are presented and compared with simulation predictions. Discussions on the measurement uncertainties that may explain the simulation/measurement discrepancies are included.

This paper is organized as the following. Section 2 introduces the antenna structures including its packaging. Section 3 describes the detailed antenna design procedures. The simulated (by finite-element electromagnetic solver HFSS) and measured antenna properties are compared in Section 4.

ANTENNA STRUCTURE/PACKAGING

Since packaging at mmWave frequency is challenging and critical to system performance [30], mmWave antenna designs should incorporate the packaging aspects as early as possible. The geometry of the entire antenna structure including its packaging is shown in Figure 1(a), which consists of a top fused silica substrate (superstrate), a bottom air cavity made of copper and dielectric sealant encapsulant with a dielectric constant of 4 (ENC4) for further integration with flip-chipped ICs [30]. Similar package structures have been applied previously for folded dipole antennas and more detailed package design information can be found in [30, 31]. The applied cavity size is optimized according to the antenna impedance bandwidth. The superstrate thickness T = 300 μm is a compromise since thinner fused silica samples would be difficult to manufacture and a thicker superstrate leads to reduced efficiency due to more energy staying in the dielectric layer as surface waves. All of the metallization (2 mm thick gold) of the patch antenna and its feed lines is on the lower side of the fused silica substrate, which has negligible loss tangent at 60 GHz and a relatively low dielectric constant of 3.8. To minimize the surface wave effects that can affect the radiation in the antenna plane, a gold ring is placed on the upper side of the fused silica. The fused silica substrate is mounted on top of an air cavity with copper walls, as shown in Figure 1(b). The cavity has an internal size of 3890 mm × 3849 mm and an external size of 4890 mm × 4849 mm. The cavity height is 500 mm, being the trade-off between the package size and the antenna bandwidth. The detailed parameters of the antenna structure and packaging are listed in Table 1.

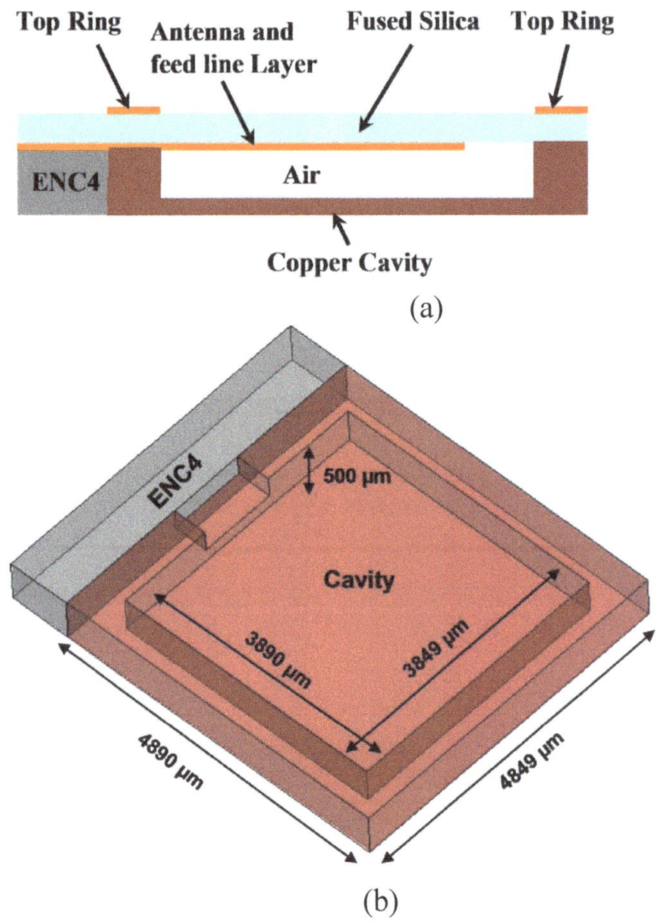

Figure 1: (a) The side view of the entire antenna structure; (b) The copper cavity.

The antenna structure shown in Figure 1 has several advantages. First, a patch antenna at the lower side of the fused silica can be thought as having an air substrate with a dielectric constant of 1 and a superstrate with a dielectric constant of 3.8. These will lead to significant bandwidth and efficiency improvements compared with a normal one substrate patch antenna. In [31], a broadband dipole antenna for 60 GHz applications using a similar cavity and superstrate was reported. Second, this packaging scheme is fully compatible with flip-chip mounted monolithic integrated circuits such as an entire SiGe transceiver [30]. A schematic of the antenna flip-chip integrated with an IC chipset is shown in Figure 2.

DETAILED ANTENNA DESIGN

For the initial design, a microstrip-fed square patch antenna with a diagonal slot is applied to achieve left-hand circular polarization [32,33], as shown in **Figure 3**. The patch size is first selected to obtain a resonant frequency around 61 GHz and the microstrip line width is tuned to match the antenna impedance. The slot length mainly controls the circular polarization performance. It is observed that the frequency at which the axial ratio is minimum increases when the slot length C increases (10 GHz when C = L/2, 51 GHz when C =

L/1.1 and 75 GHz when C = L/0.9), with the fixed patch size of 1400 μm and the slot width d kept at C/10 [34]. In addition, it is found that the slot length C also influences the antenna

	Size (μm^2)	Thickness (μm)
Fused silica ($\varepsilon_r = 3.8$, $\tan\delta = 0$)	4849 × 5555	300
ENC4 ($\varepsilon_r = 4$, $\tan\delta = 0.02$)	4849 × 665	600
Air ($\varepsilon_r = 1$, $\tan\delta = 0.0$)	3849 × 3890	500
Top ring (gold) (width = 500 μm)	4849 × 4890	2

Table 1: Configuration of the antenna structure and packaging.

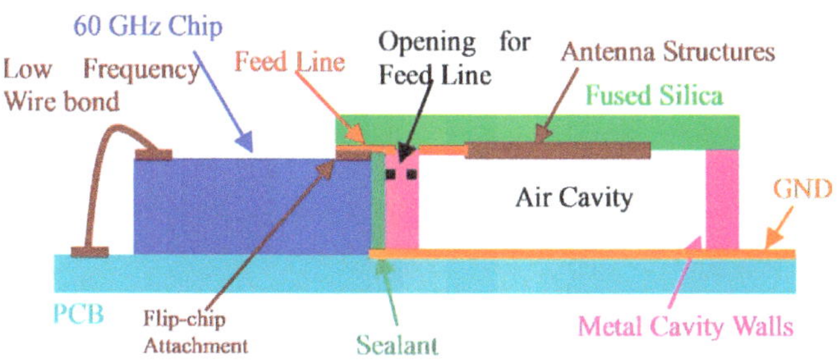

Figure 2: Schematic view of the antenna packaged with an integrated chipset.

Resonance frequency: the resonance frequency decreases when the slot length increases (with the fixed patch size and the slot width kept at C/10). Figure 4 compares the simulated antenna reflection coefficients (S_{11} in dB) for different slot lengths with the square patch size L = 1400 μm. The antenna resonance frequency shifts from 62 GHz to 57 GHz when the slot length C increases from L/2 to L/0.9. The final dimensions of the patch antenna are optimized to make both the resonant frequency and the frequency with the minimum axial ratio at 61 GHz. The finalized design parameters are listed in Table 2.

The simulated reflection coefficients (S_{11} in dB) and axial ratio of the optimized antenna are plotted in Figure 5. It can be seen that S_{11} is less than –10 dB from 59 GHz to 75 GHz, with a bandwidth of about 24%. The axial ratio is less than 6 dB from 58 GHz to 72 GHz, with a bandwidth of 21.5%.

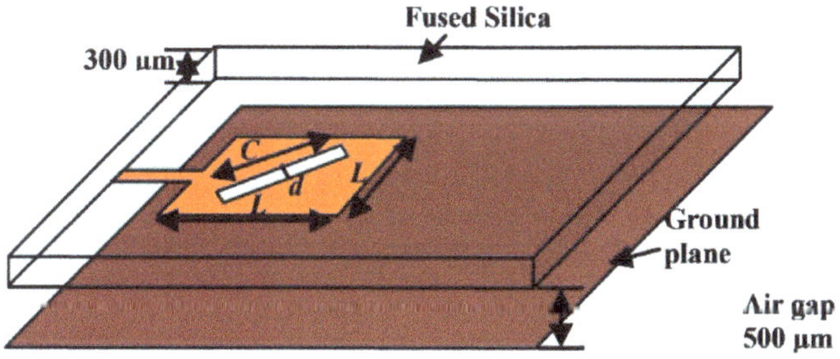

Figure 3: The microstrip-fed square patch antenna with a diagonal slot.

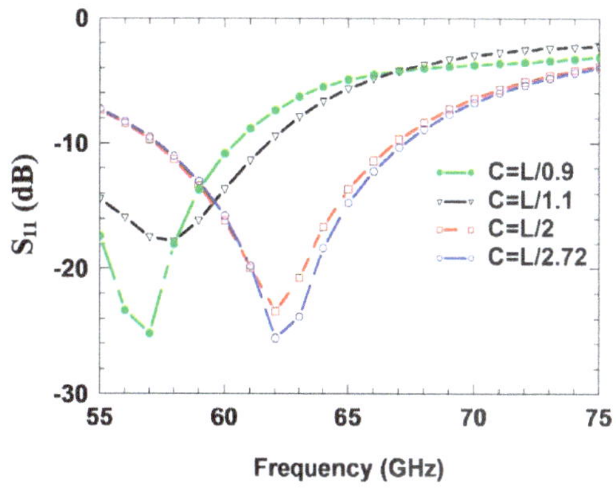

Figure 4: Simulated reflection coefficients (S_{11} in dB) of the square patch antenna (L = 1400 μm) with different slot lengths C (slot width d = C/10).

Antenna length	$L = 1050$ μm
Antenna width	$L = 1050$ μm
Slot length	$C = 1141$ μm
Slot width	$d = 114$ μm
Microstrip line width	$W_0 = 400$ μm
Microstrip impedance	$Z_0 = 102\ \Omega$

Table 2: The dimension of the finalized patch antenna design.

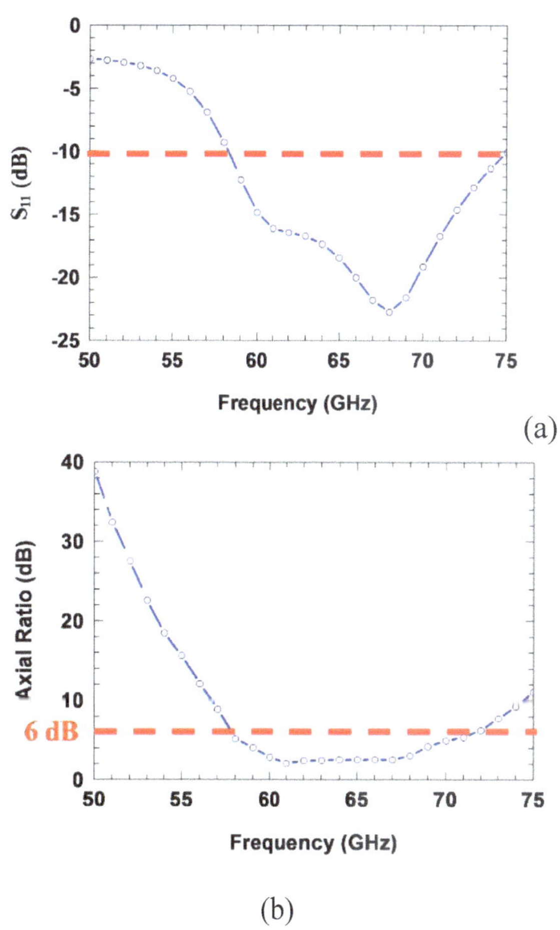

Figure 5: Simulated microstrip-fed square patch antenna performance: (a) Reflection coefficients (S_{11} in dB); (b) Axial ratio.

Comparing with conventional patch antennas, the bandwidth of the optimized antenna is much greater due to the utilization of the air cavity and the fused silica superstrate. The simulated antenna efficiency is higher than 90% for the entire frequency range from 50 to 75 GHz.

A CPW to microstrip transition is then designed to satisfy the CPW interface of the antenna packaging and enable convenient measurements using a CPW probe. The transition section is composed of a tapered microstrip line and a CPW line (labeled CPW4) with two approximately λ/4 open-stubs to achieve the desired grounding at the transition [35] around the center frequency of 61 GHz, as shown in Figure 6(a) with straight open stubs and Figure 6(b) with bent open stubs. The CPW4 section is designed to have the same impedance (102 Ω) as the microstrip. Two additional CPW lines (labeled CPW3 and CPW2) are used to transform the 102 Ω CPW line (CPW4) to the input 50 Ω CPW line (labeled CPW1).

To evaluate the impact of the CPW packaging, the HFSS simulated performances of the antennas with direct microstrip feeding, the CPW feeding with the straight open-stubs and the bent open-stubs are compared in

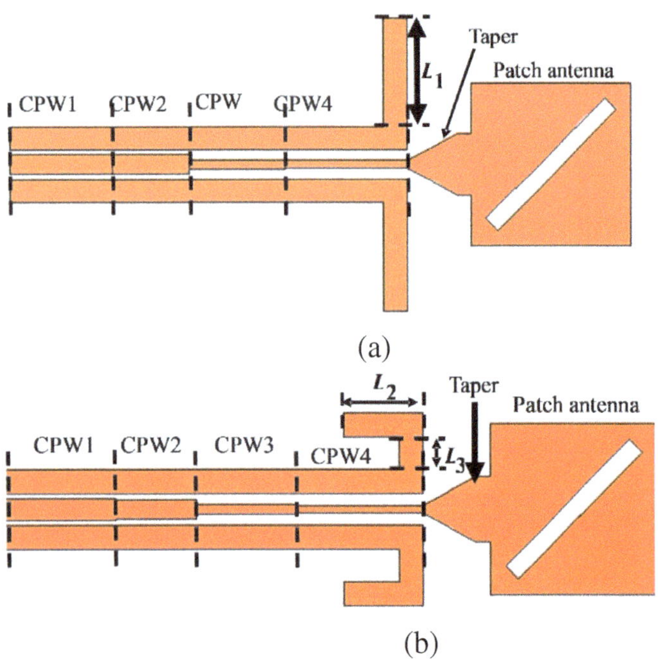

Figure 6: The top view of the antenna layer including (a) The CPW feed with straight open-stubs; (b) The CPW feed with bent open-stubs.

Figure 7. The simulated antenna efficiencies are greater than 90% for all the cases. The reflection coefficients and axial ratios of the three cases are similar at frequencies around 61 GHz (58 GHz - 64 GHz), as expected. With the CPW feeding sections, the antenna has a wider bandwidth and a narrower axial ratio bandwidth compared to those of the direct microstrip feeding. The S_{11} of the antennas with the CPW feedings are less than 10 dB from 50 GHz to 70 GHz with a bandwidth of 33%, which are larger than the 17% bandwidth of the antenna with the microstrip feeding. The 6 dB axial ratio bandwidths of the antennas with the microstrip feeding, the CPW feeding with straight open-stubs and bent open-stubs are 18.7%, 9%, and 9.9%, respectively. In combination, the bandwidth with both S_{11} less than –10 dB and axial ratio less than 6 dB drops from 17% to 9% and 9.9% when the CPW feeding sections are added to match the CPW interface, which is still sufficient for the applications at 60 GHz. In addition, the axial ratios of the antenna with bent open-stubs are around 1 dB lower than the case with straight open-stubs at frequencies around 61 GHz. Therefore, to reduce the impact of the λ/4 open-stubs on the radiation characteristics (especially the axial ratio) of the antenna, the bent λ/4 open-stubs are used for the final design. The final dimensions of the feeding sections are listed in **Table 3**, where L is the transmission line length, W is the transmission line width, g is the CPW gap, W_1 and W_2 are the widths of the two ends of the tapered mirostrip. The widths of the two CPW grounds are 150 um each.

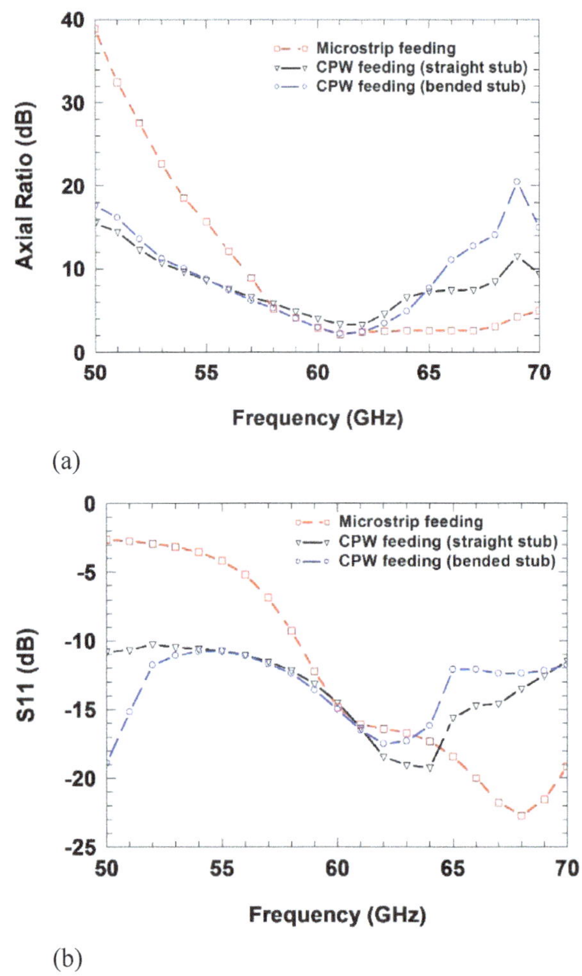

Figure 7: Comparison of antennas with different feedings (a) the axial ratio (b) the reflection coefficient (S_{11} in dB).

CPW1	L = 665 μm, W = 128 μm, g = 31 μm
CPW2	L = 500 μm, W = 124 μm, g = 33 μm
CPW3	L = 631 μm, W = 56 μm, g = 67 μm
CPW4	L = 800 μm, W = 50 μm, g = 70 μm
λ/4 open-stub	L_1 = 700 μm, L_2 = 500 μm, L_3 = 350 μm
Taper	L = 340 μm, W_1 = 50 μm, W_2 = 400 μm
Microstrip	L = 100 μm, W = 400 μm

Table 3: The geometry of the CPW to microstrip transition.

COMPARISON OF MEASUREMENT AND SIMULATION

To validate the performances of the designed 60 GHz left-hand circular polarized antenna, a prototype is fabricated by a thin-film resolution metal patterning process. The detailed fabrication procedure can be found in [30, 31]. **Figure 8** shows a photo of the fabricated and packaged antenna. Note that the antenna layer on the bottom side of the fused silica can be clearly seen because of the transparency of the fused silica superstrate.

The antenna performances including return loss, axial ratio, gain and radiation pattern are characterized from 50 GHz to 65 GHz using a CPW probe based measurement setup in an anechoic chamber, as it is described in detail in [29]. To measure the axial ratio of the fabricated antenna, both a straight and a twisted waveguide adapter are used to feed the receiving standard gain horn. The challenge of the axial ratio measurement at 60 GHz is that it is very sensitive to the phases of the two polarizations. For example, when the straight waveguide adapter is replaced by the twisted one or vice versa, small misalignment would lead to the inaccuracy of the measured axial ratio.

Figure 9 plots the simulated and measured S_{11} of the fully packaged antenna (reference point at the beginning of the CPW line). The measured result indicates a 10 dB bandwidth greater than 26% (limited by the experimental frequency range), agreeing well with the simulation result. The measured axial ratio is plotted in **Figure 10**, together with the simulated result.

Figure 8: A photo showing the top view of the fabricated and packaged 60 GHz left-hand circular polarized patch antenna.

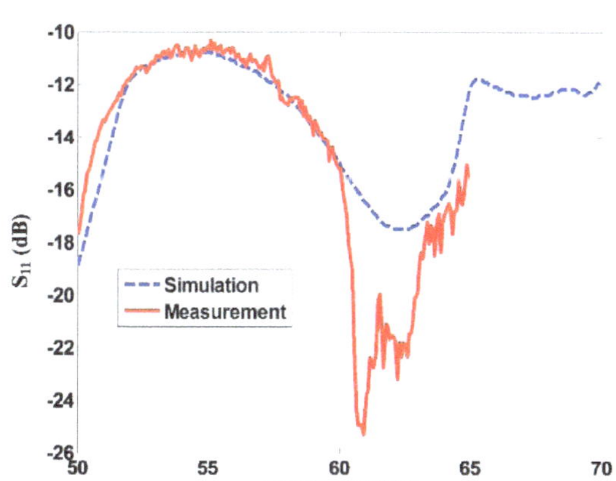

Figure 9: The simulated and measured reflection coefficients (S_{11} in dB) of the fully packaged antenna.

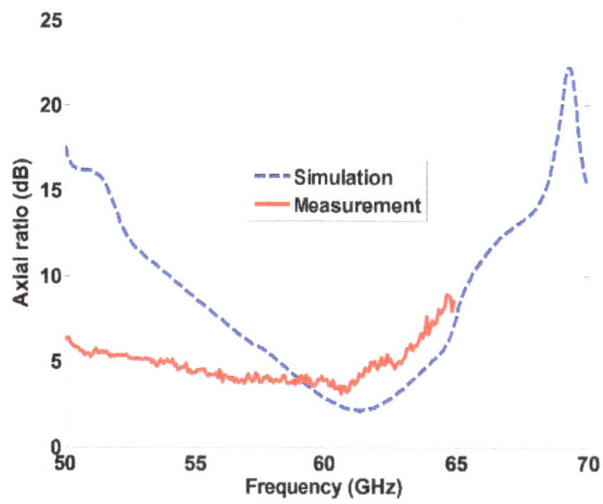

Figure 10: The simulated and measured axial ratio of the fully packaged antenna.

The simulated 6 dB axial ratio bandwidth is 9.9%, with a minimum value of 2.7 dB at 61 GHz. The measured minimum axial ratio (3.0 dB) also occurs around 61 GHz and the 6 dB axial ratio bandwidth is 22.7%. The measurement and simulation results agree well above 60 GHz but the measured values are significantly smaller at the lower frequencies. The discrepancy between the simulated and measured axial ratios is likely due to the measurement uncertainties discussed previously.

The radiation characteristics of the antenna are measured in both x-z and y-z planes (see Figure 8). The simulated and measured antenna gains (x-z plane) as a function of frequency are shown in Figure 11(a). They match reasonably well with each other near the center frequency (at 61 GHz, the measured and simulated gains are 6.4 dB and 7.3 dB, respectively). The measured gain in the y-z plane (not shown) also has similar shape and range but is distorted a bit due to the interference of the antenna feeds in the measurement. Our simulation shows that the antenna efficiency is greater than 90% within the simulated frequency range, as the dashed curve plotted in Figure 11(b). The measured antenna efficiency η_{mea} is then estimated as: $\eta_{mea} = \eta_{sim} * G_{mea}/G_{sim}$ (with the assumption that the measured and simulated directivities are consistent), where η_{sim} is the simulated antenna efficiency, G_{sim} and G_{mea} are the simulated and measured antenna gains, respectively. It can be seen from Figure 11(b) that the measured efficiency (solid curve) varies from 75% to 92% from 57 to 64 GHz.

Finally, reasonable agreements between the simulated and measured radiation patterns in both the x-z and y-z planes are observed, as shown in Figure 12. The measured patterns away from the broadside in the x-z plane show some discrepancy, probably due to measurement uncertainties. The measured co polarized and cross-polarized radiation patterns at 61 GHz are compared in Figure 13. One thing to notice is that the broadside gains in the y-z and x-z planes are different by about 1.4 dB. This discrepancy is also believed to be caused by experimental uncertainties.

CONCLUSION

A fully packaged left-hand circularly polarized antenna for 60 GHz wireless communications is demonstrated. Wide bandwidth and high efficiency are achieved by utilizing an air cavity and a fused-silica superstrate. The antenna packaging with a CPW interface is compatible to semiconductor integrated circuits. The measured antenna properties including return loss, axial ratio, gain and radiation patterns agree reasonably well with the simulation results. The demonstrated antenna achieved an impedance bandwidth greater 26%, a 6 dB axial ratio bandwidth of 22.7% and efficiency above 75% for the entire frequency range.

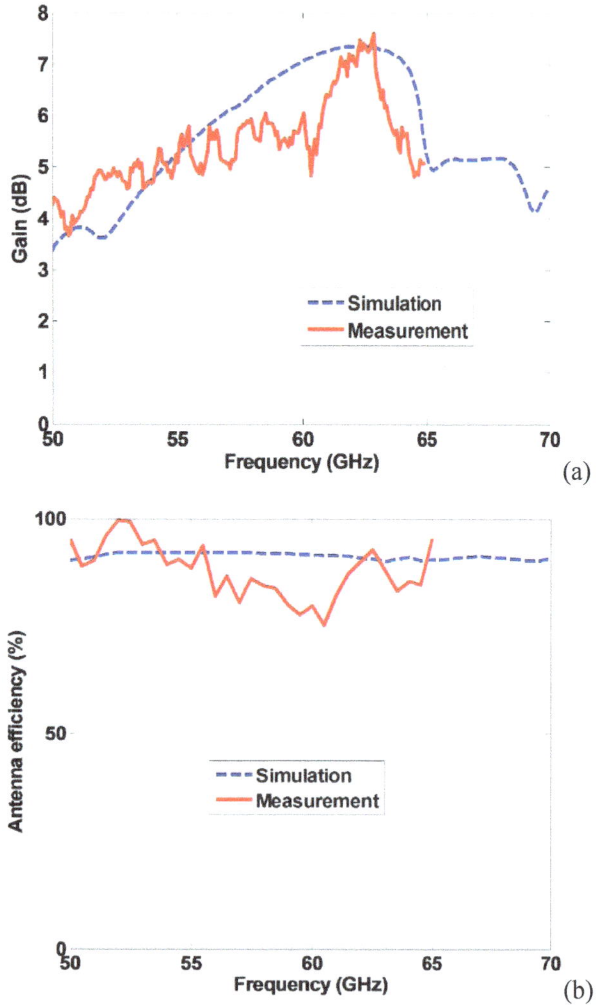

Figure 11: The simulated and measured antenna gain (a) and antenna efficiency (b) of the fully packaged antenna.

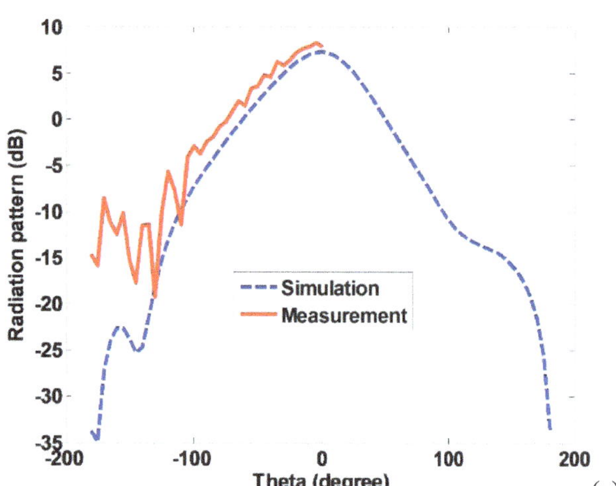

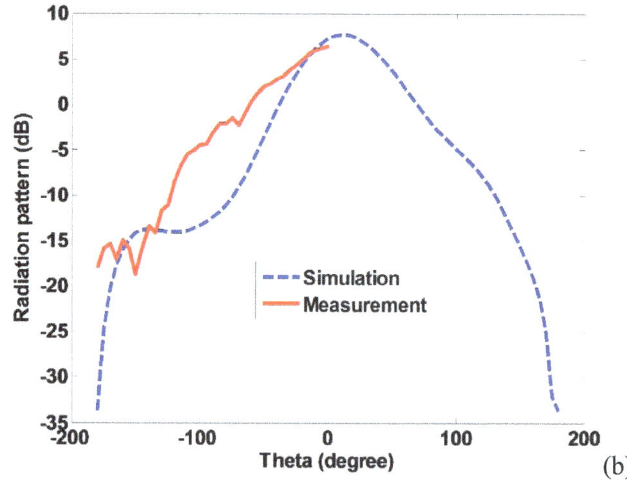

Figure 12: The simulated and measured radiation patterns of the fully packaged antenna at 61 GHz in: (a) y-z plane; (b) x-z plane.

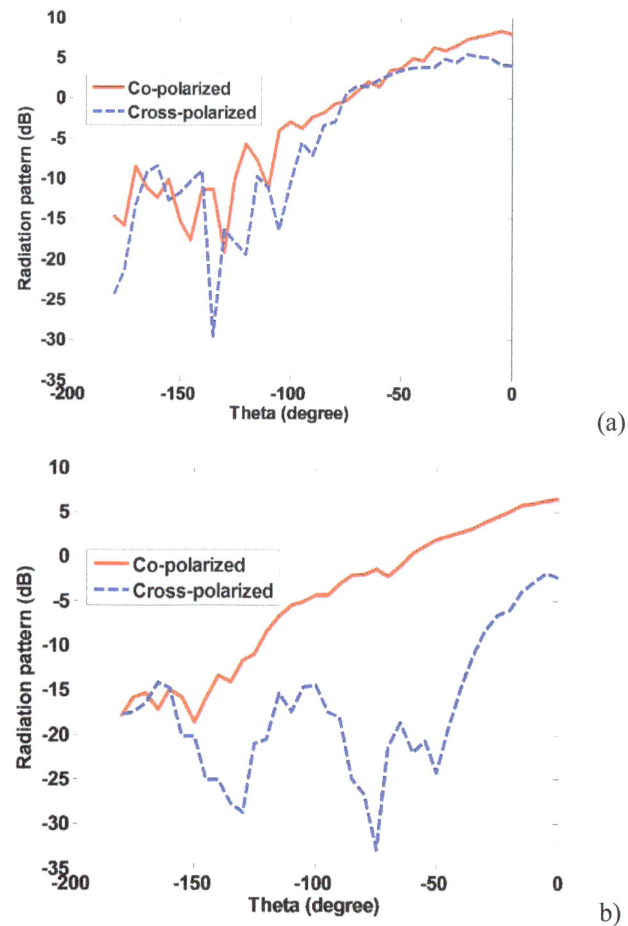

Figure 13: The measured co-polarized and cross-polarized radiation patterns of the fully packaged antenna at 61 GHz in: (a) y-z plane; (b) x-z plane.

ACKNOWLEDGEMENTS

This work was supported in part by the US Army Research Laboratory and by the US Army Research Office under agreement number W911NF-1-01-0285.

REFERENCES

1. J. F. Buckwalter, et al., "An Integrated Subharmonic Coupled-Oscillator Scheme for a 60 GHz Phased-Array

Transmitter," IEEE Transaction on Microwave Theory Technology, Vol. 54, No. 12, 2006, pp. 4271-4280. doi:10.1109/TMTT.2006.885581

2. T. S. Rappaport and D. A. Hawbaker, "Wide-Band Microwave Propagation Parameters Using Circular and Linear Polarized Antennas for Indoor Wireless Channels," IEEE Transaction on Communication, Vol. 40, No. 2, 1992, pp. 240-245.doi:10.1109/26.129185

3. T. Manabe, et al., "Polarization Dependence of Multipath Propagation and High-Speed Transmission Characteristics of Indoor Millimeter-Wave Channel at 60 GHz," IEEE Transaction on Vehicle Technology, Vol. 44, No. 2, 1995, pp. 268-274.doi:10.1109/25.385918

4. C. Loyez, N. Rolland, P. A. Rolland and O. Lafond, "Indoor 60 GHz Radio Channel Sounding and Related T/R Module Considerations for High Data Rate Communications," Electronics Letter, Vol. 37, No. 10, 2001, pp. 654- 655. doi:10.1049/el:20010439

5. M. Lye, R. B. Waterhouse, D. Novak, F. Zavosh and J. T. Aberle, "Design and Development of Printed Antenna Remote Units for Optically Distributed Mobile Communications," IEEE Microwave Guided Wave Letter, Vol. 8, No. 12, 1998, pp. 432-434. doi:10.1109/75.746766

6. N. C. Karmakar, "Investigations into a Cavity-Backed Circular-Patch Antenna," IEEE Transaction on Antennas and Propagation, Vol. 50, No. 12, 2002, pp. 1706-1715. doi:10.1109/TAP.2002.807427

7. N. Herscovici, "A Wide-Band Single-Layer Patch Antenna," IEEE Transaction on Antennas and Propagation, Vol. 46, No. 4, 1998, pp. 471-474. doi:10.1109/8.664109

8. M. M. Faiz and P. F. Wahid, "A High Efficiency L-Band Microstrip Antenna," Proceeding on IEEE International URSI Conference, Orlando, 11-16 July 1999, pp. 272-275.

9. W. Choi, C. Pyo, Y. H. Cho, J. Choi and J. Chae, "High Gain and Broadband Microstrip Patch Antenna Using a Superstrate Layer," IEEE Antennas and Propagation Society International Symposium Digest, Columbus, Vol. 2, 22-25 June 2003, pp. 292-295.

10. L. Bernard, R. Loison, R. Gillard and T. Lucidarme, "High Directivity Multiple Superstrate Antennas with Improved Bandwidth," IEEE Antennas and Propagation Society International Symposium Digest, San Antonio, Vol. 2, 16-21 June 2002, pp. 522-525.

11. N. G. Alexopoulos and D. R. Jackson, "Fundamental Superstrate (Cover) Effects on Printed Circuit Antennas," IEEE Transaction on Antennas and Propagation, Vol. 32, No. 8, 1984, pp. 807-816. doi:10.1109/TAP.1984.1143433

12. J. Lee, N. Kidera, G. DeJean, S. Pinel, J. Laskar and M. M. Tentzeris, "A V-Band Front-End with 3-D Integrated Cavity Filters/Duplexers and Antenna in LTCC Technologies," IEEE Transaction on Microwave Theory Technology, Vol. 54, No. 7, 2006, pp. 2925-2936.doi:10.1109/TMTT.2006.877440

13. X. Tang, S. Xiao, B. Wang and J. Wang, "A 60 GHz Wideband Slot Antenna Based on Substrate Integrated Waveguide Cavity," International Journal of Infrared and Millimeter Waves, Vol. 8, No. 4, 2007, pp. 275-281. doi:10.1007/s10762-007-9206-z

14. M. Sun, Y. P. Zhang, K. M. Chua, L. L. Wai, D. Liu and B. P. Gaucher, "Integration of Yagi Antenna in LTCC Package for Differential 60 GHz Radio," IEEE Transaction on Antennas and Propagation, Vol. 56, No. 8, 2008, pp. 2780-2783. doi:10.1109/TAP.2008.927577

15. S. Hu, Y. Xiong, L. Wang, R. Li and T. Lim, "A Millimeter-Wave Wideband High-Gain Antenna and Its 3D System-in-Package Solution in a TSV-Compatible Technology," Electronic Components and Technology Conference, Lake Buena Vista, 31 May-3 June 2011, pp. 869- 872.

16. S. Hu, Y. Xiong, L. Wang, R. Li, J. Shi and T. Lim, "Compact High-Gain mmWave Antenna for TSV-Based System-in-Package Application," IEEE Transactions on Components, Packaging and Manufacturing Technology, Vol. 2, No. 5, 2012, pp. 841-846.doi:10.1109/TCPMT.2012.2188293

17. R. Sauleau and P. Coquet, "Input Impedance of Electromagnetic Bandgap Resonator Antennas," Microwave and Optical Technology Letter, Vol. 41, No. 5, 2004, pp. 369- 375.doi:10.1002/mop.20144

18. S. J. Franson and R. W. Ziolkowski, "Gigabit per Second Data Transfer in High-Gain Metamaterial Structures at 60 GHz," IEEE Transaction on Antennas and Propagation, Vol. 57, No. 10, 2009, pp. 2913-2925. doi:10.1109/TAP.2009.2029277

19. E. I. Lamminen, A. R. Vimpari and J. Säily, "UC-EBG on LTCC for 60-GHz Frequency Band Antenna Applications," IEEE Transaction on Antennas and Propagation, Vol. 57, No. 10, 2009, pp. 2904-2912. doi:10.1109/TAP.2009.2029311

20. Khromova, R. Gonzalo, I. Ederra, J. Teniente, K. Esselle and P. De-Hon, "Novel All-Dielectric mmWave Horn Antennas Based on EBG Structures," Proceedings of the 5th European Conference on Antenna and Propagation, Rome, 11-15 April 2011, pp. 2959-2961.

21. A.-C. Bunea, F. Craciunoiu and G Sajin, "28 GHz CRLH Antenna on Silicon Substrate," Proceedings of the 41st European Microwave Conference, Manchester, 10-13 October 2011, pp. 579-582.

22. G. M. Rebeiz, "Millimeter-Wave and Terahertz Integrated Circuit Antennas," Proceedings of the IEEE, Vol. 80, No. 11, 1992, pp. 1748-1770. doi:10.1109/5.175253

23. D. Nesic, A. Nesic and V. Brankovic, "Circular Polarised Printed Antenna Array with Broadband Axial Ratio," IEEE Antennas and Propagation Society International Symposium Digest, Columbus, 22-25 June 2003, pp. 912- 915.

24. Kim, S. Pinel, S. Laskar and J. Yook, "Circularly & Linearly Polarized Fan Beam Patch Antenna Arrays on Liquid Crystal Polymer Substrate for V-Band Applications," Asia Pacific Microwave Conference proceedings, Suzhou, 4-7 December 2005.

25. H. Uchimura, N. Shino and K. Miyazato, "Novel Circular Polarized Antenna Array Substrates for 60 GHz-Band," IEEE MTT-S International Microwave Symposium

Digest, Long Beach, 14-18 June 2005, pp. 1875-1878. doi:10.1109/MWSYM.2005.1517099

26. M. Barakat, C. Delaveaud and F. Ndagijimana, "Circularly Polarized Antenna on SOI for the 60 GHz Band," Proceedings of the 2nd European Conference on Antenna and Propagation, Edinburgh, 11-16 November 2007, pp. 1-6.

27. R. Zhou, D. Liu and H. Xin, "Design of Circularly Polarized Antenna for 60 GHz Wireless Communications," Proceedings of the 3rd European Conference on Antenna and Propagation, Berlin, 23-27 March 2009, pp. 3787- 3789

28. Balcells, Y. Damgaci, B. A. Cetiner, J. Romeu and L. Jofre, "Polarization Reconfigurable MEMS-CPW Antenna for mmWave Applications," Proceedings of the 4th European Conference on Antenna and Propagation, Barcelona, 12-16 April 2010, pp. 1-5.

29. T. Zwick, C. Baks, U. Pfeiffer, D. Liu and B. Gaucher, "Probe Based MMW Antenna Measurement Setup," IEEE Antennas and Propagation Society International Symposium Digest, Monterey, 20-26 June 2004, pp. 747- 750.

30. U. R. Pfeiffer, et al., "A Chip-Scale Packaging Technology for 60 GHz Wireless Chipsets," IEEE Transaction on Microwave Theory Technology, Vol. 54, No. 8, 2006, pp. 3387-3397.doi:10.1109/TMTT.2006.877832

31. J. Grzyb, D. Liu, U. Pfeiffer and B. Gaucher, "Wideband Cavity-Backed Folded Dipole Superstrate Antenna for 60 GHz Applications," IEEE Antennas and Propagation Society International Symposium Digest, 5-10 June 2006, pp. 3939-3942.

32. P. C. Sharma and K. C. Gupta, "Analysis and Optimized Design of Single Feed Circularly Polarized Microstrip Antennas," IEEE Transaction on Antennas and Propagation, Vol. 31, No. 6, 1983, pp. 949-955. doi:10.1109/TAP.1983.1143162

33. J. L. Kerr, "Microstrip Polarization Techniques," Proceeding of Antenna Applications Symposium, 20-22 September 1978, 19 p.

34. C. A. Balanis, "Antenna Theory, Analysis and Design," 1st Edition, John Wiley & Sons, New York, 1982.

35. J. P. Raskin, G. Gauthier, L. P. B. Katehi and G. M. Rebeiz, "Mode Conversion at GCPW-to-Micros-Trip-Line Transition," IEEE Transaction on Microwave Theory Technology, Vol. 48, No. 1, 2000, pp. 158-161. doi:10.1109/22.817486

CHAPTER 11

DEVELOPMENT OF 16-CHANNELS COMPACT EEG SYSTEM USING REAL-TIME HIGH-SPEED WIRELESS TRANSMISSION

B. R. Myung, and S. K. Yoo

Keywords: Electroencephalography, High-speed Wireless Transmission, Real-time

INTRODUCTION

The brainwave(EEG: Electroencephalography) means the measurement of the displacement using the electrode that appears as a microscopic signal on the surface of the brain in forms of synthesized electrical signal which have occurred by the numerous nerve cells in the brain. The brainwave signal has temporal and spatial variation depending on the brain activity, condition of the body at the time of measurement and brain function. The measurement of the brainwave is essential for the diagnosis of brain function and dysfunction. Currently used mainly in hospitals and doctors will measure and analyze the brainwave for the diagnosis of diseases such as epilepsy and brain damage. In the recent years, research has been actively conducted in order to utilize the brainwave with the intelligent brain-computer interface (BCI) and human-computer interface (HCI). For each part of the cerebral cortex where most of the nerve cells exist, functions that appear physically and mentally are different and estimating the location of the activated nerve cells can be used in a variety of research. In the past, it was verified by opening the skull but currently a noninvasive method of medical equipments such as X-ray, CT and MRI is used. But MRI and CT cause a behavioral constraint to the patients when measuring because it requires over 30 minutes of measuring time and it is in adequate to measure in daily life. Whereas the brainwave measurement equipment can be manufactured inexpensively and by wearing the helmet with attached electrodes can lessen the constraints on behavior enabling a continuous health monitoring. EEG measurement system used for heath status monitoring systems such as BCI and HCI are to be used during a daily life therefore it should be simple to wear and provide maximum activity.[3,5,6,7]

In this study, while ensuring the user's activity, a miniaturized 16-channel EEG system that uses Wi-Fi module for transferring the measured EEG data at high speed will be developed. This EEG system will be de- signed with 16 channels for estimating the location of the various brainwave pattern category and nerve cells and will be able to transmit measured brainwave data within the channel without loss using 802.11 compliant high speed Wi-Fi modules. Also it will be designed with an analog front-end IC having a single-chip configuration with digital filters to enable portability and wear. Also in this article, based on 10-20 system to reference A1 will verify the system by observing the changes in the EEG signal by attaching electrodes.

MATERIALS

EEG system which was developed in this study uses 802.11 compliant Wi-Fi modules for high speed wireless communications and was designed by using Analog front-end IC and digital filter for miniaturization. It is important for wireless EEG system for BCI and HCI health status monitoring to have low power consumption, so devices with low power consumption such as Wi-Fi module, A/D converter, memory, micro controller, Analog front-end IC that includes UART module which can operate under a low power was used. [6,7] The Figure 1 is a block diagram of the 16 channel EEG system. This system is configured with EEG measurement part (Analog front-end IC), microprocessor part and Wi-Fi module part.

The electrodes that are attached to the user's head for EEG measurement will be disc electrodes which are attached to the area by paste. Signals received from the electrodes will be amplified through the Analog Frontend IC and converted to a

Citation: B. R. Myung and S. K. Yoo, "Development of 16-channels Compact EEG System Using Real-time High-speed Wireless Transmission," Engineering, Vol. 5 No. 5B, 2013, pp. 93-97. doi: 10.4236/eng.2013.55B019.

digital signal through A/D converter which will be transferred to the micro processsor using the SPI communication. A signal processing will take place within the microprocessor with an incomeing signal through the digital filter. This signal through the UART communication will be transferred as Wi-Fi communication packets and the Wi-Fi module will use its wireless communication to transfer the data to the PC and the program will display the measured EEG signal from the user's head by channel. Transferring of data between the microcontroller and Wi-Fi module is made possible through UART communication and the transfer rate is 115,200 bps. Wi-Fi can have maximum transfer rate (802.11 b) of 11 Mbps. 16 channel EEG system has maximum power consumption of 80 mA, A/D's resolving ability of 24 bit and sampling frequency of 512 Hz.[4,5,7]

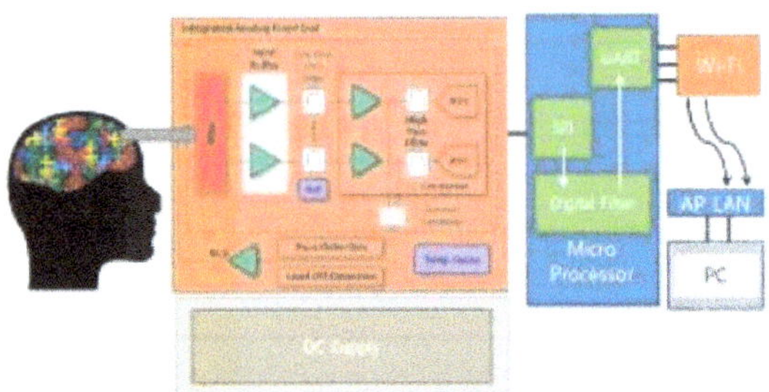

Figure 1: 16 channel EEG system block diagram.

Analog front-end

Analog Front-end IC used in this study is developed with 8 channel 24 bit dedicated for EEG/ECG which is integrated version of Analog Front-end product. It has power consumption of only 1mW per channel and can save up to 95% of power with high channel density and equipment portability which is very suitable IC for this study. The Figure 2 is an internal block diagram for Analog Front-end IC

Analog Front-end IC has 8 input channels. In this study, EEG is measured by arranging the electrodes in common reference method thus each input channel's negative input is binding with reference and EEG's GND channel is connected to the Right-Leg-Driven(RLD) circuit built in the Analog Front-end IC. EEG input signal measured through the electrodes will go through the internal amplifier within Analog Front-end IC to be amplified 12 times which then the signal will be transferred through the Delta-sigma A/D converter to the microprocessor through the built in SPI communications module. At such time, A/D converter's sampling frequency is 512 Hz.

Microprocessor

The microprocessor uses SPI communications to control and receive data from the Analog Front-end IC and transfers the data through Wi-Fi module using UART communication.

For the microprocessor, STM32F103 series of ST Microelectronics was used. STM32F103 series provides 16 MHz, 3.3 v with typical 25 uA current consumption which has low power characteristics and includes SPI and UART communications module and suitable to be utilized for this study. Also with JTAG emulation mode built in, without the need of surrounding circuits for the emulation, software upgrade and debugging is possible even with the chip mounted on the target board. The Figure 3 is a program flowchart for the KEIL uvision 4 compiler coded microprocessor.

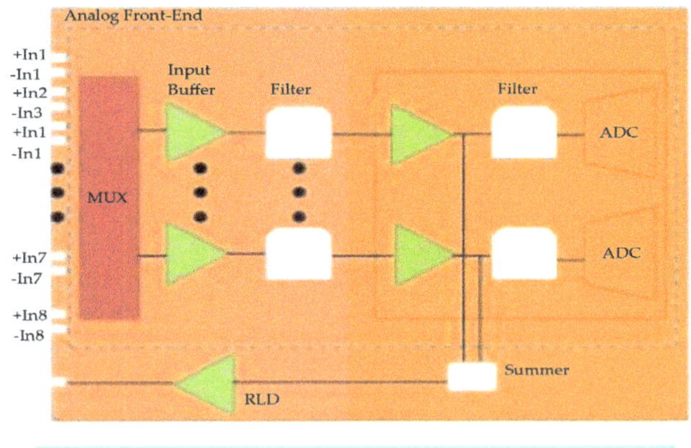

Figure 2: Analog front-end block diagram.

Digital Filter

Signal received through the SPI communication is stored in the internal buffer then through digital filter signal processing is performed. The brainwave has the frequency band of 0.1 - 100 Hz and the Table 1 shows each divided frequencies of the brainwave.[1]

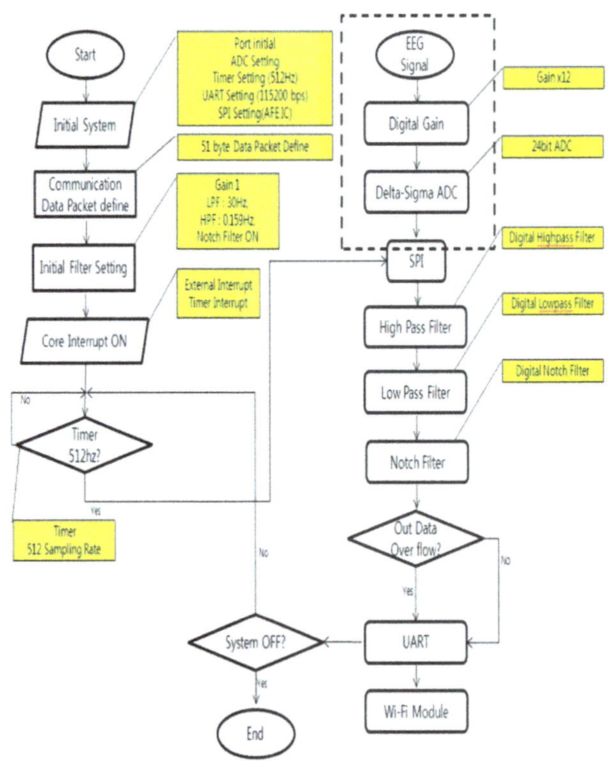

Figure 3: Micro processor flow chart.

Table 1: EEG wave.

Wave	Frequency band
Delta	Below 4 Hz
Theta	4 - 7 Hz
Alpha	8 - 13 Hz
Beta	14 - 30 Hz
Gamma	Over 30 Hz

When measuring the EEG, in order to see the frequency band of preference, EEG system is equipped with Low Pass Filter (LPF), High Pass Filter (HPF) and Notch Filter. Using a program, filter setting value can be selected at the microprocessor.(Table 2) Each filter is configured with 4th LPF, 2nd HPF, 6th Notch Filter and designed with the Direct Form-II IIR Filter.[2]

Wi-Fi Module

Wi-Fi is grafting of a wireless technology onto a Hi-Fi, which is a wireless LAN technology enabling high performance communication. It uses the frequency band of 2.4 Ghz and using this technology, a notebook can be used around the house with a wireless connection. Also using this high speed technology, 5 PC's can be connected to share files, graphics, videos and audios. Wi-Fi can provide 11Mbps of speed and at optimal environment can be used anywhere in the radius of 500m. Wi-Fi module used in this study is WizFi210 by Wiznet. Wizfi210 is Serial to Wi-Fi product that can easily and rapidly change the previous serial application to Wi-Fi capable solution. WizFi210 provides the AT Command Set which allows all functions of the module to be selected using a serial interface. Therefore, not only the serial equipment but for microprocessor as well, Wi-Fi can be easily set-up using the UART. WizFi210 is 801.11b compliant and supports up to 11Mbps of speed during a wireless interface. Also uses low power wireless communications technology with an optimal hardware provides stability and also provides Serial to Wi-Fi solution with dramatically improved power consumption, therefore it is suitable for this study.[8] The Figure 4 is an internal block diagram of WizFi- 210.

Wi-Fi module receives EEG data from the microprocessor and transfers the data to the PC through wireless communication. Wi-Fi module operates with the Limited AP Mode where the wireless communication was possible by connecting through the AP set by the Wi-Fi module using a PC.

PC EEG Viewer Program

To check the changes in EEG signal measured from the EEG System developed through this study, PC Viewer Program was realized using the C# and this program was created by Visual Studio 2010.

Table 2: EEG frequency setting value

Filter	Frequency value
High Pass Filter	0.159 Hz, 0.53 Hz, 1.09 Hz, 5.31 Hz
Low Pass Filter	30 Hz, 50 Hz, 70 Hz, 100 Hz
Notch Filter	60 Hz

The Viewer program function receives the EEG data sent from the Wi-Fi module and displays the received signal by channel. After initiating this program, clicking the button on top left corner, Wi-Fi module and connection will be conducted and it will analyze the header and tail information of the transferred data packet to display the EEG signal by channel. Signals that are displayed can be stored for EEG signal analysis and the stored signals in file format can be analyzed by using the Tool. Also High Pass Filter, Low Pass Filter, Notch Filter and Gain can be adjusted to select the desired frequency band. The Figure 5 shows the EEG Viewer Program developed by this study.

METHODS

The following is the 16 channel miniaturized EEG system developed by this study

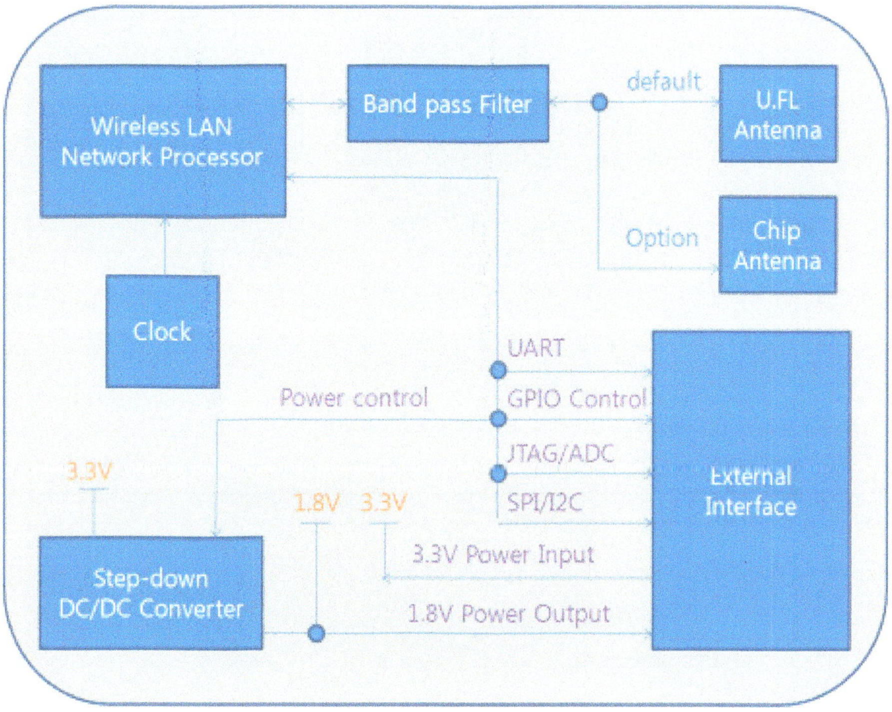

Figure 4: Wizfi210 block diagram.

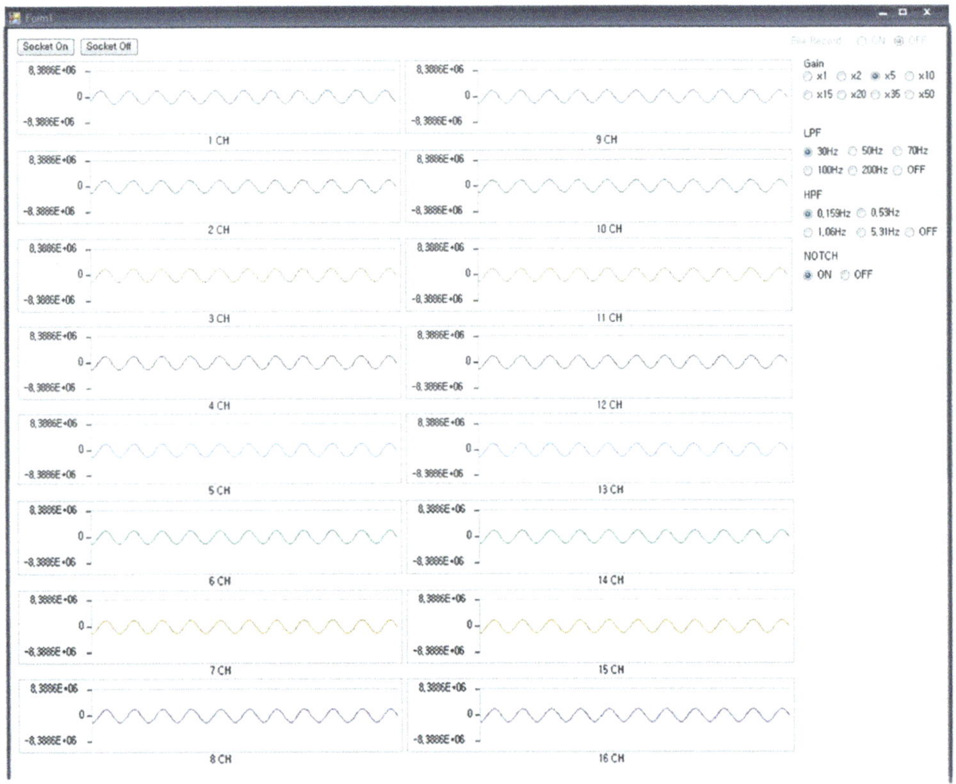

Figure 5: PC Viewer program.

Similar to the Figure 6 above, using the EEG system with confirmed Sine Wave through the simulator, have measured the EEG by having a stable normal person with its eye open. A paste was attached to the area where the measurement of the Glass Corporation's Disc electrodes was necessary. International 10 - 20 system was based to attach the electrodes and the EEG was measured as seen in the Figure 7, Figure 8 and the measured signal is shown on the Figure 9. Then, the reference electrodes and the ground electrodes were attached to the A2.[7]

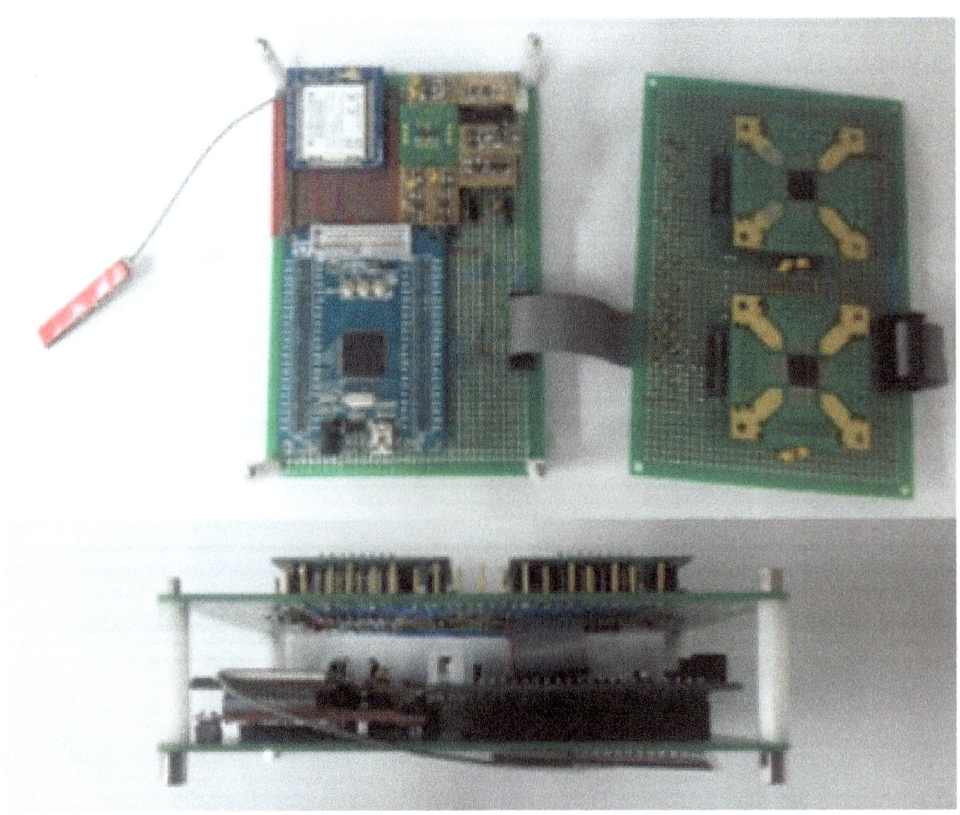

Figure 6: 16 channels EEG system using Wi-Fi.

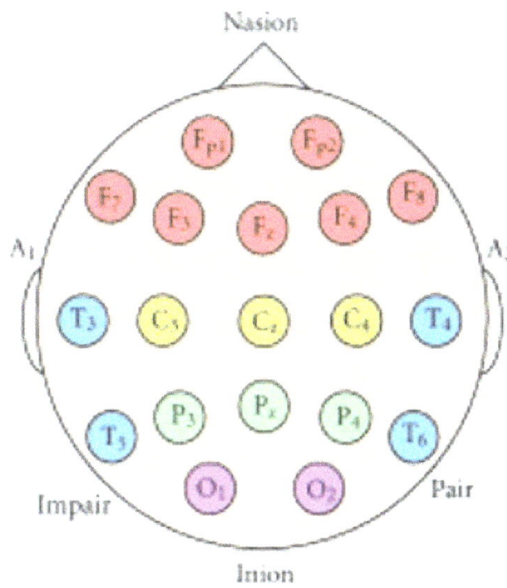

Figure 7: International 10 - 20 System

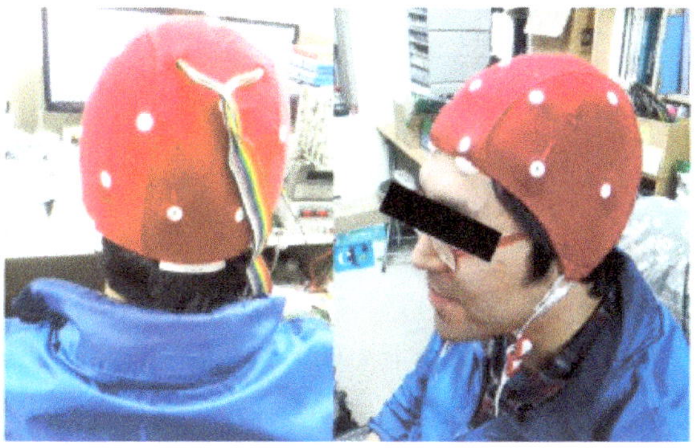

Figure 8: Setup Electrode using EEG Cap.

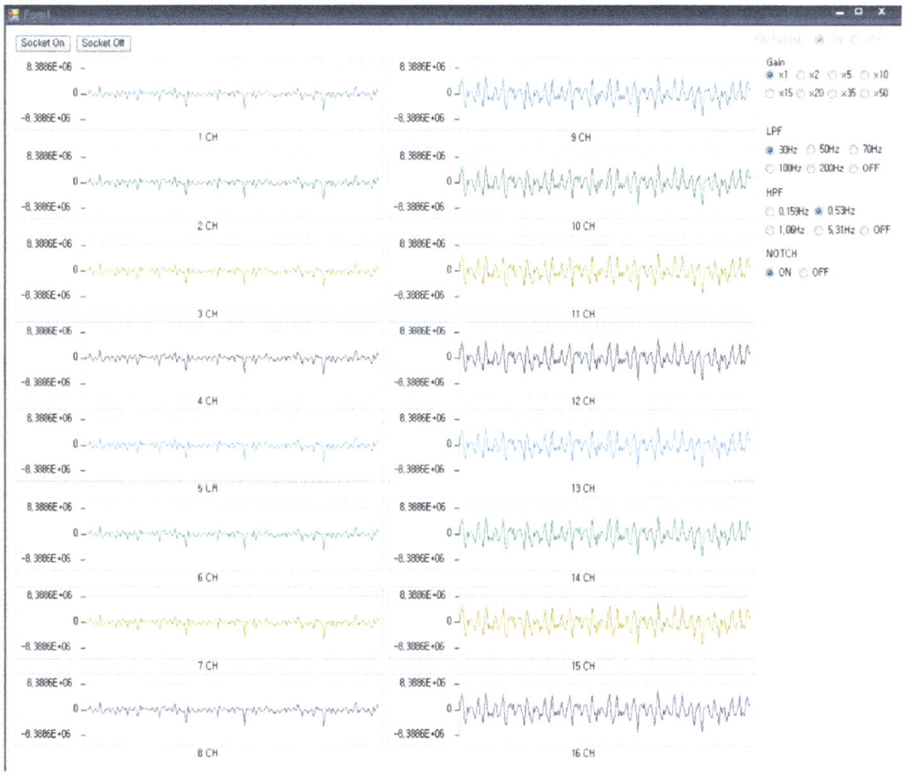

Figure 9: Measurement EEG Signal.

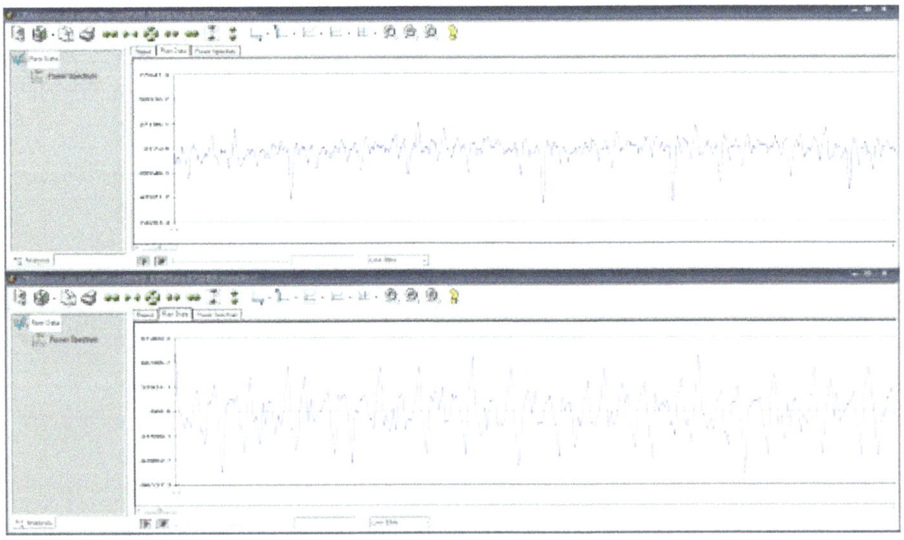

Figure 10: Stored EEG Data.

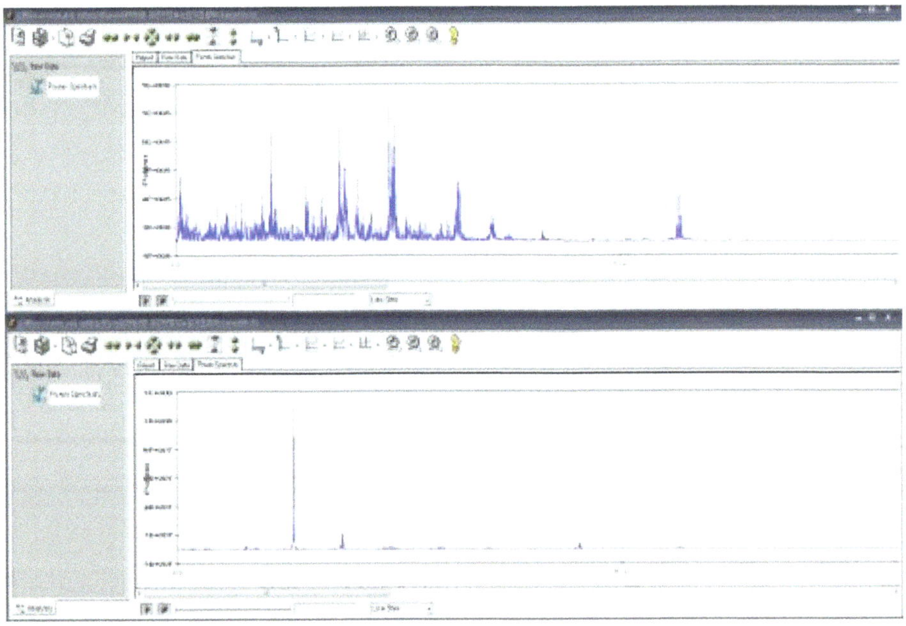

Figure 11: Power Spectrum of EEG Signal.

In order to analyze the signal by storing it as the data, the data was brought from the Complexity. Complexity is a program for analyzing the measured signal. Stored EEG data can be seen in the Figure 10.

As in Figure 11, a power spectrum for the stored signal was checked. The Figure 11 is a measurement of EEG Signal(top) from Fz and EEG signal(bottom) from O2. Both graphs were measured with an eyes closed and at O2 a signal was measured near the 13Hz and thus Alpha Wave measurement was checked. [2]

RESULTS

In this study, 16 channel miniaturized EEG system using the Wi-Fi module's real-time high speed wireless communication was designed and produced. And using this system to measure the EEG we have conducted an Eye-blocking test. Measuring the brainwave to be used with BCI and HCI health monitoring, it was necessary to develop a system that is easy to wear having the mobility during a daily life. To achieve this, in this study, a wireless module was used to configure the system to reduce the activity constraint and also we have selected a single chip for configuration in order to miniaturize the system and to reduce power consumption controlling the single chip using a microprocessor. Also, using a high speed Wi-Fi module, a lot of data within the 16 channel was transferred without loss to the PC.

ACKNOWLEDGEMENTS

This work was supported by the National Research Foundation of Korea(NRF) grant funded by the Korea government(MEST) (No.2010-0026833) and supported by the Industrial Strategic Technology Development Program (10031977-2012-23) funded by the Ministry of Knowledge Economy (MKE, Korea)

REFERENCES

1. R. Cooper, J. W. Osselton and J. C. Shaw, "EEG Technology- Third edition", Butterworths&Co Ltd, 1980
2. L. Sornmo and P. Laguna, "Bioelectrical Signal Processing in Cardiac and Neurological Applicaions," Elsevier academic press.
3. J. D. Bayliss, "A Flexible Brain-Computer Interface," University of Rochester, 2001
4. Gilsup Song, "A Study Compact EEG Measurment System Development Using Bluetooth and EEG Pattern Classification," Chonnam national university, Gwangju, Korea.
5. L. Brown, J. van de Molengraft and C. Van Hoof, "A Low-Power, Wireless, 8-Channel EEG Monitoring Headset," 32nd Annual International Conference of the IEEE EMBS, 2010
6. Ching-Sung Wang, "Design of a 32-Channel EEG System for Brain Control Interface Applications," Journal of Biomedicine and Biotechnology, Vol. 2012, Article ID 274939, 2012. doi :10.1155/2012/274939
7. M. Modarreszadeh and R. N. Schmidt, "Wireless, 32-Channel, EEG and Epliepsy Monitoring System," Proceedings of 19th International Conference, IEEE/EMBS 1997, Chicago, IL. USA.
8. Wiznet, "WizFi210 User Manual V1.14," 2013, www.wiznet.co.kr.

CHAPTER 12 REAL-TIME AUDIO & VIDEO TRANSMISSION SYSTEM BASED ON VISIBLE LIGHT COMMUNICATION

Yingjie He, Liwei Ding, Yuxian Gong, and Yongjin Wang

Keywords: Visible Light Communications, LED, Real-Time Video and Audio Broadcast System, Light Source Arrangement, Illuminance Distribution

INTRODUCTION

Visible Light Communication (VLC) was first proposed in 2004 by Toshihiki Komine and has progressed rapidly ever since with the development of solid state light sources, especially light emitting diodes (LEDs) [1]. The great popularity of VLC owes largely to the advantages of LED such as high brightness, low cost, small size, low power consumption, long lifetime and low heat radiation. VLC explores the unregulated visible light portion of the electromagnetic spectrum and acts as a supplement rather than substitute of established RF systems. With complementary strengths of both technologies, excess capacity demands of RF channels can be off-loaded to VLC networks, which enable users to seamlessly access to the Internet while keeping high Qos levels and avoiding network congestions and delays. The concept of a fullduplex multi-access system for LED-based wireless communications is proposed by Nakagawa laboratories [1]. Simplex and duplex transceiver prototypes have been demonstrated in [2] which also prove the effectiveness of replacing existing illumination systems with LED lighting devices. With 300 THz of bandwidth available for VLC, multi- gigabit-per-second data rates could be provided over short distances, for example, using array of LEDs in a multiple-input multiple-output (MIMO) system [3,4]. Besides wireless connectivity in home environment, combining the function of illumination and transmission is also attractive in specific scenarios where healthy, secured and non-interfered communication is necessary, such as hospital, underground mine, underwater, airplane, etc. In this paper, we develop a VLC prototype with large increase in transmission distance and improvement in channel capacity. The optical link which performs as the substitute of connector wire can be widely applied in video conference, real-time video frequency monitoring, smart traffic system and various practical scenarios where illumination and data transmission is in joint demand to serve modern daily life. MATLAB program is used to simulate the illuminance distribution for two practical light source deployments. The rest of the paper is organized as follows: In section 2, a prototype of the VLC real-time video/ audio transmission system is presented. The experimental result is shown and analyzed in section 3. Section 4 illustrates the correlation between light source layout and illuminance distribution based on MATLAB simulation within a model room. Finally, Section 5 concludes the article.

VLC SYSTEM DESCRIPTION

Figure 1 shows the block diagram of the real-time video and audio broadcast system, which consists of a VLC wireless LOS link with two optical channels.

VLC is free space optical communication, and line of sight (LOS) is the common link between two points in optical wireless communication system, where the transmitter directs the visible light beam in a straight and unobstructed path to the receiver [5]. Intensity modulation and direct detection are adopted by the prototype for data transmission. The system is constructed as Figure 2. The light emitter is a simple, off-the shelf LED obtained from Cree (XLamp MX-6) which provides a luminous flux of 107 lm at the current of 350 mA. A Reflection cup is mounted on each LED for the purpose of gaining light that is more parallel and concentrated. Video and audio signal captured by video camera are amplified by a self-designed amplifier and then superimposed onto two LED lamps respectively by the aid of a bias-T circuit. Thus, the output light rays changes in intensity corresponding to the

Citation: Y. He, L. Ding, Y. Gong and Y. Wang, "Real-time Audio & Video Transmission System Based on Visible Light Communication," Optics and Photonics Journal, Vol. 3 No. 2B, 2013, pp. 153-157. doi:10.4236/opj.2013.32B037.

variation in signal, which however is insensitive to human eyes due to the rapid frequency response of LED devices. The distance between two LEDs was about 10 cm in order to avoid mutual interference caused by light sources. The transmitter circuit is given in Figure 3. At the receiver, two highly sensitive Si PIN photodiodes (Hamamtsu S6968) are used to detect light transmitted over two separate optical channels. And the directionality of the PDs is required to be aligned with the most intense portion of the emitted light beams. After detection, optical signals are converted into photo electric current proportional to the variation of incident light which then is amplified and filtered by a low pass filter (LPF). The receiver circuit is shown in Figure 4. An audio and video capture module with USB interface is designed to convert the received analog signal into digital signal [6]. Finally, the video and audio are played real time on the screen of a video editor for further processing such as simple cutting, filtering or storing. In the prototype, transmission distance of nearly 2 m can be achieved without any light focusing measures and can be increased to 3 m with a focusing lens inserted between the transmitter and receiver.

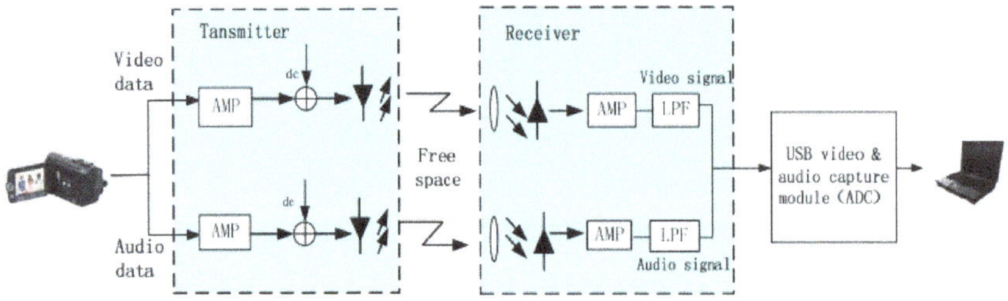

Figure 1: Block diagram of real-time video/audio VLC transmission system

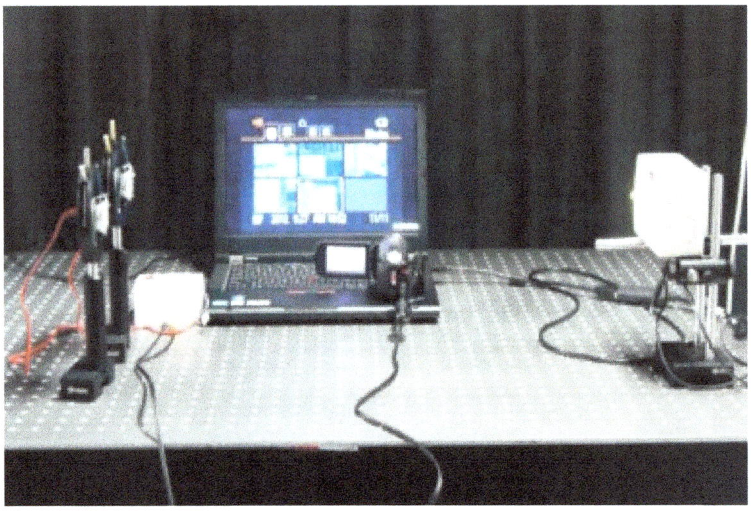

Figure 2: Picture of the real-time video/audio transmission prototype using LEDs

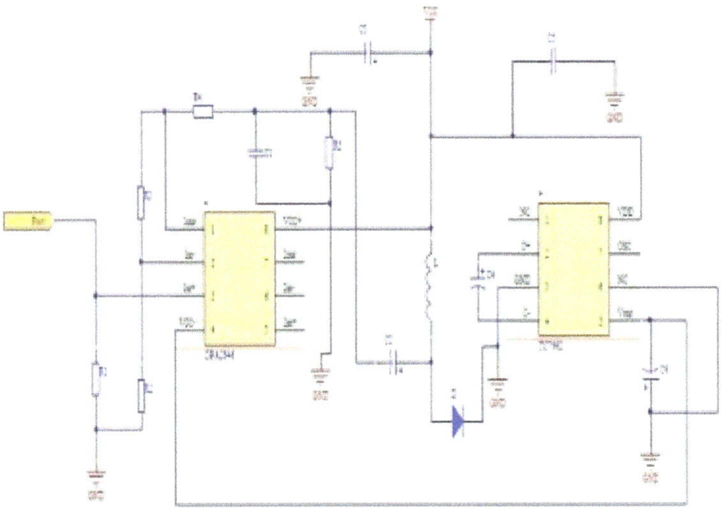

Figure 3: Transmitter circuit.

RESULTS AND ANALYSIS

To test the system performance, we have transmitted a video signal captured by a video camera. Video and audio signals were extracted from the AV output port of the camera. Through optical transmission, high quality video and audio were recovered and played on the laptop screen by the aid of self-designed ADC module with USB 2.0 interface. High-speed rendering, wireless transmission and real time performance mean less time waiting and more time creating. Figure 5 shows the waveforms of video signal before and after transmission over the optical channel. And the picture shown in Figure 6 clearly indicates that signal transmitted by optical channel can be recovered real time at 2 m without lens through the air in the laboratory. Some distortions still exist comparing the original to the received image which maybe results from the process of photoelectric conversion, signal amplification or the introduction of background noise.

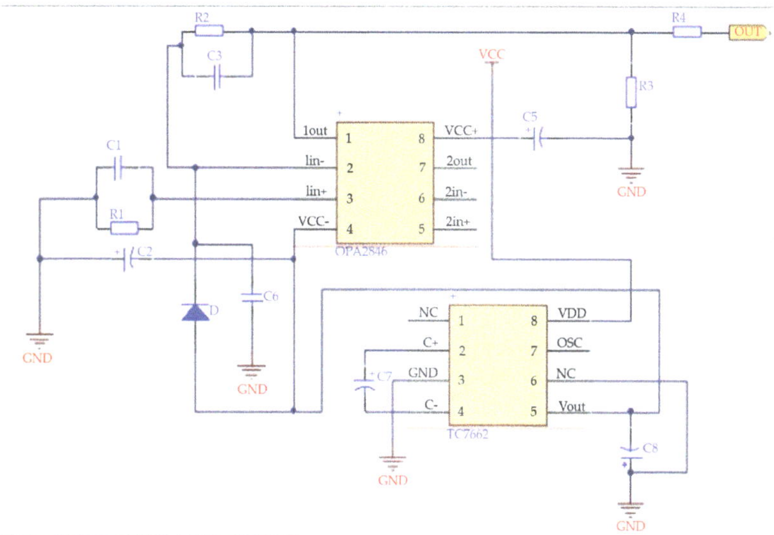

Figure 4: Receiver circuit.

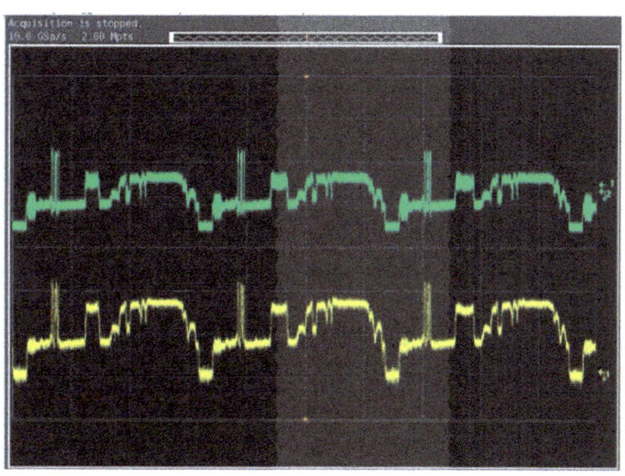

Figure 5: Experimental test of VLC prototype, received signal at the distance of 2 m, emitted (up) and received signal (bottom).

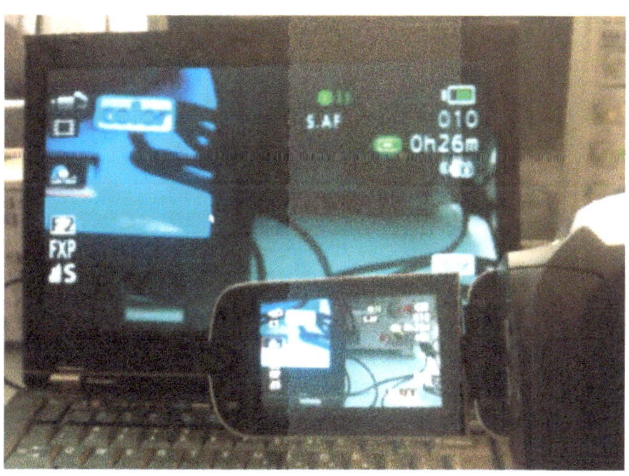

Figure 6: Comparison between original image and received image.

To improve the system performance, analog signal from web camera should be converted into digital signal before optical wireless transmission.

MODELING AND DESIGNING OF AN IN-DOOR VLC SYSTEM

The final objective of VLC development is the application of off-the-shelf LEDs in home environment wireless network to satisfy the needs of both illumination and data transmission. The prototype presented above with single LED replaced by LED arrays, is capable of supporting lighting, data broadcast as well as monitoring. In the paper, we modeled a room with the size of 5*5*3 (m). To study the illuminance distribution of LED system in practical scenarios, two types of light source arrangements located on the ceiling were simulated and discussed as shown in Figure 7. The coordinate positions of LED arrays for square and round deployments are (1.05:0.1:1.55, 1.05:0.1:1.55, 3) and (2.5+r*cos((0:1:17+ 0.5)*pi/36), 2.5-r*sin((i+0.5)*pi/36), 3） respectively. Here, 1.05:0.1:1.55 indicates that the value on coordinates increases with the step size of 0.1 m, and 0:1:17 means that the step size is assumed to be 1 m. The value of i is 1:1:18 and the radius are set to be R1=2.15 m and R2=2.35 m. Assuming that radiation of LEDs conforms to the Lambertian pattern and light reflections on the wall are not considered. The horizontal light illuminance at the point (x, y, z) is given by

$$E(x,y,z) = \frac{I(0)\cos^{m}(\phi)}{D^{2}\cdot\cos(\psi)} \quad (1)$$

where, I(0) is the center luminous intensity of LEDs, f is the angle of irradiance, y is angle of incidence, D is the distance between the LED transmitter and the receiver, and m is the order of Lambertian emission given by

$$m = -\ln 2 / \ln(\cos\varphi_{1/2}) \quad (2)$$

where, $\varphi_{1/2}$ is the semi-angle at half power [7,8]. MATLAB simulation specifications are depicted in Table 1.

The simulation of illuminance distribution for two different light source arrangements is shown in Figure 8. The minimum and maximum values for the square layout are 310.59 lx and 823.42 lx respectively. Figure 8(b) gives the simulation results for round deployment with the maximum value of 321.79 lx and minimum value of 592.92 lx. Considering that the lighting illumination requirements of the International Organization for Standardization is from 300lx to 1500 lx [7], the two arrangements of light sources proposed here can meet the joint demands of illumination and communication based on the simulation results.

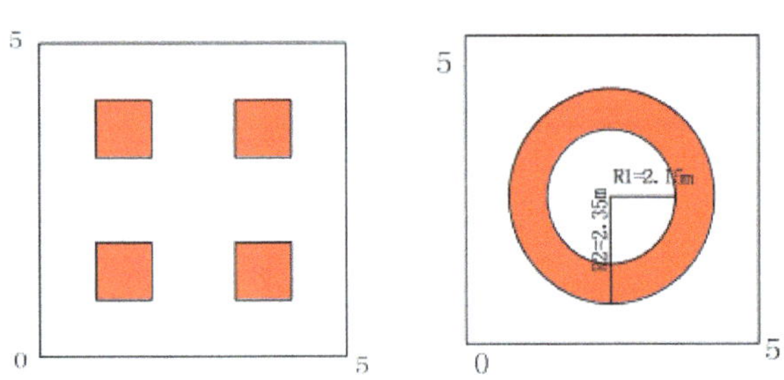

Figure 7: Square array (Left) & Round array (right).

Table 1: Parameters for MATLAB simulation

Parameters	Values(square)	Values(round)
Room size	5*5*3(m)	5*5*3(m)
Number of LED arrays	4*36	4*36
Semi-angle at half power	45°	45°
Luminous intensity	10(cd)	10(cd)
Distance between LEDs	10(cm)	20(cm)

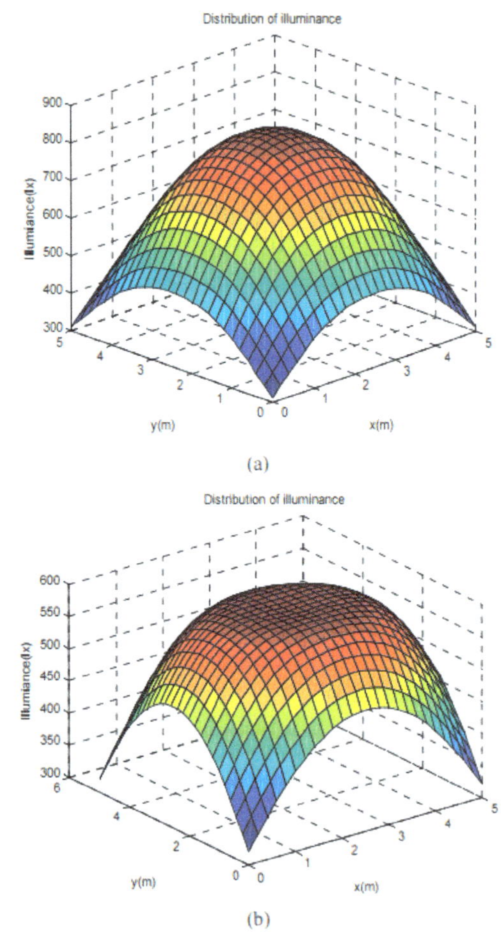

Figure 8: Illuminance distribution with the LED semi-angle of 45 degree (a) Illuminance distribution for 4x4 square deployment (b) Illuminance distribution for round deployment.

Furthermore, illuminance for round LED arrangement is more evenly distributed than that of the square arrangement judging from the simulation diagram, so that lighting effects for round deployment should be better in this case

CONCLUSIONS

In this paper, we proposed a real-time video and audio broadcast prototype using commercial LED lamps. System design is illustrated in detail and experimental results are presented. It is shown that transmission of high quality video/audio images with the distance of 3 m can be achieved and improvements can be made by adding focusing lens between the transmitter and the receiver. Although distortions still exist comparing images before and after transmission, it demonstrates that high quality wireless optical transmission using LED is successful. Illuminance distribution is simulated and compared for two light sources arrangements in a model room. Future work will concentrate on real-time digital video/audio transmission and communication link that can be connected to a TCP/IP network [9]

ACKNOWLEDGEMENTS

This work is jointly supported by NSFC (11104147), Jiangsu 973 project (Bk2011027), and research project (NY211001, BJ211026).

REFERENCES

1. Xin LIN, Kenichi IKAWA, Kazutoshi, HIROHASHI, Nakagawa Laboratories, Inc. "High-speed Full-Duplex Multiaccess System for LED-Based Wireless Communications Using Visible Light", the 2nd International MultiConference on Engineering and Technological Innovation, 2009.
2. T. D. C. Little, P. Dib, K. Shah, N. Barraford and B. Gallagher, "Using LED Lighting for Ubiquitous Indoor Wireless Networking," IEEE International Conference on Wireless and Mobile Computing, pp. 373-378, 2008. doi:10.1109/WiMob.2008.57
3. Y. A. Alqudah and M. Kavehrad, "MIMO Characterization of Indoor Wireless Optical Link Using a Diffuse-Transmission Configuration," IEEE Transactions on Communications, Vol. 51, No. 9, pp. 1554-1560, 2003. doi:10.1109/TCOMM.2003.816945
4. Z. Lubin, D. O'Brien, M. H. Le, G. E. Faulkner, L. Kyungwoo, J. Daekwang, O. YunJe and W. E. Tae, "High Data Rate Multiple Input Multiple Output (MIMO) Optical Wireless Communications Using White LED Lighting," IEEE Journal On Selected Areas In Communications, Vol. 27, No. 9, pp. 1654-1662, 2009. doi:10.1109/JSAC.2009.091215
5. D. K. Son, E. B. Cho and C. G. Lee, "Demonstration of Visible Light Communication Link for Audio and Video Transmission," Photonics Global Conference (PGC), pp. 1-4, 2010. doi:10.1109/PGC.2010.5706094
6. J. Rufo, J. Rabadan, F. Delgado, C. Quintana and R. Perez-Jimenez, "Experimental Evaluation of Video Transmission through LED Illumination Devices," Consumer Electronics, IEEE Transactions, Vol. 56, No. 3, pp. 1411-1416. doi:10.1109/TCE.2010.5606277
7. H. Q. Nguyen, J. H. Choi, M. Kang, Z. Ghassemlooy, D. H. Kim, S. K. Lim, T. G. Kang and C. G. Lee, "A MATLAB-based Simulation Program for Indoor Visible Light Communication System," Communication Systems Networks and Digital Signal Processing, pp. 537-541, 2010.
8. J. M. Kahn and J. R. Barry, "Wireless Infrared Communications," Proceedings of the IEEE, pp. 265-298, 1997. doi:10.1109/5.554222 [9] J. Rufo, F. Delgado, C. Quintana, A. Perera, J. Rabadan and R. Perez-Jimmenez, "Visible Light Communication Systems for Optical Video Transmission," Microwave and Optical Technology Letters, Vol. 52, No. 7, pp. 1572-1576, 2010. doi:10.1002/mop.25236

CHAPTER 13 DC PROGRAMMING IN COMMUNICATION SYSTEMS

Hoai An Le Thi, and Tao Pham Dinh

Keywords: Communication systems, DC programming, DCA, Exact penalty, DC constraint, Mixed 0–1 DC programs

INTRODUCTION

Optimization plays a key role in communication systems since most of issues of this domain are related to optimization problems. Convex programming has been studied for about a century. It has provided both a powerful tool and an intriguing mentality for the analysis and design of communication systems over several years in the past. During the last decade, H. A. Le Thi (B) Laboratory of Theoretical and Applied Computer Science, UFR MIM, University of Lorraine, Ile du Saulcy, 57045 Metz, France e-mail: hoai-an.le-thi@univ-lorraine.fr T. Pham Dinh Laboratory of Mathematics, LMI, National Institute for Applied Sciences, Rouen, Avenue de l'Université, 76801 Saint-Etienne-du-Rouvray Cedex, France e-mail: pham@insa-rouen.fr an increasing amount of effort has been put into nonconvex optimization to deal with challenge problems appeared in many applications of this filed (in fact, most real-life problems are of nonconvex nature). The absence of convexity creates a source of difficulties of all kinds, in particular, the distinction between the local and global minima, the nonexistence of verifiable characterizations of global solutions, etc., that causes all the computational complexity while passing from convex to nonconvex programming. In general, unlike the convex programming, there is no iterative algorithm converging directly to a global solution of a nonconvex program. Finding a global solution of a nonconvex program, especially in the large-scale setting, is the holy grail of optimizers.

The special context of practical problems in CS, along with the dramatic progress of novel technologies, requires well-adapted optimization techniques. For example, solution methods to network management in the context of mobile service should take into account the following questions:

- The topology of networks is dynamic and real-time data transmissions are needed. Hence, real-time algorithms are expected.
- Self-organization and self-configuration require all protocols in mobile networks to be distributive and collaborative. By the way, distributed algorithms are necessary.
- Location/tracking management, in addition to the handover management and routing. In the case of hybrid communication networks, the choice of the best access gateway among a number of available access technologies becomes one of the important considerations. Routing in hybrid networks should be handled by finding a suitable mathematical model and efficient algorithms.
- Multi-user communications service involves large scale setting optimization problems. Therefore, the algorithms should be able to solve large-size problems.

Many challenging issues arise from nonconvex optimization in communication systems, especially how to design suitable models and develop efficient, fast, scalable and distributed algorithms to tackle large-scale practical problems in the areas of wireless networking, internet engineering, mobile services in self-organized hybrid networks

A variety of nonconvex optimization techniques have been recently developed by researchers in optimization on one side and in communications systems on another side for studying communication theory as well as for solving practical problems in CS (see e.g, [1–9,12,14–21,24–33,38–54]. Generally, there are two different, but complementary approaches: global approaches such as Cutting Plane (CP), Branch and Bound (B&B), Branch and Cut (B&C) can guarantee the globality of the solutions but they are very expensive, and cannot handle problems of high dimensionality; and local approaches, on the contrary, are much faster while only local minima are available. Many current approaches are not generally effective for practical large-scale problems. Finding efficient algorithms

Citation: Hoai An Le Thi, Tao Pham Dinh, DC programming in communication systems, DOI: 10.1007/s40595-013-0010-5.

that realize a compromise to overcome these drawbacks is a challenge of nonconvex programming. Such algorithms must exploit domain-specific structures of the problems being considered.

As an innovative approach to nonconvex programming, Difference of Convex functions (DC) programming and DC Algorithms (DCA) are increasingly used by researchers in CS (see e.g. [1,2,16,22,45,54] and references therein). The objective of this paper is to show that many challenging problems in CS can be modeled as DC programs and solved by DCA-based algorithms. We offer the community of researchers in CS promising approaches in a unified DC programming framework to tackle various applications such as routing, power control, congestion control of the Internet, resource allocation in networks, etc

Nonconvex optimization problems in CS

In terms of optimization, nonconvex problems appeared in

CS can be divided into three classes:

- Minimizing a nonconvex function on a convex set;
- Minimizing a convex/nonconvex function on a nonconvex set;
- Minimizing a convex/nonconvex function on a convex/ nonconvex set with integer variables.

The reader will see that these classes of nonconvex programs in CS can be formulated or reformulated as DC programs and solved by DCA.

Why DC programming and DCA?

DC programming is an extension of convex programming which is sufficiently large to cover almost all nonconvex optimization problems, but not much to still allow using the arsenal of powerful tools in convex analysis and convex optimization. DC programming and DCA constitute the backbone of nonconvex programming and global optimization. The use of DCA for solving nonconvex optimization problems in CS is motivated by the following facts:

- DCA is a philosophy rather than an algorithm. For each problem, we can design a family of DCA-based algorithms. The flexibility of DCA on the choice of DC decomposition offers DCA schemes having the potential to outperform standard methods.
- By exploiting the nice effect of DC decomposition of the objective function we can build distributed algorithms. This issue is very important in communication networks that involve multi-users, in particular in the purpose of personalized mobile services.
- Convex analysis provides powerful tools to prove the convergence of DCA in a generic framework. Hence, any DCA-based algorithm enjoys (at least) general convergence properties of the generic DCA scheme that are already available.
- DCA is an efficient, fast and scalable method for smooth/ nonsmooth nonconvex programming. To the best of our knowledge, DCA is actually one of the rare algorithms for nonsmooth nonconvex programming which allows to solve large-scale DC programs. DCA was successfully applied to a lot of different and various nonconvex optimization problems to which it quite often gave global solutions and proved to be more robust and more efficient than related standard methods. In particular, DCA has already efficiently solved large-scale DC programs in network optimization (see [1, 2, 24–27, 29–31, 45–51, 54] and the list of references in [22]).

We will show how to solve these three classes of problems in CS by DC programming and DCA. For beginning, let us give in Sect. 2, a brief introduction of DCA programming and DCA. The solution methods of each class of problem will be presented in Sects. 3, 4, and 5 where, in addition to development of generic models and algorithms, methods for typical applications in CS will be illustrated. In Sect. 6, we mention another issue in CS for which DCA can also be investigated. Section 7 concludes the paper.

DC PROGRAMMING AND DCA

A Brief Introduction

We are working with the space $X = \mathbb{R}^n$ which is equipped with the canonical inner product $\langle \cdot, \cdot \rangle$ and the corresponding

Euclidean norm $\| \cdot \|$ thus the dual space Y of X can be identified with X itself. We follow [13,37], for definitions of usual tools in modern convex analysis, where functions could take the infinite values $+\infty$. A function $\theta : X \to \mathbb{R} \cup \{+\infty\}$ is said to be proper if it is not identically equal to $+\infty$. The effective domain of θ, denoted by dom θ, is

$\operatorname{dom} \theta = \{x \in X : \theta(x) < +\infty\}$.

The indicator function χC of a nonempty closed convex C set is defined by $\chi C(x) = 0$ if $x \in C$, $+\infty$ otherwise. The set of all lower semicontinuous proper convex functions on X is denoted by $\Gamma_0(X)$. Let $\theta \in \Gamma_0(X)$, then the conjugate function of θ, denoted θ^*, is defined by

$\theta^*(y) = \sup\{x, y - \theta(x) : x \in X\}$.

We have $\theta^* \in \Gamma_0(Y)$ and $\theta^{**} = \theta$.

Nonsmooth convex functions are handled using the concept of sub differentials. For $\theta \in \Gamma_0(X)$ and $x_0 \in \operatorname{dom} \theta$, $\partial\theta(x_0)$ denotes the sub differential of θ at x0, and is defined by

$$\partial\theta(x_0) := \{y \in Y : \theta(x) \geq \theta(x_0) + \langle x - x_0, y \rangle, \forall x \in X\}. \quad (1)$$

Each $y \in \partial\theta(x_0)$ is called a sub gradient of θ at x0. The sub differential $\partial\theta(x_0)$ is a closed convex set in Y. It generalizes the derivative in the sense that θ is differentiable at x0 if and only if $\partial\theta(x_0) \equiv \{\nabla\theta(x_0)\}$. Recall the well-known properties related to sub differential calculus of $\theta \in \Gamma_0(X)$:

$$y_0 \in \partial\theta(x_0) \Leftrightarrow x_0 \in \partial\theta^*(y_0) \Leftrightarrow \langle x_0, y_0 \rangle = \theta(x_0) + \theta^*(y_0); \quad (2)$$

$$\partial\theta^*(y_0) = \operatorname{argmin}\{\theta(x) - \langle x, y_0 \rangle : x \in X\}. \quad (3)$$

A function $\theta \in \Gamma_0(X)$ is said to be polyhedral convex if

$$\theta(x) = \max\{\langle a_i, x \rangle - \beta_i : i = 1, \ldots, m\} + \chi_C(x) \, \forall x \in X,$$

where C is a nonempty polyhedral convex set in X. A DC program is of the form

$$(P_{dc})\,\alpha = \inf\{f(x) := g(x) - h(x) : x \in X\}, \quad (4)$$

with $g, h \in \Gamma_0(X)$. Such a function f is called a DC function, and g − h, a DC decomposition of f, while the convex functions g and h are DC components of f. In (Pdc) the nonconvexity comes from the concavity of the function— h (except the case h is affine since (Pdc) then is a convex program). It should be noted that a convex constrained DC program can be expressed in the form (4) by using the indicator function on C, that is

$$\inf\{f(x) := g(x) - h(x) : x \in C\}$$
$$= \inf\{\varphi(x) - h(x) : x \in X\}, \text{with}\, \varphi := g + \chi_C.$$

Hence, throughout this paper, DC program of the form (4) is referred to as "standard DC program".

Polyhedral DC programs(P_{dc}) (i.e., when g or h are polyhedral convex) play a key role in nonconvex programming (see [25,34,35] and references therein), and enjoy interesting properties related to local optimality and DCA's convergence.

The DC duality is based on the conjugate functions and the fundamental characterization of a convex function $\theta \in \Gamma_0(X)$ as the pointwise supremum of a collection of affine minorants:

$$\theta(x) = \sup\{\langle x, y\rangle - \theta^*(y) : y \in Y\}, \quad \forall x \in X. \quad (5)$$

That associates the primal DC program (4)(P_{dc}) with its dual DC program (D_{dc}) defined by

$$(D_{dc}) \quad \alpha = \inf\{h^*(y) - g^*(y) : y \in Y\}, \quad (6)$$

and investigates their mutual relations. We observe the perfect symmetry between primal and dual DC programs: the dual to (D_{dc}) is exactly (P_{dc}).

It is worth noting the wealth of the vector space DC(X) $= \Gamma_0(X) - \Gamma_0(X)$ spanned by the "convex cone" $\Gamma_0(X)$ [34, 35]: it contains most realistic objective functions and is closed under operations usually considered in optimization.

The complexity of DC programs resides, of course, in the lack of verifiable globality conditions. Lets recall the general local optimality conditions in DC programming (subdifferential's inclusion): if x^* is a local solution of (P_{dc}) then

$$\emptyset \neq \partial h(x^*) \subset \partial g(x^*). \quad (7)$$

The condition (7) is also sufficient (for local optimality) in many important classes of DC programs (see [34,35]).

A point x^* is said to be a critical point of g—h (or generalized KKT point for (Pdc)) if

$$\partial h(x^*) \cap \partial g(x^*) \neq \emptyset. \quad (8)$$

Note that, by symmetry, the dual part of (7) and (8) are trivial.

DC Programming and DCA were introduced by Pham Dinh Tao in their preliminary form in 1985. These theoretical and algorithmic tools are extensively developed by Le Thi Hoai An and Pham Dinh Tao since 1994 to become now classic and increasingly popular. DCA is a continuous primal dual subgradient approach based on local optimality and duality in DC programming for solving standard DC programs (P_{dc}).

DCA's philosophy

The key idea behind DCA is to replace in (Pdc), at the current point x^k , the second component h with its affine minorant defined by

$$h_k(x) := h(x^k) + \langle x - x^k, y^k\rangle,\ y^k \in \partial h(x^k)$$

to give birth to the primal convex program of the form

$$(P_k) \quad \inf\{g(x) - h_k(x) : x \in X\}$$
$$\Longleftrightarrow \inf\{g(x) - \langle x, y^k\rangle : x \in X\}$$

whose solution set is $\partial g^*(y^k)$.The next iterate x^{k+1} is taken in $\partial g^*(y^k)$.

Dually, a solution x^{k+1} of (P_k) is then used to define the dual convex program (D_{k+1}) obtained from (D_{dc}) by replacing g^* with its affine minorant

$$(g^*)_k(y) := g^*(y^k) + \langle y - y^k x^{k+1}\rangle$$

to obtain the convex program

$$(D_{k+1}) \inf\{h^*(y) - [g^*(y^k) + \langle y - y^k, x^{k+1}\rangle] : y \in Y\}$$
$$\Leftrightarrow \inf\{h^*(y) - \langle y, x^{k+1}\rangle : y \in Y\}$$

whose solution set is $\partial h(x^{k+1})$. The next iterate y^{k+1} is chosen in $\partial h(x^{k+1})$.

DCA scheme

Initialization: Let $x^0 \in \mathbb{R}^n$ be a guess, k ← 0. Repeat

– Calculate $y^k \in \partial h(x^k)$

– Calculate $x^{k+1} \in \partial g^*(y^k)$, which is equivalent to $x^{k+1} \in$ argmin $\{g(x) - x, y^k : x \in \mathbb{R}^n\}$ (P_k)

– k ← k + 1

Until convergence of $\{x^k\}$

DCA's convergence properties

Convergence properties of DCA and its theoretical basis can be found in [25,34,35]. For instance, it is important to mention that

- DCA is a descent method without linesearch but withglobal convergence: the sequences $\{g(x^k) - h(x^k)\}$ and $\{h*(y^k) - g*(y^k)\}$ are decreasing.
- If the optimal value α of problem (P_{dc}) is finite and the infinite sequences $\{x^k\}$ and $\{y^k\}$ are bounded, then everylimitpoint $x*$ (resp. $y*$)ofthesequence$\{x^k\}$(resp. $\{y^k\}$) is a critical point of $g - h$ (resp. $h^* - g*$), i.e. $\partial h(x^*) \cap \partial g(x^*) \neq \emptyset$ (resp. $\partial h^*(y^*) \cap \partial g^*(y^*) \neq \emptyset$).
- DCA has a linear convergence for DC programs.
- DCA has a finite convergence for polyhedral DC programs.

For a complete study of DC programming and DCA, the reader is referred to [25,34,35] and the references therein. Without going into details, let us mention the key properties of DCA.

Key properties of DCA

a. Flexibility the construction of DCA is based on g and h but not on f itself, and there are as many DCA as there are DC decompositions. This is a crucial fact in DC programming. It is important to study various equivalent DC forms of a DC program, because each DC function f has infinitely many DC decompositions which have crucial implications for the qualities (speed of convergence, robustness, efficiency, globality of computed solutions, etc.) of DCA.

b. DCA is a descent method without linesearch, with global convergence (i.e. DCA converges from an arbitrary initial point).

c. Return from DC programming to convex programming DCA consists in an iterative approximation of a DC program by a sequence of convex programs that will be solved by appropriate convex optimization algorithms. This property is called successive convex approximation (SCA) in some recent works in CS.

d. Versatility With suitable DC decompositions DCA generates most standard algorithms in convex and nonconvex optimization. Hence DCA offers a wide framework for solving convex/nonconvex optimization problems. In particular, DCA is a global approach to convex programming, i.e., it converges to optimal solutions of convex programs reformulated as DC programs. Consequently, it can be used to build efficient customized algorithms for solving convex programs generated by DCA.

How to apply DCA for solving practical problems

It would be wrong to think that using DCA for solving a practical problem is a simple procedure. Indeed, the generic DCA scheme is an overall philosophical idea rather than a single algorithm. There is not only one DCA but a family of DCAs for a considered problem. While DC decompositions exist for a very large class of functions, there are no general procedure for determining such DC decompositions. The design of an efficient DCA for a concrete problem is an art which should be based on theoretical tools and on its special structure. It consists of the five steps :

a. Find a DC formulation of the considered optimization problem: this can be done if the feasible domain C is a convex set and the objective function f is DC, otherwise we must use the approximation or reformulation technique based on relevant theoretical tools.

b. Design a DCA scheme for (P_{dc}). This consists of i) computing a subgradient of h and ii) solving the convex program of the form (P_k). In the ideal case (what is not always possible, especially for nondifferentiable DC programs) an optimal solution of (P_k) is explicitly determined, which corresponds to the explicit computation of a subgradient of g^*. Otherwise one should find efficient convex optimization algorithms suitably adapted to (Pk)'s specific structures in order to save computation time.

c. Search for "good" DC decompositions. If the computations in i) and ii) are not satisfactory (costly, computed solutions by DCA are not sufficiently good ...) then one has to find more suitable DC decomposition. That is a difficult issue and should be done by exploiting distinctive features of the class of DC programs at hand. Here reformulation techniques play a key role to obtain suitable models. Reformulation techniques should be diversified and have recourse to good mathematical backgrounds in numerical analysis and optimization.

d. Search for "good" starting points. That can be done by combining with other approaches (heuristic or local search methods, solutions of convex relaxation problems in global optimization methods). Another efficient way in finding a convex minorant consists of the objective function on the feasible set C and solving the resulting convex program whose solution is used to initialize DCA. This strategy must be developed in depth and specifically, by exploiting the structure of the problem (P_{dc}). e. Globalize DCA: to guarantee globality of sought solutions or to improve their quality it is advised to combine DCA with global optimization techniques

It goes without saying that the two steps (c) and (d) and solution methods for convex programs (P_k) (if necessary) constitute the key issues for successful applications of DC programming and DCA

We show below how to use DCA for solving the three classes of nonconvex problems in CS.

MINIMIZING A NONCONVEX FUNCTION UNDER A CONVEX CONSTRAINT SET

The mathematical formulation of this class of problem is given by

$$(P)\ \inf\{f(x) : x \in C\} \qquad (9)$$

where $C \subset \mathbb{R}^n$ is a closed nonempty convex set and f is a nonconvex function on $\mathbb{R}^n$. Many applications in CS can be formulated in the form of (P) whose typical examples are Network Utility Maximization (NUM) [9,30], power control problem [2,6,48,54]), dynamic spectrum management in DSL systems [27,53] MIMO relay optimization [15], sum-rate maximization, proportional-fairness and max–min optimization of SISO/MISO/MIMO ad hoc networks [14,17,43,44].

We will consider two examples of applications of DCA on NUM [30] and DSL [27]. Complete works on NUM and power control using DCA can be found in [24, 48, 31].

Can one get a DC formulation for any problem in this class?

The answer is "yes", one can always formulate (P) as a standard DC program of the form (P_{dc}). In fact, as indicated above, the vector space DC(X) contains most realistic objective functions. In the rare cases where f is not DC (for example, when f is a discontinuous function), one can approximate f by a DC function or use reformulation techniques to get an equivalent DC program

Useful DC decompositions and corresponding DCA

To illustrate the way to construct DC decompositions and design the resulting DCA, let us show the two useful DC decompositions of the problem (P) in (9) and discuss on their effectiveness

Assume that there exists a nonnegative number η (resp. ρ) such that the function $\frac{1}{2}\eta\|x\|^2+f(x)$ (resp. $\frac{1}{2}\rho\|x\|^2-f(x)$) is convex. We can now write (P) in the form of DC program (P_{dc}) with, for example, two following DC decompositions:

$$g(x) := \chi_C(x) + \frac{1}{2}\eta\|x\|^2 + f(x); \quad h(x) := \frac{1}{2}\eta\|x\|^2 \tag{10}$$

and

$$g(x) := \chi_C(x) + \frac{1}{2}\rho\|x\|^2; \quad h(x) := \frac{1}{2}\rho\|x\|^2 - f(x). \tag{11}$$

The DCA applied to (P) with decomposition (10) and/or (11) can be described as follows

Algorithm DCAP1 Let x^0 be given in $\mathbb{R}^n$. Set k ← 0. Repeat

– Calculate x^{k+1} by solving the convex program

$$\min\left\{\frac{1}{2}\eta\|x\|^2 + f(x) - \langle x, \eta x^k\rangle : x \in C\right\}, \tag{12}$$

– k ← k + 1

Until convergence of $\{x^k\}$.

– Calculate $y^k \in \partial\left(\frac{1}{2}\rho\|.\|^2 - f(.)\right)(x^k)$

– Calculate x^{k+1} by solving the convex program

$$\min\left\{\frac{1}{2}\rho\|x\|^2 - \langle x, y^k\rangle : x \in C\right\},$$

$$\text{i.e. } x^{k+1} = \text{Proj}_C(y^k/\rho).$$
$$- \; k \leftarrow k+1 \tag{13}$$

Until convergence of $\{x^k\}$.

Here, ProjC stands for the orthogonal projection on C. As indicated above, we are greatly interested in the choice of DC decompositions: what is "the best' among (10) and (11)? The answer depends on C and f. In fact, the performance of the DCA depends upon that of the algorithm for solving convex programs (12) and (13). For certain problems, for example, box constrained quadratic programming and ball constrained quadratic programming, Algorithm DCAP2 is greatly less expensive than Algorithm DCAP1, because the orthogonal projection onto C in these cases is given in explicit form (see for example [35]). In practice, when f is differentiable and the computation of its gradient is not difficult, and the projection on C can be inexpensively determined, the use of DCAP2 is very recommended

For using the above DC decompositions the crucial question is how to determine a nonnegative number η (resp. ρ) such that the function $\frac{1}{2}\eta\|x\|^2+f(x)$ (resp. $\frac{1}{2}\rho\|x\|^2-f(x)$) is convex. In many practical problems such η and ρ exist and can be computed according to the properties of the function f. For example, when f is a smooth function with Lipschitz continuous gradient ρ is nothing but the Lipschitz constant of ∇ f.

DCA for network utility maximization

Network utility maximization (NUM) has many applications in network rate allocation algorithms and Internet congestion control protocols. Consider a communication network with L links, each with a fixed capacity of cl bps, and S sources (i.e., end users), each transmitting at a source rate of x_s bps. Each source s emits one flow, using a fixed set L(s) of links in its path, and has a utility function $U_s(x_s)$. Each link l is shared by a set of sources denoted S(l), (the set of users using link l). NUM, in its basic version, consists of maximizing the total utility of the network $\sum_s U_s(x_s)$ over the source rates x, subject to linear flow constraints for all links l:

$$\begin{aligned} &\text{maximize} && \sum_{s\in\mathcal{S}} U_s(x_s) \\ &\text{s.t} && \sum_{s\in S(l)} x_s \le c_l \quad \forall l,\ x_s \ge 0 \quad \forall s \in \mathcal{S}, \end{aligned} \tag{14}$$

where S denotes the set of users. Here the vector variable is $x = (x_s)_{s\in\mathcal{S}} \in \mathbb{R}^S$ and the constraint set is a well-defined convex polytope. There are many nice properties of the basic NUM model due to several simplifying assumptions of the utility functions and flow constraints, which provide the mathematical tractability of problem (14) but also limit its applicability. In particular, the utility functions Us are usually assumed to be concave increasing. In such a case, the optimization problem (14) is a convex program, and so far is easy to solve. In the past, maximization of concave utility functions and the resulting distributed rate allocation for elastic traffic have gained extensive attention. Based on the concavity and continuity assumptions on utility functions and the elasticity assumption on application traffic, rigorous mathematical frameworks for standard price-based distributed algorithms have been investigated.

However, it is known that for many multimedia applications, user satisfaction may assume nonconcave shape as a function of the allocated rate. Furthermore, in some other models of utility functions, the concavity assumption on $U_s(x_s)$ is also related to the elasticity assumption on rate demands by users. When demands for x_s are not perfectly elastic, $U_s(x_s)$ may not be concave. In this case, the resulting NUM becomes nonconvex and significantly harder to be analyzed and solved. Since inelastic flows with nonconcave utility functions represent important applications in practice, today solving the NUM problem with nonconcave utility function is a challenge of the analysis and design of communication systems by nonconvex optimization techniques.

As an illustrative example, we consider the NUM problem with Sigmoidal-like utility functions [30] that are used in many multimedia applications and Internet congestion control (for example, the utility for voice applications is modeled by a Sigmoidal function with a convex part at low rate and a concave part at high rate). Other useful utility functions can also be solved by DCA-based algorithms (see [24]).

Consider sigmoidal utilities in a standard form:

$$U_s(x_s) = \frac{1}{1 + e^{-(a_s x_s + b_s)}},$$

where as > 0, $b_s < 0$ and a_s, b_s are integers. The Sigmoidal function is neither convex nor concave, but it is DC (difference of convex functions). Then the resulting NUM problem is a DC program. We are going to present a DC decomposition for the sigmoidal function.

We have

$$U_s(x_s) = e^{(a_s x_s + b_s)} - \frac{e^{2(a_s x_s + b_s)}}{1 + e^{(a_s x_s + b_s)}}.$$

It is easy to verify that the functions

$$h_s(x_s) := e^{(a_s x_s + b_s)} \quad \text{and} \quad g_s(x_s) := \frac{e^{2(a_s x_s + b_s)}}{1 + e^{(a_s x_s + b_s)}}$$

are convex (their derivative is increasing). Therefore, U_s is a DC function, and so is $-U_s$.

Let K be the set defined by the constraints of Problem (14), say

$$K := \left\{ \sum_{s \in S(l)} x_s \le c_l \quad \forall l \in \{1 \ldots L\},\ x_s \ge 0 \quad \forall s \in \mathcal{S} \right\}$$

and denote by χK the indicator function on K. Then the Sigmoidal NUM problem can be expressed as

$$\begin{aligned} &\max_{x \in K} \left\{ U(x) := \sum_{s \in \mathcal{S}} U_s(x_s) \right\} \\ &= -\min_{x \in K} \left\{ \sum_{s \in \mathcal{S}} \frac{e^{2(a_s x_s + b_s)}}{1 + e^{(a_s x_s + b_s)}} - e^{(a_s x_s + b_s)} \right\} \\ &= -\min \{ g(x) - h(x) : x \in K \} \text{ where } g(x) \\ &:= \sum_{s \in \mathcal{S}} g_s(x_s),\ h(x) := \sum_{s \in \mathcal{S}} h_s(x_s). \end{aligned}$$

Since gs and hs are convex functions, the function g and h are convex too (note also that g and h are differentiable). Hence the Sigmoidal NUM problem is a DC program that can be written in the standard form as

$$\min \left\{ [\chi_K(x) + g(x)] - h(x) : x \in \mathbb{R}^S \right\}. \tag{15}$$

According to the general DCA scheme, applying DCA to (15) amounts to computing two sequences $\{y^k\}$ and $\{x^k\}$ in the way that

$$y^k = \nabla h(x^k),$$
$$x^{k+1} \in \operatorname{argmin} \left\{ [\chi_K(x) + g(x)] - \langle x, y^k \rangle : x \in \mathbb{R}^S \right\}.$$

Hence the algorithm can be described as follows.

Algorithm: DCA for Sigmoidal utility maximization:

1. Choose $x^0 \in \mathbb{R}^S$ as the initial point. Let > 0 be sufficiently small, k ← 0.
2. Repeat

Set $y^k = (a_1 e^{a_1 x_1^k + b_1}, \ldots, a_s e^{a_s x_1^k + b_1})$.

Set x^{k+1} as an optimal solution of the convex program

$$\min \left\{ \sum_{s \in \mathcal{S}} \frac{e^{2(a_s x_s + b_s)}}{1 + e^{(a_s x_s + b_s)}} - \langle x, y^k \rangle : x \in K \right\}$$

$k \leftarrow k + 1$

until $\|x^{k+1} - x^k\| \le \epsilon(1 + \|x^k\|)$. (16)

Note that the convex problem (16) can be distributedly implemented using the Lagrangian dual decomposition method as shown in [30].

Spectrum Management Problem (SMP)

Discrete multitone (DMT) [42] has been adopted as standard in various DSL applications such as asymmetric DSL (ADSL) and more recently for very-high-bit-rate digital subscriber line (VDSL) by International Telecommunication Union (ITU). For a sufficiently large number of subcarriers, DMT transmission over a frequency-selective fading channel can be modeled as a set of K parallel independent flat-fading sub-carrier AWGN channels (Additive white Gaussian noise). Under this Gaussian assumption, the achievable bit-loading rate of user n on tone k is

$$\begin{aligned} r_k^n &= \log_2 \left(1 + \frac{1}{\Gamma} \frac{|g_k^{n,n}|^2 p_k^n}{\sum_{m \ne n} |g_k^{n,m}|^2 p_k^m + \omega_k^n} \right) \\ &= \log_2 \left(1 + \frac{1}{\Gamma} \frac{p_k^n}{\sum_{m \ne n} h_k^{n,m} p_k^m + \sigma_k^n} \right). \end{aligned}$$

where p_k^n denotes user n's transmit power spectral density (PSD) on tone k; ω_k^n denotes user n's transmit noise power on tone k; $g_k^{n,m}$ is the channel path gain from user m to user n on tone k; $h_k^{n,m} = \frac{|g_k^{n,m}|^2}{|g_k^{n,n}|^2}$ is the normalized interference path power gain from user m to user n on tone k; $\sigma_k^n = \frac{\omega_k^n}{|g_k^{n,n}|^2}$ is the noise variance of user n on tone k, and Γ is the SNR-gap to capacity. The data rate of user n is $R_n = f_s \sum_{k=1}^{K} r_k^n$, where f_s is the DMT symbol rate.

The goal of spectrum management problem is to achieve best possible user rates tradeoff among users in the network, i.e., to find the boundary of rate region. Assume that each user is subject to an individual total transmission power constraint. One way to define the SMP in the literature is consider the following optimization problem

$$\max_{p_1, p_2, \ldots, p_K} \left\{ R_1 : R_n \ge T_n,\ \forall n \ge 1;\ \sum_k p_k^n \le P_n, \forall n, k; \right.$$
$$\left. p_k^n \le p_k^{n,\text{mask}}, \quad \forall n, k \right\},$$

where T_n is minimum target rates of user n and P_n is maximum total transmission power of user n. The SMP (17) aims to maximize the rate of user 1 while guarantees the achievable rates of other users higher than their required minimum target rates Tn. P_n denotes the maximum total transmission power of user n. Spectral mask constraints $p_k^{n,\text{mask}}$ may also be applied if needed.

Among various Dynamic Spectrum Management (DSM) techniques, centralized Optimal Spectrum Balancing (OSB) achieves the maximum data rates by computing the optimal PSDs (power spectral density) for all modems in DSL systems.

The centralized algorithm based on dual decomposition for OSB, proposed in [3], decouples joint optimization across all tones to make the problem solvable per-tone basis. If the rate region is convex (the assumption that the rate region is convex is justified in [3] for two-user in DSL system, and the same logic for two-user can be applied to justify the convexity of rate region for multiple-user case), solving the problem (17) amounts to solving the following weighted sum rate optimization problem [5]:

$$\max_{p_1,\dots,p_K} \left\{ \sum_n \omega_n R_n : \sum_k p_k^n \leq P_n \quad \forall n;\ 0 \leq p_k^n \leq p_k^{n,\text{mask}} \quad \forall k, n \right\}, \tag{18}$$

where the weight for user 1, ω_1, is set to unity, resulting in the maximization of the rate of user 1; whereas $\omega_n \geq 0,\ n \neq 1$ can be adjusted to guarantee the target rate of user n. In [27], we have investigated DC programming and DCA for solving (18).

A nice DC formulation of SMP (18)

First, we write Problem (18) in the form of a minimization program

$$\min_{p=(p_1,\dots,p_K)} \left\{ f(p) := -\sum_{n=1}^{N} \omega_n R_n : \sum_{k=1}^{K} p_k^n \leq P_n\ \forall n;\ 0 \leq p_k^n \leq p_k^{n,\text{mask}}\ \forall k=1\dots K, n=1\dots N \right\}. \tag{19}$$

A natural DC decomposition of f (easily deduced from the definition of r_k^n) has been given in [27]. However, as indicated in [27], from numerical point of views, the DCA scheme corresponding to this DC decomposition is not interesting because it requires an iterative algorithm for solving a convex program at each iteration. In an elegant way we introduced in [27] a nice DC reformulation of the problem (19) (based on the second DC decomposition discussed in Sect. 2) for which the resulting DCA is explicitly determined via a very simple formula. Such a DC decomposition of f is inspired by the following result.

Theorem 1 There exists ρ > 0 such that the function

$$h(p) := \frac{1}{2}\rho||p||^2 - f(p) \tag{20}$$

is convex on C, the feasible set of (19), say $C := \{p \in \mathbb{R}^{K\times N} \mid \sum_{k=1}^{K} p_k^n \leq P_n, 0 \leq p_k^n \leq p_k^{n,\text{mask}}\ \forall n = 1, 2, \dots, N;\ k = 1, 2, \dots, K\}$.

Proof See [27].

Using the theorem above, we get the next DC decomposition of f :

$$g(p) = \frac{1}{2}\rho||p||^2,\quad h(p) = \frac{1}{2}\rho||p||^2 - f(p), \tag{21}$$

and Problem (19) can be now written in the form min{ f (p) := g(p) − h(p) | p ∈ C}

or again, in the standard form of DC program: $\min\{\chi_C(x) + f(p)) \mid p \in R^{N\times K}\}$.

Then, DCA apply to Problem (19) is described as follows.

DCA-SMP

Initialization: Let $\epsilon > 0$ be given, $p^{(0)} \in C$ be an initial point, set r := 0;

Repeat: set

$$\mathbf{q}^{(r)} = \nabla h(\mathbf{p}^{(r)}) = \rho\,\mathbf{p}^{(r)} - \nabla f(\mathbf{p}^{(r)})$$

and calculate $p^{(r+1)} \in \partial(g+\chi_C)^*(q^{(r)})$ by solving the linear constrained quadratic program

$$\min\left\{\frac{1}{2}\rho||p||^2 - \langle p, \mathbf{q}^{(r)}\rangle \mid p \in C\right\} \tag{22}$$

Set $r + 1 \leftarrow r$

until either $||\mathbf{p}^{(r+1)} - \mathbf{p}^{(r)}|| \leq \epsilon(||\mathbf{p}^{(r)}||+1)$ or $|f(\mathbf{p}^{(r+1)}) - f(\mathbf{p}^{(r)})| \leq \epsilon(|f(\mathbf{p}^{(r)})| + 1)$.

The advantage of the DC decomposition (21) is that the resulting DCA-SMP requires, at each iteration, the computation of the projection of a point on the set C (having very specific structure) for which efficient algorithms are available (see [27]).

NONCONVEX CONSTRAINT SET

A typical application of this class of problems is Internet routing (see for example [5]). Mathematically, the nonconvex constraint set is expressed as

$$E := \{x \in C,\ f_i(x) := g_i(x) - h_i(x) \leq 0,\ i = 1, \dots, m\}$$

where C is a closed nonempty convex set in $\mathbb{R}^n$, $g_i, h_i \in \Gamma_0(\mathbb{R}^n)$, $i = 0, \dots, m$, and $f_i(x) \leq 0$ are called DC constraints. The generic formulation of this class of problems takes the form

$$\alpha = \inf\{f_0(x) := g_0(x) - h_0(x) : x \in E \quad (P_{dcg}) \tag{23}$$

where E is assumed to be nonempty.

This class of nonconvex programs (called general DC programs) is the most general in DC Programming and, a fortiori, more difficult to treat than that of standard DC programs (P_{dc}) because of the nonconvexity of the constraints. It is not new and has been addressed in [34]. Its renewed interests is due to the fact that this class appears, increasingly, in many models of nonconvex variational approaches.

We can solve (P_{dcg}) by DCA via penalty techniques. First, we transform (P_{dcg})-(23) into (P_{dc}) via penalty techniques in DC programming.

Let the functions p and p^+ be defined by

$$p(x) := \max\{f_i(x) : i = 1, \dots, m\};$$
$$I(x) := \{i \in \{1, \dots, m\} :\ f_i(x) = p(x)\}$$
$$p^+(x) := \max\{0, p(x)\},$$

which are DC functions with the following DC decompositions (in case g_i, h_i are finite on C for i = 1,..., m,(see, e.g., [34]), obtained directly from those of $f_i, i = 1,\dots, m$.

$$p(x) = \max_{i=1,\dots,m}\left\{g_i(x) + \sum_{j=1,j\neq i}^{m} h_j(x)\right\} - \sum_{j=1}^{m} h_j(x) \tag{24}$$

$$p^+(x) = \max_{i=1,\dots,m}\left\{\sum_{j=1}^{m} h_j(x), g_i(x) + \sum_{j=1,j\neq i}^{m} h_j(x)\right\} - \sum_{j=1}^{m} h_j(x). \tag{25}$$

The general DC program (P_{dcg})-(23) can then be formulated as

$$\alpha = \inf\{f_0(x) := g_0(x) - h_0(x) \ : \ x \in C,\ p^+(x) \leq 0\} \tag{26}$$

and its penalized is a standard DC program

$$\alpha(\tau) = \inf\{\varphi_\tau(x) := f(x) + \tau p^+(x) : x \in C\}.\ (P_\tau) \tag{27}$$

Let DC decompositions of f_0 and p^+ be given by

$$f_0(x) = g_0(x) - h_0(x); \tag{28}$$

$$p^+(x) = p_1(x) - p_2(x), \tag{29}$$

where g_0, h_0, p_1, p_2 are convex functions defined on the whole space. Then, we have the following DC decomposition for ϕ_τ

$$\varphi_\tau(x) = g_\tau(x) - h_\tau(x), \tag{30}$$

where,

$$g_\tau(x) := g_0(x) + \tau p_1(x); \quad h_\tau(x) := h_0(x) + \tau p_2(x). \tag{31}$$

Exact penalty (relative the constraint $p^+(x) \leq 0$) for (26) means that there is $\tau_0 \geq 0$ such that for all $\tau > \tau_0$ both DC programs (P_{dcg})-(23) and (P_τ)-(27) are equivalent in the sense that $\alpha(\tau) = \alpha$ and (P_{dcg})-(23) et (P_τ)-(27) have the same (global) solution set. In this case, the solution of (P_{dcg})-(23) can be achieved by applying DCA to a standard DC program (P_τ)-(27) with $\tau > \tau 0$.

Exact penalty techniques in DC programming have been widely investigated in our works [25, 28, 34]. However, from a computational point of view, an inconvenience of this exact penalty method is that the penalty parameter is generally unknown. Moreover, there are practical optimization problems for which the exact penalization is not satisfied. In [28], we proposed to develop the DCA for solving general DC program (P_{dcg})-(23) by using a penalty technique with updated parameter.

The generalized DCA can be deduced from DCA as follows: instead of fixing the penalty parameter τ , DCA is applied to the sequence of ($P_{\tau k}$) with an increasing sequence of penalty parameters $\{\tau k\}$ given by a updating rule from the current iteration x^k such that x^{k+1} is the next iteration of DCA applied to ($P_{\tau k}$) from x^k . Our work consists in the statement of appropriate updating rules for the sequence $\{\tau_k\}$ and the refinement of constraint qualifications used, in order to ensure global convergence (to a critical point of (P_{dcg})- (23)) and efficiency of DCA1. It is also important that the sequence $\{\tau_k\}$ is constant after a certain rank. The penalty introduced uses l_∞ norm, but we can also consider the l_1 -norm where $q(x) := \sum_{i=1}^{m} f_i^+(x)$ replaces $p^+(x)$. Some other DCA-based algorithms for (P_{dcg})-(23) have been developed in [36].

INTEGER VARIABLES

Several applications in CS can be formulated as an optimization problem with (mixed) integer variables. Here we mention some classes of problems which have been successfully solved by DCA

Cross-layer optimization in multi-hop time division multiple access (TDMA) networks

Efficient design of wireless networks is a challenging task due to the interference nature of shared wireless medium. Recently, the concept of cross-layer design has been investigated extensively. In [26, 29] a cross-layer optimization framework, i.e., joint rate control, routing, link scheduling and power control for multi-hop TDMA networks, has been considered. Particularly, we studied a centralized controller that coordinates the routing process and transmissions of links such that the network lifetime is maximized [29] and the quality of-service (QoS) constraints on the minimum source rates are satisfied. Alternatively, the energy consumption is an important design criterion for a multi-hop wireless network. In [26], we considered the energy minimization-based crosslayer design problem. We will show below that the aforementioned problems can be formulated as mixed integer-linear programs (MILP) and then efficiently solved by DCA.

In the considered TDMA network, time is partitioned into fixed-length frames, and each frame is further divided into J time slots with unit duration. Since the resource allocation is the same in all frames, we concentrate our design on a single frame. A node may need to transmit in one or more slots for its own traffic and/or relay traffic from other nodes. If a node transmits in a slot, while its transmission power can be varied from $[0, P_{max}]$, its transmission rate is fixed at a unit rate. In the TDMA-based network, a channel is specified by two elements (j,l), $j \in J$, $l \in L$, where $J = \{1, 2,\dots, J\}$. For the channel, the resource allocation is denoted by (s_j^l, P_j^l), where $s_j^l = 1$ means link l is active at slot j while $s_j^l = 0$ otherwise, and $P_j^l > 0$ denotes the transmission power of link l at slot j if $s_j^l = 1$, $P_j^l = 0$ otherwise

At each node, the difference of its outgoing traffic and its incoming traffic should be the traffic generated by itself, i.e.,

$$\sum_{l\in\mathcal{O}(n)}\sum_{j=1}^{J} s_j^l - \sum_{l\in\mathcal{I}(n)}\sum_{j=1}^{J} s_j^l = r_n, \quad n \in \mathcal{N} \tag{32}$$

where $\mathcal{O}(n)$ and $\mathcal{I}(n)$ are the set of outgoing links and incoming links at node n, respectively. The values of s_n for the non-source nodes are set to zero, or equivalently all the traffic entering such nodes must be routed.

The energy consumption at node $n \in N$ can be written as

$$\mathcal{E}_n = \sum_{l\in\mathcal{O}(n)}\sum_{j=1}^{J} P_j^l + \sum_{l\in\mathcal{O}(n)}\sum_{j=1}^{J} \epsilon_l s_j^l + \sum_{l\in\mathcal{I}(n)}\sum_{j=1}^{J} \varepsilon_l s_j^l, \quad n \in \mathcal{N} \tag{33}$$

where ϵ_l and εl denote the energy needed to transmit and receive a unit of traffic over link l, respectively. Note that ϵ_l, ε_l include the energy consumed by the signal processing blocks at the link ends.

Interference Model

Wireless channel is a shared medium and interference limited where links contend with each other for channel use. Moreover, interference relations among the nodes and/or links can be modeled in various ways, for example, by using the signal-to-interference-plus-noise-ratio (SINR)- based model [32, 52]. Specifically, if the link l ∈ L is active at slot j (i.e., $s_j^l = 1$), the following inequality should hold so as to guarantee the transmission quality of the link

$$\mathrm{SINR}_j^l = \frac{P_j^l h_{ll}}{\sum_{k \neq l} P_j^k h_{kl} + \eta_l} \geq \gamma^{\mathrm{th}} \tag{34}$$

where SINR_j^l is the SINR for link l at slot j, h_{kl} is the path gain from the transmitter of link k to the receiver of link l , ηl is the noise power at receiver of link l, and γ th is the required SINR threshold for accurate information transmission.

We assume that all wireless nodes are low-mobility devices and/or the topology of the network is static or changes slowly allowing enough time for computing the new scheduler. An example of such networks is a wireless sensor network for environmental monitoring with fixed sensor locations. In this case, the need for distributed implementation is not necessary. From the preceding discussions, the energy minimization based cross-layer design, i.e., joint rate control, routing, link scheduling, and power allocation problem can be mathematically formulated as

$$\min_{r_n, P_j^l, s_j^l} \quad \sum_{n \in \mathcal{N}} \mathcal{E}_n \tag{35a}$$

subject to:

$$\sum_{l \in \mathcal{O}(n)} \sum_{j=1}^{J} s_j^l - \sum_{l \in \mathcal{I}(n)} \sum_{j=1}^{J} s_j^l = r_n, \; n \in \mathcal{N}$$

$$r_n \geq r_n^{\min}, \; n \in \mathcal{N} \tag{35b}$$

$$r_n \geq r_n^{\min}, \; n \in \mathcal{N} \tag{35c}$$

$$\sum_{l \in \mathcal{I}(\hat{n})} \sum_{j=1}^{J} s_j^l = \sum_{n \in \mathcal{N}} r_n \tag{35d}$$

$$\sum_{l \in \mathcal{O}(n)} s_j^l + \sum_{l \in \mathcal{I}(n)} s_j^l \leq 1, \quad \forall n \in \{\mathcal{N} \cup \hat{n}\}, \; j=1, \ldots, J \tag{35e}$$

$$\forall l \in \mathcal{L}, \; j = 1, \ldots, J \tag{35f}$$

$$0 \leq P_j^l \leq P_{\max} s_j^l, \quad \forall l \in \mathcal{L}, \; j = 1, \ldots, J \tag{35g}$$

$$s_j^l \in \{0, 1\}, \quad \forall l \in \mathcal{L}, \; j = 1, \ldots, J \tag{35h}$$

where $\hat{n}$ denotes the common sink node for all data generated in the network, D is a very large positive constant. The objective function is the energy consumption in the network.1 Constraints (35b) ensure that the data generated by source nodes are routed properly. Constraints (35c) guarantee that the rate for each node is no less than a minimum rate. The minimum rates are possibly different for nodes and are usually determined by the network QoS. Nodes which do not generate traffic have $r_n = r_n^{\min} = 0$. Constraint (35d) is the flow conservation at the traffic destination for all the sources. Constraints (35e) state that a node cannot receive and transmit simultaneously in one particular time slot. Constraints (35f) make sure the SINR requirement is met: if a link l is active in time slot j, then the SINR at receiver of link l must be larger than the given threshold γ th which also depends on the system implementation. Constraint (35f) is automatically satisfied if link l is not scheduled in time slot j. Constraint (35g) states that if a link l is scheduled for time slot j, i.e., $s_j^l = 1$, then the corresponding power value P_j^l must be less than $P_{\max}$. Otherwise, P_j^l obviously equals to zero. We also impose binary integer constraints on s_j^l .

It can be seen that the cross-layer optimization problem (35a)–(35h) belongs to a class of well-known mixed-integer linear programs (MILPs). The combinatorial nature of the optimization (35a)–(35h) is not surprising and it has been shown in some previous works, albeit with different objective functions and formulations [7, 32, 52]. Theoretically, MILPs are NP-hard which is clearly inviable for practical scenarios when the dimension is large. It has been shown in [26] that, at optimality, the source rate constraints (35c) must be met with equalities for all sources

Note that by considering (35a), one aims to minimize the total energy consumption, it may cause some particular nodes spending more energy than the other nodes, and thus, running out of energy quicker. Therefore, equal energy distribution among nodes is not optimal. Another design objective which may help to prevent such situation is as follows

$$\min_{r_n, P_j^l, s_j^l} \quad \max_{n \in \mathcal{N}} \mathcal{E}_n \tag{36a}$$

Subject to: The constraints (35b)–(35h) (36b)

The optimization problem (36a)–(36b) aims at minimizing the maximum energy consumed at nodes(s). As a result, more nodes are likely to be involved in the routing algorithm, i.e., relaying information for other nodes. For simplicity, the optimization problem (35a)–(35h) is often considered in the literature.

The cross-layer optimization problem (35a)–(35h) has worst case exponential complexity when BnB methods are used to compute the solution. Moreover, when modeling practical networks and depending on the number of links, nodes and time slots, problem with large sizes may arise. As a result, it is extremely difficult to schedule links optimally. Most research in literature is based on heuristic at the cost of performance degradation, for example, see [7, 8, 52]. In [26], we investigated a DCA scheme to solve the mixed 0–1 linear program (35a)–(35h) efficiently

The network lifetime maximization problem [29] is similar to (35a)–(35h) in which (35a) is replaced by network lifetime maximization.

Quality of service (QoS) routing problems

The Unicast (resp. Multicast) QoS routing emphasizes to find paths (resp. a set of paths) from a source node to a destination node (resp. a set of destination nodes) satisfying the QoS requirements. The Routing problems become more complex as far as we consider mobile networks or hybrid networks, because of dynamic topology and real time routing procedure.

As an example, we consider a scenario in Multicast routing problem, such as we are staying in a car parking place. The mobile services are provided in each moving car, equipped with a mobile device, via a car service center likes in the car parking place. There are m cars sending their requests to a mobile car service center, they need help to find the route to go to their destinations under the travel time constraint, the less latency traffic jam, the jitter time delay constraint, the travel cost (same sources, different destinations, considering local constraints to each mobile vehicle). Therefore, based on the temporary update data of the network state, the mobile car service system has to calculate the route and given the answer for each car in a few seconds. In this context, we need a centralized and efficient algorithm to calculate the routes.

The problem of finding a path in network with multiple constraints (the MCP problem) is NP-complete. We reformulated the MCP [46] and MCOP (multi-constrained optimal path problem) [47, 51] problem as Binary Integer Linear Programs (BILP) and investigated DCA-based algorithms for solving them. The DCA is fast and furnished an optimal solution in almost all cases, and a near-optimal solution in the remaining cases. For large scale problems we investigated the proximal decomposition technique to solve convex subprograms at each iteration of DCA. Computational results show that this approach is efficient, especially for large-scale settings where the powerful CPLEX fails to be applicable.

The partitioning-hub location-routing problem

The Partitioning-Hub Location-Routing Problem (PHLRP) is a hub location problem involving graph partitioning and routing features. PHLRP consists of partitioning a given network into sub-networks, locating at least one hub in each sub-network and routing the traffic within the network at minimum cost. There are various important applications of PHLRP, such as the deployment of network routing protocol problems and the planning of freight distribution problems. In [50] we formulated this problem as a Binary Integer Linear Programming (BILP) and then investigate DCA for solving it. Preliminary numerical results are compared with the well-known commercial solver CPLEX, they show the efficiency and the superiority of DCA.

The car pooling problem

Carpooling is a well-known transport solution that consists of sharing a car between a driver and passengers sharing the same route, or part of it. The challenge is to minimize both the number of required cars and the additional cost in terms of time for the drivers. To solve the problem, several tasks should be performed: choosing drivers and passengers, allocating passengers to cars, computing an optimal route for the cars. As such, the carpooling transport problem may be described as some kind of fleet management problem. In [49], we formulated this problem as a Mixed Integer Linear Program for which DCA has been efficiently applied. In order to globally solve the problem, we combine DCA with classical Branch and Bound algorithm. DCA is used to calculate upper bound while lower bound is obtained from a linear relaxation problem. Preliminary numerical results are compared with CPLEX. They show the efficiency and the superiority of DCA-based algorithms.

The minimum m-dominating set problem

Let G = (V, E) be a graph, where V is the set of nodes and E is the set of edges of G. A dominating set D of a graph G = (V, E) is a subset of nodes $D \subseteq V$ such that every vertex not in D is joined to at least one member of D by some edge. The domination number γ (G) is the number of vertices in a smallest dominating set for G. The dominating set problem is a classical NP-complete decision problem [10] and has various applications in CS. A classical network application for this problem would be to choose a set of locations to install relay antennas. In ad hoc networks, creating a dominating set is a way to organize the network and is generally used as a first step for generating a connected dominating set [11]. This problem is formulated as a BILP for which DCA is investigated in [39]. Numerical results show that the DCA is efficient even for very large instances of problem. Moreover, our algorithm obtained better solution in significantly less time than CPLEX.

In a general framework, we show below how to solve optimization problems with integer variables by DCA.

From combinatorial optimization to DC programming: reformulation

Mixed zero-one concave minimization programming problem

Let $D \neq \emptyset$ be a bounded polyhedral convex set in $\mathbb{R}^n$ and let $J \subset \{1 \dots n\}$.

$$\min\{f(x) : x \in D,\ x_i \in \{0,1\},\quad \forall i \in J\}, \tag{37}$$

where f is a finite concave function on D.

Let $K := \{x \in D: 0 \leq x_i \leq 1,\ \forall i \in J\}$ and define $p(x) = \sum_{i \in J} x_i(1-x_i)$ clearly, p is a concave function with nonnegative values on K and

$$\{x \in D,\ x_i \in \{0,1\}\} = \{x \in K : p(x) = 0\}$$
$$= \{x \in K : p(x) \leq 0\}.$$

Hence (37) can be reformulated as (by Theorem below)

$$\min\{ f(x) : x \in K,\ p(x) \leq 0\}$$

which is equivalent to, for any $t > t_0$

$$\min\{ f(x) + tp(x) : x \in K\}. \tag{38}$$

Theorem 2 [23] Let K be a nonempty-bounded polyhedral convex set in $\mathbb{R}^n$ and f, p be finite concave on K. Assume the feasible set of (P) be nonempty and p be nonnegative on K. Then there exists $t_0 \geq 0$ such that for every $t > t_0$ the following problems have the same solution sets:

(Pt) $\alpha(t) = \inf\{ f(x) + tp(x) : x \in K\}$,

(P) $\alpha = \inf\{ f(x) : x \in K,\ p(x) \leq 0\}$.

Furthermore

(i) if the vertex set of K , denoted by V(K), is contained in $\{x \in K : p(x) \leq 0\}$, then $t_0 = 0$.

(ii) if V(K) is not contained in $t_o \leq \frac{f(x^o)-\alpha(0)}{S}$, then $t_0 \leq f(x^o) - \alpha(0)$ S for every $x^o \in K$, $p(x^o) \leq 0$, where $S := \min\{p(x) : x \in V(K),$

$p(x) > 0\}$.

DCA for solving (Pt)

$$(P_t)\quad \alpha(t) = \inf_{x\in K}\{f(x) + tp(x)\}$$
$$= \inf_{x\in\mathbb{R}^n}\{F_t(x) := \chi_K(x) + f(x) + tp(x)\}.$$

Assuming that a subgradient of − f is computable. One DC decomposition of F_t can be chosen as

$$F_t(x) := g(x) - h(x) \text{ with } g(x) := \chi_K(x),$$
$$h(x) := -f(x) - tp(x). \tag{39}$$

In this case, (P_t) is a polyhedral DC program because χK is a polyhedral convex function, and the general DCA scheme becomes:

$$y^k \in \partial(-f(x^k) - tp(x^k));$$
$$x^{k+1} \in \operatorname{argmin}\left\{-\langle x, y^k\rangle : x \in K\right\}. \tag{40}$$

Besides the computation of subgradients of − f and set $\nabla(-p)(x) = \sum_{i\in J}(2x_i - 1)$, the algorithm requires one linear program at each iteration the convergence properties can be stated as follows:

Theorem 3 i) DCA generates a finite sequence $x^1,\ldots, x^{k*}$ contained in V(K) such that $f(x^{k+1}) + tp(x^{k+1}) \le f(x^k) + tp(x^k)$, $p(x^{k+1}) \le p(x^k)$ for each k , and x^{k*} is a critical point of g − h.

ii) If, in addition, h is differentiable at x^{k*}, then x^{k*} is actually a local minimizer to (Pt).

iii) Let $t > t_1 := \max\left\{\frac{f(x)-\alpha(0)}{S} : x \in V(K) \cap \{x \in K : p(x) = 0\}\right\}$ if at an iteration q one has $p(x^q) = 0$, then $p(x^k)=0$ and $f(x^{k+1}) \le f(x^k)\ \forall k \ge q$.

Extension cases

Based on new results related to exact penalty and error bounds in DC programming [28], the same reformulation technique via exact penalty can be used for

- Linear constrained mixed zero-one DC programming problems.
- Linear constrained mixed integer DC programming problems.

ANOTHER ISSUE: SOLVING CONVEX PROGRAMS BY DCA

Another issue which is also important in CS but were not discussed in this paper is how to solve large-size convex programs. Although convex programming has been studied for about a century, an increasing amount of effort has been put recently into developing fast and scalable algorithms to deal with large scale problems. While some convex regularizations involve convex quadratic programs (QP) for which standard QP solvers can be certainly used, many first-order methods have been developed in the last years for large scale convex problems. Since DC programming and DCA encompass convex programming and convex programs can be recast as (infinitely many) DC programs to which DCAs become global, (i.e. providing optimal solutions), one can make use of these theoretical and algorithmic tools to better reformulate and solve convex programs.

CONCLUSION

We have presented DC programming and DCA for modeling and solving three challenging classes that cover most nonconvex programs in communication systems. These theoretical and algorithmic tools have been outlined in an appropriate way to make them understandable to the reader. They highlight the distinctive features (flexibility, versatility, inexpensiveness, scalability, efficiency and globality) of DC programming and DCA. It is desirable that our approaches will help researchers and practitioners tackle efficiently their nonconvex programs, especially in the large-scale setting.

REFERENCES

1. Alvarado, A., Scutari, G., Pang, J.S.: A New Decomposition Method for Multiuser DC-Programming and its Applications, arXiv: 1308.3521v2 [cs.IT] (2013). http://arxiv.org/pdf/1308.3521. pdf
2. Al-Shatri, H., Weber, T.: Achieving the maximum sum rate using D.C. programming in cellular networks. IEEE Trans. Signal Process. 60(3), 1331–1341 (2012)
3. Cendrillon, R., Yu, W., Moonen, M., Verlinden, J., Bostoen, T.: Optimal multiuser spectrum management for digital subscriber lines. IEEE Trans. Comm. 54, 922–933 (2006)
4. Chiang, M.: Geometric programming for communication systems. Found. Trends Commun. Inf. Theory 2, 1–154 (2005)
5. Chiang, M.: Nonconvex optimization of communication systems. In: Gao, D., Sherali, H. (eds.) Advances in Mechanics and Mathematics, Special Volume on Strang's 70th Birthday, vol. 3, pp. 136– 196. Springer, Berlin (2008)
6. Chiang, M., Hande, P., Lan, T., Tan, C.W.: Power control in wireless cellular networks. Found. Trends Netw. 2, 1–156 (2008)
7. Commander, C.W., Pardalos, P.M.: A combinatorial algorithm for the TDMA message scheduling problem. Comput. Optim. Appl. 43(3), 449–463 (2009)
8. ElBatt, T., Ephremides, A.: Joint scheduling and power control for wireless ad hoc networks. In: Proc. IEEE INFOCOM'02, pp. 976–984, New York, USA (2002)
9. Fazel, M., Chiang, M.: Nonconcave network utility maximization through sum of squares method. Proc. IEEE Control and Decision Conference, Seville, Spain (2005)
10. Garey, M.R., Johnson, D.S.: Computers and Intractability. A Guide to the Theory of NP-Completeness. W. H. Freeman Co, San Franciso (1990)
11. Guha, S., Khuller, S.: Approximation algorithms for connected dominating sets. Algorithmica Number 4(20), 374–387 (1998)
12. Hande, P., Zhang, S., Chiang, M.: Distributed Rate Allocation for Inelastic Flows. IEEE/ACM Transactions on Networking, vol. 15, No 6 (2007)

13. Hiriart-Urruty, J.B., Lemaréchal, C.: Convex Analysis and Minimization Algorithms. Parts I&II. Springer, Berlin (1991)

14. Hong, M., Luo, Z.-Q.: Signal processing and optimal resource allocation for the interference channel. Elsevier e-Reference-Signal Processing (2013). http://arxiv.org/pdf/1206.5144v1.pdf

15. Khabbazibasmenj, A., Roemer, F., Vorobyov, S., Haardt, M.: Sumrate maximization in two-way AF MIMO relaying: polynomial time solutions to a class of DC programming problems. IEEE Trans. Signal Process. 60(10), 5478–5493 (2012)

16. Kha, H.H., Tuan, H.D., Nguyen, H.H.: Fast global optimal power allocation in wireless networks by local D.C. programming. IEEE Trans. Wireless Commun. 11(2), 510–512 (2012)

17. Kim, S.-J., Giannakis, G.B.: Optimal resource allocat ion for MIMO ad hoc cognitive radio networks. IEEE Trans. Inf. Theory 57(5), 3117–3131 (2011)

18. Kelly, F.P., Maulloo, A., Tan, D.: Rate control for communication networks: shadow prices, proportional fairness and stability. J. Op. Res. Soc. 49(3), 237–252 (1998)

19. Julian, D., Chiang, M., ONeill, D., Boyd, S.: QoS and fairness constrained convex optimization of resource allocation for wireless cellular and ad hoc networks. Proc. IEEE INFOCOM (2002)

20. La, R.J., Anantharam, V.: Utility-based rate control in the Internet for elastic trafic. IEEE/ACM Trans. Netw. 10(2), 272–286 (2002)

21. Lee, J.W., Mazumdar, R.R., Shroff, N.: Non-convex optimization and rate control for multi-class services in the Internet. Proc. IEEE Infocom, Hong Kong, China (2004)

22. Le Thi, H.A., DC Programming and DCA. http://lita.sciences. univ-metz.fr/~lethi

23. Le Thi, H.A., Pham Dinh, T., Le, D.M.: Exact penalty in DC programming. Vietnam J. Math. 27(2), 169–178 (1999)

24. Le Thi, H.A.: Network Utility Maximisation: a unified DC programming approach. Technical Report, LITA (2012)

25. Le Thi, H.A., Pham Dinh, T.: The DC (difference of convex functions) Programming and DCA revisited with DC models of real world nonconvex optimization problems. Ann. Oper. Res. 133, 23– 46 (2005)

26. Le Thi, H.A., Nguyen, T.K., Phan, T.K., Pham Dinh, T.: Energy minimization-based cross-layer design in wireless networks. In The Proceedings of the 2008 High Performance Computing & Simulation Conference (HPCS 2008) Nicosia, Cyprus, June 3–6, pp 283–289 (2008)

27. Le Thi, H.A., Ta, A.S., Pham Dinh, T., Le, N.T.: Optimal Spectrum Balancing in Multi-User DSL Network by DC programming and DCA, Technical report, LITA-UPV-M (2009)

28. Le Thi, H.A.: Pham Dinh, T., Huynh, V.N.: Exact penalty and error bounds in DC programming. J. Global Optim. 52(3), 509– 535 (2012)

29. Le Thi, H.A., Nguyen, Q.T., Phan, T.K., Pham Dinh, T.: DC Programming and DCA Based Cross-layer Optimization in Multihop TDMA Networks. Intelligent Information and Database Systems. Lecture Notes in Artificial Intelligence LNCS/LNAI (2013, to appear)

30. Le Thi, H.A., Pham Dinh, T.: Network Utility Maximisation: a DC programming approach for Sigmoidal Utility function. Proceedings of IEEE conference Advance Techonogies for Communications ATC'13 Ho Chi Minh city october 16–18, 2013, 978–1-4799- 1089-2/13/31.00. IEEE, pp. 50–54 (2013)

31. Le Thi, H.A., Ta, A.S., Pham Dinh, T.: DC programming and DCA for some resource allocation problems in Communication, Networks (2013, submitted)

32. Madan, R., Cui, S., Lall, S., Goldsmith, A.: Cross-layer design for lifetime maximization in interference-limited wireless sensor networks. IEEE Trans. Wireless Commun. 5(11), 3142–3152 (2006)

33. Palomar, D.P., Chiang, M.: A tutorial on decomposition methods for network utility maximization. IEEE J. Select. Areas Commun. 24(8) (2006)

34. Pham Dinh, T., Le Thi, H.A.: Convex analysis approach to dc programming: theory, algorithms and applications. Acta Math. Vietnam. 22(1), 289–357 (1997)

35. Pham Dinh, T., Le Thi, H.A.: DC optimization algorithms for solving Mthe trust region subproblem. SIAM J. Optim. 8, 476–505 (1998)

36. Pham Dinh, T., Le Thi, H.A.: Recent advances on DC programming and DCA. Transactions on Computational Collective Intelligence, Springer, Berlin (2013, to appear)

37. Rockafellar, R.T.: Convex Analysis. Princeton University Press, N.J. (1970)

38. Schleich, J., Bouvry, P., Le Thi, H.A.: Decentralized Fault-tolerant Connected Dominating Set Algorithm for Mobile Ad hoc Networks. Proceedings of the: International Conference on Wireless Networks, World Congress in Computer Science Computer Engineering, and Applied Computing, July 13–16, 2009. Las Vegas, USA, ICWN 2009, 354–360 (2009)

39. Schleich, J., Le Thi, H.A., Bouvry, P.: Solving the Minimum mDominating Set problem by a Continuous Optimization Approach based on DC Programming and DCA. J. Combin. Optim. 24(4), 397–412 (2012)

40. Srikant, R.: The Mathematics of Internet Congestion Control. Birkhauser, Basel (2004)

41. Song, K.B., Cheung, S.T., Ginis, G., Cioffi, J.M.: Dynamic spectrum management for next-generation dsl systems. IEEE Commun. Mag. 40, 101–109 (2002)

42. Starr, T., Cioffi, J.M., Silverman, P.: Understanding Digital Subscriber Line Technology. Prentice Hall, Upper Saddle River (1999)

43. Schmidt, D., Shi, C., Berry, R., Honig, M., Utschick, W.: Distributed resource allocation schemes: Pricing algorithms for power control and beamformer design in interference networks. IEEE Signal Process. Mag. 26(5), 53–63 (2009)

44. Scutari, G., Facchinei, F., Song, P., Palomar, D., Pang, J.: Decomposition by partial linearization: Parallel optimization of multi-agent systems, IEEE Trans. Signal Process (2013, submitted). http:// arxiv.org/pdf/1302.0756v1.pdf

45. Ta, A.S.: Contributions aux développements des nouvelles technologies de communication en transport multimodal par des techniques d'optimisation, thèse de doctorat soutenue au LMI-INSA de Rouen Juin (2012)

46. Ta, A.S., Le Thi, H.A., Khadraoui, D., Pham Dinh, T.: Solving QoS routing problems by DCA. In Intelligent Information and Database Systems. Lecture Notes in Artificial Intelligence (LNAI), vol. 5991, pp. 460–470. Springer, Berlin (2010)

47. Ta, A.S., Le Thi, H.A., Khadraoui, D., Pham Dinh, T.: Solving Multicast QoS Routing Problem in the context V2I Communication Services using DCA, 9th IEEE/ACIS ICIS: August 18–20, 2010, pp. 471–476. Yamagata, Japan (2010)

48. Ta, A.S., Le Thi, H.A., Pham Dinh, T.: Power control by DC programming and DCA. In: The proceedings of International Conference on Industrial Engineering and Systems Management IESM (2011)

49. Ta, A.S., Le Thi, H.A., Arnould, G., Khadraoui, D., Pham Dinh, T.: Solving car pooling problem using DCA. Proceedings of IEEE conference Global Information Infrastructure Symposium (GIIS 2011), Danang 4–6 (2011) (published by IEEE Xplore)

50. Ta, A.S., Le Thi, H.A., Khadraoui, D., Pham Dinh, T.: Solving Partitioning-Hub Location-Routing Problem using DCA. J. Ind. Manage. Optim. 8(1), 87–102 (2012)

51. Ta, A.S., Le Thi, H.A., Pham Dinh, T., Khadraoui, D.: Solving many

52. to many multicast QoS routing problem using dca and proximal decomposition technique, In Proc. IEEE International Conference on Computing, Networking and Communications, pages 809–814, Hawaii, American, 30 January-2 February (2012) (published by IEEEXplore)

53. Tang, J., Xue, G., Chandler, C., Zhang, W.: Link scheduling with power control for throughput enhancement in multihop wireless networks. IEEE Trans. Vehicular Tech. 55(3), 733–742 (2006)

54. Tsiaflakis, P., Diehl, M., Moonen, M.: Distributed spectrum management algorithms for multiuser dsl networks 56(10), 4825–4843 (2008)

55. Vucic, N., Shi, S., Schubert, M.: DC programming approach for resource allocation in wireless networks, pp. 380–386. In International Symposium on Modeling and Optimization in Mobile, Ad Hoc and Wireless Networks (WiOpt) (2010)

CHAPTER 14 HOME NETWORKING AT 60 GHZ

Zulkuf Genc, Bao Linh Dang, Jing Wang, and Ignas Niemegeers

Keywords: Home networking60 GHz, Future home networks, Wireless

INTRODUCTION

The arrival of broadband communication into homes and the continuing evolution of multimedia and communication devices are increasing user demands for networked homes. The Consumer Electronics Association expects that more than 50% of US homes will have home networks by 2008. A home network of today is typically oriented to sharing of data, internet access, and peripherals. The future networks, however, will be dominated by advanced multimedia applications like high-definition television (HDTV), Internet protocol television (IPTV), multiplayer gaming, voice-over Internet protocol, etc. The occurrence of currently unpredicted and highly demanding future applications is also quite possible. This kind of applications will demand huge amount of bandwidth and pose strict quality of service (QoS) requirements on the home networking. For example, the data rates required for raw and MPEG2 compressed HDTV streams are about 2 Gbps and 20 Mbps, respectively. In case of multiple HDTV streams being concurrently played, the bandwidth to support the required data rates will be much higher.

Additional to bandwidth, mobility also appears as another major factor shaping the future home networks. According to a recent survey, laptops will overtake the domination of PCs all over the world by 2011 [1]. The same survey also reveals the fact that 88% of laptop users have a wireless network at home. Looking at this picture, it can be expected that homes will mostly be equipped with wireless networks in the future, and the expectations from these networks will fairly be high.

This paper reviews the 60-GHz wireless technology from a future home networking perspective. It discusses the challenges and the research issues that need to be addressed in order to successfully deploy this millimeter wave band for wireless local area networks (WLAN). Moreover, potential solutions for addressing the presented challenges are also discussed in this paper. We show that it is possible to design and deploy a full size WLAN operating at 60 GHz in a home environment.

The rest of the paper is organized as follows. Section 2 elaborates on the motivations and background information for the 60-GHz band. Section 3 explains a typical use case scenario for a future home network operating at 60 GHz. Two architectural approaches for home networking at 60 GHz will be presented in Section 4. Section 5 discusses the challenges and research issues. Finally, Section 6 concludes the paper.

MOTIVATION AND BACKGROUND

Today's WLAN technologies are not capable of supporting the high expectations for the future home networks. These expectations even do not look achievable on the roadmaps of existing 802.11x and UWB [2]. The IEEE 802.11n technology, which is still under development as the latest version of 802.11 series, is expected to offer only 100–200 Mbps of actual throughput [3]. On the other hand, the standardization activities of WirelessHD group towards the next generation wireless digital network interface for consumer electronics and PC products aim to achieve data rates from 2 to 20 Gbps [4]. SiBeam Inc. recently introduced a new technology for the production of WirelessHD compliant chips for multi gigabit high-definition video and audio distribution in short range [5]. A research group from GIT has already achieved data transfer rates of 15 Gbps at a distance of 1 m and 5 Gbps at 5 m [6]. All these achievements have been made by the utilization of 60-GHz radio band. It has a 5-GHz block of bandwidth in the globally license-free spectrum between 59 and 64 GHz as seen in Fig. 1.

Citation: Zulkuf Genc, Bao Linh Dang, Jing Wang, Ignas Niemegeers, Home networking at 60 GHz, DOI 10.1007/s12243-008-0047-0.

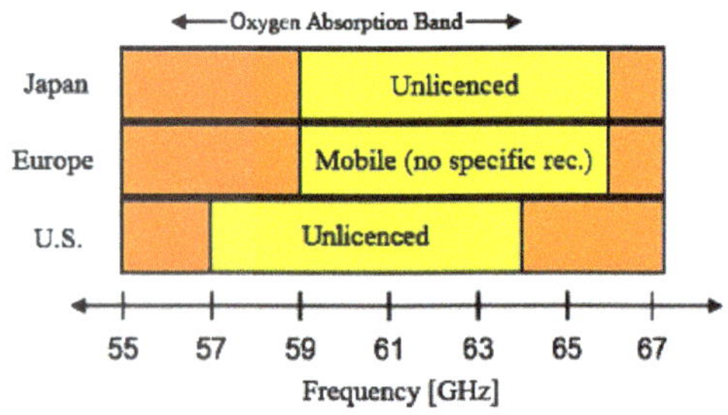

Figure 1: 60-GHz spectrum allocation.

The unique oxygen absorption band at 60 GHz also enables the transmission at higher power levels, up to 40 dBm in the USA and Europe and 10 dBm in Japan, than the IEEE 802.11 and UWB standards. This huge amount of available spectrum and high power levels allow the data transfer at the rate of gigabits per second over indoor distances. The transmission distance, however, is mainly determined by the location of the walls in a typical home. Sixty-gigahertz radio signals cannot penetrate walls due to the heavy attenuation. This characteristic turns each room into a separate cell in which the use of the whole system capacity is possible. Similar to cellular networks, frequencies can be reused here to help the network achieve higher throughput. The naturally isolated cells created by high oxygen absorption and heavy attenuation also form a very secure network environment. Additionally, radio signal transmissions at the higher frequencies inherently become more directional and require smaller antennas. These antennas can better focus their energy in the transmission direction by minimizing the interference and interception possibility. As for the health concerns, 60-GHz radio does not pose a risk to human health, since its signals cannot penetrate through human skin into the body [6]. The allowable radiation power limits for 60 GHz are also much lower than the safety levels determined in the studies [7]. Some early critics of 60-GHz technology were generally pointing out its implementation cost. However, recent advances showed that it is possible to produce low-cost CMOS circuits operating at 60 GHz [6, 8]. Indeed, a research group from Melbourne University has even demonstrated a working transceiver integrated on a CMOS chip that operates at 60 GHz [9]. It is expected to cost less than $10 to the manufacturers. These advances, together with promising features of 60 GHz, have led the standardization activities. IEEE has formed 802.15.3 c study group to develop a millimeter-wave-based alternative physical layer (PHY) for wireless personal area network (WPAN) [10]. Ecma International has also started to develop an international standard for 60-GHz short range communication to utilize it at bulk data transfer, high-definition multimedia streaming, and WPAN applications [11].

USE CASE SCENARIO

HDTV and the ultra-high-definition video are leading a revolution of home entertainment experience. The trend is being pushed towards such an emergent future home environment where people are going to be surrounded by high capacity multimedia devices. Those devices can either be the fixed household devices like LCD and DVR or wall-mounted HDTV plasmas or the personal portable devices such as laptop, video game consoler, and DVD player for the next generation DVD technologies like Blue-Ray. The establishment of 60-GHz multi-gigabit links between these devices will enable the easy and quick delivery of high-definition content without the confusion and unaesthetic view of cables. Moreover, these multi-gigabit links will eliminate the need for compression of the high-definition content. The uncompressed data transfer will enable the production of lower cost higher quality video products which are free of compression and decompression equipment's.

One highly possible use case scenario could be a simple network for high-definition content sharing and distribution between end-user devices such as the storage device, HDTV, laptop, handset devices, and speakers as seen in Fig. 2. Thanks to 60-GHz links, the user can transfer the HD video content of a football game, which he downloaded into his handset from a public kiosk in the stadium right after the game, to the storage device in seconds. While the high-definition video content is shown on the HDTV via 60-GHz link, the audio is distributed to speakers via 60 GHz or a lower frequency radio. At the same time, another user can download this content into his laptop in a few seconds.

Figure 2: A use case scenario demonstrating the high-definition content sharing and distribution via 60-GHz links

Apart from this specific scenario, in the presence of wireless connectivity, people will be free to move inside the home with their mobile devices by keeping the active multimedia sessions on them. The multimedia sessions will also be able to follow a mobile user from one device to another around the user. Besides the home networking capabilities, the high-speed wireless connectivity will allow seamless, easy, and uninterrupted access to the bandwidth-intensive broadband services as well. People will enjoy these services in the comfort of their homes by ensuring the privacy and security of their home network.

The realization of such scenarios described above will certainly draw more consumers into home networking. In response to this interest, more advanced services and applications can be expected to growingly take place in the market. The key point in this phase will be the availability of a future-proof wireless technology, which will meet the requirements of the desired home network, like enormous bandwidth need (multi-Gbps), simplicity, tight QoS necessities, and security. The existence of such a technology can also help overcome the problem of interoperability. We believe that 60-GHz technology can play this role in future home networks once it becomes standardized and gains more maturity.

COMMUNICATION INFRASTRUCTURES

In order to guarantee the full radio coverage at 60 GHz within the entire house, the home network communication infrastructure should specially be designed by ensuring the dependability of highly sensitive 60-GHz links all over the house. Considering the flexibility and economy of architectural and interior design, the candidate communication infrastructure should, at the best, eliminate the enormous amount of cabling in the house and allow itself to be integrated into the user ecosphere in an unobtrusive and aesthetic manner.

We envision two possible infrastructure approaches for 60-GHz home networks by taking into account the radio characteristics of this band. One is to use a wired and wireless infrastructure mix, as shown in Fig. 3a, which deploys fibers to connect rooms, and put simple antennas at the end of the fiber to emit radio signals to form individual 60-GHz radio cells. The other approach is to deploy purely wireless infrastructure, as in Fig. 3b, where the network devices are connected in ad hoc manner, with dedicated devices relaying the 60-GHz radio signals through the walls. We believe that these two candidates are complementary in timeframe for market introduction and capabilities.

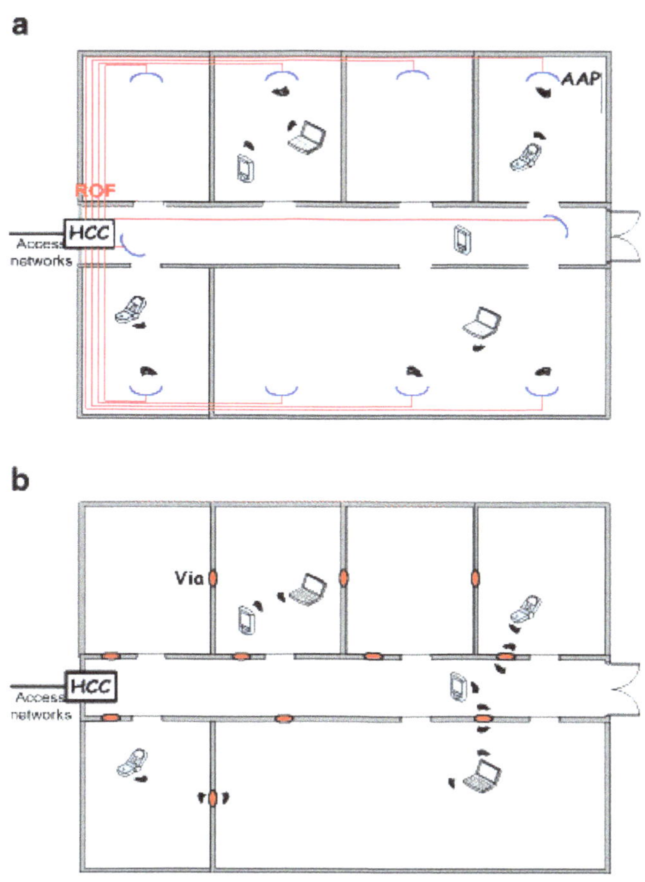

Figure 3: **a** Cell-based home network communication infrastructure. **b** Ad-hoc-based home network communication infrastructure

Cell-based communication infrastructure

This approach is based on the novel radio-over-fiber (RoF) technique [12] which requires a low-cost plastic optical fiber infrastructure to be installed in new buildings or retrofitted to existing buildings. As illustrated in Fig. 3a, there are two essential infrastructural elements. The first one is the home communications controller (HCC), a concept which evolves from the prevailing residential gateways such as a DSL access gateway or a cable set-top box. Besides the gateway functions of interfacing between the home network and the access networks, it will be implemented with the full network stack functionality and additional control and management units as the central intelligence of the network. The other infrastructural element is the antenna access point (AAP). It is a wall socket access point which can be considered as a dumb antenna device only with layer-2 functionality for interfacing the radio signals.

The fiber infrastructure forms a tree with the HCC at the root and with the AAPs at the tips. In each room, at least one AAP is deployed, and thus, at least one short-range 60-GHz radio cell is formed potentially superimposed to offer 60-GHz connectivity. For uplinks, 60-GHz signal from the source device is picked up by the AAP in the close vicinity, modulated as ROF microwave signals, and guided to the HCC. At the HCC, the incoming signal is converted to the proper optical frequency and routed to the AAP close to the destination device where the signal is modulated once again into 60-GHz radio as the downlink traffic. The inter-room connectivity provided by this infrastructure can thus, in principle, provide 60-GHz radio connectivity crossing the entire house.

Ad-hoc-based communication infrastructure

In order to completely free the future home networks from cabling, the ad-hoc-based approach aims to investigate a new type of 60-GHz radio relaying device. This device is considered as two pieces which can be attached to each side of a wall to relay the 60-GHz signals from one side to other one. The idea here is based on the fact that a loss of 10–20 dB through the wall can be compensated by attaching a transmitter on one side of the wall and increasing the transmission power with either a power amplifier or a phased antenna array. Referring to the experimental results in [13], 10- to 20-dB gain is able to compensate the penetration loss of 60-GHz radio through most common wall materials (e.g., brick, glass, wooden board, etc.). Thus, by attaching a receiving antenna on the other side of the wall, it is feasible to catch the signal with enough signal strength. In this approach, the only infrastructural elements required are the HCC and the wall-mounted radio transponders, which we call as "via" devices (Via-s). A 60-GHz home network with the Via-s does not need a backbone network to interconnect the room-based cells. Instead, it basically forms an in-house ad hoc network with all the Via-s by connecting the devices in different rooms over them. In such a network, the HCC serves to connect home network to the access networks. However, HCC is not considered being the only device here having control and management functionalities as in the cell-based approach. We envision a solution distributing these functionalities with network intelligence among, if not all, a subset of Via devices to better adapt to the dynamic and distributed nature of such an ad hoc network.

CHALLENGES AND RESEARCH ISSUES

Despite the provision of multi-Gbps data rates, networking at the 60-GHz band exposes some serious challenges that prevent the adoption of this frequency band in the range of LAN. In this section, the challenges of networking at the 60-GHz band and their research issues are discussed.

Connectivity

As mentioned in previous sections, in an indoor environment, the propagation of signals at 60 GHz is strongly weakened by surrounding obstacles and walls. Concrete walls can be considered as reliable cell boundaries since they can attenuate signals as much as 40 dB. A person standing in between a line-of-sight connection can also take 20 dB away from the link budget. To further analyze the coverage at 60-GHz band, a simulation study has been carried out using a race-tracing tool called radio-wave propagation simulator [14]. This simulator has been shown to be accurate in terms of statistical properties [15]. Figure 4 illustrates the simulated in-home environment containing multiple small rooms (5×8 m), a large room (15×8 m), and a corridor. Immobile people and objects such as sofas, tables, etc. were placed onto the floor. In the first simulation scenario, an omnidirectional antenna with 0 dBm of transmission power is placed in the center of each room at the height of 3 m. The antenna placed in the big room has 3-dBm antenna gain. Two others omni-antennas are placed in the corridor. Signal strength is recorded at a height of 1.5 m.

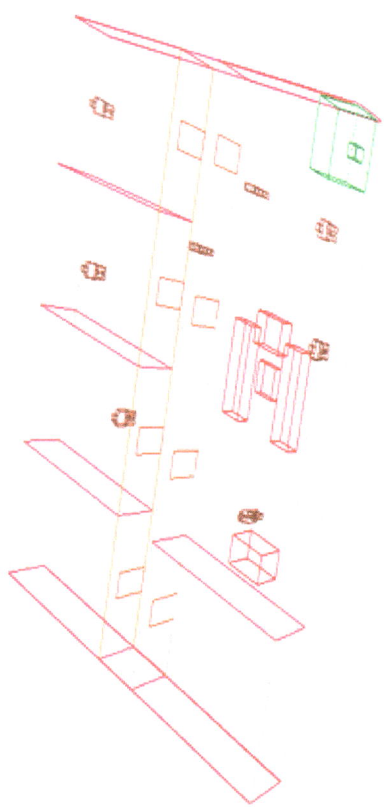

Figure 4: The simulated in-home environment

Figure 5 shows the simulated signal coverage of the first scenario. As predicted, the transmission of 60-GHz signals is severely effected by shadowing. Not only can large objects such as trees, furniture, etc. cause shadowing, but even a person can completely block the signal. Thus, to achieve the full connectivity in a crowded room is rather challenging by this deployment. Moreover, a number of antennas are required to cover the long corridor. To improve the signal coverage in the big room, two dielectric lens antennas are deployed for the big room in the second scenario (Fig. 6). As a result, the signal coverage is significantly improved. A large part of the shadowed areas behind the people has been removed. In this paper, the use of multiple antennas to improve signal coverage is referred by the term "antenna redundancy". Another option to overcome the line-of-sight limitation of 60-GHz propagation is to make use of smart antennas. Since the antenna dimension is inversely proportional to the operating frequency, it is possible to deploy many antennas in a small fixed area. The antenna array formed by these small antennas can intelligently adjust the gain, the transmission power, and the beam direction according to the link conditions. For example, if any obstacle blocks the line-of-sight path of the radio wave, a smart antenna array can use the reflections by redirecting the energy onto the reflecting objects [16]. However, these techniques are not enough to create a reliable WLAN covering whole home, since walls are still impassable barriers for 60-GHz radio signals. New methods and devices are required to overcome this problem.

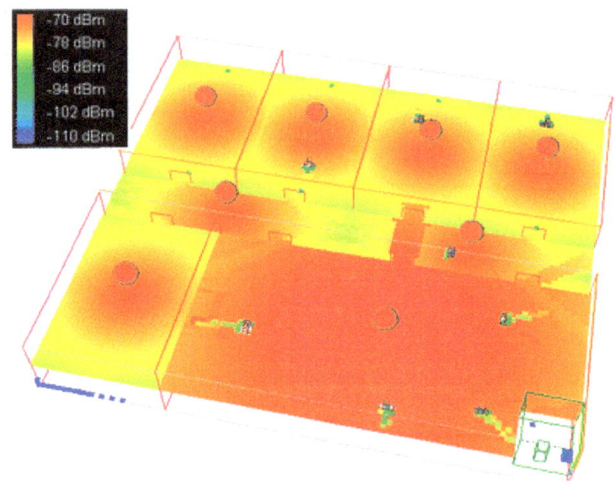

Figure 5: Signal coverage at the 60-GHz frequency band

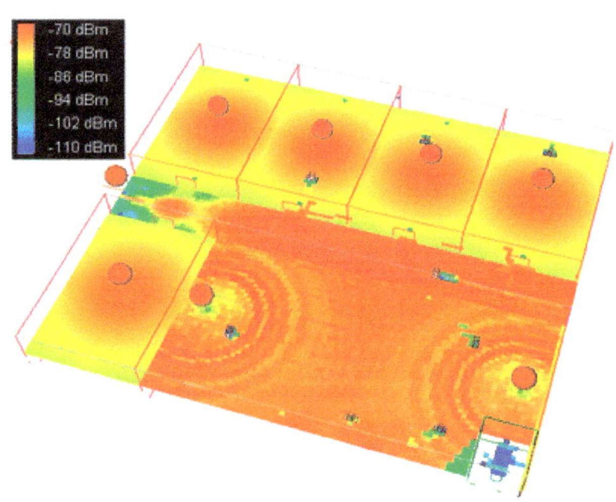

Figure 6: Signal coverage at the 60-GHz frequency band—multiple antennas

Mobility

Another problem that arose is the directional overlapping areas between cells. As can be seen in Fig. 7, overlapping only exists around the opened areas, i.e., opened doors and windows.

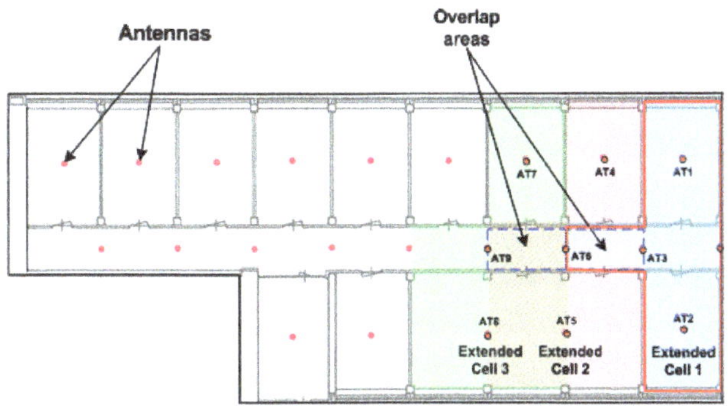

Figure 7: The extended cell (*EC*) concept

Consequently, overlapping areas are often narrow and directional. In a multichannel communication system where handovers (HO) are required when a mobile station (MS) roams from one cell to another, these overlapping areas might be too small to allow an MS sufficient time to trigger and complete a handover. In this paper, we use the term *corner effect* to refer to this problem. The directional overlapping poses difficulties on the design of the mobility handling mechanism, since the transition between two cells is too sharp and short to trigger a handover. This problem gets worse as mobile users in an in-home environment are much likely to perform sharp turns while moving from one room to another. To perform a radio-signal-measurement-based handover, a good decision has to be made based on the signal strength averaging and hysteresis margin. In order to guarantee a seamless communication environment, it therefore requires a minimum overlapping area between two adjacent cells. In other words, a mobile station must be able to listen to both APs so it can determine which one is better to connect to. In [17], the relation between the minimum overlapping area and the handover frequency defined by the ratio between a mobile user's velocity and the cell size has been carried out. As calculated in the paper, the minimum overlapping area for an indoor environment is 20% of the cell's diameter (2 m). It is therefore crucial that large enough overlapping areas are created in the system in order to guarantee a seamless communication environment.

With this target in mind, an infrastructure for broadband in-building networks at millimeter-wave GHz band was presented in [12, 18]. It is proposed to group multiple adjacent antennas into an extended cell (EC) and to allow the antennas to transmit the same content over the same frequency channel. Moreover, an EC is designed to cover several adjacent rooms and a part of a transitional area. By doing so, an overlapping area between two ECs can always be created in the transitional zone, and thus, seamless communication can be achieved, as mobile users always have to pass a transitional area in order to move from one room to another (Fig. 7). Using this concept, the corner effect is mitigated, as a mobile user does not have to perform a HO as long as it is still in the EC. The number of HOs will therefore be substantially decreased. Furthermore, a form of spatial diversity can also be achieved with the EC concept since multiple copies of a signal are concurrently sent by all the antennas in an EC. Shadowing is reduced since there is a better chance that a mobile station receives a good signal. To illustrate the effectiveness of the EC concept, simulations have been carried out. The simulation setup was described in [12]. The simulation results in Fig. 8 clearly show the improvement in terms of the average number of drop calls when the EC concept is employed. However, since a call lasts longer, the average number of HOs per call also increases. As a result, a trade-off should be made between the average EC size and the required quality of service in terms of drop calls.

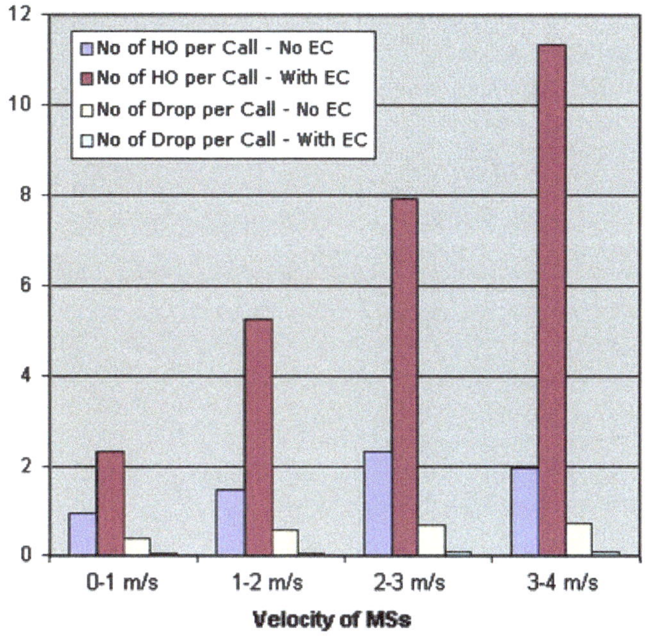

Figure 8: Average number of HOs and call drops

This EC concept connects well with the antenna redundancy ideas and the RoF infrastructure approach presented in Section 4.1. New antennas can be easily connected to the optical infrastructure to improve the signal coverage. Moreover, these new antennas will be included into the existing extended cell.

Self-configuration

Self-configuration refers to a means of automatically setting up networking-capable devices and organizing them in a network in the way that the intended tasks requested by users are optimally performed. For home networks, self-configuration is of great importance, as the network owners are most likely to have little or none expertise of network administration. It is also realistic to assume that they are not willing to have a third party administration.

Another important drive for seeking a sophisticated self-configuration strategy in home networks is fuelled from the characteristics of 60-GHz radio. While offering a great opportunity to prosper versatile bandwidth-demanding home applications, it raises greater challenges to realize a self-configurable home networking. The short radio range characteristics and the vulnerability for line-of-sight (LOS) obstruction of 60-GHz links result in frequent changing on link quality and network topology. This situation causes a large amount of reconfiguration processes. As the 60-GHz radio promises for high data rate and real-time applications, the timeliness of the reconfiguration processes is especially demanded, as it is directly tied to the user perception quality.

Self-configuration is essentially an important issue of network control and management at different connectivity levels (i.e., physical and link, network and service level) to enable distributed application entities that spread over the home network components to cooperate for a particular application.

While the network and service levels self-configuration processes (i.e., IP configuration, service discovery, etc.) have great generality for all the radio technologies, the novelty of 60-GHz radio calls for special concerns on physical and link level self-configuration process. An example with the cell-based communication infrastructure is when the number of users or terminals in a room (e.g., a meeting room) becomes so large that the picocell capacity cannot guarantee the required service quality, the self-configuration process should automatically add extra resources, e.g., establish an extra picocell by turning on an additional wavelength in the fiber to that room and handing off part of the load to that new picocell. For ad-hoc-based approach, frequent disturbances of LOS communication links (e.g., people standing in LOS links) require fast resource discovery strategy to reestablish connectivity. Self-configuration is thus required to efficiently organize devices into piconets and scatternets to support multihop connections. Such process must be established for supporting the cooperation between distributed devices and be tailored for 60-GHz radio with respect to the use of directional antenna, LOS links, and the specialty of the MAC protocol.

Although most current contributions related to self-configuration issue have been focusing on developing and optimizing isolated layer-dependent protocols, we share a vision with the new networking paradigm such as autonomic communications [19] and are seeking a promising way to cooperate different levels of configuration under a cross-layer optimization strategy in the sense that the configuration information at different levels may be mutually beneficial. For example, it could be feasible and desirable to take into account 60-GHz channel state information available at the physical layer when making decisions at higher layer levels concerning channel allocation, cell formation, route selection, etc. However, as pointed out in [20], cross-layer approach should not only focus on the layer-wise performance enhancement blindly but rather be guided by the high-level goal required by the applications.

To solve the potential confliction among configuration actions at different layers, a promising way is to employ a general management module to conciliate self-configuration processes in parallel threads rather than merely increase complexity to achieve local sophistication within individual configuration protocols. Such a self-configuration management module is expected to incorporate three basic functions: (1) information abstraction to aggregate the contextual information from different connectivity levels and to define configuration states of the network; (2) configuration coordination to optimize configuration processes under an overview of the configuration states and under joint consideration and decision of all connectivity levels; and (3) *protocol interface* to exert the decision from the configuration coordinator to adjust and operate the layer-based configuration protocols. This management module should be implemented in the capable devices in terms of computation, storage, and power resources while taking into account the location and mobility. By deploying the cell-based communication infrastructure, as all the traffic goes through the HCC, it should at the best take the role of the self-configuration manager and form a centralized management structure. For the ad-hoc-based approach, besides the HCC, other capable devices should also be implemented fully or partially in the module to provide distributed management. In this way, hierarchical self-configuration management can be organized.

Cognitive networking

As the deployment of 60 GHz results in more dynamic and distributed networking environment, control and management of such a network stringently require the efficiency of the processes. That means that on one hand, the timeliness is highly desired for configure and adapt the 60-GHz applications to a user desired state. In particular, for personal applications, the maximum allowed configuration time is determined by the patience of the user (often of the order of a second). On the other hand, necessity of reconfiguration and adaptation processes must be decided on occurrence of the networking state changes, in other words, whether and when a reconfiguration process should be triggered. This is critical to achieve a stable networking system, which shields the network dynamics from its users. With further concerns on building home network to be a "self-" capable system, the network should be able to identify its goals centered on its users' need and manage on its own to realize those goals regardless the heterogeneous, distributed, and dynamic network environment. This requires new mechanisms to evoke the expert-like intelligence on the network.

One promising solution to tackle such challenge is to incorporate the novel concept of cognitive networking into play. Being specified in [21], a cognitive network has a cognitive process that can perceive current network conditions and then plan, decide, and act on those conditions. The network can learn from these adaptations and use them to make future decisions, all while taking into account the user experience. With the ability to learn from the consequences of the network operations and to accumulate experience on executing such operations effectively and efficiently, the cognitive network is expected to switch network control and management processes from the reactive to the proactive manner. More specifically, the experience can help, for example, to predict ahead of time when a reconfiguration/adaptation needs to start. The idea is based on the observation that in-home usage of the network resource exhibits predictable patterns over time. An example would be the daily routine of inhabitants of a house who repeat the same patterns of behavior and movement from room to room.

Inspired by the idea of constructing a knowledge plane for the Internet [22], the architecture of the cognitive home network is expected to be implemented with a cognitive plane (CP). In our opinion, the CP is best to be positioned vertically crossing all the other three existing planes (i.e., data, control, and management planes) as shown in Fig. 9.

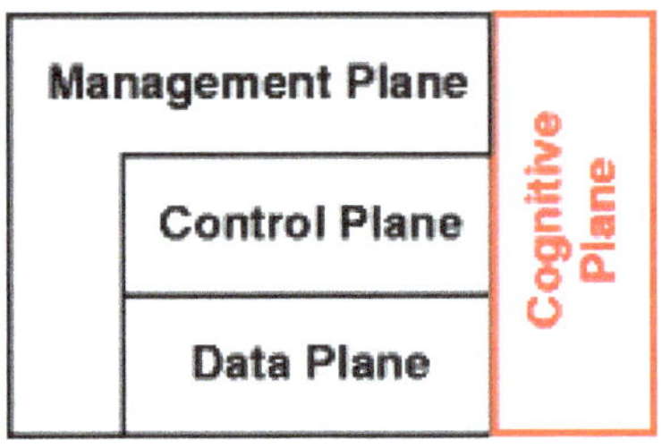

Figure 9: FHN architecture: a view of plane division

The purpose of this design is twofold. One is to enable

sophisticated decision making for high level instruction. This mainly works for the management plane and control plane to realize network self-management. The other one is for knowledge circulation. Knowledge aggregation and distribution need to be carried out by data plane, while the control and management planes should be responsible for admission of knowledge accessing and provision. Within the CP, four basic functionalities are proposed and will bear further deliberation: (1) a *monitoring* function perceives environment and collects meta-information from the users, the network, and the environment in which the network operates from which user patterns and context can be derived; (2) a *learning* function processes and converts the aggregated information, extracting, adjusting, and correcting patterns and deriving context to form the basis for predicting the next services demanded from the network operations; (3) a *decision* function initiates proactive configuration and reconfiguration based on learned patterns and context through specific cognitive techniques; (4) an *execution* function takes decisions made from the decision function and exerts them into actions upon network operation.

A key question is how to identify the most applicable cognitive techniques for specific network control and management operations. This should be answered from a joint effort of telecommunication and artificial intelligence expertise. Further, we need to identify which information needs to be sensed and what context should be consisted of. We can build here on the researches that have been done in context-aware computing and networking [23] and in ambient intelligence domain [24]. Although the concept of cognitive networking has illuminated a potential research direction for realizing network intelligence, the additional cost (e.g., system complexity, power consumption, etc.) by introducing such a new network structure should not be ignored. Introduced in [21], the prices of anarchy, ignorance, and control can be considered as further evaluation of the necessity and the fitness of adopting cognitive approach as the solution for the complex network problems.

CONCLUSION

In this paper, we review and present the main aspects of the 60-GHz radio regarding future home networks. We demonstrated two approaches to create a future-proof home network utilizing the 60-GHz band. The cell-based approach connects the rooms with fiber-optic cables and transmits the data at 60 GHz over these cables using the radio-over-fiber technique. The ad-hoc-based approach deploys special Via devices to enable the penetration of 60-GHz signals through the walls by forming a full wireless network all over the house. We also introduce the fundamental challenges with the research issues we are currently working on. With regards to connectivity, antenna redundancy can be a solution for the shadowing caused by immobile and mobile obstacles. This solution is not costly when combined with ROF, since the antennas are connected to the existing optical fiber infrastructure and they are very simple. Another promising solution to this problem looks like deployment of smart antennas. To overcome the corner effect and guarantee a seamless communication, the EC concept can be used. For the self-configuration concerns, we consider to employ a general management module operating across the layers. Finally, we investigate the cognitive networking paradigm and its application to the problems of future home networking.

As the future work, we will keep working on the integration of 60-GHz technology into the home networks. We will explore the challenges in more detail towards finding concrete solutions.

ACKNOWLEDGMENT

This research was carried out in the "Future home network" and "Broadband In-home Networks employing Radio over Fiber" projects within IOP GenCom program and Adaptive Ad-Hoc Free Band Wireless Communications project within Freeband program, both funded by the Dutch Ministry of Economic Affairs.

REFERENCES

1. Online BBC article. http://news.bbc.co.uk/1/hi/technology/6474581.stm
2. Interview. http://www.gearlog.com/2007/07/gearlog_interview_joy_laskar_l.php
3. SiBEAM White Paper. The future of wireless technology achieving gigabit data rates
4. The WirelessHD specification. http://www.wirelesshd.org/company/about.html
5. OmniLink60ä technology. http://www.sibeam.com/products/omnilink.html
6. GIT news release. http://www.gatech.edu/news-room/release.php?id = 1431
7. Kues H et al (1999) Absence of ocular effects after either single or repeated exposure to 10 mW/cm^2 from a 60 GHz CW source. Bioelectromagnetics 20:463–473
8. Doan CH, Emami S, Niknejad AM, Brodersen RW (2005) Millimeter-wave CMOS design. IEEE J Solid-State Circuits 40(1):144–155
9. NICTA 2008 Media Releases (2008) NICTA develops a world first in semiconductor technology for the wireless home and office of the future. http://www.nicta.com.au/news, February
10. TG3cwebpage. http://www.ieee802.org/15/pub/TG3c.html
11. Ecma International 60 GHz standardization activity. http://www.ecma-international.org/activities/Communications/ga-2007-015.pdf
12. Dang BL, Prasad V, Niemegeers I, Larrode MG, Koonen AMJ (2006) Toward a seamless communication architecture for in-building networks at the 60 GHz band. Proceedings of the 31st IEEE Conference on Local Computer Networks (LCN2006), November
13. Langen B, Lober G, Herzig W (1994) Reflection and transmission behavior of building materials at 60 GHz. IEEE PIMRC '94, The Hague, The Netherlands, September
14. Radioplan. Radiowave Propagation Simulator. www.radioplan.com
15. Smulders P, Li C, Yang H, Martijn E, Herben M (2004) 60 GHz indoor radio propagation comparison of simulation and measurement results. Proceedings of the 11th IEEE Symposium on communications and

vehicular technology in the Benelux. Belgium

16. SiBEAM White Paper. 60 GHz architecture for wireless video display.
17. Emmelmann M (2005) Influence of velocity on the handover delay associated with a radio-signal-measurement-based handover decision. Proceedings of IEEE Vehicular Technology Conference (VTC 2005 Fall), Dallas, TX, USA, September
18. Dang BL, Garcia Larrode M, Prasad V, Niemegeers I, Koonen AMJ (2007) Radio over fiber based architecture for seamless wireless indoor communication in the 60 GHz band. Elsevier Computer Communications 30:3598–3613 (18 December)
19. Dobson S et al (2006) Survey of autonomic communications. ACM Transactions on Autonomous and Adaptive Systems (TAAS) 1(2):223–259
20. Kawadia V, Kumar PR (2005) A cautionary perspective on cross-layer design. IEEE Wireless Communications/IEEE Personal Communications 12(1):3–11 (February)
21. Thomas RW, Friend DH, Dasilva LA, Mackenzie AB (2006) Cognitive networks: adaptation and learning to achieve end-to-end performance objectives. IEEE Commun Mag 44(12):51–57 (December)
22. Clark D, Partridge C, Ramming C, Wroclawski J (2003) A knowledge plane for the Internet. ACM SIGCOMM 2003, Karlsruhe, August
23. Schilit BN, Hilbert DM, Trevor J (2002) Context-aware communication. IEEE Wireless Commun 9(5):46–54 (October)
24. Stajano F (2004) Security for ubiquitous computing. Wiley, 2002 Technology in the Benelux, November

CHAPTER 15

PERFORMANCE OF BLUETOOTH 2.X + EDR TECHNOLOGY IN NOISY ENVIRONMENTS

Jose-Rafael Luque, Maria-Jose Morón, and Eduardo Casilari

Keywords: Bluetooth, Enhanced data rate, RFCOMM, Packet delay

INTRODUCTION

Bluetooth (BT) communication standard has been widely adopted for the deployment of personal area network (PAN) and body area network (BAN) networking applications.

Although it was initially created to avoid cables in short range networks, BT interfaces have been integrated into a high diversity of commercial devices, from handheld products (smartphones, laptops, electronic pads, tablets) to medical sensors, automobile electronics, music players, headsets or gaming consoles such as Sony PlayStation 3 or Nintendo Wii. This expansion of BT-enabled hardware fostered the release in 2004 of a version of the standard (BT 2.0 + EDR) aimed at providing a bandwidth higher than the basic speed supported by BT 1.1 and 1.2 versions. In BT 1.1 and 1.2 this basic bit rate of 1 Mbps was achieved with a Gaussian frequency shift keying (GFSK) modulation. On the contrary, the main novelty of BT 2.0 + EDR and 2.1 + EDR standards [1, 2] (hereinafter referred to as BT 2.x + EDR) is the adoption of two new modulation schemes. These schemes allow a faster data transfer through the so-called enhanced data rates (EDR). These novel differential phase shift keying (DPSK) based schemes (π/4-DQPSK and 8-DPSK) enable the transmission of the BT packet payload at a bit rate of 2 Mbps or 3 Mbps, respectively.

BT physical layer operates in the Industrial, Scientific and Medical (ISM) 2.4 GHz unlicensed band. As a result, BT radio connections are prone to interferences from other wireless transmitters (Bluetooth or not) working in the same ISM band. Because of the popularity of some of these technologies coexisting in the same ISM band (especially 802.11 but also 802.15.4), most real transmission of BT-capable applications will take place in the presence of a strong environmental noise.

Common non real-time data are transmitted by BT through asynchronous connection-oriented links (ACL). Bluetooth ACL packets must be positively acknowledged by the receiver if they are properly received. In other case, the packet must be retransmitted, which increases the transmission delay and degrades the performance of BT. Noise can induce unrecoverable bit errors in the packets and provoke these retransmissions. In this sense, the impact of the noise on the bit errors suffered by the BT signal highly depends on the modulation scheme that is being employed. In fact, EDR allows higher bit data rates at the cost of a higher sensitivity to the noise.

This paper analytically characterizes and evaluates the effects of noise on the packet delay experienced by transmissions for the different modulations contemplated by BT. The study permits to assess the feasibility of BT communications in the presence of a certain noise level, characterized as a value of signal-to-noise ratio (SNR) per bit.

This paper is organized as follows: Sect. 2 reviews the literature related to the modelling and performance evaluation of the different modulations contemplated by the BT 2.x + EDR specification (GFSK, π/4-DQPSK and 8-DPSK). Basing on existing studies, Sect. 3 presents a model to compute the delay of BT packets in ideal conditions (i.e. with no packet retransmissions). In Sect. 4 this model is extended to characterize the non-ideal case where noise introduces an extra delay due to retransmissions. This analytical model is utilized in Sect.5 to evaluate the performance of the different modulations for different noise levels. Section 6presents an empirical evaluation of the analytical model for actual transmissions using the basic rate. Finally Sect. 7 recapitulates the main conclusions of the paper.

2 Related work

Many works in the literature have provided insights on the analysis of the existing modulation techniques for RF communications. In the case of the three modulating schemes

Citation: Jose-Rafael Luque, Maria-Jose Morón, Eduardo Casilari, performance of Bluetooth 2.x + EDR technology in noisy environments, DOI: 10.1007/s11276-015-0888-1.

employed by BT 2.x + EDR, extensive studies have been devoted to characterize the relationship between the SNR and the bit error ratio (BER). For example the performance of GFSK is modelled in [3], π/4-DQPSK in [4–11] and 8-DPSK in [12–14].

Some of these abstract models are particularized to improve or evaluate specific characteristics of the BT transceivers. For example, the study in [15] offers an analysis of the GFSK modulation with non-coherent detection for systems that employ frequency hopping as the spread spectrum technique (which is the case of BT). Assuming a Rayleigh-type radio channel, the authors propose a model to relate the BER to the SNR ratio and the modulation index.

The work in [16] details the design of a GFSK transceiver with carrier frequency offset correction. The design incorporates a mechanism to simplify the implementation of a multi-symbol receiver. The study shows the results that could be obtained if this design is applied to a BT receiver.

Similarly, the paper in [17] proposes a scheme for a digital GFSK transceiver that includes a new method to estimate the carrier frequency. The scheme is evaluated through simulations analysing the obtained BER. Simulations assume an additive white Gaussian noise (AWGN) channel.

An analogous mechanism for carrier synchronization in a GFSK receiver is offered in [18]. By means of simulations the mechanism (which is not implemented) is contrasted against the behaviour (in terms of the obtained BER) of the typical DFT-based method. Noise is also modelled by an AWGN channel.

After proposing a new radio channel model (based on Markov chains), the performance of BT piconets is evaluated in [19].

Other works in the literature specifically focus on characterizing the impact of noise on BT transmissions. Authors in [20] study the performance of BT 1.1 in Gaussian and Rayleigh-type channels. The paper provides an analytical formulation (with some mathematical mistakes) to calculate the throughput that can be obtained by the different BT packet types as a function of the SNR. The proposed expressions allow to compute the packet error rate (PER) from the BER experienced by a GFSK modulation. Finally the study discusses the convenience of the different packet types for the considered channel types (Raleigh or AWGN).

A RF simulation system for BT transceivers is presented in [21]. The system emulates the channel by combining an AWGN and the effect of interfering signals.

Authors in [22] propose a set of rules to select the BT packet type basing on the estimated SNR. The study suggests employing new modulations (not considered by BT 2.x + EDR) and new algorithms to calculate the SNR. However, the viability of the implementation of these algorithms at the LM (Link Manager) layer of an actual BT host controller is not analysed.

The performance of BT transmissions in Nakagami-type channels is studied in [23] through simulations by observing the evolution of the BER as a function of the SNR. Analyses are conducted for different values of the fading parameter of the Nakagami distribution (as well as for an AWGN channel).

The study in [24] expands the model in [20] (conceived for BT 1.1) to include the EDR modes. As in [20] the benefits of using the different sizes for the new packet types—under an AWGN channel—are evaluated. The paper proposes a set of rules for selecting the packet type as a function of the SNR. Authors suggest that this selection must be performed at the L2CAP layer although they do not clarify how this decision must be implemented at the baseband layer.

A MATLAB simulation model for the physical layer of a BT 2.0 + EDR transceiver is presented in [25]. The BER computed with this model for different values of SNR is compared to that obtained with theoretical models which also assume an AWGN channel.

The operation of Wireless Application Protocol (WAP) applications over BT is evaluated in [26]. The modelling of the communications again assumes that bit errors in a simulated BT piconet are caused by an AWGN channel.

Zanella [27] presents a Markovian model of the BT 2.0 + EDR transmissions in fading channels. In [28] the same author offers theoretical expressions to calculate the bit error probability for BT 2.0 transmissions for both AGWM and Rice fading channels.

The recent study in [29] describes different GFSK demodulation techniques in BT framework. In addition, the authors propose a novel scheme for a digital GFSK demodulator that is able to improve the performance of BT without introducing any complexity in the hardware implementation. The analysis of the behaviour is carried out for an AWGN channel.

Other references dealing with the optimization and implementation of BT demodulators for GFSK, π/4-DQPSK and 8 DPSK modulations can be found in [30–38].

The throughput of the recent version of BT for low energy devices is modelled in [39]. The model, conceived for error-prone links, offers an expression to compute the maximum throughput as a function of the BER.

As aforementioned, bit error rate has been analytically characterized in some studies (such as those authored by Zanella). However, to the best of our knowledge, the packet delay has not been analytically modelled in the existing literature about the performance of BT 2.x + EDR in noisy environments. Data delay is an important metric to evaluate the performance of any networking application. In fact packet delay is a key factor to determine the quality of service that is actually perceived by the end-users. Moreover, if a finite period ('flush timeout') is set for the proper transmission and reception of the BT packets, data can be discarded if not successfully acknowledged within this interval. Additionally ACL links are disconnected if no packets are received during a certain supervision timeout period (whose default value is 20 s). The timeouts set by the upper layers should be also considered to evaluate the impact of packet delay. Consequently, delay induced by environmental noise may cause applications to become not viable if a certain SNR is not guaranteed. The goal of the present work is to offer an analytical characterization of the delay of BT 2.x + EDR standards for the different modulations (packet type) and under diverse conditions of the noise. The model considers the effects of the retransmissions provoked by the noise together with the procedures (packetization, segmentation, header overhead, etc.) operated at the different layers in the whole BT protocol stack for a typical serial port profile (SPP) connection. By employing this model, the performance of the different available rates of BT 2.x + EDR can be predicted as a function of the measured SNR. Thus the utilized modulation at the BT host controller could be dynamically adjusted to minimize the delay.

Hence, the feasibility of using the EDR modulations will be investigated basing on a realistic metric of the behaviour of a BT connection.

CHARACTERIZATION OF THE DELAY FOR THE IDEAL CASE

The specification of BT 2.0 + EDR enables the BT host controller to use diverse types of Baseband packets. The maximum allowed payload sizes for the different packet types are summarized in Table 1. The acronyms DH (data high rate) and DM (data medium rate) indicate the protection utilized by each type for the payload bits. In particular, DM packets incorporate a forward error correction (FEC) field to increase the robustness of the transmission but at the cost of transporting less user data than DH-type. In BT, both DM and DH packets may consist of 1, 3 or 5 BT slots. Thus, in the identifiers of the packet type in Table 1, the number (1, 3 or 5) after DM or DH defines the number of BT slots of the packet. Similarly, the digit before the name indicates the modulation technique and consequently the utilized data rate (1, 2 or 3 Mbps). It should be remarked that EDRs are only conceived for DH-type packets. Thus, DM packets are necessarily transmitted at 1 Mbps.

Table 1: Maximum payload and utilized rate for the different packet types of Bluetooth 2.x + EDR

Packet type	Bit rate (Mbps)	Maximum payload (bytes)	Notation in the text
DM1	1	17	L_1^M
DM3	1	121	L_3^M
DM5	1	224	L_5^M
1-DH1	1	27	$L_{1,1}^H$
1-DH3	1	183	$L_{1,3}^H$
1-DH5	1	339	$L_{1,5}^H$
2-DH1	2	54	$L_{2,1}^H$
2-DH3	2	367	$L_{2,3}^H$
2-DH5	2	679	$L_{2,5}^H$
3-DH1	3	83	$L_{3,1}^H$
3-DH3	3	552	$L_{3,3}^H$
3-DH5	3	1,021	$L_{3,5}^H$

Aiming at estimating the time required to transmit a BT packet, we have to consider the influence of the existing layers in the BT protocol stack: mainly the BT Baseband, L2CAP (Logical Link Control & Adaptation Protocol) and the protocol selected by the BT profile. BT profiles, which are intended to guarantee the interoperability between different vendors, specify standard interfaces to access particular services. SPP, which is the basis for other BT profiles, is the most commonly implemented profile in commercial BT devices, ranging from handheld devices (BlackBerry units, Smartphones, etc.) to medical wireless sensors or peripherals such as keyboards or GPS.

SPP uses RFCOMM protocol to emulate RS232 cable communications. RFCOMM conveys the user data (organized in frames) to the lower BT stack layers by means of L2CAP. L2CAP multiplex, segment and reassemble the data flowing from/to the upper layers. This fragmentation is executed so that only one single RFCOMM frame is encapsulated into every L2CAP frame. L2CAP frames are in turn layered over BT Baseband. As a consequence, L2CAP frames are divided into a series of BT baseband packets which are transmitted to the radio medium by the physical layer.

To calculate the time which is needed to transmit a certain number of user data bytes, we have to take into account the fragmentation and the packet overhead applied by each protocol layer. In the absence of retransmissions, this time can be computed [40] as:

$$t_R(N_U) = n_{nff}(N_U) \cdot t_{ACK}(L_R + O_R(L_R) + H_L) + t_{TX}(L_{ff}(N_U) + O_R(L_{ff}(N_U)) + H_L) \quad (1)$$

being:

- N_U: the number of user data bytes,
- L_R: the maximum size of the RFCOMM frames. The value of the size is limited by two bounds: the maximum frame size (N_1) of RFCOMM and the maximum transfer unit of L2CAP for RFCOMM (M_R):

$$L_R = \min(N_1, M_R - O_R(N_1)) \quad (2)$$

where $O_R(x)$ is the overhead that RFCOMM introduces in each frame: 5 or 4 bytes depending if the data size is higher than 127 bytes or not.

H_L: the size of the L2CAP header (4 bytes).

$n_{nff}(N_U)$: the number of non-final RFCOMM frames which are required to transport the N_U user data bytes. This number is directly computed from N_U and L_R, in the way:

$$n_{nff}(N_U) = \left\lceil \frac{N_U}{L_R} \right\rceil - 1 \quad (3)$$

where ⌈x⌉represents the lowest integer higher than x.

$L_{ff}(N_U)$: the number of data bytes of the last RFCOMM frame, also computable from N_U and L_R as:

$$L_{ff}(N_U) = ((N_U - 1) \bmod L_R) + 1 \tag{4}$$

$t_{ACK}(x)$: the delay required to transmit and acknowledge all the Baseband BT packets corresponding to the first and the intermediate RFCOMM frames. This variable has to compute all the one-slot packets (ACK) which must be sent by the receiver to acknowledge every received packet. The term $t_{ACK}(x)$ can be calculated with the recursive expression in Eq. (5). This formula takes into consideration that a RFCOMM frame may require more than one BT packet. In that case, as the last condition in the expression indicates (when $x > L_{5s}$), the frame will be segmented and transmitted as a series of 5-slot BT packets followed by one 1, 3 or 5-slot packet. The formula assumes the optimal case in which no retransmission occurs, so that the transmission of an n-slot packet just requires (n + 1) slots (n for the packet plus 1 slot for the acknowledgment).

$$t_{ACK}(x) = \begin{cases} 0 & x = 0 \\ 2 \cdot T_S & 0 < x \leq L_{1s} \\ 4 \cdot T_S & L_{1s} < x \leq L_{3s} \\ 6 \cdot T_S & L_{3s} < x \leq L_{5s} \\ 6 \cdot T_S \cdot \left\lfloor \frac{x}{L_{5s}} \right\rfloor + t_{ACK}(x \bmod L_5) & x > L_{5s} \end{cases} \tag{5}$$

In the previous expression the operator ⌊x⌋⌊x⌋ represents the highest integer lower than x, T_S defines the duration of a BT slot (625 µs), while L_{1s}, L_{3s} and L_{5s} are the maximum sizes of the payload of a 1, 3 and 5-slot BT packet, respectively. These sizes are 17, 121 and 224 bytes for DM-type packets (see Table 1). In the case of DH-type packets, most vendors' BT 2.x chipset tries to minimize the number of employed BT packets. Consequently, packets with a higher maximum payload (i.e., those used with a rate of 3 Mbps) are chosen. Thus, L_{1s}, L_{3s} and L_{5s} correspond to $L_{3,1}{}^H$, $L_{3,3}{}^H$ and $L_{3,5}{}^H$ (83, 552 and 1,021 bytes respectively).

$t_{TX}(x)$ defines the time to transmit the final RFCOMM frame (of x bytes). As the reception of user data finishes when the last bit of the final frame arrives at the receiver, the term does not add the final acknowledgement slot of the last BT packet by only considering the time needed to convey the bytes in the last frame. Again, this transmission delay essentially depends on the employed packet-type and rate. For the DM-type, we have that:

$$t_{TX}^M(x) = \begin{cases} 0 & x = 0 \\ d_1(x) & 0 < x \leq L_5^M \\ t_{TX}^M(x \bmod L_5^M) + t_{ACK}(L_5^M) \lfloor x/L_5^M \rfloor & x > L_5^M \end{cases} \tag{6}$$

where $d_1(x)$, defined in (8), is the time required to transmit a BT packet with a data payload of x bytes at the basic rate of 1 Mbps.

As in the case of t_{ACK}, this expression for t_{TX} also contemplates the segmentation that takes place if the final L2CAP frame needs more than one DM5 packet (which occurs if $x > L_5{}^M$). So, the acknowledgment slots of the corresponding intermediate 5-slot BT packets are also computed.

For DH-type packets, the time to transmit the last packet of the final RFCOMM frame depends on the modulation that is being employed. The selection of the rate in turn relies on the implementation of BT 2.x + EDR. Basing on the tests performed with actual interfaces using chipsets [41] from CSR (the most popular vendor of BT technology), we have concluded in a previous study [40] that the firmware selects the rate following the next criteria:

1. If DH-type packets are considered, the firmware only employs enhanced rates (1-DH1, 1-DH3 and 1-DH5 packets are not utilized).
2. The packet size (that is to say, the number of required slots) is tried to be minimized. Thus, the rate of 3 Mbps is preferred if the use of a rate of 2 Mbps obliges to increase the number of utilized slots.
3. If the data can be transported at 2 Mbps within the same number of slots than at 3 Mbps, the rate of 2 Mbps is chosen as it incorporates a more robust modulation technique.

Assuming these criteria, the term $t_{TX}(x)$ for DH packets can be computed as it follows:

$$t_{TX}(x) = \begin{cases} 0 & x = 0 \\ d_2(x) & 0 < x \leq L_{2,1}^H \\ d_3(x) & L_{2,1}^H < x \leq L_{3,1}^H \\ d_2(x) & L_{3,1}^H < x \leq L_{2,3}^H \\ d_3(x) & L_{2,3}^H < x \leq L_{3,3}^H \\ d_2(x) & L_{3,3}^H < x \leq L_{2,5}^H \\ d_3(x) & L_{2,5}^H < x \leq L_{3,5}^H \\ t_{ACK}\left(L_{3,5}^H\right) \cdot \left\lfloor \frac{x}{L_{3,5}^H} \right\rfloor + t_{TX}\left(x \bmod L_{3,5}^H\right) & x > L_{3,5}^H \end{cases} \tag{7}$$

The previous expression incorporates the changes in the used modulation (between 2 and 3 Mbps) that occur when the data exceed the maximum payload of the different packet types.

In Eqs. (6) and (7), the variable $d_i(x)$ defines the time that the baseband requires to transmit x user data at a rate (i) of 1, 2 or 3 Mbps. This time can be estimated from the number of symbols, $n_{symi}(x)$, that must be sent to transport the data:

$$d_i(x) = t_{sym} \cdot n_{sym_i}(x) + \Delta_{EDR_i} \quad \text{with } i \in \{1, 2, 3\} \tag{8}$$

where t_{sym} is a constant describing the time to transmit a symbol (1 µs, as BT transmits 1 Megasymbol per second for any the employed rate) while Δ_{EDRi} defines the extra-delay caused by the use of the enhanced rates of 2 and 3 Mbps. In particular, under EDR, BT introduces additional control and timing information to allow the radio module to synchronize to the new modulation format (see Fig. 1 for a better understanding of the structure of a BT packet utilizing EDR).

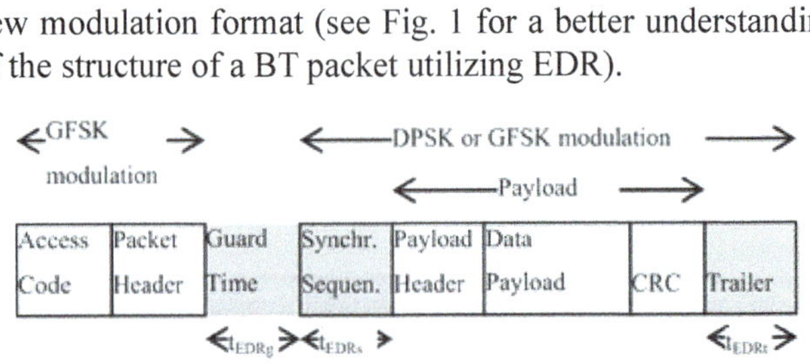

Figure 1: Bluetooth EDR packet format (shadowed fields are specific of EDR packets)

According to the BT specifications [1, 2], this extra delay (Δ_{EDRi}) results from summing three components:

$$\Delta_{EDR_i} = \begin{cases} 0 & i = 1 \\ t_{EDRg} + t_{EDRs} + t_{EDRt} & (i = 2) \vee (i = 3) \end{cases} \tag{9}$$

where:

t_{EDRg} is the guard time between the last symbol of the packet header (encoded with GFSK) and the EDR Synchronization Sequence. This short period enables the BT radio module to prepare for the change in the modulation. The standard defines a variable range for this value from 4.75 to 5.25 μs. For our model t_{EDRg} is rounded to 5 μs.

t_{EDRs} is the constant time of 11 μs required to transmit the 11 symbols of the enhanced data rate synchronization sequence. The goal of this sequence is to allow the synchronization of the symbol timing and phase for the new modulation type.

t_{EDRt} represents the 2 μs corresponding to the specific 2 symbol trailer that must be added after the packet payload to mark the end of the packet.

On the other hand, the number of required symbols $n_{symi}(x)$ can be directly computed from the type of the utilized rate (i) and the number of bits in the packet payload [$n_d(x)$] and in the associated overhead of the BT packet:

$$n_{sym_i}(x) = O_{BB} + \left\lceil \frac{n_d(x)}{i} \right\rceil \tag{10}$$

where O_{BB} indicates the 126 bits of overhead of the BT packet (72 bits of the initial access code plus 54 bits of the packet header). These two fields are always transmitted under the basic GFSK modulation (at 1 Mbps, with 1 bit per symbol). Thus, the number of symbols required to transport the overhead coincides with the number of bits O_{BB}. Conversely, as the rest of the packet can make use of the EDRs, the number of symbols that the payload demands depends both on the number of bits $n_d(x)$ and the number of bits per symbol i (1, 2 or 3 for a rate of 1, 2 and 3 Mbps, respectively).

To calculate the value of $n_d(x)$, we have to consider that DM packets support FEC 2/3 protection, so that 5 redundancy bits are added for every 10 information bits (if the number of bits is not a multiple of 10, the packet is filled with extra bits):

$$n_d(x) = \begin{cases} (x + H_P(x) + O_{CRC}) \cdot 8 & \text{DH packets} \\ 15 \cdot \left\lceil \frac{(x + H_P(x) + O_{CRC}) \cdot 8}{10} \right\rceil & \text{DM packets} \end{cases} \tag{11}$$

where O_{CRC} (2 bytes) is the overhead incorporated to the CRC field while $H_p(x)$ describes the size of the payload header: 1 byte for DM1 and 1-DH1 type packets, and 2 bytes for the rest of packet types:

$$H_p(x) = \begin{cases} 1 & x \leq L_{1s} \\ 2 & x > L_{1s} \end{cases} \tag{12}$$

where L_{1s} is the maximum size of the payload of a 1-slot packet for DM1 and 1-DH1 types: $L_{1s} = L_1^M$ (17 bytes) and $L_{1s} = L_{1,1}^H$ (27 bytes), respectively. As abovementioned, CSR Bluetooth firmware does not employ 1-DH1 packets. Thus $H_p(x)$ is always 2 bytes for DH packets.

The strategy for the rate selection may change for the firmware of other vendors (for example, ISSC BT interfaces [42] only use 3-DH1, 3-DH3 and 3-DH5 packets). However, the previous equations could be easily adapted for these variations.

CHARACTERIZATION OF THE DELAY WITH PACKET RETRANSMISSIONS

The model presented in the previous section is conceived for the optimal case in which all the BT packets are properly received and acknowledged, so that no retransmission is required. However, as aforementioned, environmental noise can cause packet retransmissions, which will be reflected in a variable packet delay? The goal is now to define the analytical relationship between the mean packet delay of the BT transmissions and the existing SNR. As it has been stated in the state-of-the-art (in Sect. 2), the characterization of the transmission channel as an AWGN channel has been widely considered by the literature for the analysis of the performance of BT connections and even for the design of BT radio transceivers [17–21, 23–26, 28–30, 32,35, 36]. In studies on BT technology, several reasons have been argued to justify the use of this Gaussian characterization of the channel. BT communications are usually deployed in static (indoor) office scenarios. Therefore the transmission model can assume that short term (fast) fading has a marginal effect on BT operation. Typical BT connections consist in low range communications where line-of-sight (LOS) from the receiver to the transmitter normally exists. In this sense, indoor environments are usually determined by the existence of other interfering technologies (such as Wi-Fi) which operate in the same 2.4 GHz ISM band. These interfering sources (with different spectral properties) can be reasonably modelled by a noise with a constant spectral density [43–46]. Authors in [47] experimentally show that the in-band interference power of Wi-Fi can be modelled as a band-limited AWGN, at least for the central part of the active Wi-Fi channels.

An AWGN channel adds a white (flat power) noise to the whole considered signal bandwidth. In the next sub-sections, assuming this AWGN channel, a formula for estimating the mean packet delay is calculated.

Computation of the Bit Error Probability

The first step is to analytically relate the bit error rate (BER) of the BT transmissions to the existing SNR for the different modulations employed by BT, that is to say, GFSK, π/4-DQPSK and 8-DPSK modulations.

For an AWGN channel, a noise with constant energy is added in the whole signal bandwidth. So, the SNR ratio can be expressed [48] as:

$$SNR = \frac{P_r}{N_O \cdot B} \tag{13}$$

where:

P_r is the power (in W) at the receiver.

$N_o/2$ is the power spectral density (PSD) of the noise introduced by the channel, defined in W/Hz.

2·B symbolizes the bandwidth (in Hz) of the signal at the receiver.

Aiming at comparing the impact of noise on the digital modulations utilized by BT 2.x + EDR, we characterize the SNR in terms of the energy emitted per bit or per symbol:

$$SNR = \frac{P_r}{N_0 \cdot B} = \frac{E_S}{N_0 \cdot B \cdot T_S} = \frac{E_b}{N_0 \cdot B \cdot T_b} \tag{14}$$

where E_s and T_s denote the energy (in W) per symbol and the symbol transmission period, respectively, while E_b and T_b are the energy per bit (in W) and the bit transmission period.

The terms B, T_S and T_b are pre-defined values by the BT standard. Thus, the SNR can be directly referred to the SNR ratios per symbol (γ_s) and bit (γ_b), as it follows [48]:

$$\gamma_s = \frac{E_s}{N_0} \tag{15}$$

$$\gamma_b = \frac{E_b}{N_0} \tag{16}$$

The relationship between these two variables can be calculated as:

$$\gamma_b = \frac{\gamma_s}{log_2 M} \tag{17}$$

where M represents the number of symbols of the modulation constellation.

On the other hand, to compute the BER or bit error probability, P_b, as a function of γ_b, we have to analyse the particularities of every modulation employed by BT 2.x + EDR. For the case of multi-level symbol constellations, P_b can be obtained from the probability of symbol error (P_s) and the number (M) of existing constellation symbols. Assuming that bit errors are uncorrelated, for low values of P_s (e.g. below 0.1) P_b can be reasonably approximated [48] as:

$$P_s = 1 - (1 - P_b)^{log_2 M} \underset{P_b \to 0}{\approx} (log_2 M) P_b \to P_b \approx \frac{P_s}{log_2 M} \tag{18}$$

The relationship between P_b and γ_b for the three modulations employed by BT 2.x is obtained in the next three subsections:

GFSK modulation

GFSK scheme can be assimilated to a FSK modulation. To determine the bit error probability, we have to take into account that BT was conceived as a low-cost technology. At the expense of a lower performance, non-coherent detection is preferred to simplify the hardware architecture of the demodulator. When this type of detection is utilized, the bit error probability can be computed [49] as:

$$P_b = \frac{1}{2} e^{\left(-\frac{\gamma_b}{2}\right)} \tag{19}$$

The previous expression describes a lower bound for P_b which assumes an orthogonal modulation scheme with a modulation index h of 0.5. However, BT specification imposes the index h to have a value in the interval [0.28–0.35], lower than 0.5. Under this non-orthogonal modulation, P_b can be estimated [50] from:

$$P_b = Q_1(\alpha, \beta) - \frac{1}{2} e^{\left(-\frac{\alpha^2 + \beta^2}{2}\right)} I_0(\alpha, \beta) \tag{20}$$

where $Q_1(\alpha, \beta)$ represents the Marcum Q-function, which is defined as follows:

$$\begin{aligned} Q_1(\alpha, \beta) &= \int_{\beta}^{\infty} x e^{\left(-\frac{x^2 + \alpha^2}{2}\right)} I_0(\alpha x) dx \\ &= e^{\left(-\frac{\alpha^2 + \beta^2}{2}\right)} \cdot \sum_{n=0}^{\infty} \left(\frac{\alpha}{\beta}\right)^n I_n(\alpha, \beta); \\ (\beta > \alpha > 0) \end{aligned} \tag{21}$$

where $I_n(x)$ expresses the n-order modified Bessel function while α and β are two parameters that can be calculated from γ_b [50]:

$$\alpha = \sqrt{\frac{\gamma_b}{2}\left(1 - \sqrt{1 - \rho^2}\right)} \tag{22}$$

$$\beta = \sqrt{\frac{\gamma_b}{2}\left(1 + \sqrt{1 - \rho^2}\right)} \tag{23}$$

In the previous expressions the term ρ is the so-called correlation coefficient, which can be calculated from the modulation index h:

$$\rho = \frac{\sin(2\pi h)}{2\pi h} \tag{24}$$

Therefore, to calculate the bit error probability, P_b, for transmissions with the basic rate of BT, we just have to consider that the modulation index h is in the range [0.28–0.35]. In any case, this computation neglects the effect of the Inter-Symbol Interference (ISI) which is introduced by the Gaussian filter implemented by GFSK.

ϖ/4-DQPSK *modulation*

As it refers to the performance of the radio receiver, π/4-DQPSK (Differential Quaternary PSK) modulation has an identical behaviour to DQPSK (4-DPSK) for an ideal lineal AWGN channel [51]. Thus, for π/4-DQPSK, the symbol error probability, P_s, can be approximated [51] as:

$$P_s = \frac{\sin\frac{\pi}{M}}{2\pi} \int_{-\pi/2}^{\pi/2} \frac{exp(\gamma_s \cdot (1 - \cos\frac{\pi}{M} \cdot \cos\theta))}{(1 - \cos\frac{\pi}{M} \cdot \cos\theta)} d\theta \tag{25}$$

The π/4-DQPSK constellation defines 4 different symbols (M = 4), so that 2 bits are codified per symbol. On the other hand, once the value of M is known, P_b and γ_b can be easily derived fromP_s and γ_s by applying Eqs. (18) and (17), respectively. So we have that the expression for the calculus of P_b under π/4-DQPSK is:

$$P_b = \frac{1}{log_2 M} \cdot \frac{\sin\frac{\pi}{M}}{2\pi} \int_{-\pi/2}^{\pi/2} \frac{exp((\gamma_b \cdot log_2 M) \cdot (1 - \cos\frac{\pi}{M} \cdot \cos\theta))}{(1 - \cos\frac{\pi}{M} \cdot \cos\theta)} d\theta \tag{26}$$

8-DPSK modulation

In this case, the Eq. (26) also permits to calculate the bit error probability, P_b, if M is set to 8 (as 8-DPSK codifies $log_2 8 = 3$ bits per symbol). However, as it is remarked in [50], the exact computation of this formula for P_b under M-DPSK becomes very complex for values of M higher than 2. Authors in [51] provide an upper bound for the value of P_s which can be computed as:

$$P_s \leq 2.06 \cdot \sqrt{\frac{1+\cos\frac{\pi}{M}}{2\cdot\cos\frac{\pi}{M}}} \cdot Q\left(\sqrt{\gamma_s\left(1-\cos\frac{\pi}{M}\right)}\right) \quad (27)$$

From this expression and Eq. (18), P_b can be easily derived.

Computation of the mean packet delay

After determining the bit error probability (P_b) as a function of the SNR ratio per bit (γ_b), the model developed in Sect. 3 has to be extended to compute the packet delay in scenarios where bit errors and, consequently, packet retransmissions can occur. For this purpose, the components of the delay (now defined as the mean expected values $\overline{t_{ACK_i}(x)}$ and $\overline{t_{TX_i}(x)}$) are recalculated depending on both the employed modulation and the PER, that is to say, the probability that a packet has to be retransmitted due to uncorrectable bit errors. So, the first step to extend the analytical model of Sect. 3 is to determine the relationship between this PER and the bit error probability (P_b). As previously commented, DM packet type includes specific mechanisms to correct bit errors in the payload and reduce packet transmissions. Conversely, the payload of DH packets is unprotected. On the other hand, the packet header presents the same protection with independence of the packet type (the packet header is FEC encoded for both types). As a result, the calculus of this relationship between PER and P_b has to be particularized for each packet type and each packet field (payload and header).

Retransmission Probability Due To Errors in the Payload of the DH Packets

As long as DH packet type does not include any correction mechanism for the payload, the probability that a DH packet is correctly received is the probability that no bit error occurs during the transmission of the packet.

Assuming that bit errors are uncorrelated, the probability of transmitting i correct bits can be characterized as a binomial distribution in the following way:

$$P(X=i) = \binom{n}{i} \cdot (1-P_b)^i \cdot P_b^{n-i} \quad (28)$$

where the random variable X represents the number of correct bits in a received packet of n bits while P_b is the bit error probability.

Thus, if a DH packet with a payload of x bytes contains $n_d(x)$ bits, the probability $P_{DHd}(x)$ of receiving this packet without any bit error is:

$$P_{DH_d}(x) = P(X = n_d(x)) = (1-P_b)^{n_d(x)} \quad (29)$$

This equation is valid for both the basic transmission mode and the EDRs.

Retransmission Probability Due To Errors in the Payload of the DM Packets

The payload of DM packets is encoded with a 2/3 FEC protection. In particular, a shortened Hamming (15, 10) code is adopted. This code, which utilizes 15 bits to encode 10 data bits, can correct one bit in every group of 15 bits. Assuming the binomial distribution used for DH packets, the probability of not suffering an unrecoverable error within a group (P_g) can be calculated from the probability of receiving at least 14 correct bits as it follows:

$$P_g = \sum_{i=14}^{15} \binom{15}{i} \cdot (1-P_b)^i \cdot P_b^{(15-i)} \quad (30)$$

The retransmission of a DM packet is required when two or more bit errors occur in a group. So, the success probability (i.e. the probability $P_{DMd}(x)$ that a DM packet is properly received) can be derived from P_g by taking into account the number of $n_g(x)$ 15-bit groups in the packet:

$$P_{DM_d}(x) = P(X = n_g(x)) = (P_g)^{n_g(x)} \quad (31)$$

Both DM and DH packets incorporate a Cyclical Redundancy Code of 16 bits (CRC-16), which enables the detection of any error altering an odd number of bits [52]. Therefore, this model presumes that the probability of not detecting an error at the Baseband layer is negligible. So, all packets with unrecoverable errors are assumed to be retransmitted.

Retransmission Probability Due To Errors in the Header of the Packets

The header of BT Baseband packets is protected with FEC 1/3. For every bit in the 18-bit header, two redundancy bits are added, so that 18 groups of 3 bits (54 bits) are transmitted in the header of every packet. Under FEC 1/3 protection, a single bit error in any 3-bit group can be corrected. Thus, the probability (P_H) of not suffering an unrecoverable error in the header can be computed as the probability of having one or no error per group:

$$P_H = \left(3\cdot(1-P_b)^2 \cdot P_b + (1-P_b)^3\right)^{18} \quad (32)$$

Global Retransmission Probability

Assuming again than errors in the payload and in the header are uncorrelated, the global probability of a packet not being retransmitted can be directly determined by multiplying P_H by$P_{DMd}(x)$ for DM packets (or by $P_{DHd}(x)$ for DH packets). Then the packet retransmission (or packet loss) probability P_l is computable as:

$$P_l(x) = \begin{cases} 1 - P_{DH_d}(x)\cdot P_H & \text{DH packets} \\ 1 - P_{DM_d}(x)\cdot P_H & \text{DM packets} \end{cases} \quad (33)$$

Once that the packet retransmission probability is known, the next step is to calculate the mean number of times that packets are transmitted. The hardware of the BT host controller can be configured to retransmit a packet up to a certain number of times which is set by a specific timer (flushTO). In most actual implementations, the default value of this timer does not set any limit. As a consequence, the mean number of transmission attempts $(\overline{N_{Tx}})$ of a packet can be directly derived from the retransmission probability:

$$\overline{N_{Tx}} = (1 - P_l) \cdot \sum_{i=1}^{\infty} i \cdot P_l^{(i-1)} = \frac{1}{1 - P_l} \tag{34}$$

Taking into account this new variable, we can now calculate the mean time to transmit and confirm x-byte user data when retransmissions take place. As in Sect. 3, the times needed by non-final and final RFCOMM frames are separately estimated.

The mean time $(\overline{t_{ACK_l}}(x))$ to transmit and confirm a non-final RFCOMM frame not requiring more than one 5-slot packet can be computed as:

$$\begin{aligned}\overline{t_{ACK_l}}(x) &= \frac{P_l}{1 - P_l} \cdot t_{RTx}(x) + t_{ACK}(x) \\ &= (\overline{N_{Tx}} - 1) \cdot t_{RTx}(x) + t_{ACK}(x)\end{aligned} \tag{35}$$

where $t_{ACK}(x)$ is again the time required to transmit and confirm the last (and successful) retransmitted packet, while $t_{RTX}(x)$ represents the mean time required to retransmit a packet.

Remark that, in the previous formula, $(\overline{N_{Tx}} - 1) \cdot t_{RTx}(x)$ represents the extra time caused by unsuccessful attempts of transmitting a packet. This delay component $t_{RTx}(x)$ due to the retransmission of a packet must be in the range:

$$t_{RTx_{min}}(x) \leq t_{RTx}(x) \leq t_{RTx_{max}}(x) \tag{36}$$

The inferior limit of this range $[t_{RTXmin}(x)]$ can be computed as:

$$t_{RTx_{min}}(x) = \begin{cases} 0 & x = 0 \\ 2 \cdot T_S & 0 < x \leq L_1 \\ 4 \cdot T_S & L_1 < x \leq L_3 \\ 6 \cdot T_S & L_3 < x \leq L_5 \end{cases} \tag{37}$$

where L_1, L_2 and L_5 are the maximum sizes of a BT packet of 1, 3 or 5 slots respectively. Table 1defines the values of these sizes depending on the packet type (DM or DH) and the employed data rate (basic rate for DM packets or up to 3 Mbps for DH packets). In particular, these values are 17, 121 and 224 bytes if DM-type packets are being employed. In opposition, when DH-types (and EDR) are considered, the values for L_1, L_2 and L_5 are determined by the maximum sizes of 3-DH1, 3-DH3 and 3-DH5 packets (83, 552 and 1,021 bytes, respectively).

Note that Eq. (37) presumes the best case in which all packets are retransmitted immediately after a bit error is detected and a negative acknowledgment (NAK) is received at the transmitter to indicate the error. Consequently this delay just computes the slots (1, 3 or 5) to transmit the erroneous packet plus the corresponding slot for the NAK packet. On the other hand, the transmission of two consecutive PDU can be separated up to T_{poll} seconds (n_{poll}slots). So, the maximum delay $t_{RTXmax}(x)$ provoked by a retransmission is:

$$t_{RTx_{max}}(x) = T_{poll} \tag{38}$$

The value of $t_{RTx}(x)$ within this range depends on the hardware implementation of the BT host controller and on the number of slaves that form the BT piconet. In the rest of this analysis of this article we assume the worst case in which every retransmission introduces a delay of T_{poll}.

When the data in an intermediate frame exceed the maximum capacity of the payload of a single BT packet, $\overline{t_{ACK}}(x)$ can be again recursively calculated as:

$$\overline{t_{ACK_l}}(x) = \begin{cases} 0 & x = 0 \\ (\overline{N_{Tx}} - 1) \cdot t_{RTx}(x) + t_{ACK}(x) & 0 < x \leq L_5 \\ ((\overline{N_{Tx}} - 1) \cdot t_{RTx}(L_5) + t_{ACK}(L_5)) \cdot \left\lfloor \frac{x}{L_5} \right\rfloor + \overline{t_{ACK_l}}(x \bmod L_5) & x > L_5 \end{cases} \tag{39}$$

where L_5 is the maximum size of a 5-slot BT packet.

The time required by the final RFCOMM frame $(\overline{t_{TX_l}}(x))$ can be obtained extending the previous equations for $\overline{t_{ACK_l}}(x)$. Thus, this delay $\overline{t_{TX_l}}(x)$ can be recursively computed as:

$$\overline{t_{TX_l}}(x) = \begin{cases} 0 & x = 0 \\ (\overline{N_{Tx}} - 1) \cdot t_{RTx}(x) + t_{TX}(x) & 0 < x \leq L_5 \\ \overline{t_{ACK_l}}(L_5) \cdot \left\lfloor \frac{x}{L_5} \right\rfloor + \overline{t_{TX_l}}(x) \cdot (x \bmod L_5) & x > L_5 \end{cases} \tag{40}$$

As in the case without retransmissions (see Eq. 7), the previous expression also considers the time to transmit and confirm the intermediate packets of the final frame by adding the term $\overline{t_{ACK_l}}(x)$ (being $\left\lfloor \frac{x}{L_5} \right\rfloor$ the number of intermediate packets).

After determining $\overline{t_{ACK_l}}(x)$ and $\overline{t_{TX_l}}(x)$, the global mean packet time required to transmit N_U user data bytes when retransmissions (losses) occur $(\overline{t_{R_{M_l}}}(N_U))$ is obtained by substituting in Eq. (1) $t_{ACK}(x)$ and $t_{TX}(x)$ by these two new terms. In the sense master–slave, this mean delay is:

$$\begin{aligned}\overline{t_{R_{M_l}}}(N) = \frac{T_{poll}}{2} &+ n_{nff}(N_U) \cdot \overline{t_{ACK_l}}(L_R + O_R(L_R) + H_L) \\ &+ \overline{t_{TX_l}}(L_{ff}(N_U) + O_R(L_{ff}(N_U)) + H_L)\end{aligned} \tag{41}$$

In the sense slave–master, the mean delay is practically the same. We only have to add the slot (T_s) in which the slave is addressed by the master. So, this mean delay $\overline{t_{R_{S_l}}}(N)$ is:

$$\overline{t_{R_{S_l}}}(N_U) = T_S + \overline{t_{R_{M_l}}}(N_U) \tag{42}$$

DISCUSSION OF THE PERFORMANCE OF BLUETOOTH 2.X + EDR

This section discusses the performance of the different BT modulations in the presence of a certain SNR, that is to say, in the non-ideal case where an existing noise can induce packet losses and, consequently, packet retransmissions. The performance is evaluated in terms of the mean transmission delay, which is calculated by using the previous analytical model implemented through Python scripts [53]. MATLAB computing and simulation environment [54] was also employed. In our study, we consider two scenarios depending on the size of the user data that have to be transmitted. To isolate the effect of the retransmissions, the two considered sizes were 70 and 200 bytes. The limit of 200 bytes is set taking into consideration the lowest maximum capacity of the different BT packet types. The maximum size of the data payload of a single 5-slot DM-type packet is 224 bytes while 5-slot packets of the DH type (1-DH5, 2-DH5 and 5-DH5), which do not include any FEC protection field for the data, can transport more user data in a single packet (see Table 1).

Remark that the comparison would not be adequate for user data that, after adding the headers of L2CAP and RFCOMM layers, require a payload with a size higher than this value of 224 bytes. In that case the transmissions with DM packets would demand to divide the data into several BT packets. This fragmentation implies an intrinsic extra delay which is not directly related to the impact of the noise on the delay experienced by the different modulations employed by BT.

For both scenarios (with N = 70 and 200 bytes), Table 2 presents the mean computed delay ($\overline{t_{TX_i}(N)}$) obtained for the ideal case without noise ($\gamma_b = \infty$) and for three different values of the SNR per bit ($\gamma_b = 10.5$ dB, $\gamma_b = 12$ dB and $\gamma_b = 15$ dB). The delay is calculated by using the formula in Eq. (41), taking into consideration the overhead introduced by the headers of RFCOMM and L2CAP. Table 2 shows the results for both the basic bit rate of 1 Mbps (which can use DM or DH-type packets) and when the EDRs (which only employ DH-type packets) are utilised. In this second case, the selection of the rate and the packet rate follows the algorithm described in Sect. 3. Accordingly, with the EDR modulations, a 3-DH1 packet is employed to transmit 70 bytes while a 2-DH3 packet is considered to transport 200 user data bytes. In the case of DM packets (always transmitted at the basic rate of 1 Mbps), 70 user data bytes need one DM3 packet (a 3 slot packet) while 200 bytes only fit in a DM5 packet (comprising five slots). If DH type is used at the basic rate, a DH3 and a DH5 packets are required to transport 70 and 200 bytes, respectively.

Table 2: Comparison of the PER, BER and mean transmission delay obtained for the existing packet types and different values of the SNR per bit and diverse packet

γ_b(dB)	*N* (user data bytes)	BER	Packet type	PER	$\overline{t_{TX_i}(N)}$ (ms)
∞	70	0	DM3	0	1.116
∞	70	0	DH3	0	0.782
∞	70	0	3-DH1	0	0.360
∞	200	0	DM5	0	2.691
∞	200	0	DH5	0	1.830
∞	200	0	2-DH3	0	0.996
15	70	5.426×10^{-6}	DM3	2.056×10^{-7}	1.116
15	70	5.426×10^{-6}	DH3	3.533×10^{-3}	0.808
15	70	9.632×10^{-3}	3-DH1	9.983×10^{-1}	3,570.000
15	200	5.426×10^{-6}	DM5	5.301×10^{-7}	2.691
15	200	5.426×10^{-6}	DH5	9.203×10^{-3}	1.911
15	200	9.264×10^{-6}	2-DH3	1.566×10^{-2}	1.115
12	70	1.106×10^{-3}	DM3	8.432×10^{-3}	1.180
12	70	1.106×10^{-3}	DH3	5.163×10^{-1}	8.786
12	70	4.117×10^{-2}	3-DH1	≈1.000	∞
12	200	1.106×10^{-3}	DM5	2.160×10^{-2}	2.884
12	200	1.106×10^{-3}	DH5	8.484×10^{-1}	50.790
12	200	1.282×10^{-3}	2-DH3	8.877×10^{-1}	60.300
10.5	70	5.648×10^{-3}	DM3	1.915×10^{-1}	2.892
10.5	70	5.648×10^{-3}	DH3	9.757×10^{-1}	301.910
10.5	70	6.554×10^{-2}	3-DH1	≈1.000	∞
10.5	200	5.648×10^{-3}	DM5	4.219×10^{-1}	9.076
10.5	200	5.648×10^{-3}	DH5	$\approx 9.999 \times 10^{-1}$	136,200
10.5	200	5.767×10^{-3}	2-DH3	$\approx 9.999 \times 10^{-1}$	143,100

For every case, Table 2 also includes the estimated BER and PER. In the table, the obtained transmission delay is marked as 'infinite' to indicate an extremely high and unrealistic value that would provoke the connection loss in an actual BT communication.

In the ideal case ($\gamma_b = \infty$), results obviously demonstrate the advantages of employing the EDRs, which reduce the delay by more than 50 %. However, the table shows that under the 8-DPSK modulation (utilized by 3-DH1) the packets are very sensitive to the presence of noise. Even for the highest considered SNR, the mean transmission delay is about several seconds (which is unaffordable for most practical applications of BT). As the values of the SNR decrease, the performance of EDR and even DH packets (at the basic rates) severely degrades while DM-type packets still experience delays under 10 ms. For a SNR of 10.5 dB and 200 user data bytes, for example, only DM type packets guarantee the feasibility of the BT connection, as far as the use of DH-type or EDRs induce an intolerable delay.

These conclusions are corroborated in Figs. 2 and 3, which represent again the mean delay $\overline{T_{TX_i}(M(N))}$ required to transmit 70 and 200 user data bytes, respectively. The delay is calculated as a function of the SNR per bit (γ_b). In order to compare the performance of the two EDR modulations individually, the curves are now computed imposing a fixed packet type. Commercial implementations of the BT Baseband tend to minimize the number of BT packets that are needed to transmit the data flowing from the higher layers. Consequently, for every packet type and data rate, BT chooses (if possible) the shortest packet (i.e. that requiring less slots) that can transport the data without fragmenting them. For the DH, 2-DH and 3-DH packets types, 70 bytes will be transported in one DH3, one 2-DH3 and one 3-DH1 packets, respectively. Similarly, 200 user data bytes require one DH5, one 2-DH3 or one 3-DH3 packet, depending on the employed rate (1, 2 or 3 Mbps).

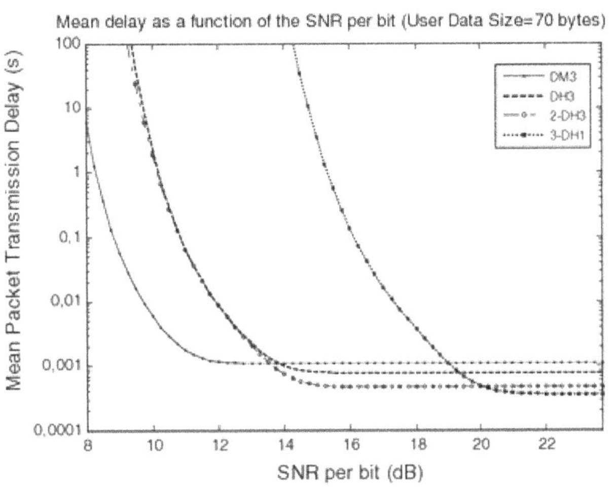

Figure 2: Mean transmission delay $\overline{T_{TX_i}(N)}$ as a function of the SNR per bit (γ_b) for the different packet types (DM and DH) and data rates (1, 2 or 3 Mbps). User data size: 70 bytes

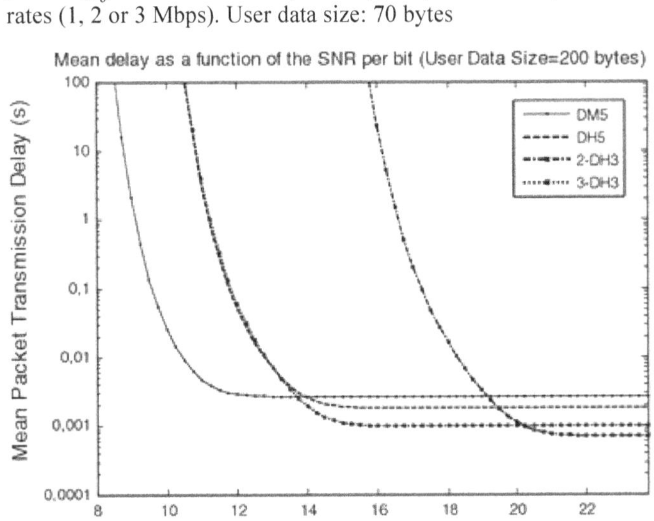

Figure 3: Mean transmission delay $\overline{T_{TX_i}(N)}$ as a function of the SNR per bit (γ_b) for the different packet types (DM and DH) and data rates (1, 2 or 3 Mbps). User data size: 200 bytes

From the graphs, we can clearly observe that the election of the most adequate packet type and data rate must be based on the existing noise conditions. For noisy environments with a low value of the SNR, the use of DH-type packets and certain high rate modulations can lead to high packet delays which will most probably cause the connection loss. In fact, the figures show the existence of different intervals of the SNR for which a specific type of packet and modulation is most suitable. In particular, we can distinguish the following ranges:

$\gamma_b \leq 8.5$ dB: Under these circumstances, independently of the employed modulation, the delay soars and the retransmissions renders the connection unfeasible. Thus, the BT link cannot be maintained for all cases.

8.5 dB $< \gamma_b \leq$ 10.0 dB: Only FEC protected DM-type packets can be utilized to guarantee the viability of the link. The use of DH-packets induces intolerable transmission delays. Consequently, enhanced rates are not possible for a SNR per bit lower than 10 dB.

10.0 dB $< \gamma_b \leq$ 13.8 dB: The use of DH and 2-DH packets is now viable while this is not the case for 3-DH type. From a SNR of 13 dB the delay with DH-2 improves the results obtained with DH packets. Nevertheless, for this range of the SNR, the delay of the transmissions with the DM-packet type (always at the basic rate of 1 Mbps) reaches its minimum so that DM-type still outperforms the behaviour of DH packets at any rate.

13.8 dB $< \gamma_b \leq$ 19.0 dB: The lowest delay is accomplished with 2-DH packets. DH-type packets at 1 Mbps also enable a better performance than the DM-type. 3-DH packets only become an option for values of SNR higher than 16 dB.

$\gamma_b >$ 19.0 dB: From this point, we can assume that the optimal performance is achieved with the highest data rate (3-DH).

EMPIRICAL EVALUATION OF THE MODEL FOR THE BASIC RATE

The main problem of evaluating the proposed model with real BT interfaces is that user cannot select the EDR that an actual BT controller employs. When the DH mode (that for which enhanced rates are available) is set, commercial BT hardware is programmed to change the modulation (i.e. the rate) as a function of the detected losses. However, from the tests that we have conducted in our laboratory (with interfaces from different manufacturers), we have concluded that present implementations of BT hosts follow a highly conservative policy to select the modulation. Thus, as soon as any loss is detected, the basic rate of 1 Mbps is set. As a consequence, it is not possible to evaluate the accuracy of the model for the EDRs in the presence of noise in a real scenario.

In any case, we have checked the validity of the model when the DM mode (and consequently the basic data rate) is utilized. For this purpose, we employed the testbed sketched in Fig. 4 and pictured in Fig. 5. To emulate a Gaussian channel in a fully controlled environment, we deployed an evaluation scenario where the transmissions between two BT units are exclusively regulated by two externally adjustable parameters: an attenuation factor and an additive Gaussian white noise. Aiming at this goal, the testbed utilizes a signal attenuator and a commercial noise generator between the two BT devices (which perform as the BT master and the BT slave). To isolate the testbed from any external interference, the communications are carried out through coaxial cables with SMA connectors (the attenuation of the cables and the connectors is lower than 0.5 dB so it can be neglected). BT interfaces were inserted in metal covered boxes for a better isolation.

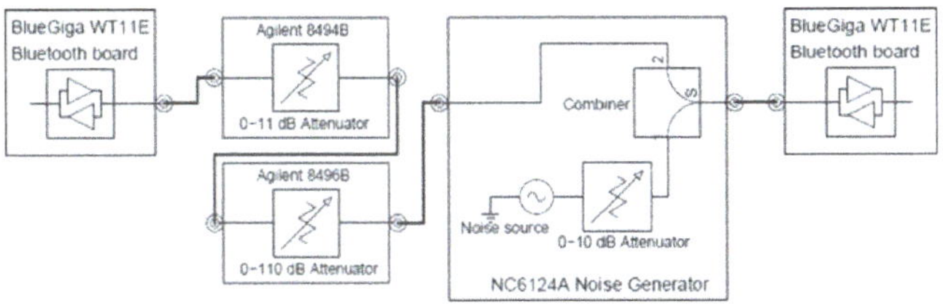

Figure 4: Connection diagram of the employed testbed

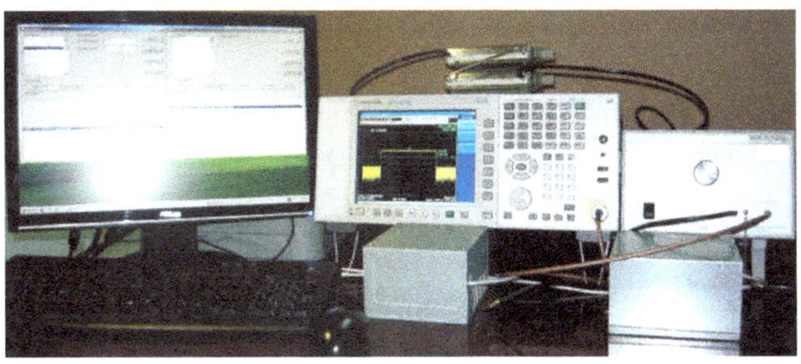

Figure 5: Picture of the global testbed employed to evaluate the accuracy of the model

As it refers to the BT interfaces, the experimental testbed includes two BT development boards from Bluegiga manufacturer [55]. These interfaces, which incorporate a serial peripheral interface (SPI) programming interface, integrate a WT11-E BT module with the Bluecore-4 chipset from CSR (the most popular vendor of BT technology).

To estimate the actual SNR (through the actual power densities of the noise and the signal at the BT receptor), we utilized an N9010A spectrum analyser from Agilent Technologies.

Figures 6 and 7 compare the mean packet delay computed by the model and the empirical results measured in the testbed for two different conditions (i.e. when two different values are set for the power of the additive Gaussian noise). The graphs show the evolution of the delay as a function of the size of the data payload. Both BT modules were connected to the same equipment (a personal computer) to avoid any synchronization problem in the measurements. The connections were accomplished by C-programmed scripts employing the Blue Z protocol stack. The scripts performed a series of periodical and systematic transmissions of user data from the master to the slave (similar results were obtained in the opposite sense). The tests were iterated by modifying the payload data size from 10 to 1,500 bytes. The delay for each data packet was estimated as the time from the start of the transmission in the master to the reception of the last bit in the slave. The delay provoked by the operating system and the USB connections with the modules were measured and detracted from the show results.

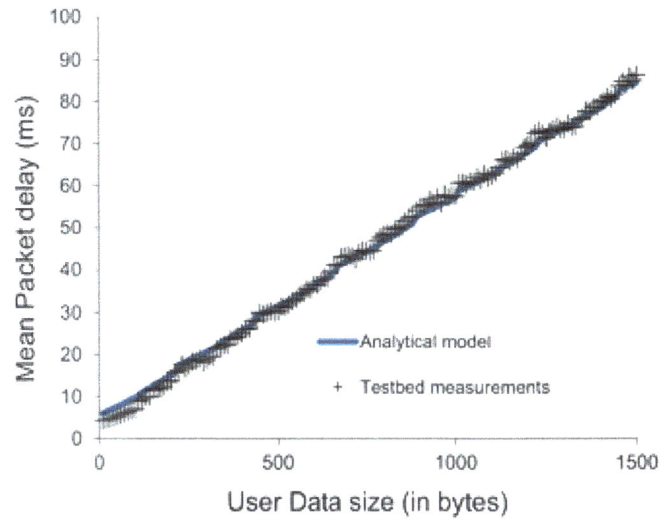

Figure 6: Comparison of the mean packet delay obtained in the real testbed and the results derived from the analytical model for a signal-to-noise ratio (SNR) of 10.95 dB (polling time, T_{poll}, of 10 ms)

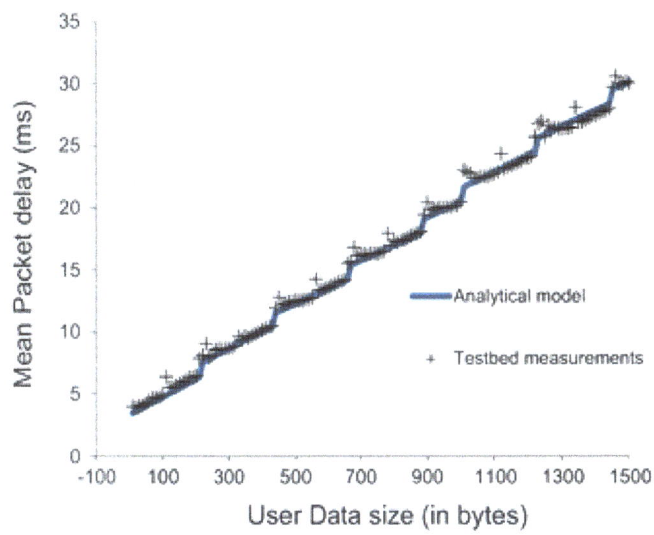

Figure 7: Comparison of the mean packet delay obtained in the real testbed and the results derived from the analytical model for a signal-to-noise ratio (SNR) of 14.95 dB (polling time, T_{poll}, of 10 ms)

Figures 6 and 7 confirm the accuracy of the model to compute the actual delay of a BT connection.

On the other hand, noisy scenarios with very higher delays are also difficult to implement because existing timeouts defined in the protocol stack force the interface to assume that the connection is lost.

We think that the conclusions of this evaluation (showing the accuracy of the model for the basic rate) could be extrapolated to the same tests with BT modules if an enhanced rate could be fixed by the user.

CONCLUSIONS

This paper has presented an analytical model that allows calculating the transmission delay of BT 2.x + EDR connections in the presence of noise. The model has been particularized for the three modulations (the basic speed of 1 Mbps and the two EDRs) which are provided by the specification. By utilizing the model, we have investigated the feasibility of the different modulations for different levels of noise (expressed in terms of the SNR per bit). As higher rates are associated to less robust signals, the performed study identifies the modulation and packet-type that offer the lowest delay for a certain SNR. This study could be employed to optimize the implementation of the BT host controller, which could select the most adequate packet type and data rate basing on the existing noise conditions.

ACKNOWLEDGMENTS

This work has been supported by European FEDER funds and the Spanish Ministry of Science and Innovation (Grants TEC2009-13763-C02-01 and TEC2013-42711-R).

REFERENCES

1. Bluetooth SIG. (2004, November 10). Bluetooth core specification version 2.0 + EDR. In Specification of the Bluetooth system. Bluetooth Special Interest Group. https://www.bluetooth.org/Technical/Specifications/adopted.htm. Accessed April 25, 2014.
2. Bluetooth SIG. (2007, 26 July). Bluetooth core specification version 2.1 + EDR. In Specification of the Bluetooth system. Bluetooth Special Interest Group. https://www.bluetooth.org/Technical/Specifications/adopted.htm. Accessed April 25, 2014.
3. Shimizu, M., Aoki, N., Shirakawa, N., Tozawa, Y., Okubo, N., & Daido, Y. (1997). New method of analyzing BER performance of GFSK with postdetection filtering. IEEE Transactions on Communications, 45(4), 429–436.
4. Chandra, A., Mandal S. C., & Bose, C. (2008). BER of π/4-DQPSK with multichannel reception: Some series solutions. In Proceedings of TENCON 2008—2008 IEEE Region 10 conference (pp. 1–6).
5. Al Falujah, I., & Prabhu, V. K. (2008). Error performance of DQPSK with EGC diversity reception over fading channels. IEEE Transactions on Wireless Communications, 7(4), 1190–1194.
6. Kiasaleh, K., & He, T. (1996). Performance of DQPSK communication systems impaired by mixer imbalance, timing error, and Rayleigh fading. In Proceedings of the IEEE international conference on communications (ICC 96) (Vol. 1, pp. 364–368).
7. Kiasaleh, K., & He, T. (1997). On the performance of DQPSK communication systems impaired by timing error, mixer imbalance, and frequency nonselective slow Rayleigh fading. IEEE Transactions on Vehicular Technology, 46(3), 642–652.
8. Miller, L. E., & Lee, J. S. (1998). BER expressions for differentially detected π/4-DQPSK modulation. IEEE Transactions on Communications, 46(1), 71–81.
9. Ng, C. S., Tjhung, T. T., Adachi, F., & Lye, K. M. (1993). On the error rates of differentially detected narrowband π/4-DQPSK in Rayleigh fading and Gaussian noise. IEEE Transactions on Vehicular Technology, 42(3), 259–265.
10. Sun, Y., Baricz, A., Zhao, M., Xu, X., & Zhou, S. (2009). Approximate average bit error probability for DQPSK over fading channels. Electronics Letters, 45(23), 1177–1179.
11. Wong, D. P. C., & Mathiopoulos, P. T. (1992). Nonredundant error correction analysis and evaluation of differentially detected π/4-shift DQPSK systems in a combined CCI and AWGN environment. IEEE Transactions on Vehicular Technology, 41(1), 35–48.
12. Chow, Y. C., Nix, A. R., & McGeehan, J. P. (1994). An error bound analysis for M-DPSK in frequency-selective Rayleigh fading channel with diversity reception. In Proceedings of theSingapore ICCS'94. Conference proceedings (Vol. 2, pp. 686–690).
13. Kostic, I. (2008). Average symbol error probability for M-ary DDPSK. Electronics Letters,44(20), 1203–1204.
14. Noneaker, D. L., & Pursley, M. B. (1994). Error probability bounds for M-PSK and M-DPSK and selective fading diversity channels. IEEE Transactions on Vehicular Technology,43(4), 997–1005.
15. Liu, H., Venkatesan, V., Nilsen, C., Kyker, R., & Magana, M. E. (2003). Performance of frequency hopped noncoherent GFSK in correlated Rayleigh fading channels. InProceedings of the IEEE international conference on communications (ICC'03) (Vol. 4, pp. 2779–2783).
16. Tibenderana, C., & Weiss, S. (2004). Low-complexity high-performance GFSK receiver with carrier frequency offset correction. In Proceedings of the IEEE international conference on acoustics, speech, and signal processing (ICASSP'04) (Vol. 4, pp. 933–936).
17. Chang, D. C., & Shiu, T. H. (2006). Digital GFSK carrier synchronization. In Proceedings of the APCCAS 2006, IEEE Asia Pacific conference on circuits and systems (pp. 1523–1526).
18. Chang, D. C. (2009). Least squares/maximum likelihood methods for the decision-aided GFSK receiver. IEEE Signal Processing Letters, 16(6), 517–520.
19. García-Armada, A., Rodríguez, B. B., Jiménez, V. P. G., & Sánchez-Fernández, M. (2007). Modelling, performance analysis and design of WPAN systems. Wireless Personal Communications, 42(3), 367–386.
20. Valenti, M. C., Robert, M., & Reed J. H. (2002). On the throughput of Bluetooth data transmissions. In Proceedings of the IEEE wireless communications and networking conference (WCNC2002) (Vol. 1, pp. 119–123).
21. Skudlarek, J.P. (2002). RadiSim—A fast digital RF behavioral simulator including bit error rate assessment for system exploration, validation, and tuning. In Proceedings of the 2002 IEEE international workshop on behavioral modeling and simulation (BMAS 2002) (pp. 61–66).
22. Yoon, J. H., Lee, S. B., & Park, S. C. (2004). Packet and modulation type selection scheme based on channel quality estimation for Bluetooth evolution systems. In Proceedings of theIEEE wireless communications and networking conference (WCNC 2004) (pp. 1014–1017).
23. Roh, J. S. (2006). Performance analysis and evaluation of Bluetooth networks in wireless channel environment. In Proceedings of the international conference on systems

and networks communications (ICSNC'06) (p. 61).

24. Hua Y., & Zou Y. (2008). Analysis of the packet transferring in L2CAP layer of Bluetooth v2.x + EDR. In Proceedings of the international conference on information and automation (ICIA 2008) (pp. 753–758).
25. Mikulka, J., & Hanus, S. (2008). Bluetooth EDR physical layer modeling. In Proceedings of the 18th international conference, Radioelektronika, 2008 (pp. 1–4).
26. Moon I. Y. (2007). Performance analysis of WAP in Bluetooth ad-hoc network system. In Knowledge-based intelligent information and engineering systems, Lecture notes in computer science (Vol. 4692/2010, pp. 390–396).
27. Zanella, A. (2009). A mathematical framework for the performance analysis of Bluetooth with enhanced data rate. IEEE Transactions on Communications, 57(8), 2463–2473.
28. Zanella. A. (2007). Analysis of the packet reception statistics of Bluetooth v2 + EDR in fading channels. Technical report 155, Department of Information Engineering, University of Padova, Italy.
29. Yu, B., Yang, L., & Chong, C. C. (2011). Optimized differential GFSK Demodulator. IEEE Transactions on Communications, 59(6), 1497–1501.
30. Han, D., & Zheng, Y. (2008). An ultra-low power GFSK demodulator for wireless body area network. In Proceedings of the 34th European solid-state circuits conference (ESSCIRC 2008) (pp. 434–437).
31. Han, U. D., & Zheng Y. (2009). A GFSK demodulator based on instant phase computation and adaptive multi-threshold quantization. In Proceedings of the IEEE Asian solid-state circuits conference (A-SSCC 2009) (pp. 249–252).
32. Ibrahim, N., Lampe, L., & Schober, R. (2007). Bluetooth receiver design based on Laurent's decomposition. IEEE Transactions on Vehicular Technology, 56(4), 1856–1862.
33. Kao, H. S., Yang, M. J., & Lee, T. C. (2007). A delay-line-based GFSK demodulator for low-IF receivers. In Proceedings of the 54th IEEE international solid-state circuits conference (ISSCC 2007) (pp. 588–589).
34. Lee, T. C., & Chen, C. C. (2006). A mixed-signal GFSK demodulator for Bluetooth. IEEE Transactions on Circuits and Systems II: Express Briefs, 53(3), 197–201.
35. Xia, B., Xin, C., Sheng, W., Valero-Lopez, A. Y., & Sánchez-Sinencio, E. (2003). A GFSK demodulator for low-IF Bluetooth receiver. IEEE Journal of Solid-State Circuits, 38(8), 1397–1400.
36. Xin, C., Xia, B., Sheng, W., Valero-Lopez, A. Y., & Sánchez-Sinencio, E. (2002). A mixed-mode IF GFSK demodulator for Bluetooth. In Proceedings of the IEEE international symposium on circuits and systems (ISCAS 2002) (Vol. 3, pp. 457–460).
37. Sheng, W., Xia, B., Emira, A. E., Xin, C., Valero-López, A. Y., Moon, S. T., & Sánchez-Sinencio, E. (2003). A 3-V, 0.35-μm CMOS Bluetooth receiver IC. IEEE Journal of Solid-State Circuits, 38(1), 30–42.
38. Hyun, S. B., Tak, G. Y., Kim, S. H., Kim, B. J., Ko, J., & Park, S. S. (2004). A dual-mode 2.4-Ghz CMOS transceiver for high-rate Bluetooth systems. ETRI Journal, 26(3), 229–240.
39. Gomez, C., Demirkol, I., & Paradells, J. (2011). Modeling the maximum throughput of Bluetooth low energy in an error-prone link. IEEE Communications Letters, 15(11), 1187–1189.
40. Luque, R., Morón, M. J., & Casilari, E. (2010). Minimum transmission delay in Bluetooth 2.0 + EDR. Electronics Letters, 46(13), 955–957.
41. CSR. Cambridge Silicon Radio Plc. BlueCore Bluetooth, chipset. http://www.csr.com.
42. ISSC (Integrated System Solutions Corp). http://www.issc.com.tw/.
43. Shin, S. Y., Park, H. S., Choi, S., & Kwon, W. H. (2007). Packet error rate analysis of ZigBee under WLAN and Bluetooth interferences. IEEE Transactions on Wireless Communications, 6, 2825–2830.
44. Shin, S. Y., Kang, J. S., & Park, H. S. (2009). Packet error rate analysis of ZigBee under interferences of multiple Bluetooth piconets. In Proceedings of the 69th IEEE vehicular technology conference (VTC Spring 2009) (pp. 1–5).
45. Yang, G., & Yu, Y. (2009). ZigBee networks performance under WLAN 802.11b/g interference. In Proceedings of the 4th international symposium on wireless pervasive computing (ISWPC 2009) (pp. 1–4).
46. Mikulka, J., & Hanus, S. (2008). Bluetooth and IEEE 802.11b/g coexistence simulation. Radio engineering, 17(3), 66–73.
47. Mangir, T., Sarakbi, L., & Younan H. (2011). Analyzing the impact of Wi-Fi interference on ZigBee networks based on real time experiments. International Journal of Distributed and Parallel Systems (IJDPS), 2(4), 1–10.
48. Goldsmith, A. (2005). Wireless communications. New York: Cambridge University Press.
49. Bruce Carlson, A. (1986). Communication systems: An introduction to signals and noise in electrical communication (3rd ed.). NewYork: McGraw-Hill Education.
50. Proakis, J. (2000). Digital communications (4th ed.). NewYork: McGraw-Hill Science.
51. Simon, M. K., & Alouini, M. S. (2005). Digital communication over fading channels (Wiley series in telecommunications and signal processing) (2nd ed.). Hoboken, NJ: Wiley-IEEE Press.
52. Moon, T. K. (2005). Error correction coding: Mathematical methods and algorithms. Hoboken, NJ: Wiley-Interscience.
53. Python programming language. https://www.python.org/.
54. MATLAB, the language of technical computing.http://www.mathworks.com/products/matlab/.
55. BlueGiga Tech. WT11 evaluation kit. http://www.bluegiga.com/. Accessed September 9, 2010.

CHAPTER 16

CONSTRAINED PRE-EQUALIZATION ACCOUNTING FOR MULTI-PATH FADING EMULATED USING LARGE RC NETWORKS: APPLICATIONS TO WIRELESS AND PHOTONICS COMMUNICATIONS

Rubens Junqueira Magalhães Afonso, Roberto Kawakami Harrop Galvão, Karl Heinz Kienitz, Hashem Alyami, Victor M. Becerra, and Sillas Hadjiloucas

Keywords: Equalization, RC networks, Percolation Mixed Integer Linear Programming, Communications, Photonics

INTRODUCTION

Multi-path propagation is a phenomenon that is often encountered when matter is probed by electrical acoustic or electromagnetic signals with wavelengths smaller than the physical dimensions of the features that need to be characterized (Lowery et al. 2012; Kaatze 2008). Interpretation of measurements where percolation of the excitatory wave through the material has taken place is never straightforward and requires adoption of solutions developed by the inverse problems community. It is also a typical problem encountered in communications where performance of wireless and mobile communication links is compromised by multi-path interference and frequency selective fading (Taub and Schilling 1986; Sari et al. 1995). The phenomenon is most prominent in indoor communication systems based on visible or IR optoelectronics based transceivers (Hashemi 1993) and limit proliferation of communications solutions at 60 GHz or THz links (Smulders 2002; Pieslewicz et al. 2005; Galvão et al. 2007a,b). In addition, it is encountered in optical fibre or waveguide based communications under multimode propagation and when photonic crystal fibres are used. It is generally accepted that irrespective of the propagation medium, most commonly adopted communication protocols such as orthogonal frequency-division multiplexing (OFDM) employing multilevel modulation schemes with non-constant amplitude (e.g. 16 QAM) generally require estimation and tracking of the fading channel parameters to perform coherent demodulation (Li et al. 1998; Edfors et al.1998; Yang et al. 2001). For particular modulation modalities, channel estimation is most commonly achieved by exploiting the correlation of the channel frequency response at different frequencies and times. Alternatively, channel estimation and multi-path fading may be based on a parametric modeling approach where the channel is modeled as a finite impulse response (FIR) filter. A channel estimator in its broader sense is often used to estimate the time delays, gains, and phases of the paths.

In the following sections, we first discuss the relevance of three-dimensional models of RCinput–output network responses to communications as this provides a novel way of modelling multi-path propagation. In the proposed approach, multi-path propagation is seen as analogous to a percolation process. We further explain that frequency selective attenuation can be parsimoniously described using system identification models which can be associated with the proposed network structure. This approach is based on the physical insight that has been recently developed by studying the underlying percolation processes discussed in the Physics literature (Stroud 1979; Soukoulis et al. 1987; Ho et al. 1990;

Citation: Rubens Junqueira Magalhães Afonso, Roberto Kawakami Harrop Galvão, Karl Heinz Kienitz, Hashem Alyami, Victor M. Becerra, Sillas Hadjiloucas, Constrained pre-equalization accounting for multi-path fading emulated using large RC networks: applications to wireless and photonics communications, DOI: 10.1007/s11082-016-0627-6

Rivas et al. 1999; Hu et al. 2004). Because of the universal relations underlining coupling of optical power between micro-resonators and dielectric waveguides (Yariv 2000), and the fact that interaction of multi-conductor transmission lines with electromagnetic radiation is usually accounted for through distributed current and voltage sources along the lines (with the current sources being related to the electric field and the voltage sources being proportional to the magnetic field), the use of percolation models for electrical networks (Desoer and Kuh 1969) should also be seen as complementing modal transmission line theory (Snyder and Love 1983; Hoefer 1985; Cangellaris 1988; Faria 1993; Castellanos et al. 1997; Rowe 1999; Akbari et al. 2000; Dabirian et al. 2005; Lin et al. 2002; Pregla 2008; Nitsch et al. 2009; Stutzman 2012; Wait 2013; Rambousky et al. 2013; Spadacini et al. 2014). As discussed in a recent review (Kafesaki and Soukoulis 2006), the propagation of EM waves after taking into consideration the transmission properties of the medium can be theoretically studied using the plane wave (PW) method, the transfer matrix method (TMM) or the finite difference time domain method (FDTD). Such simulations are normally complemented by using mode solvers such as CUDOS MOF Utilities (http://sydney.edu.au), RSoft (http://optics.synopsys.com) or COMSOL (http://www.comsol.com). The above background provides a justification for the adoption of the proposed networks to model multi-path propagation.

An advantage of the proposed approach is that it enables the development of equalization techniques that would be of relevance across the communications industry (radio, mm-wave, terahertz or photonic based), assuming coherent, incoherent or partially coherent emitter modalities and homodyne or heterodyne receiver modalities. Channel dynamics for multiple paths are written in a state space form which can be converted into an input-output realization where Laplace transforms can be used to describe frequency selective attenuation in a parsimonious manner. We then develop a pre-equalization procedure where the objective is to match the output waveform with the input function. In our case a Gaussian function is used to achieve this. The difficulty in the task lies on the fact that the proposed approach should account for constraints on instantaneous values of voltage, current and power that can be either delivered by the source or absorbed by the transmission medium. The problem is cast within a Mixed Integer Linear Programming (MILP) optimization framework that uses the developed nominal RC network model, with the excitation waveform customized to optimize signal fidelity from the transmitter to the receiver. Simulation results are then provided, showing the robustness of the proposed formulation. A general discussion of other applications in photonics and measurement science that would benefit from the proposed formulation is also provided.

CHANNEL MODELLING USING *RC* NETWORKS

In this paper, we model multi-path propagation through a transmission medium by using large three-dimensional (3D) RC networks. The rationale behind this type of model is based on the fact that for a simple RC network, with voltage input u(t) and current i(t) flowing through a resistor R and a capacitor C, the dynamic relation between the input voltage and the current can be described by a transfer function of the form:

$$G(s) = \frac{\Im[i(t)]}{\Im[u(t)]} = \frac{1}{R}\left[\frac{s}{s+(RC)^{-1}}\right] = \frac{1}{R}\left[1 - \frac{(RC)^{-1}}{s+(RC)^{-1}}\right] \tag{1}$$

where $\Im[\cdot]$ denotes the Laplace transform.

If the signals u(t) and i(t) are measured at sampling times kT, where T is a sampling period and k is the sample index, the relation between sequences u(kT) and i(kT) can be expressed by a transfer function of the form:

$$G_d(z) = \frac{Z[i(kT)]}{Z[u(kT)]} = \frac{1}{R}\left[1 - \frac{\beta}{z-\alpha}\right] \tag{2}$$

where Z[.] denotes the Z-transform. The values of α and β depend on the sampling period T, as well as the discretization method employed. If the input voltage is determined by a digital processor, then its value can be expected to remain constant over the sampling period associated to the operation of the digital-to-analog converter. In such a case, the zero-order-hold (ZOH) method can be employed (Dickinson 1991), which results in $\alpha = \exp[-T(RC)^{-1}]$. The value of β can be obtained by noting that the ZOH method preserves the DC gain of the transfer function, i.e. $G(s)|_{s=0} = G_d(z)|_{z=1}$. Since $G(s)|_{s=0} =$ =0 in this case, it follows that $\beta = 1 - \alpha = 1 - \exp[-T(RC)^{-1}]$. It can easily be seen that 0<α<1and 0<β<1. Finally, the transfer function Gd(z) can be expanded in a power series of z as:

$$G_d(z) = \frac{1}{R}\left[1 - \beta z^{-1} - \alpha\beta z^{-2} - \alpha^2\beta z^{-3} - \cdots\right] = \frac{1}{R}\left[1 - \beta\sum_{i=1}^{\infty}\alpha^{i-1}z^{-i}\right] \tag{3}$$

It is worth noting that the term z^{-i} in the Z-transform domain is associated to a delay of isampling periods in the time domain. Therefore, the dynamics of the network can be interpreted as a combination of a direct (line of sight) transmission path and multiple reflections with gains that decrease as the associated delay is increased (Reis and Galvão 2004). Such an interpretation is consistent with multipath propagation, as the attenuation tends to be stronger over longer paths.

A large 3D RC network such as the one shown in Fig. 1 is employed to represent a complex medium, comprising a multitude of propagation paths with different delays and attenuations. A typical approach to handle the distortions of the transmitted signal caused by multi-path propagation is to include a channel equalizer at the receiver side. In the present case, if u(t) and i(t) are regarded as the transmitted and received signals, respectively, a discrete-time equalizer could be implemented in the form of a finite impulse response (FIR) filter as:

$$i_{equalized}(kT) = \sum_{j=0}^{L} w_j i[(k-j)T] \tag{4}$$

with weights $w_0, w_1, \ldots w_L$ that should be adjusted according to the dynamics of the transmission medium. Such an adjustment can be carried out in an adaptive manner by using the well-known Least-Mean-Squares (LMS) algorithm e.g. (Widrow and Stearns 1985). The LMS adaptation rule can be expressed as:

$$\mathbf{w}(k+1) = \mathbf{w}(k) + \mu\,\mathbf{i}(kT)e(kT) \tag{5}$$

where $\mu > 0$ is an adaptation parameter, w(k) and i(kT) denote the vectors of weights and inputs of the FIR filter and at kth sampling time:

$$\mathbf{w}(k) = [w_0(k) \quad w_1(k) \quad \cdots \quad w_L(k)]^T \tag{6a}$$

$$\mathbf{i}(kT) = [i(kT) \quad i((k-1)T) \quad \cdots \quad i((k-L)T)]^T \tag{6b}$$

and e(kT) denotes the difference between the filter output and the desired response r(kT):

$$e(kT) = r(kT) - \sum_{j=0}^{L} w_j i[(k-j)T] \tag{7}$$

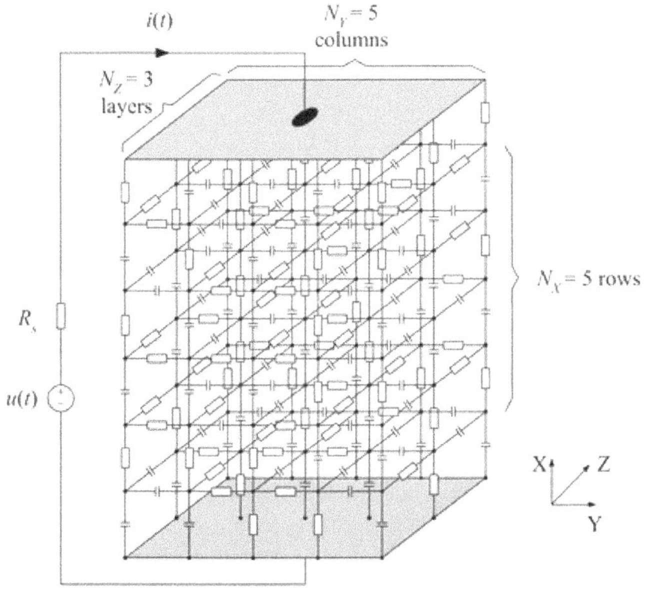

Figure 1: Schematic diagram of a 3D RC network with random allocation of R, C elements. Thegrey plates indicate a pair of electrodes employed to connect the network to a voltage source. The output resistance of the source is denoted by R_S

It is worth noting that this adaptation scheme requires the knowledge of the actual transmitted signal. In practice, the adaptation can be carried out by transmitting a known "training" signal. Alternatively, blind equalization methods may be employed (Johnson et al. 1998).

The present paper explores a different approach, where the shape of the transmitted signal is optimized by taking into account the dynamics of the medium and the capabilities of the transmitter source, in order to achieve a desired waveform at the receiver. The proposed approach can be regarded as a pre-equalization procedure, which dispenses with the need for the transmission of training signals. This has applications in RF wireless (Koenig et al. 2013) or light communications (Chow et al. 2012; Ghassemlooy et al. 2013) or in coherent optical receivers for optical fibres based communications (Savory 2010; Mori et al. 2013; Chen and Chng 2005; Liu et al. 2014). Furthermore, since such networks can be emulated in real-time using reconfigurable FPGAs, there are likely to be additional applications in FPGA based digital coherent optical OFDM signal processing which offer new opportunities for spectral efficiency, receiver sensitivity, and polarization-dispersion resilience (Yang et al. 2009). Furthermore, it is worth noting that the proposed multi-path models underlying the proposed method should be particularly useful in developing equalization for spatial mode based multiplexing schemes which are likely to provide additional capacity in future generation communication modalities (Richardson et al. 2013; van Uden et al. 2014; Mizuno et al. 2014; Huang et al. 2015; Shieh et al. 2012). Because space division multiplexing is also likely to be important in astronomical applications (Leon-Saval et al. 2010) as well as on-chip high throughput optical communications (Luo et al. 2014), the proposed modelling procedure accounting for multi-path propagation with the excitation waveform customized to optimize signal fidelity from the transmitter to the receiver should have a wide domain of technological applications.

OPTIMIZATION OF THE INPUT WAVEFORM

The problem addressed herein consists of determining an appropriate waveform for the input voltage u(t)u(t) in order to obtain a current waveform i(t)i(t) that is closest to a desired profile, respecting constraints on instantaneous values of voltage, current and power that can be either delivered by the source or absorbed by the transmission medium. More formally, we wish to minimize a 1-norm cost function of the form

$$J = \sum_{k=0}^{N} |r(kT) - i(kT)| \tag{8}$$

subject to:

$$|u(kT)| \leq u_{\max}, \; k = 0, 1, \ldots, N \tag{9a}$$

$$|i(kT)| \leq i_{\max}, \; k = 0, 1, \ldots, N \tag{9b}$$

$$|i(kT)\, u(kT)| \leq P_{\max}, \; k = 0, 1, \ldots, N \tag{9c}$$

where r(kT),k=0,1,...,N is the desired current profile, specified over N sampling times, and u_{max}, i_{max}, P_{max} are bounds on the voltage, current and power of the source. It is assumed that the RC network is initially at rest (uncharged capacitors) and that the relation between the sequences u(kT) and i(kT),k=0,1,... ,N is expressed by a model derived from the topology and R, C component values of the network. More specifically, we have adopted the modelling framework presented in (Galvão et al. 2013) to derive a linear, continuous-time state-space model, which was subsequently simplified by using an order-reduction procedure based on balanced realizations. A review of model order reduction techniques that would be suitable for the task is found in (Heydari and Pedram 2006). For use in the optimization problem, the reduced-order model was then discretized in the form:

$$x[(k+1)T] = Ax(kT) + Bu(kT) \tag{10a}$$

$$i(kT) = Cx(kT) + Du(kT) \tag{10b}$$

where $x(kT) \in \Re^n$ denotes the n-dimensional state vector at the kth sampling time, A, B, C are matrices of dimensions (n×n), (n×1), (1×n), respectively, and D is a scalar. For more details concerning the discretization of state-space models, the reader is referred to (Franklin et al.1997). The minimization of the 1-norm cost function in (8) subject to the dynamic relations expressed by the model Eq. (10a, b) and the bounds on the voltage (9a) and current (9b) values can be cast into the form of linear programming (Camacho and Bordons 1999). However, the presence of the power constraints (9c) result in a non-convex set of feasible (u(kT),i(kT)) values. In the present work, this issue is handled by using linear approximations that result in a mixed-integer linear programming (MILP) problem, as detailed below. In what follows, the indication of the sampling period T will be omitted in the equations for brevity.

CONVERSION OF THE OPTIMIZATION PROBLEM TO A MILP PROBLEM

As stated in the previous section, the inclusion of power constraints entails an optimization problem over a non-convex region of possible solutions. This issue poses a difficulty to obtain the optimal solution, as most optimization algorithms are designed to search over a convex set of feasible solutions. Therefore, it is necessary to restate the optimization problem to an equivalent form, which in turn can be solved with standard convex optimization techniques. Among the alternatives to recast the optimization problem is the conversion to a Mixed Integer Programming (MIP) problem involving both continuous and binary-valued variables.

A procedure that can be employed to convert the "or"-type constraints that result in a non-convex optimization problem to the "and"-type which entail convex optimization with the presence of binary variables, is the "big-M" method (Richards et al. 2002). A conversion of this type was presented in the formulation by Lima et al. 2010, which treated the case of power constraints in an electrical motor system by approximating the power constraints by straight line segments. In that work, the cost function was quadratic in terms of the optimization variables thus resulting in a Mixed Integer Quadratic Programming (MIQP) problem. Contrary to that work, in the optimization required in the current study, the cost function is linear in terms of the optimization variables, and should be casted as a Mixed Integer Linear Programming (MILP) problem. MILP problems are less intensive in terms of computational effort than MIQP ones, therefore, if one can obtain a linear cost whose optimization reflects the desired behavior of the system in a satisfactory manner, the computational effort to obtain the optimal solution may be reduced.

MIP problems present the drawback of exponential growth of the computational complexity with the increase of the number of binary variables. As a consequence, in obtaining a problem that can be solved in a reasonable amount of time as would be required in real-time pre-equalization applications, it is important to keep the number of binary decision variables to a minimum. In (Lima et al. 2010) the "or"-type constraints were written in terms of the bounds of the feasible region, therefore, one constraint corresponded to each half-plane of the feasible region. In contrast, in the present work the constraints are expressed in terms of convex polytopes whose union is the feasible region. In general, the number of such polytopes, Np, is less than the number of half-planes, allowing for a reduction in the number of "or"-type constraints. Figure 2 illustrates the benefits of using the polytopes for the constraints as opposed to the half-planes. As can be seen, there are 22 half-planes that need to be used with "or"-type constraints to represent the feasible region, in contrast to 21 convex polytopes. In this figure, only the power constraints are depicted, as they require an approximation by polytpes. The bounds on the voltage and current are not shown, given that their representation is a simple rectangle parallel to the axes.

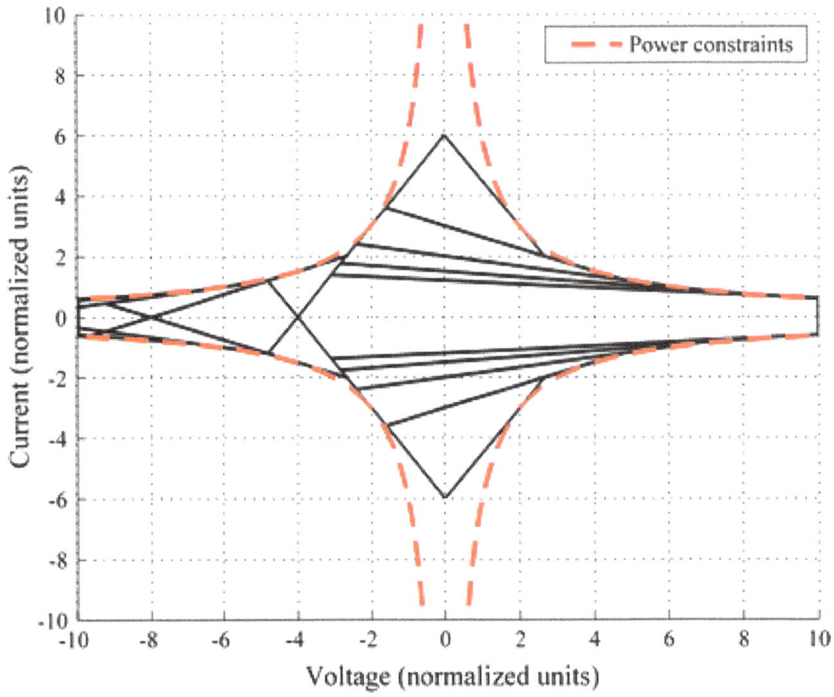

Figure 2: Illustration of number of convex polytopes versus number of half-planes used to represent the feasible region

In the traditional employment of the "big-M" method, each "or"-type constraint corresponds to one binary variable in the optimization problem. In a recent paper (Prodan et al. 2012), the authors presented an extension to that method to assign tuples of binary variables to each constraint, allowing for a number of binary variables Nb=log2 Np. By implementing this tuple association to feasible regions in this work, the number of binary variables necessary to represent the non-convex constraints is considerably reduced. The application of such method, in turn, entails a number of additional constraints on the binary variables to render the possibly unallocated tuples infeasible. This number of additional constraints is kept to a minimum by following the procedure developed in (Afonso and Galvão 2014). Therefore, the constraints over the current and voltage can be written as:

$$S_j \begin{bmatrix} u(k) \\ i(k) \end{bmatrix} \le h_j + M \cdot f_j(\lambda(k)) \cdot \begin{bmatrix} 1 \\ \vdots \\ 1 \end{bmatrix}_{F_j}, \quad 1 \le j \le N_p, 0 \le k \le N \tag{11}$$

In (11), the Np convex polytopes $\Gamma_j, 1 \le j \le N_p$, are described by the pairs of matrices and vectors Sj and hj, respectively. F_j is the number of facets of the polytope j and $[1 \ \cdots \ 1]^T_{F_j}$ is a vector of $F_j \times 1$. Each function fj is zero only if the tuple λ corresponds to the value that was assigned to render the constraints active, otherwise fj is equal to a positive integer. Since the constant scalar M is large enough to render all inequalities inactive, only one polytope constraint is enforced at each time step, as the others are relaxed:

$$f_j(\lambda) = \begin{cases} 0, \lambda \leftrightarrow \Gamma_j \\ L, otherwise \end{cases} \tag{12}$$

where L is a positive integer. To render the unassigned tuples infeasible, one constraint is appended to the optimization problem for every time instant k:

$$\sum_{l=1}^{N_b} 2^{l-1} b_l(k) \le N_p - 1, \ 0 \le k \le N \tag{13}$$

where b_l is the l-th binary variable of the tuple, in accordance to Afonso and Galvão (2014).

The cost function that was adopted in this work is the 1-norm of the vector formed by stacking the difference between the desired current value and the predicted one at every sample time, according to Eq. (8). To achieve this, auxiliary continuous variables are employed as means to penalize the absolute values of the deviations:

$$J = \sum_{k=0}^{N} \xi(k) \tag{14}$$

by imposing the following additional constraints:

$$\left. \begin{matrix} \xi(k) \ge r(k) - i(k) \\ \xi(k) \ge i(k) - r(k) \end{matrix} \right\} 0 \le k \le N \tag{15}$$

which ensure that $\xi(k)=|r(k)-i(k)|$ upon minimization of (14). The resulting optimization problem is:

$$\min_{u(k), \xi(k), b_l(k)} J = \sum_{k=0}^{N} \xi(k) \tag{16}$$

subject to:

$$x(k+1) = Ax(k) + Bu(k) \tag{17a}$$

$$i(k) = Cx(k) + Du(k) \tag{17b}$$

$$S_j \begin{bmatrix} u(k) \\ i(k) \end{bmatrix} \le h_j + M \cdot f_j(\lambda(k)) \cdot \begin{bmatrix} 1 \\ \vdots \\ 1 \end{bmatrix}_{F_j}, 1 \le j \le N_p, 0 \le k \le N \tag{17c}$$

$$\lambda(k) = [b_{N_b}(k), b_{N_b-1}(k), \ldots, b_1(k)] \tag{17d}$$

$$\sum_{l=1}^{N_b} 2^{l-1} b_l(k) \le N_p - 1, \ 0 \le k \le N \tag{17e}$$

where A, B, C, and D denote matrices as in (10a, b).

SIMULATION RESULTS

Simulations were carried out by using five realizations of 3D RC networks with $N_X = 5$ rows, $N_Y = 5$ columns, $N_Z = 3$ layers, $R_S = 0.1$, R = 1 and C = 0.5 (normalized units). These realizations were obtained by randomly varying the allocation of the R, C components within the network, as in (Galvão et al. 2013). The fraction of capacitors with respect to the total number of components (capacitors and resistors) was set to 0.5.

From one of the realizations, a reduced-order model of order three was obtained. The model reduction procedure consists of removing dynamic modes which have weak influence on the input-output behaviour of the network. This model, (henceforth termed design model) was used to describe the multi-path propagation process and was the one employed for the calculation of the voltage input sequence. The optimal sequence was then applied to five models associated to different network realizations as a means to verify the robustness of the proposed solution. The imposed constraints (in normalized units) were: $|u(k)| \le 1.5$, $|i(k)| \le 0.5$ and $|u(k)i(k)| \le 0.125$, each of them in normalized units. The sample time was set to Ts=Ts= 0.010, (also in normalized units), and N was set to 20 in the optimization problem (16), (17a–e). Following the work by Kasper (1982) on equalization of multimode optical fibre systems, the desired output signal adopted in the current simulations was a Gaussian pulse. In all figures, "Simulation model 1" is the full-order model that was used to generate the reduced-order design model. The package Multi Parametric Toolbox (MPT) (Kvasnica 2009) was used to manipulate the polytopes and IBM-CPLEX was used to solve the resulting MILP problem. Both were run under the Matlab environment.

Figure 3 illustrates the dispersion features of the RC network by using a Gaussian-shaped pulse as input voltage. As can be seen in Fig. 3a–c (time domain) and Fig. 3d–f (frequency domain), the output current waveform closely follows the input shape if the pulse is sufficiently wide. However, for narrower pulses (i.e. at higher spectral content), the pulse shape becomes distorted, deviating from the ideal Gaussian shape.

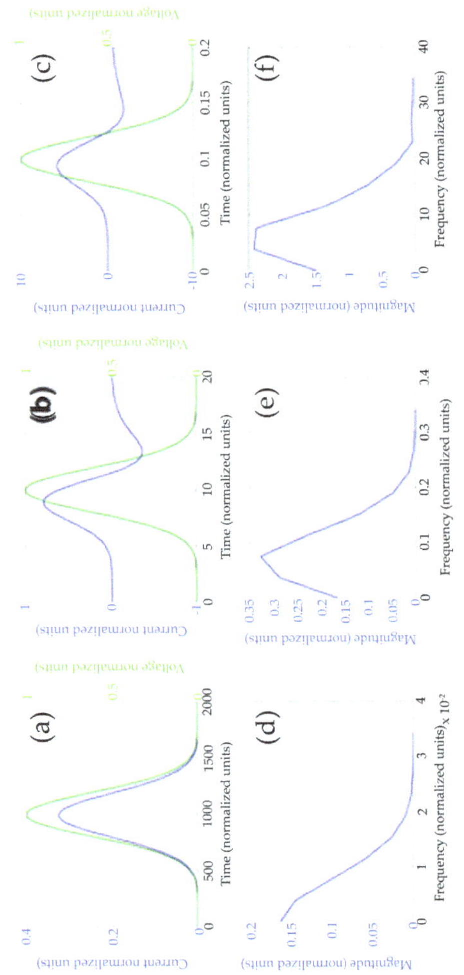

Figure 3: Top row current responses to a Gaussian voltage pulse with varying width. Bottom rowmagnitude spectra of the current responses

The results in Fig. 3 point to the necessity of a pre-equalization scheme to compensate for the network dispersion in order to obtain an output pulse closer to a Gaussian shape. For this purpose, the optimization problem (16), (17a–e) can be solved by using the desired Gaussian shape as the reference waveform. As a result, the input waveform is calculated so that the output follows the reference as close as possible, given the constraints over the voltage input, current output and power. For illustration, the optimization problem was initially solved without any of these constraints. The results are shown in Fig. 4, in which the Gaussian pulse reference with the shortest period shown in Fig. 3 was discretized to cope with the discrete-time nature of the optimization problem. It can be seen from Fig. 4a, b, that the output of the design model follows the reference perfectly in the time and frequency domains. As can be seen in Fig. 4a, there is a phase advance of the current with respect to the voltage, which is coherent with the typical response of an RC network. As for the different simulation models with random allocation of the RC components, the main characteristics of the signal are preserved. Figure 4c depicts the results in the current–voltage plane. In this case, the power constraints would be violated, i.e. the transmitter would not be able to implement the optimized voltage pulse in practice.

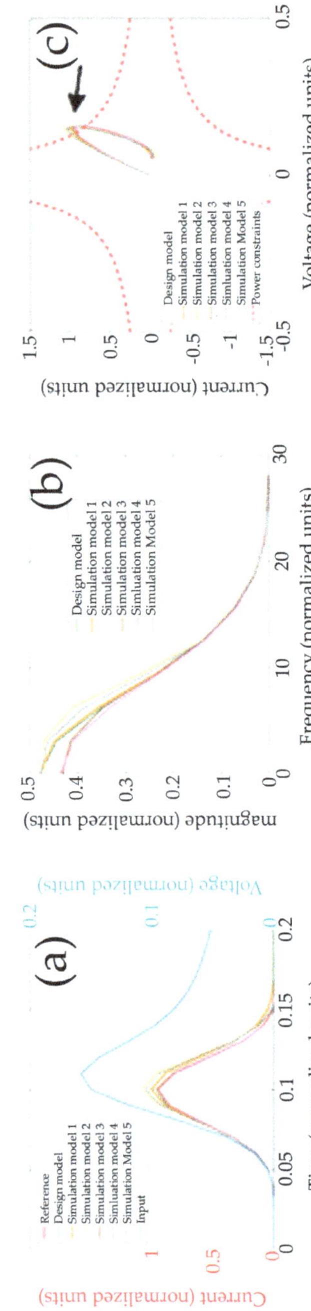

Figure 4: Response to pre-equalized signal without constraints: **a** voltage and current for different simulation models, **b** spectrum of the current responses and **c** trajectories in the current–voltage plane with the arrow indicating constraints violation

In order to cope with the source constraints, the optimization was repeated, now imposing constraints on the voltage, current and power. As can be seen in Fig. 5c, the power constraints are now properly enforced. However, the signal at the output does not follow the reference as closely as before, especially at the peak current demand, as seen in Fig. 5a. As a consequence, the output spectrum in Fig. 5b displays an elevation at high frequencies, as compared to Fig. 4b. In practice, such a distortion could be mitigated by transmitting a signal of smaller amplitude (so that the power constraint becomes inactive) at the cost of a worse signal-to-noise ratio.

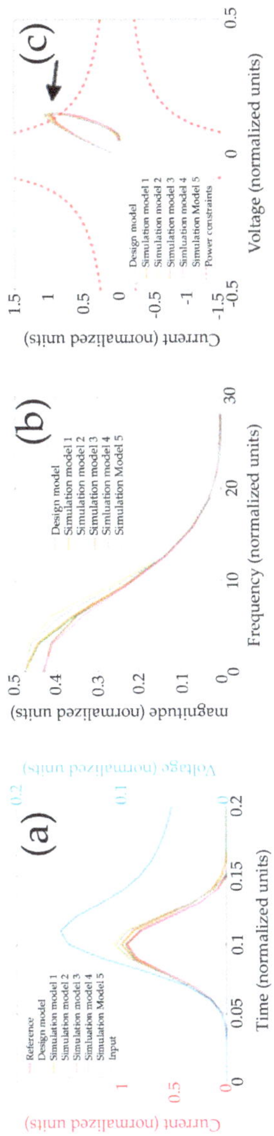

Figure 5: Response to pre-equalized signal with constraints: **a** voltage and current for different simulation models, **b** spectrum of the current responses and **c** trajectories in the current–voltage plane with the arrow indicating that there is no more constraints violation

DISCUSSION

Prospects for Pre-Equalization Applications in Communications

There is considerable scope to further explore the proposed formulation in a communications setting (Kang et al. 1999; Chen et al. 1999; Vanderveen et al. 1998). Line-of-sight links can assume a simple exponentially decaying channel model whereas diffuse links require a ceiling-bounce channel model. In the time domain, the output y(t) from the communications channel is given by: y(t)=Rh(t)∗x(t)+n(t) where h(t) is the channel impulse response, RR is the detector responsivity, n(t)n(t) is white Gaussian noise due to the lighting in the room and ∗ denotes the convolution operator. In indoors communications where multi-path propagation can be more pronounced, a ceiling-bounce model (Carruthers and Kahn 1997; Barry et al. 1993; Fernando and Balendran 2005) can be used for the channel model:

$$h(t, a) = \frac{rA_r}{3\pi L^2}\frac{6a^6}{(t+a)^7}u(t) \tag{18}$$

where u(t) is the unit step function and a depends on the room size, and the transmitter and receiver position where L is the height of the ceiling, r is the reflectivity of the ceiling above the transmitter and receiver and cc is the speed of light. The parameter a=2L/ca=2L/c is related to the multi-path rms delay spread Drms by noting that a=12Drms (13/11) −1/2. The Drms value can be directly associated to the expression in (3) defining a particular network configuration that mimics the channel response.

Furthermore, it is also worth noting that optical wireless systems commonly use simple baseband modulation schemes such as on–off keying (OOK), or pulse position modulation (PPM) (Gfeller and Bapst 1979; Barry 1994). Alternatives that are superior from a power efficiency perspective include digital pulse-interval modulation (DPIM) (Ghassemlooy et al.1998), dual header pulse-interval modulation (DHPIM) (Aldibbiat et al. 2002) and differential pulse-position modulation (DPPM) (Shiu and Kahn 1999). Furthermore, differential amplitude pulse-position modulation (DAPPM) is also a hybrid modulation technique that has attracted interest (Sethakaset and Gulliver 2005) as it incorporates aspects from both pulse-amplitude modulation (PAM) as well as DPPM. The spectral richness of the chosen input function in our simulations imply that such communications protocols can be supported by adopting a pre-equalization approach. Since demodulation of the encoded signals is commonly performed using hard-threshold decision (HTD), maximum-likelihood sequence detection (MLSD) or zero-forcing decision-feedback equalization (ZF-DFE) modalities, it seems that the piece-wise linearization adopted should not have a detrimental effect in the fidelity of the transmitted information. Generally, when intensity demodulation is performed using a direct detection technique, there should be no problem for compensating from channel distortions. In cases where a heterodyning technique would be used, our formulation would need to be augmented to account explicitly for phase.

Ultra-wide-band communications (Taylor 1994; Win and Scholtz 1998; Vaughan and Scott1999; Saleh and Valenzuela 1987; Maloney et al. 1990) are defined as communications using very short duration pulse signals of sufficiently high rise-time and fall-time having a large ratio of 3-dB bandwidth to the signal's center frequency, typically above 0.25. Since future indoor short-distance communication links are likely to use ultra-wide bandwidth radio or infrared transmitters (referred to as short-pulse or impulse radio systems) embedded in the lighting infrastructure of the buildings, it is also appropriate to also consider propagation of such signals. In this case, the channel model has the form: $r(t) = \sum_n p_n(t - \tau_n)$ where τn is the delay of the nth propagation path, and pn is the received waveform from each path so that τ(t)represents a weighted sum of time-shifted versions of the waveform p(t). In this formulation, the channel impulse response would be a function of time and azimuth angle: h(t,θ)=h(t)h(θ)and independent descriptions of the multipath time-of arrival h(t) and angle-of-arrival h(θ)expressions would need to be associated with the RC network structure. A greater sensitivity to frequency-selective fading in multi-path propagation will normally be associated in the UWB implementation of the technique.

It is important to further note that the proposed formulation should also have a wide applications domain in free space or in-fibre propagation using coherent sources (Asif et al.2013). This includes free-space adaptive cohort secure communication where the generated waveform can take advantage of selective frequency fading to minimize eavesdropping. Pre-equalization of soliton pulses launched into optical fibers and dispersion flattening in non-linear fibres have many applications in the optical fibre communications industry and should be possible with the proposed methodology. Re-casting the optimization procedure in the spectral domain would also enable control of wavelength dependent birefringence in optical fibres. The technique also offers new prospects for other more exotic forms of free-space communications such as through spatial hole burning in the atmosphere where controlled filamentation processes would enable the opening of ad hoc communications channels (Hadjiloucas et al.2013).

Finally, a further advantage of adopting the proposed RC networks to describe multi-path frequency selective fading is the fact that they display transfer functions where fractional Laplace transforms can be used to describe frequency selectivity in a parsimonious manner.

The work also opens the possibility for pre-distortion equalization in communications on the basis of identification using fractional order calculus techniques (Jacyntho et al. 2015).

Prospects for Pre-Equalization Applications in Photonic Networks

In previous works (Galvão et al. 2013) we have already discussed that the descriptor based formulation and input/output transfer function derivation associated to the adopted RCnetworks can be used to model attenuation in different dielectric media (e.g. nano-dielectrics, amorphous or porous materials) which are typically characterized using electrical impedance spectroscopy techniques. Although the optimization procedure was implemented assuming voltages and currents, the methodology is generic and can be implemented in a wide range of propagating media across different parts of the electromagnetic spectrum. Such approach makes the proposed propagation models of use well beyond the confines of existing communications modalities and enables us to take into consideration advances in the femtosecond pulse laser community (Wise et al. 2008; Nathan et al. 2005; Keller 2003; Limpert et al. 2003; Xu et al. 2006; Sardesai et al. 1998). In amplified femtosecond systems for example, spectral management at the input port of the amplifier is necessary to ensure that optical components operate within tolerances and will not be damaged by large peak pulse power transients. This is commonly achieved using hyper-Michelson interferometers, rf-excited acousto-optic modulators or liquid crystal masks in 4f optical systems which are programmed to perform a pre-equalization process. Pulse management is currently performed on a trial–error basis using a spectral phase interferometer or frequency resolved optical gating module which monitors the amplifier output after the excitation waveform has passed through the dispersion-compensated optical components of the system with a spectrometer monitoring the spectral input to the amplifier. Essentially this is a black-box approach which monitors inputs and outputs and inferring the processes in the amplifying medium without providing any understanding of the time domain waveform reflected and refracted through the system optical components. As a result amplitude and phase pulse shaping is always ultra-conservative. With the proposed technique, it is possible to adopt a less conservative approach at the laser input port and more importantly perform pulse management in non-linear optical components, this would have applications in quantum optics (photonic crystal sources of correlated photon pairs to be used in polarization entangled photons applications), high harmonic generation (frequency doubling or tripling crystals), parametric amplification at longer wavelengths, THz generation, four-wave mixing and electromagnetically induced transparency experiments. It should also be of relevance to the advancement of ultra-high power microwave and laser technologies e.g. gyrotron research for plasma excitation and target implosion experiments for fusion as performed at the National Ignition Facility (US).

The formulation also paves the way for pre-equalization of optical pulses coupled to photonic crystal fibres (Reeves et al. 2003; Konorov and Zheltikov 2003; Ouzounov et al. 2005). Photonic crystal fibres offer significant promise for ultra-high bandwidth terabyte communications as well as for the generation of atto-second pulses to perform time resolved X-ray studies (Takashi et al. 2004; Drescher et al. 2001; Christov et al. 1997). One of the main reasons for not widely adopting photonic crystal fibres to generate atto-second pulses through non-linear conversion of carrier envelope stabilized femtosecond pulses is that the fibre output does not remain sufficiently stable over a sufficiently long period required to perform several experiments with the generated atto-second pulses. There is a particularly high sensitivity of the non-linear conversion process to laser beam walk-off. The assumption of slowly varying dynamics in theRC network adopted in this study should enable the development of a time-varying pre-equalization scheme for dynamic compensation leading to a more stable output.

Future simulations related to the specific case of photonic crystal fibres will concentrate on the application of the modal transmission line theory as has been developed for modelling multilayered periodic media as well as to perform modal analysis of arbitrary shape optical waveguides. In order to achieve this one needs to consider the temporal harmonic electromagnetic fields in a dielectric medium as solutions of the source-free Maxwell equations which can be transformed into a system of differential equations with the electric and magnetic field vectors in the x and y dimensions related to capacitive and inductive responses (Karpisz et al. 2015; Dabirian et al. 2005) with expressions analogous to those found in the telegraph equations. Such approach requires the extension of the current three-dimensional RCframework to incorporate inductive components, with the descriptor equations associated with that structure and its boundary conditions converted into state space form. This is a topic of current research by our group.

Another interesting domain of applications is in polaritonic integrated circuits (Werley et al.2010) which currently promise significant potential to advance THz spectroscopy. In these systems, the pulses are generated within the crystal and phonons from crystal edge reflections introduce additional frequency selective attenuation to the input waveform. The pre-equalization procedure proposed can ensure spectral richness of the excitation waveform reaching the sample, accounting for phonon band-gap interference, thus enabling persistent sample excitation.

Finally, it is worth noting that although the formulation discussed in this paper was confined to electrical excitation and was explored within a communications setting, it is straightforward to assume a far more general structure for the adopted RC networks with nodes composed of Gibb's free energy expressions related to acoustic, acousto-optic or chemical processes. Such approach would enable the formulation to provide optimal excitation in acousto-optic crystals or even to more complex physico-chemical systems e.g. ferroelectrics. It may, therefore, be concluded that the proposed formulation is of general interest across the Physical Sciences.

Future refinement of the proposed RC network topology will enable the derived models to account for propagation in complex photonic crystal structures (Datta et al. 1993; Sigalas et al.1994; Benisty et al. 2002; Kafesaki et al. 2002; Foteinopoulou et al. 2007; Sigalas et al. 1999; Reboud et al. 2010; Vasilantonakis et al. 2012; Tasolamprou et al. 2010). Such developments would have applications in studies of dispersion compensation in fibre based communication systems (Knight et al. 1998; Várallyay et al. 2010). Furthermore, with the addition of non-linear components in the network it would be possible to simulate propagation in metamaterials (Koschny et al. 2005) providing an alternative approach to their modelling as presented in recent work (Liaskos et al. 2015). Finally, it is worth noting that experiments have already been proposed to characterize in both space and time the propagation of an ultrashort pulse in a multimode optical fiber where the optical field amplitude and phase at the output of the fiber can be estimated as a function of delay using time-gated spatial heterodyne interferometry (Rokitski and Fainman 2003). It would therefore be particularly useful to use spectral phase interferometry to experimentally verify the validity of the proposed RC modeling approach to emulate multimode amplitude attenuation and phase delay.

CONCLUSION

This contribution provides a new way of interpreting multi-path propagation in a dielectric medium as well as in communication channels by assuming an analogy to multi-layer RCnetworks with randomly allocated resistors and capacitors being responsible for the delay and attenuation properties of the medium. The network responses can describe frequency-selective attenuation encountered in path-dependent propagation. A pre-equalization procedure for a Gaussian pulse input that takes into account the capabilities of the transmission source as well as the transmission properties of the medium is developed. The problem is cast within a Mixed Integer Linear Programming (MILP) optimization framework that uses the developed nominalRC network model, with the excitation waveform customized to generate the desired function in a least squares sense at the receiver. Simulations are carried out with different network realizations in order to evaluate the sensitivity of the solution with respect to changes in the transmission medium associated with multi-path propagation. The proposed approach is of relevance where channel identification is ill-conditioned and equalization techniques are difficult to implement (e.g. displaying fractional order dynamics). It is applicable to both established communication modalities across the EM spectrum as well as emergent indoor communications assuming various modulation protocols or UWB schemes. It is anticipated that such formulations will have a much wider applicability to measurement science, the inverse problem and the photonics modelling community.

ACKNOWLEDGMENTS

R.J.M.A., R.K.H.G. and K.H.K. acknowledge the support of CNPq (Grants 303714/2014-0 and 302566/2012-0) and FAPESP (Grants 2011/13777-8, 2011/17610-0 and 2011/18632-8). In addition, H. A. would like to acknowledge his embassy for providing a research scholarship.

REFERENCES

1. Afonso, R.J.M., Galvão, R.K.H.: Comments on "Enhancements on the hyperplanes arrangements in mixed-integer programming techniques".: association method for unallocated tuples to allow for a single additional constraint. J. Optim. Theory Appl. 162, 996–1003 (2014)
2. Akbari, M., Schünemann, K., Burkhard, H.: A full-wave three-dimensional analysis of open dielectric waveguides and periodic structures using an equivalent network model. Opt. Quantum Electron. 32, 991–1004 (2000)
3. Aldibbiat, N.M., Ghassemlooy, Z., McLaughlin, R.: Dual header pulse interval modulation for dispersive indoor optical wireless communication systems. IEE P-Circ. Dev. Syst. 149(3), 187–192 (2002)
4. Asif, R., Shahid, H., Arshad, F., Saleem, R.: Scalable nonlinear equalization in high-bit-rate optical transmission systems. Photon. Res. 1(3), 130–135 (2013)
5. Barry, J.R., Kahn, J.M., Krause, W.J., Lee, E.A., Messerschmitt, D.G.: Simulation of multipath impulse response for indoor wireless optical channels. IEEE J. Select. Areas Commun. 11(3), 367–379 (1993)
6. Barry, J.R.: Wireless Infrared Communications. KluwerAcademic Publishers, Norwell (1994), 181p
7. Benisty, H., Olivier, S., Weisbuch, C., Agio, M., Kafesaki, M., Soukoulis, C.M., Qiu, M., Swillo, M., Karlsson, A., Jaskorzynska, B.: Models and measurements for the transmission of submicron-width waveguide bends defined in two-dimensional photonic crystals. IEEE J. Quantum Electron. 38, 770–785 (2002)
8. Camacho, E.F., Bordons, C.: Model Predictive Control. Springer, London (1999), 280p
9. Cangellaris, A.C.: Distributed equivalent sources for the analysis of multiconductor transmission lines excited by an electromagnetic field. IEEE Trans. Microw. Theory Tech.36(10), 1445–1448 (1988)
10. Carruthers, J.B., Kahn, J.M.: Modeling of nondirected wireless infrared channels. IEEE Trans. Commun. 45(10), 1260–1268 (1997)
11. Castellanos, F., Marti, J.R., Marcano, F.: Phase-domain

multiphase transmission line models. Int. J. Electron Power. 19(4), 241–248 (1997)

12. Chen, J.T., Paulraj, A., Reddy, U.: Multichannel maximum likelihood sequence estimation (MLSE) equalizer for GSM using a parametric channel model. IEEE Trans. Commun. 47, 53–63 (1999)
13. Chen, S., Chng, E.S.: Fractionally spaced blind equalization with low-complexity concurrent constant modulus algorithm and soft decision directed scheme. Int. J. Adapt. Control Signal Process. 19, 471–484 (2005)
14. Chow, C.W., Yeh, C.H., Liu, Y., Liu, Y.F.: Digital signal processing for light emitting diode based visible light communication. IEEE Photon. Soc. Newslett. 26(5), 9–13 (2012)
15. Christov, I.P., Murnane, M.M., Kapteyn, H.C.: High-harmonic generation of attosecond pulses in the "single-cycle" regime. Phys. Rev. Lett. 78, 1251–1254 (1997)
16. Dabirian, A., Akbari, M.: Modal transmission-line theory of optical waveguides. J. Electromagn. Wave 19(7), 891–906 (2005)
17. Dabirian, A., Akbari, M., Mortensen, N.A.: Propagation of light in photonic crystal fibre devices. J. Opt. A Pure Appl. Opt. 7, 663–668 (2005)
18. Datta, S., Chan, C.T., Ho, K.M., Soukoulis, C.M.: Effective dielectric constant of periodic composite structures. Phys. Rev. B 48(20), 14936–14943 (1993)
19. Desoer, C.A., Kuh, E.S.: Basic Circuit Theory. Ch. 14. McGraw-Hill, New York (1969), 876p
20. Dickinson, B.W.: Systems: Analysis, Design, and Computation. Ch. 3 (Discretization of Continuous Time Systems), 124–158. Prentice-Hall, Englewood Cliffs (1991)
21. Drescher, M., Hentschel, M., Kienberger, R., Tempea, G., Spielmann, C., Reider, G.A., Corkum, P.B., Krausz, F.: X-ray pulses approaching the attosecond frontier. Science 291, 1923–1927 (2001)
22. Edfors, O., Sandell, M., van de Beek, J.J., Wilson, S.K., Börjesson, P.O.: OFDM channel estimation by singular value decomposition. IEEE Trans. Commun. 46, 931–939 (1998)
23. Faria, J.A.B.: Multiconductor Transmission-Line Structures-Modal Analysis Techniques. John Wiley & Sons Inc, New York (1993), 203p
24. Fernando, X.N., Balendran, B.: Adaptive denoising and equalization of infrared wireless CDMA system. EURASIP J. Wirel. Commun. 1, 20–29 (2005)
25. Foteinopoulou, S., Kafesaki, M., Economou, E.N., Soukoulis, C.M.: Backwards surface waves in photonic crystals. Phys. Rev. B 75, 245116-1–245116-6 (2007)
26. Franklin, G.F., Powell, J.D., Workman, M.L.: Digital Control of Dynamic Systems, 3rd edn. Addison-Wesley, Boston (1997), 742p
27. Galvão, R.K.H., Hadjiloucas, S., Izhac, A., Becerra, V.M., Bowen, J.W.: Wiener system subspace identification for mobile wireless mm-wave networks. IEEE Trans. Veh. Technol. 56(4), 1935–1948 (2007a)
28. Galvão, R.K.H., Izhac, A., Hadjiloucas, S., Becerra, V.M., Bowen, J.W.: MIMO Wiener model identification for large scale fading of wireless mobile communications links. IEEE Commun. Lett. 11(6), 513–515 (2007b)
29. Galvão, R.K.H., Hadjiloucas, S., Kienitz, K.H., Paiva, H.M., Afonso, R.J.M.: Fractional order modeling of large three-dimensional RC networks. IEEE Trans. Circuits Syst. I Reg. Pap. 60(3), 624–637 (2013)
30. Gfeller, F.R., Bapst, U.H.: Wireless in-house data communication via diffuse infrared radiation. Proc. IEEE 67(11), 1474–1486 (1979)
31. Ghassemlooy, Z., Hayes, A.R., Seed, N.L., Kaluarachchi, E.D.: Digital pulse interval modulation for optical communications. IEEE Commun. Mag. 36(12), 95–99 (1998)
32. Ghassemlooy, Z., Popoola, W., Rajbhandari, S.: Optical Wireless Communications: System and Channel Modelling with MATLAB. CRC Press, Taylor and Francis Group, Boca Raton (2013), 575p
33. Hadjiloucas, S., Hashem, A., Becerra, V.M., Galvão, R.K.H.: Channel equalization for indoor lighting communications networks. Dielectr. J. Phys. Conf. Ser. 450, 012051 (2013)
34. Hashemi, H.: The indoor radio propagation channel. Proc. IEEE 81, 943–968 (1993)
35. Heydari, P., Pedram, M.: Model-order reduction using variational balanced truncation with spectral shaping. IEEE Trans. Circuits Syst. I Reg. Pap. 53(4), 879–891 (2006)
36. Ho, K.M., Chan, C.T., Soukoulis, C.M.: Existence of a photonic gap in periodic dielectric structures. Phys. Rev. Lett. 65(25), 3152–3155 (1990)
37. Hoefer, W.J.R.: The transmission-line matrix method-theory and applications. IEEE Trans. Microw. Theory Tech. 33(10), 882–893 (1985)
38. Hori, T., Takayanagi, J., Nishizawa, N., Goto, T.: Flatly broadened wideband and low noise supercontinuum generation in highly nonlinear hybrid fiber. Opt. Express 12(2), 317–324 (2004)
39. http://optics.synopsys.com/rsoft/rsoft-passivedevice-beamprop.htm. Accessed 17 Feb 2016
40. http://sydney.edu.au/science/physics/cudos/docs/research/UsingFibre.pdf. Accessed 17 Feb 2016
41. http://www.comsol.com/products/wave-optics/. Accessed 17 Feb 2016
42. Hu, L., Hecht, D.S., Grülner, G.: Percolation in transparent and conducting carbon nanotube networks. Nano Lett. 4, 2513–2517 (2004)
43. Huang, H., Milione, G., Lavery, M.P., Xie, G., Ren, Y., Cao, Y., Ahmed, N., Nguyen, T.A., Nolan, D.A., Li, M.-J., Tur, M., Alfano, R.R., Willner, A.E.: Mode division multiplexing using an orbital angular momentum mode sorter and MIMO-DSP over a graded-index few-mode optical fiber. Sci. Rep. 5, 14931 (2015)
44. Jacyntho, L.A., Teixeira, M.C.M., Assunção, E., Cardim, R., Galvão, R.K.H., Hadjiloucas, S.: Identification of fractional-order transfer functions using a step excitation. IEEE Trans. Circuits Syst. II Exp. Briefs 62, 896–900 (2015)

45. Johnson Jr., R., Schniter, P., Endres, T.J., Behm, J.D., Brown, D.R., Casas, R.A.: Blind equalization using the constant modulus criterion: a review. Proc. IEEE 86(10), 1927–1950 (1998)

46. Kaatze, U.: Perspectives in dielectric measurement techniques for liquids Meas. Sci. Technol.19, 112001 (2008)

47. Kafesaki, M., Soukoulis, C.M.: A historical perspective and a review of the fundamental principles in modeling 3D periodic structures with an emphasis on volumetric EBGs. In: Engheta, N., Ziolkowski, R. (eds.) Metamaterials: Physics and Engineering Explorations, pp. 215–234. Wiley-VCH, New York (2006)

48. Kafesaki, M., Agio, M., Soukoulis, C.M.: Waveguides in finite-height two-dimensional photonic crystals. J. Opt. Soc. Am. B 19, 2232–2240 (2002)

49. Kang, I., Fitz, M.P., Gelfand, S.B.: Blind estimation of multi-path channel parameters: a modal analysis approach. IEEE Trans. Commun. 47, 1140–1150 (1999)

50. Karpisz, T., Salski, B., Szumska, A., Klimczak, M., Buczynski, R.: FDTD analysis of modal dispersive properties of nonlinear photonic crystal fibers. Opt. Quantum Electron. 47, 99–106 (2015)

51. Kasper, B.L.: Equalization of multimode optical fiber systems. Bell Syst. Tech. J. 61(7), 1367–1388 (1982)

52. Keller, U.: Recent developments in compact ultrafast lasers. Nature 424, 831–838 (2003)

53. Knight, J.C., Broeng, J., Birks, T.A., Russell, P.S.J.: Photonic band gap guidance in optical fibers. Science 282, 1476–1478 (1998)

54. Koenig, S., Lopez-Diaz, D., Antes, J., Boes, F., Henneberger, R., Leuther, A., Tessmann, A., Schmogrow, R., Hillerkuss, D., Palmer, R., Zwick, T., Koos, C., Freude, W., Ambacher, O., Leuthold, J., Kallfass, I.: Wireless sub-THz communication system with high data rate. Nat. Photonics 7, 977–981 (2013)

55. Konorov, S.O., Zheltikov, A.M.: Frequency conversion of subnanojoule femtosecond laser pulses in a microstructure fiber for photochromism initiation. Opt. Express 11(19), 2440–2445 (2003)

56. Koschny, T., Markoš, P., Economou, E.N., Smith, D.R., Vier, D.C., Soukoulis, C.M.: Impact of inherent periodic structure on effective medium description of left-handed and related metamaterials. Phys. Rev. B. 71, 245105 (2005)

57. Kutz, J.N., Lyngå, C., Eggleton, B.J.: Enhanced supercontinuum generation through dispersion-management. Opt. Express 13(11), 3989–3998 (2005)

58. Kvasnica, M.: Real-Time Model Predictive Control via Multi-Parametric Programming. Verlag Dr. Müller, Saarbrücken (2009), 272p

59. Leon-Saval, S.G., Argyros, A., Bland-Hawthorn, J.: Photonic Lantern Photonic lanterns: a study of light propagation in multimode to single-mode converters. Opt. Express 18(8), 8430–8439 (2010)

60. Li, Y., Cimini, L.J., Sollenberger, N.R.: Robust channel estimation for OFDM systems with rapid dispersive fading channels. IEEE Trans. Commun. 46, 902–915 (1998)

61. Liaskos, C., Tsioliaridou, A., Pitsillides, A., Akyildiz, I.F., Kantartzis, N.V., Lalas, A.X., Dimitropoulos, X., Ioannidis, S., Kafesaki, M., Soukoulis, C.M.: Design and development of software defined metamaterials for nanonetworks. IEEE Circ. Syst. Mag. 15, 12–25 (2015)

62. Lima, T.P., Galvão, R.K.H., Yoneyama, T.: Predictive control with power constraints using mixed-integer quadratic programming. In: XIV Congresso Latinoamericano de Control Automático, Santiago, Chile, 2010 (paper LCA-033)

63. Limpert, J., Clausnitzer, T., Liem, A., Schreiber, T., Fuchs, H.-J., Zellmer, H., Kley, E.-B., Tnnermann, A.: High-average-power femtosecond fiber chirped-pulse amplification system. Opt. Lett. 28, 1984–1986 (2003)

64. Lin, C.H., Leung, K.M., Tamir, T.: Modal transmission line theory of three-dimensional periodic structure with arbitrary lattice configurations. J. Opt. Soc. Am. A 19, 2005–2017 (2002)

65. Liu, X., Chandrasekhar, S., Winzer, P.J.: Digital signal processing techniques enabling multi-Tb/s superchannel transmission. IEEE Signal Process. Mag. 31(2), 16–24 (2014)

66. Lowery, N., Potthast, R., Vahdati, M., Holderbaum, W.: On discrimination algorithms for ill-posed problems with an application to magnetic tomography. Inverse Probl. 28(6), 065010 (2012)

67. Luo, L.-W., Ophir, N., Chen, C.P., Gabrielli, L.H., Poitras, C.B., Bergmen, K., Lipson, M.: WDM-compatible mode-division multiplexing on a silicon chip. Nat. Commun. 5, 3069 (2014)

68. Maloney, J., Smith, G.S., Scott, W.R.: Accurate computation of the radiation from simple antennas using the finite-difference time-domain method. IEEE Trans. Antennas Propag.38(7), 1059–1068 (1990)

69. Mizuno, T., Kobayashi, T., Takara, H., Sano, A., Kawakami, H., Nakagawa, T., Miyamoto, Y., Abe, Y., Goh, T., Oguma, M., Sakamoto, T., Sasaki, Y., Ishida, I., Takenaga, K., Matsuo, S., Saitoh, K., and Morioka, T.: 12-core x 3-mode dense space division multiplexed transmission over 40 km employing multi-carrier signals with parallel MIMO equalization. In: Optical Fiber Communication Conference: Postdeadline Papers, (Optical Society of America, 2014), paper Th5B.2

70. Mori, T., Sakamoto, T., Yamamoto, T., Tomita, S.: Modal dispersion compensation by using digital coherent receiver with adaptive equalization in multi-mode fiber transmission. Opt. Fiber Technol. 19, 132–138 (2013)

71. Nitsch, J., Gronwald, F., Wollenberg, G.: Radiating Nonuniform Transmission-Line Systems and the Partial Element Equivalent Circuit Method. John Wiley & Sons Inc., New York (2009), 348p

72. Ouzounov, D.G., Hensley, C.J., Gaeta, A.L., Venkateraman, N., Gallagher, M.T., Koch, K.W.: Soliton pulse compression in photonic band-gap fibers. Opt. Express 13(16), 6153–6159 (2005)

73. Piesiewicz, R., Jemai, J., Koch, M., Kurner, T.: THz channel characterization for future wireless gigabit

indoor communication systems. In: Proceedings of SPIE—Terahertz and Gigahertz Electronics and Photonics IV, 2005, 5727, 166–176 (2005)

74. Pregla, R.: Analysis of Electromagnetic Fields and Waves: The method of lines. Research Studies Press Limited, John Wiley & Sons Inc., New York (2008), 522p
75. Prodan, I., Stoican, F., Olaru, S., Niculescu, S.I.: Enhancements on the hyperplanes arrangements in mixed-integer programming techniques. J. Optim. Theory Appl. 154, 549–572 (2012)
76. Rambousky, R., Nitsch, J.B., Garbe, H.: Application of the transmission-line super theory to multi-wire TEM-waveguide structures. IEEE Trans. Electromag. Compat. 55(6), 1311–1319 (2013)
77. Reboud, V., Kehagias, N., Kehoe, T., Leveque, G., Mavidis, C., Kafesaki, M., Sotomayor-Torres, C.: Nanoimprinted plasmonic crystals for light extraction applications. Microelectron. Eng. 87, 1367–1369 (2010)
78. Reeves, W.H., Skryabin, D.V., Biancalana, F., Knight, J.C., Russell, P.S.J., Omenetto, F.G., Efimov, A., Taylor, A.J.: Transformation and control of ultra-short pulses in dispersion-engineered photonic crystal fibres. Nature 424, 511–515 (2003)
79. Reis, N.O.G., Galvão, R.K.H.: A platform for didactic experiments of adaptive equalization. In: Proceedings of the 15th Brazilian Automatic Control Conference, Gramado. Brazil (in Portuguese) (2004)
80. Richards, A., Schouwenaars, T., How, J.P., Feron, E.: Spacecraft trajectory planning with avoidance constraints using mixed-integer linear programming. J. Guid. Control Dyn. 25(4), 755–764 (2002)
81. Richardson, D.J., Fini, J.M., Nelson, L.E.: Space-division multiplexing in optical fibers. Nat. Photonics 7, 354–362 (2013)
82. Rivas, J.G., Sprik, R., Soukoulis, C.M., Busch, K., Lagendijk, A.: Optical transmission through strong scattering and highly polydisperse media. Europhys. Lett. 48(1), 22–28 (1999)
83. Rokitski, R., Fainman, S.: Propagation of ultrashort pulses in multimode fiber in space and time. Opt. Express 11(13), 1497–1502 (2003)
84. Rowe, H.E.: Electromagnetic Propagation in Multi-Mode Random Media. John Wiley & Sons Inc., New York (1999), 233p
85. Snyder, W., Love, J.D.: Optical Waveguide Theory. Chapman & Hall, London (1983), 738p
86. Saleh, A.A.M., Valenzuela, R.A.: A statistical model for indoor multipath propagation. IEEE J. Select. Areas Commun. 5, 128–137 (1987)
87. Sardesai, H.P., Chang, C.C., Weiner, A.M.: A femtosecond code-division multiple-access communication system test bed. J. Lightw. Technol. 16(11), 1953–1964 (1998)
88. Sari, H., Karam, G., Jeanclaude, I.: Transmission techniques for digital terrestrial TV broadcasting. IEEE Commun. Mag. 33, 100–109 (1995)
89. Savory, S.J.: Digital coherent optical receivers: algorithms and subsystems. IEEE J. Sel. Topics Quantum Electron. 16(5), 1164–1179 (2010)
90. Sethakaset, U., Gulliver, T.: A differential amplitude pulse-position modulation for indoor wireless optical communications. EURASIP J. Wirel. Commun. 1, 3–11 (2005)
91. Shieh, W., Li, A., Al, A.A., Chen, X.: Space-division multiplexing for optical communications. IEEE Photonics Soc. Newslett. 26(5), 4–12 (2012)
92. Shiu, D.-S., Kahn, J.M.: Differential pulse-position modulation for power-efficient optical communication. IEEE Trans. Commun. 47(8), 1201–1210 (1999)
93. Sigalas, M.M., Biswas, R., Ho, K.M., Soukoulis, C.M.: Waveguides in three-dimensional metallic photonic band-gap materials. Phys. Rev. B 60(7), 4426–4429 (1999)
94. Sigalas, M.M., Soukoulis, C.M., Chan, C.T., Ho, K.M.: Electromagnetic-wave propagation through dispersive and absorptive photonic-band-gap materials. Phys. Rev. B 49(16), 11080–11087 (1994)
95. Smulders, P.: Exploiting the 60 GHz band for local wireless multimedia access: prospects and future directions. IEEE Commun. Mag. 40(1), 140–147 (2002)
96. Soukoulis, C.M., Economou, E.N., Grest, G.S.: Localization in quantum percolation: transfer-matrix calculations in three dimensions. Phys. Rev. B 36(16), 8649–8654 (1987)
97. Spadacini, G., Grassi, F., Marliani, F., Pignari, S.A.: Transmission-line model for field-to-wire coupling in bundles of twisted-wire pairs above ground. IEEE Trans. Electromagn. Compat.56(6), 1682–1690 (2014)
98. Stroud, D.: Percolation effects and sum rules in the optical properties of composites. Phys. Rev. B 19, 1783–1791 (1979)
99. Stutzman, W.L., Thiele, G.A.: Antenna Theory and Design, 3rd edn. John Wiley & Sons Inc., New York (2012), 848p
100. Tasolamprou, A.C., Zhang, L., Kafesaki, M., Koschny, T., Soukoulis, C.M.: Frequency splitter based on the directional emission from surface modes in dielectric photonic crystal structures. Opt. Express 23(11), 13972–13982 (2010)
101. Taub, H., Schilling, D.L.: Principles of Communication Systems, 2nd edn. McGraw-Hill, New York (1986), 759p
102. Taylor, J.D.: Introduction to Ultra-Wideband Radar Systems. CRC Press, Boca Raton (1994), 688p
103. van Uden, R.G.H., Correa, R.E., Lopez, E.A., Huijskens, F.M., Xia, C., Li, G., Schülzgen, A., de Waardt, H., Koonen, A.M.J., Okonkwo, C.M.: Ultra-high-density spatial division multiplexing with a few-mode multicore fiber. Nat. Photonics 8, 865–870 (2014)
104. Vanderveen, M.C., van der Veen, A.J., Paulraj, A.: Estimation of multipath parameters in wireless communications. IEEE Trans. Signal Process. 46, 682–690 (1998)
105. Várallyay, Z., Saitoh, K.: Photonic crystal fibre for dispersion control. In: Pal, B. (eds.) Frontiers in Guided Wave Optics and Optoelectronics, Ch. 10, pp. 185–208. INTECH, Croatia (2010)

106. Vasilantonakis, N., Terzaki, K., Sakellari, I., Purlys, V., Gray, D., Soukoulis, C.M., Vamvakaki, M., Kafesaki, M., Farsari, M.: Three-dimensional metallic photonic crystals with optical bandgaps. Adv. Mater. 24, 1101–1105 (2012)
107. Vaughan, R.G., Scott, N.L.: Super-resolution of pulsed multipath channels for delay spread characterization. IEEE Trans. Commun. 47, 343–347 (1999)
108. Wait, J.R.: Electromagnetic Waves in Stratified Media: Revised Edition Including Supplemented Material. Pergamon Press, Oxford (2013), 620p
109. Werley, C.A., Wu, Q., Lin, K.-H., Tait, C.R., Dorn, A., Nelson, K.A.: Comparison of phase-sensitive imaging techniques for studying terahertz waves in structured $LiNbO_3$. J. Opt. Soc. Am. B 27(11), 2350–2359 (2010)
110. Widrow, B., Stearns, S.D.: Adaptive Signal Processing. Prentice Hall, Englewood Cliffs (1985), 528p
111. Win, M.Z., Scholtz, R.A.: Impulse radio: how it works. IEEE Commun. Lett. 2(2), 36–38 (1998)
112. Wise, F.W., Chong, A., Renninger, W.H.: High-energy femtosecond fiber lasers based on pulse propagation at normal dispersion. Laser Photonics Rev. 2, 58–73 (2008)
113. Xu, B., Coello, Y., Lozovoy, V.V., Harris, D.A., Dantus, M.: Pulse shaping of octave spanning femtosecond laser pulses. Opt. Express 14(22), 10939–10944 (2006)
114. Yang, B., Letaief, K.B., Cheng, R.S., Cao, Z.: Channel estimation for OFDM transmission in multipath fading channels based on parametric channel modeling. IEEE Trans. Commun.49(3), 467–479 (2001)
115. Yang, Q., Chen, S., Ma, Y., Shieh, W.: Real-time reception of multi-gigabit coherent optical OFDM signals. Opt. Express 17(10), 7985–7992 (2009)
116. Yariv, A.: Universal relations for coupling of optical power between micro-resonators and dielectric waveguides. Electron. Lett. 36, 321–322 (2000)

CHAPTER 17

BROADBAND TRAFFIC IN NEXT-GENERATION NETWORKS

P. Loskot, M. A. M. Hassanien, F. Farjady, M. Ruffini, and D. Payne

Keywords: Broadband services, Next-generation networks, Network operator, Traffic growth drivers, Traffic forecasting, Video streaming

INTRODUCTION

Businesses are increasingly dependent on the state-of-the-art communication systems. Therefore, it is important that the rural as well as remote areas have the same services and capabilities as the urban areas. Provided that the businesses are not restricted by the availability and capabilities of the communication infrastructure, for instance, provided that the network operators can flexibly assign and reassign the network resources to subscribers based on their current demand, the businesses can spring up and flourish anywhere, and even the sparsely populated areas are likely to attract the inward investments. GDP growth, inward investments, stimulation of new entrepreneurships and businesses, and other measures of economic activity are known to be important economic and commercial drivers of deploying next-generation networks (NGN). The reports by the Organisation for Economic Co-operation and Development (OECD) in Europe and by the Office of Communications (Of com) in the UK show high correlations between offerings of the broadband services in a geographical area and its GDP growth [22, 31].

In general, there is interplay between the content providers and the bandwidth (i.e. network) providers; each of them represents a distinct market. Youtube is a good example of co-evolution between the content and network providers whereas the iPlayer from BBC proves that there exist occasional disruptions of the natural growth. The broadband networks are a long-term investment [41]; for instance, it will take at least next 10–15 years to fibre the whole UK. The network providers invest into the network expansion just ahead of the (permanently) increasing bandwidth demand. Hence, forecasting of the future network services [5] is extremely valuable for the network operators who are especially concerned to identify important drivers of traffic in NGN.

In general, we need to study future broadband services to show that the subscribers can make good use of the extra bandwidth provided by the new NGN technologies such as fibre-to-the-cabinet (FTTC) and especially fibre-to-the-premises (FTTP) and that these technologies can be also justified by the expected benefits to the community [6, 10]. The large bandwidth offered by NGN will provide new service monetization opportunities. For example, renting the processing power within the cloud computing service model motivates deployment of fast broadband connections to the customer premises while allowing to exploit the high bandwidth which will have a positive impact on the revenues for the network operator. In addition, the available bandwidth can be partitioned and flexibly rented on demand for remote processing and cloud computing applications. The dedicated portions of the available bandwidth can be utilized to support different service classes, to simplify traffic management and monitoring by isolating different types of traffic and to create new service bundling options that can be offered to the subscribers. For instance, local loop unbundling (LLU) offered by BT enables other communication service providers to install their equipment at the BT's exchanges or nearby in order to connect the customers' lines or local loops to the communication provider's own network.

Citation: P. Loskot, M. A. M. Hassanien, F. Farjady, M. Ruffini, D. Payne, Broadband traffic in next-generation networks, DOI: 10.1007/s12243-014-0424-9

Even though many broadband services do not generate any revenues for the network providers, identifying opportunities for monetization of services in NGN is very valuable. The models of broadband services have become more complex over time [19], and currently, these models incorporate issues such as bonding, vectoring, and packaging of services. Thus, the offered services drive development of traffic models which are then used to map potential revenues [17]. For instance, open service models drive regulatory models which in turn drive business models including monetization of broadband services.

Traffic models can be used to generate artificial traffic data, for example, for evaluation of different network architectures by computer simulations. Traffic volume is often driven by a single emerging factor which usually weakens over time, and traffic saturates. Thus, traffic models should produce the logistic curves rather than follow the exponential growth curves [39]. Public sources of broadband traffic data are very useful for developing and validating traffic models. Some recommended sources of publicly available traffic data are as follows: in the UK, the House of Lords government report by the Head of Strategy; in Europe, the reports provided by OECD and Of com; and also the reports usually published by the national network operators and regulators. The latter contain traffic data at network exchanges which can be used to accurately predict traffic growth in the whole country. However, the key for understanding these traffic data reports are the assumptions and measurement conditions that were used to generate the reported data. Traffic models are usually built assuming some key parameters and factors such as analysis of historical data, the number of broadband connections per exchange, adoption and usage rates of applications, statistical behaviour of subscribers, processing capabilities of end-hosts, access speeds of broadband connections and so on. In addition, it is useful to classify values of the underlying traffic model parameters as being either conservative, optimistic or an upper bound.

There are also other public sources of broadband traffic data. In particular, the MINTS [28] project collated the Internet traffic growth rates for different regions around the world including Western Europe in years 2002–2009. The year-over-year (YoY) traffic growth rates are predicted by linear least squares fitting of historical traffic data [26]. However, since such analysis assumes aggregated (i.e. average) traffic data, extrapolation of these results to different countries is limited. Assuming traffic data of the Internet service providers (ISP) in Japan, Cho et al. [8] suggest that video services will not significantly affect overall traffic in the near future whereas these services will eventually reshape traffic in broadband networks in a longer term. The traffic reports produced by FTTH Council Europe [16] contain forecasted traffic data for Europe for the following 5 years. The paper [21] predicts streaming video traffic over the period 2010–2015.

Cisco Visual Networking Index (VNI) [9] provides detailed traffic forecasts over the next 5 years for different regions and countries. These traffic forecasts are used extensively in many published research works. In the first step, the number of users and broadband connections and their types are estimated using traffic data from broadband regulators. The application adoption and usage rates are then estimated for each application and service type using traffic data from the service providers and using the customers' surveys. The average traffic data are calculated by multiplying the bitrates of applications and services with their anticipated usage times and the number of subscribers. The traffic estimates for the core Internet, the managed ISP and the mobile subscribers are reconciled by assessing traffic migration between the segments of the broadband network. Finally, in the last step, the obtained traffic forecasts are validated using the actual YoY traffic growth data following by possible readjustments of the traffic estimates in each step. It should be noted that Cisco VNI only reports the sources of traffic data used; however, the actual data used are not provided.

For example, the Ministry of Japan estimates the total annual broadband traffic to be 0.5 exabyte in 2010 while Cisco VNI traffic forecast is 0.6 exabyte for Japan in 2010. For the UK, the Cisco VNI forecasts have not been verified. The Cisco traffic data would have to be rescaled to account for specifics of the different UK regions since traffic usage patterns in these regions are different. For instance, Wales has a medium take-up of broadband services which is nevertheless higher than that in Scotland, but lower than those in Northern Ireland and England. However, the reported take-up growth rates in Wales in 2009–2010 are higher than those in England and comparable to those in Northern Ireland while most of this take-up is happening in the rural areas. In addition, the main Internet protocol television (IPTV) provider in the UK Tiscali did not provide its services in Wales until 2009 although it already did in other parts of the UK.

The main aim of this paper is to explore evolution of broadband traffic over a 20-year perspective into the future even though the current network technologies are likely to be obsolete by then [41]. We will also attempt to identify disruptive services that will spur new bandwidth demands. The long-term traffic predictions can be obtained by evaluating the limits of human senses (especially sighting) and then assuming that these limits will be achieved by future services. Alternatively, we will consider the likely contents to be transferred over the future broadband networks to form the upper bounds of traffic.

BUILDING LONG-TERM TRAFFIC MODELS

Broadband traffic is usually characterized by the sustained data rates, whereas the traffic peak values are a good indication of the potential network capacity demands for the broadband network dimensioning purposes. We can aim at characterizing the traffic dynamics either at the network core or at the network edges [7]. The latter is more challenging since traffic at the network edges experiences more variations, and thus, it is more difficult to forecast than traffic at the network core.

It is useful to distinguish between short-term and long-term traffic modelling and analysis [40]. The short-term traffic analysis usually employs statistical methods such as regression [1, 26], whereas the long-term traffic analysis is dominantly affected by the network technology used as well as by the various economic and demographic factors [2]. There are different methodologies to forecast broadband traffic [24, 32, 43]. The models commonly used in the relevant projects and in the open literature are judgemental, i.e. they assume that past traffic is a good indication of future traffic [39, 40, 46]. Other often used traffic models are the time-series models that assume periodic traffic measurements in order to update the judgemental models. A great challenge common to all these models is the availability of traffic data [3] which is usually rather limited due to political, legal, business and other reasons [40]. However, the publicly available traffic data from the government and the regulator sites can give good indication about the broadband traffic trends over the past several years.

Our approach to long-term traffic forecasting assumes the upper bounds of future traffic. The traffic upper bounds can be constructed by considering how much information content an average person is able to absorb at most. Traffic in future networks will be dominated by 3D video (currently, there is no 3D standard, but it is very likely that eventually such standard will be adopted), the Internet storage and other services in the cloud. The future multimedia delivery is likely to be point-to-point (Pt-Pt), thus moving traffic from the network core to its edges. The Pt-Pt services and other non-multimedia applications will be driving the bandwidth requirements especially in the upstream even though the downstream and upstream traffic volumes have a tendency to become more balanced over time.

The video-based applications and services are expected to be the strongest driver of the future traffic growth and even exceed the bandwidth demands of the file-sharing services. Moreover, some of the existing services are likely to be enhanced with embedded video; the upcoming merge of MSN with Skype is one such example. The human eye perception and emulation of the real-life vision are the ultimate and challenging limits to reach. Hence, the traffic upper bounds coupled with the anticipated levels of data compression can provide good estimates of the required broadband network capacity. It is also important to understand how many broadband devices we may find in an average household in the future. It is reasonable to assume that there will likely be only one large screen with a very high resolution in the household which will then become the main driver of traffic to the residential subscribers. However, many other devices at home will also have broadband connection and will be used for video viewing.

We have to consider that, in the future, video streaming may be, at least in some instances, replaced by a full initial download of the whole multimedia file before its replay. The compression levels and progress in the state-of-the-art compression methods have significant impact on the required network bandwidth. The large bandwidth of NGN will require less compression and, at the same time, provide better video quality. The current trend is to make applications more interactive and personalized. The existing compression methods such as MPEG-4 are insufficient for the stringent latency requirements of the interactive video applications, so new compression algorithms will have to be developed. Moreover, in future multimedia applications with video streaming being the key integral component, the users will be in charge to choose what to see, zoom-in on demand, change viewing angles and focal distances, and go back in time, and all this will be happening in real time with fast response and negligible latency.

The future video will be multi-view (the exact definition is unclear at the moment) with large content redundancy among the views which will allow good levels of compression. However, since the bandwidth is not for free, the users are unlikely to try to maximize their bandwidth use. Furthermore, the multimedia broadcast channels are likely to be shared by the viewers. The experience shows that the number of such channels rarely exceeds the number of users per household. Furthermore, due to large content redundancy and correlations, the traffic peak values do not add up linearly with the number of views. A simple adding up of traffic flows is a very loose upper bound, and more realistic is to use statistical summation (i.e. expectation or the average weighted by the probabilities). It is unlikely that parallel video streams would be sent at the same time to any household except at times of channel switching when the new channel has to be buffered first, so that at this time traffic will peak. The response times of channel switching need to be comparable to the time it takes for the user to realize that the channel has changed (about 1 s).

DRIVERS OF FUTURE APPLICATIONS AND SERVICES

The important networking topics relevant to long-term traffic modelling are the services for user-created contents, the targeted advertising, and the related economies of the Internet. Some examples of the current rapidly expanding services are as follows: network virtualization, multicast and Pt-Pt delivery of multimedia (e.g. IPTV), online gaming, virtual reality and immersive environments, and localization-aware services. The future of gaming will likely change since new client (gaming) terminals are now being introduced to the market. Human-to-machine (H2M), M2H, M2M and H2H communications are another emerging service type.

The current trend to providing services in the broadband networks is to use bandwidth partitioning with dedicated bandwidth allocations. One example of this trend is Flexgrid [14]: the network bandwidth is dynamically allocated based on the current demand. In the area of wireless networks, small cells are currently being aggressively deployed with the number of femtocells expected to increase the number of macrocell base stations early in 2013. Since small-cell base stations such as femtocells are heavily relying on backhauling via broadband networks, the mobile traffic off-load may significantly contribute to overall broadband traffic. Another significant and rapidly increasing contributor to broadband traffic volumes are the software downloads and upgrades. It is therefore of interest to explore methods for optimizing the software delivery that can reduce traffic in the core network. For example, considering the concept of cloud computing, the computing power can be alternatively distributed at the network edges. Thus, cloud computing and data storage are likely to evolve to be provided partly centrally in the cloud and partly distributed at the network edges. These services will affect the upstream and downstream traffic symmetry.

Enhanced Personal Communications

At present, H2H communications over broadband networks rely mainly on email, online chat (e.g. MSN) and audio/video calls (e.g. Skype). It is expected that these services will be gradually combined into one integrated and highly flexible application for online (real-time) as well as offline (leave-a-message) communications. Moreover, everyone will be assigned a universal unique ID or a personal account. Most

H2H messages are likely to be multimedia with embedded video and collections of high-resolution images and sounds. The current video technologies are being evolved towards being capable to fully exploit the potential limits of human sighting such as stereoscopic and 3D viewing and the capability to sense the direction the person is looking at and at which object he/she is focusing attention. Ultimately, the holographic video of the scene may be transmitted. Furthermore, the availability of additional senses will enhance and change the applications for human communications as we know them today. The broadband network users will be able to experience full-extent virtual reality with holographic images and sounds, touch and smell feelings, perhaps even enriched with the taste. Such new rich multimedia communications will be used extensively by social networking and gaming applications, and more importantly, they will incur much higher bandwidth requirements than their current versions.

Web Content Customization

Majority of contents in 20 years' time are going to be generated by the broadband subscribers in their premises. Many applications and services on the broadband networks are becoming gradually highly personalized which will contribute to increased traffic variations. The search engines have evolved to be more intelligent with optimized presentation of the search results. The multimedia will become searchable to the extent that is now only possible for the text-based contents. The applications will be getting a lot smarter and provide a high level of interactivity. Future traffic will be much more affected by attributes such gender, age, emotional state, ethnical background and other such factors. This development is already evident in Japan where multiple sensors are exploited to collect and analyse data for advertising purposes [25]. These developments are further supported by the progress in artificial intelligence and discovery informatics [13] which are playing increasingly important roles in the creation of fully customized contents. Furthermore, it is expected that the broadband networks will be used much more efficiently, for example, by providing the contents in several multi-resolution descriptions chosen for given context and scenario. These developments require accurate predictive models of the users' behaviour and their intention.

Internet of Things (IoT) and Machine Communications

Privacy is one of the top research, ethical and legal issues of the current Internet. The trend is to assign an ID not only to all network hosts but also to all network processes, activities and transactions. Home automation has been established to control lighting, heating, ventilation, air conditioning, security systems, home robots and appliances, and many other home systems including the home networks for distribution of audio and video in the households [37]. There is a strong emerging trend to embed broadband network connections into all home devices to enable their remote control over the Internet. The IoT and M2M communications (where devices and/or software components communicate) [11] are expected to increase the bandwidth demand. However, it is unlikely that the IoT and M2M communications will significantly affect the broadband traffic outward/inward of the households since most of this traffic will remain within the home network boundaries. The M2M communications can be also found in cars where sensors are frequently used to aid the car maintenance, to inform the developers [35] and to control car settings through dedicated applications [15]. In addition, the M2M communications are used for the electricity, gas and water utility metering in building the smart grids [33], and there are many other emerging applications. On the other hand, the video surveillance applications for increased safety and security are becoming very attractive, and, unlike the locally constrained M2M communications (used at home and in-car networks), they are likely to require significant additional bandwidth. Such applications are expected to grow rapidly and will include different kinds of sensors for automated control of machines. One of the distinct features of using sensors is that they will become almost ubiquitous, but more importantly, they will represent permanent traffic sources possibly requiring dedicated broadband connections which can generate large aggregated traffic over time. This will change requirements on the broadband network design in order to accommodate such dedicated links.

Cloud Computing and Remote Processing

There is a noticeable increase of demand for cloud computing applications [20], which is also driven by proliferation of portable devices with limited computing power and storage [12]. Today, cloud computing includes solutions for delivering hardware and software resources as a service or platform over the broadband networks [29]. For instance, the handheld devices may be used as terminals to access extensive centralized computing resources. The future terminals will have enhanced human-computer interfaces; at minimum, they will be equipped with touch screen, high-resolution display, cameras and other sensors. Alternatively, the computing resources can be deployed or distributed at the network edges which may have some salient benefits. Consequently, cloud computing will change the structure of broadband networks. The related standardization efforts and developments in improving the security and trust of cloud computing are contributing to its acceptance. Furthermore, it is already observed that online storage services are generating a large amount of traffic which can be expected to only increase. Large amounts of data from ubiquitous sensors, other measuring campaigns and, particularly, scientific experiments generate truly massive amounts of data to be stored online in large databases. The data storage and access to such databases is likely to generate large traffic volumes. Thus, the bandwidth requirements of cloud computing and remote processing services are expected to far exceed those of other daily used services (e.g. WWW browsing).

Education and Work Online

Classroom-based lecturing is becoming an obsolete form of educating students. Traditional classroom lectures are now being enhanced or even, in some cases, replaced by online learning methods. There are many case studies clearly demonstrating this trend. We envision that high-quality interactive video streaming links to join students from anywhere around the world into virtual classrooms and virtual labs will be a norm in 20 years' time. Online education can be further enhanced by the use of remote controllers including robots. A very similar discussion is applicable to working online supported by emergence of virtual offices and even virtual enterprises. Hence, education and work online will be significant drivers of future broadband traffic.

Network Architectures

The broadband networks today are mainly concerned with delivering traffic to residential users. In 20 years, the demand for broadband services will become truly ubiquitous. In addition to the residential (home) and enterprise access, the broadband access will be required in public areas and spaces such as parks, streets, bus stops, shopping malls, restaurants, in cars and buses, etc. Majority of this broadband access will be realized wirelessly; however, it is very likely that backhaul traffic will have to be collected by wired networks in order to be cost-effective and in order to provide the sufficient bandwidth for such extremely large aggregated traffic. We believe that this is the main research direction when considering opportunities for significant investments in future broadband networks, especially in 20 years' time. Inevitably, such network developments will be happening in the new market environments and with new business models. The future broadband networks will require different topology and architecture designs as well as different network management strategies in order to provide truly ubiquitous broadband access everywhere. For example, traffic will be automatically re-routed to most efficiently exploit the available instantaneous network transport capacity. In addition, new applications and services can be developed that are more aware of the broadband network state and that can adapt their parameters. In summary, the broadband network infrastructure is a trade-off between the transport capacity and its implementation and operational cost. It is expected that the capacity will remain the main driver in the near future; however, at some point within the next 20 years, the bandwidth demand may saturate, and the cost factors may dominate the broadband network developments [41].

TRAFFIC VOLUMES OF VIDEO STREAMING

As discussed above, applications and services with embedded video streaming will be the dominant driver of traffic growth in future broadband networks. The key parameters of video streaming are as follows: screen size, resolution, color depth, frame rate, achievable compression and statistical multiplexing of streams [4]. Our investigation will consider high-resolution (definition) video services (i.e. the number of horizontal pixels being 4K or even 8K) and the opportunities for significantly improving video quality.

There are several methods how to provide 3D video resulting in 30 80 % overhead compared to the base 2D video. The expected peak rates of future video are as large as 200 megabits/s (Mb/s). For example, YouTube, Netflix and similar other services constitute highly compressed video streams (rates about 1 Mb/s) whereas future IPTV will be of much higher quality (8–16 Mb/s or even 22–30 Mb/s). The take-up rate (i.e. rate of adding new customers) is another key parameter to consider not only for evaluations of traffic but also for cash-flow modelling, e.g. Version reports 25–35 % take-up rates. Assuming the logistic curve [39], the take-up rate grows in the first 2–3 years followed by a linear growth in the intermediate years and the final decline in the last few years. Next, we will summarize the existing and expected future video streaming methods and their bandwidth requirements.

2d Video Streaming

Most of today's 2D video streaming methods use either standard definition (SD) or high definition (HD) format. The minimum and maximum bandwidth requirements of the SD and HD video streaming with MPEG-2 and MPEG-4 compressions, respectively, are given in Table 1[23].

Table 1: The bandwidth requirements for 2D video streaming

2D video format	Minimum bandwidth (Mb/s)	Maximum bandwidth (Mb/s)
SD MPEG-2	2	5
SD MPEG-4	1.5	2
HD MPEG-2	15	20
HD MPEG-4	5	10

3D video streaming

The 3D video technology is still under active development and standardization activities. There are different methods for displaying and encoding 3D video. Some of them are already established while others have just been introduced. The ultimate aim of 3D video is to produce static and, later on, also moving holographic pictures. The main factors influencing 3D video development are the implementation and computing complexity, availability of network bandwidth and ultimately adoption by the consumers. Thus, the main players in adoption of 3D video services are the ISP, network infrastructure providers, TV equipment manufacturers, content providers and subscribers. For example, Philips started to develop an auto-stereoscopic TV to have been launched at the end of 2011, but then, this project was cancelled due to slow adoption of 3D video by the customers [27]. On the other hand, Toshiba, Samsung, Sony and Panasonic launched their auto-stereoscopic TV in the past 2 years [34, 36]. Toshiba makes further investments into developing higher resolution 3D TV sets aiming to provide simultaneous 3D views which are more than any other available TV content of today [44].

The 3D video technologies can be classified into two main classes: 3D video with added depth perspective and 3D multi-view video. The first approach is relatively well established by now, and it uses stereoscopic display with viewing glasses or with parallax barriers. The second approach uses auto-stereoscopic display without any need for viewing glasses. We next review the common 3D video encoding formats.

2D-plus-depth (2D + z) coding

2D + z coding is one of the established encoding formats for 3D stereoscopic displays. The 3D view is achieved by adding depth map details into the conventional 2D images. This 3D encoding represents about 10–20 % overhead compared to the conventional 2D format.

Multi-viewpoint coding (MVC)

MVC is a relatively new 3D encoding format with the initial standard launched by ITU in 2009 as H.264/AVC. The main idea is to record and send multiple viewpoints together and then reconstruct these viewpoints as single 3D images. Assuming that there is no compression involved, the required bandwidth is upper-bounded as a product of the number of multi-views (recommended to be eight for industrial uses) and the bandwidth required for a single view [45]. The required bandwidth for the MVC encoding is increased by about 30–80 % per viewpoint over the conventional 2D single-viewpoint images.

Multi-viewpoint plus depth (MVD)

Since the MVC technique requires a relatively large number of multi-views, and thus, it has large bandwidth consumption, the MVD technique has been introduced to combine a depth map with MVC to reduce the required number of multi-views. Typically, the required number of multi-views for MVD is three in order to achieve the same viewing experience. This 3D video encoding format is still under active research [30, 38, 46].

Assuming the HD MPEG-4 video encoding, Table 2 provides the upper and lower bounds of the bandwidth requirements for different 3D video technologies. Due to very large bandwidth consumption of MVC with eight views, the number of views for MVC is expected to grow gradually, so that it can be accommodated in the available bandwidth.

Table 2: The bandwidth consumption per future 3D encoding stream

Coding scheme	Minimum bandwidth (Mb/s)	Maximum bandwidth (Mb/s)
2D+z	6	12
MVC (8 views)	52	144
MVD (3 views+depth)	23	60

It is expected that the bandwidth requirements due to introduction of 3D video will be suddenly increased, and then this demand will be reduced with development of better video encoding formats and improved compression methods. This situation occurred previously with introduction of 2D video. However, there is still lack of 3D video standards, so the actual bandwidth requirements may be eventually different from those presented in Table 2. The 3D technology is expected to also have a major impact on online gaming, (enhanced) social networking, virtual worlds and other applications that are using video streaming [18]. Thus, video streaming will dominate and possibly reshape traffic of many applications and services. In the near future, the impact of 3D video on average traffic will be moderate; however, its impact on peak values will be much more noticeable. These developments encourage display manufacturers to develop new 3D viewing equipment and technologies.

20+YEARS BROADBAND TRAFFIC FORECAST

We consider a case study of broadband traffic of residential subscribers per household between the cable modem and the optical splitter just outside the customer premises. Aggregated traffic to a local exchange or to a metro node then scales directly with the number of premises and the number of splitter branches. Since the aggregation effectively corresponds to averaging, such traffic will have smaller variations. As discussed in the previous section, we can assume that 3D video streaming (increasingly embedded in other applications) will dominate the bandwidth demands in NGN which will be estimated separately for the downstream and the upstream.

Download Bandwidth Requirements

The 3D video encoding standards are launched every 3 years [42]. It takes about one additional year for the 3D video codec to become considered by wider markets, so we assume that a new 3D video format is introduced every 4 years. Although the number of parallel video streaming channels is likely to be increasing in the future with the introduction of new multimedia applications, we assume that at any time each household is most likely consuming one single video stream. The traffic upper and lower bounds are obtained assuming video at high quality (e.g. high-definition TV) and at minimum quality (e.g. small-screen TV), respectively. These bounds are given in Tables 3 and 4 including the calculated YoY traffic growth rates which are the key parameter for dimensioning the broadband networks. The upper bound shows good approximation of the logistic curve of traffic demand which saturates in the last 4 years of the 20-year period considered. The spread between the upper and lower bounds could be reduced by assuming more realistic conditions.

Table 3: The bandwidth demand upper bounds

Year	HD MPEG-4	2D+z	MVC: 2 views	MVC: 4 views	MVC: 8 views	MVD: 3 views+depth	Total (Mb/s)	YoY growth (%)
2012	4						40	
2016		4					48	5
2020			4				144	32
2024				4			288	19
2028					4		576	19
2032					4	1	636	3

Table 4: The bandwidth demand lower bounds

Year	SD MPEG-4	HD MPEG-4	2D+z	MVC: 2 views	MVC: 4 views	MVD: 3 views+depth	Total (Mb/s)	YoY growth (%)
2012	1						1.5	
2016		1					5	35
2020			1				6	5
2024				1			13	21
2028					1		26	19
2032					1	1	49	17

Upload Bandwidth Requirements

Upstream traffic is expected to grow due to development of new interactive multimedia as well as non-interactive and/or non-multimedia applications; however, the expected growth rate in the upstream will be smaller than that in the downstream. The multimedia applications requiring significant uplink bandwidth are those for interactive communications with embedded video. The cameras capable of recording 3D video are likely to become available within the 20-year horizon. The upload of 3D videos to multimedia or storage servers will contribute significantly to the uplink bandwidth demands even though these uploads could be scheduled for times when the uplink bandwidth becomes available (e.g. overnight or in early morning hours). Furthermore, as discussed above, new M2M communications and other emerging applications and services will make broadband traffic generated at the subscriber's premises to appear peakier with large instantaneous variations. Consequently, the ratio of the downlink to uplink traffic volumes in today's broadband connections being 10:1 will be much more balanced in future broadband networks [21]; this ratio may eventually become 1:1 even though it is difficult to predict how soon this will happen.

In order to gain a first insight into traffic variations in the residential households, we assess how likely or how often traffic delivered to the households is going to reach the previously computed upper bounds. We assume the average usage hours for different applications and user groups as shown in Table 5 [31]. We then calculate the probabilities P_1, P_2 and P_3 corresponding to the probability of accessing the broadband network, the probability of using some multimedia service and the probability of using other non-multimedia applications, respectively (see Table 6). In particular, the probability that all family members (in all age groups) use multimedia with embedded 3D video streaming at the same time and independently from each other is only 0.7 %. The probability that a single adult user consumes a single multimedia stream at any time is 29 %. In other words, the largest bandwidth peak due to multiplexing multiple independent video streams will occur with the probability 0.7 % (i.e. approximately 1 out of 100 households at any time or, equivalently, for any household, for 1 h every 100 h). On the other hand, approximately three out of ten households will generate traffic corresponding to a single video stream at any time, or, equivalently, each household will consume, on average, a single video stream for approximately one third of the day (i.e. 7.2 out of 24 h).

Table 5: The assumed average usage hours per day

Age	Video streaming	Virtual worlds	Online gaming	Online education and working	Other traffic	Total usage hours/day
>18	1.5	1	0.5	4	4	11
14–18	2	2	1	5	5	15
6–10	1	0.5	2.5	0.5	0.5	5

Table 6: The usage probabilities of broadband applications

Age	P_1	P_2	P_3
>18	0.458	0.636	0.364
14–18	0.625	0.67	0.33
6–10	0.208	0.9	0.1

CONCLUSIONS

The long-term traffic trends as well as the disruptive services that will spur new bandwidth demand in NGN were explored. The main public sources of broadband traffic data were provided. The important services and application drivers of traffic growth in future broadband networks were identified. These drivers included both multimedia (2D and 3D video) as well as non-multimedia (software upgrades and download, Internet storage) applications. Long-term traffic models to forecast traffic in the 20+-year horizon from today were investigated. In particular, the upstream and downstream traffic volumes in future NGN were obtained assuming upper bounds of broadband traffic between customer premises and the local exchange or metro node and assuming the most bandwidth-demanding services such as applications with embedded interactive 2D and 3D video streaming. The upstream versus downstream traffic volume ratio of 1:1 is likely to be reached in 20 years' time; however, it is difficult to predict whether this may occur much earlier. The developed long-term traffic models can be connected to socio-economic models of NGN to evaluate the actual economic benefits, to map the potential revenues and to obtain the cash-flow curves of the network deployment employing different competing optical (or other) technologies.

ACKNOWLEDGEMENT

This work was funded by the Welsh Government and British Telecom plc as the Digital Wales Research Hub (DWRH) pilot project in the broadband traffic modelling and evaluation methods. The authors would like to also acknowledge the guidance provided by D. Payne and suggestions provided by N. Doran, D. Nesset and J. Seton.

REFERENCES

1. Al-Tamimi AK, Jain R, So-In C (2012) High-definition video streams analysis, modeling, and prediction. Adv Multimed 539396:1–13
2. Andrés L, Cuberes D, Diouf M, Serebrisky T (2010) the diffusion of the Internet: a cross-country analysis. Telecommun Policy 34(5–6):323–340
3. Bauer S, Clark D, Lehr W (2009) Broadband microfoundations: the need for traffic data. In: Proceedings of beyond broadband access conference. New America Foundation, Washington, DC
4. Bo G (2011) How much bandwidth does Broadcast HD video use? Ciena Insights Blog.http://www.ciena.com/connect/blog/How-much-bandwidth-does-Broadcast-HD-video-use.html
5. British Telecom plc (2007) BT's view on future services. Video podcast.http://www.youtube.com/watch?v=VN3hF8dX8TM
6. British Telecom plc (2011) BT's public statements on the real-world challenges of getting fibre to households.http://www.theregister.co.uk/2011/05/25/bt_fttp_trial_rollout_delayed/ andhttp://www.theregister.co.uk/2011/03/03/bt_milton_keynes_fibre_to_the_premises_trial/
7. Brown P, Petrovic S (2011) A new statistical approach to estimate global file populations from local observations in the eDonkey P2P file sharing system. Ann Telecommun 66(1–2):5–16
8. Cho K, Fukuda K, Esaki H, Kato A, Murai J (2008) Observing slow crustal movement in residential user traffic. In: Proceedings of CAIDA-WIDE Workshop, Marina del Rey, CA, USA. http://www.caida.org/workshops/wide-casfi/0808/slides/residential_user_traffic.pdf
9. Cisco-VNI: Cisco Visual Networking Index (VNI). http://www.cisco.com/en/US/netsol/ns827/networking_solutions_sub_solution.html
10. Crandall RW, Eisenach JA, and Ingraham AT (2013) The long-run effects of copper-loop unbundling and the implications for fiber. Telecommun Policy 37(4–5):262–281CrossRef
11. Crosby T (2013) How machine-to-machine communication works.http://computer.howstuffworks.com/m2m-communication.htm
12. Dediu H (2012) when will tablets outsell traditional PCs? Asymco.http://www.asymco.com/2012/03/02/when-will-the-tablet-market-be-larger-than-the-pc-market/
13. Delgass WN (2004) Hollywood's 'minority report' becomes reality. Eng Manag J 6:5
14. Finisar (2013) Flexgrid technology for optical networks. http://www.finisar.com/technical-information
15. Ford Motor Company (2011) Ford teams with AT&T to wirelessly connect new Ford Focus Electric. http://corporate.ford.com/news-center/press-releases-detail/pr-ford-teams-with-at26t-to-34274
16. FTTH-CE (2013) Fibre to the Home Council Europe. http://www.ftthcouncil.eu
17. García AE, de Lope Rodríguez L, Hackbarth KD (2010) Application of cost models over traffic dimensioning with QoS restrictions. Ann Telecommun 65(3–4):135–144
18. George J (2006) Start thinking about 3 to 30 Gbps by 2030! Broadband Properties, No 42–47. http://www.broadbandproperties.com
19. Grote W, Grote A, Delgado I (2008) IEEE 802.11 goodput analysis for mixed real-time and data traffic for home networks. Ann Telecommun 63(9–10):463–471
20. Harris D (2011) Cisco: Cloud traffic fast approaching the zettabyte era.http://gigaom.com/2011/11/29/cisco-cloud-traffic-fast-approaching-the-zettabyte-era/
21. Harstead E (2011) Future bandwidth demands favors TDM PON, not WDM PON. In: Proceedings of optical fiber communication conference and exposition, Los Angeles
22. Hoernig S, Jay S, Neumann KH, Peitz M, Plückebaum T, Vogelsang I (2012) The impact of different fibre access network technologies on cost, competition and welfare. Telecommun Policy 36(2):96–112
23. ICF News (2013) How much bandwidth? Newsletter of International Cablemakers Federation. http://www.icf.at/en/6000/how_much_bandwidth.html
24. Karagiannis T, Molle M, Faloutsos M (2004) Long-range dependence ten years of Internet traffic modeling. IEEE Internet Comput 8(5):57–64
25. Lah K (2010) Is 'Minority Report' becoming reality? CNN International Business News.http://business.blogs.cnn.com/2010/03/11/is-minority-report-becoming-reality
26. Madden G, Tan J (2007) Forecasting telecommunications data with linear models. Telecommun Policy 31(1):31–44
27. Martin N (2008) 3D television without the special glasses. The Telegraph (30 May).http://www.telegraph.co.uk/news/uknews/2054047/3D-television-without-the-specialglasses.html

28. MINTS: Minnesota Internet Traffic Studies (MINTS). Project homepage:http://www.dtc.umn.edu/mints

29. Monaco A (2012) A view inside the cloud: why the time is right for cloud computing to soar. The Institute, the IEEE news source. http://theinstitute.ieee.org/technology-focus/technology-topic/a-view-inside-the-cloud

30. Na ST, Oh KJ, Lee C, Ho YS (2008) Multi-view depth video coding using depth view synthesis. In: Proceedings of IEEE international conference on circuits and systems, pp. 1400–1403

31. Ovum Europe (2012) The benefit to the Canadian economy from the wireless telecommunications industries: an economic impact assessment. Technical report for CWTA, pp. 1–50

32. Papagiannaki K, Taft N, Zhang ZL, Diot C (2005) Long-term forecasting of Internet backbone traffic. IEEE Trans Neural Netw 16(5):1110–1124

33. Rogers Communications (2011) Hydro-Quebec chooses Rogers for wireless connectivity of Smart Meter collectors. http://www.newswire.ca/en/story/731359/hydro-quebec-chooses-roger-for-wireless-connectivity-of-smart-meter-collectors

34. Samsung (2010) Samsung learning resources: 3D TV.http://www.samsung.com/us/learningresources3D

35. Schwartz E (2003) When machines speak. InfoWorld Vol. 11.17.03, No 45

36. Shimbun Y (2010) Toshiba to launch 3-D TVs that don't require glasses, Daily Yomiuri Online. http://www.yomiuri.co.jp/dy/business/T100824003774.htm

37. Smarthouse (2013) Online retail store for home automation.http://www.smarthousestore.co.uk

38. Smolic A, Kimata H, Vetro A (2005) Development of MPEG standards for 3D and free viewpoint video. In Proc SPIE Conf Opt East: Commun, Multimed Display Technol 6014:262–273

39. Stordahl K (2010) Market development up to 2015. Celtic project CP5–013, Deliverable D3.4, Telenor

40. Stordahl K (2011) Long-term penetration and traffic forecasts for the Western European fixed broadband market. In: Proceedings of 22nd European regional ITS conference, Budapest

41. Stordahl K (2008) Broadband demand and the role of new technologies. In: Proceedings of 13th international telecommunications strategy and planning symposium, Budapest

42. Sullivan GJ (2005) Overview of international video coding standards (preceding H.264/AVC). ITU-T VICA Workshop, Geneva, Switzerland. http://www.itu.int/ITU-T/worksem/vica/docs/presentations/S0_P2_Sullivan.pdf

43. Thilakawardana D, Moessner K (2009) Traffic modelling and forecasting using genetic algorithms for next-generation cognitive radio applications. Ann Telecommun 64(7–8):535–543CrossRef

44. Toshiba 55ZL2 (2005) Review of Toshiba 3D TV. http://reviews.cnet.co.uk/tvs/toshiba-55zl2-review-50005001

45. Tytgat D (2009) New applications and opportunities in 3DTV IP distribution. In: Proceedings of Workshop VI: 3DTV: a distant perspective? Alcatel Lucent

46. Vetro A, Sehoon Y, Smolic A (2008) Towards a 3D video format for auto-stereoscopic displays. Technical report TR2008–057, Mitsubishi Electric Research Laboratories

CHAPTER 18 TCP PERFORMANCE IN MOBILE ENVIRONMENTS

Deguang Le, Xiaoming Fu and Dieter Hogrefe

Keywords: TCP, Performance, Cross-layer, Mobility management, Mobile IPv6

INTRODUCTION

With the development of wireless access technologies, such as IEEE 802.11x, GPRS, 3G cellular networks, and Bluetooth, there is a trend converging these systems towards an IP-based infrastructure. To allow users with mobile devices to stay connected while moving to a new location, various mobility proposals have been developed [19]. One of the most mature solution is the IETF Mobile IPv6 (MIPv6) protocol [15], which handles mobility in the network layer. Various extensions for MIPv6 have been also developed, including the IETF Hierarchical Mobile IPv6 (HMIPv6) [27], Fast Handovers for Mobile IPv6 (FMIPv6) [18] and their cross-layer enhancements such as link layer triggers [21].AlthoughtheMIPv6-based mobility approach makes the transport layer transparent from mobility, it may cause performance degradation in upper layers [2], especially for TCP [25], the key transport layer protocol.

TCP is a reliable transport protocol tuned to perform well in traditional fixed networks, where the route for each active TCP connection is relatively invariable and network congestion is the primary factor of affecting its performance. However, in mobile environments, it is common that the route of active connection changes more frequently due to communicating endpoints moving to a different location with different network characteristics. Without considering mobility, TCP will perform normal network congestion and trigger congestion control (e.g. exponential backoff of retransmission timeout or reduction of its congestion window) as done in fixed network cases, which may not be desired in mobile environments. This may result in significant end-to-end transport delay deviation and packet loss.

Existing works on TCP over wireless networks [5,8] focused on issues associated with wireless link features such as Bit Error Rate (BER) and loss, and usually do not consider the issue of mobility. Two broad approaches have been widely investigated: (1) link layer-based approach, e.g. SNOOP TCP [4]; (2) split connection approach, e.g. I-TCP [3], M-TCP [6]. These works assume the TCP over wireless problem as local problem and as a result attempt to address it locally.

Since support for mobility in IP networks has emerged as a critical issue, TCP performance issues over mobile environments using MIPv6 have been recently studied. Hsieh etal. [12] Show that FMIPv6 and FMIPv6 in the network layer alone can bring better TCP performance than standard MIPv6. In the transport layer based approach, Fast Retransmission (FR) [7] and Freeze TCP [9] attempt to improve TCP performance by distinguish packet losses due to the host mobility from those due to the network congestion. FR [7] handles this by fast retransmission of the earliest unacknowledged data segments, immediately after completing the handover, while Freeze TCP [9] freezes the connection state as soon as the Mobile Node (MN) detects an impending handover. However, the trigger of reacting mechanisms to mobility in most of these approaches is rather dependent on the feedback from the transport layer itself, or the link signal strengths. None of these approaches could effectively and timely enough to dctcct nctwork situation change (e.g. route change) in existence of any underlying mobility mechanism.

Demo-Vegas [11] is a TCP extension for performance improvements over Mobile IPv4 (MIPv4) networks, finely tuned for TCP Vegas without considering other possible TCP versions. Matsushita etal. [20] Proposed modifications in both the mobile node and the access gateway, basically by applying the split connection approach to HMIPv6. Both approaches improve TCP performance to some extent, but

Citation: Deguang Le, Xiaoming Fu, Dieter Hogrefe, TCP Performance in Mobile Environments, DOI: 10.1007/s11277-008-9624-9.

do not sufficiently offer a general solution to improve TCP performance over mobile environments.

In this paper, based on a qualitative investigation on TCP over MIPv6 networks, we propose TCP-CM, a cross-layer approach utilizing the cooperation between network layer and transport layer for performing TCP reactively in mobile environments. TCP-CM can be easily extended for different versions of TCP (e.g. Vegas, Reno and Tahoe) and MIPv6 extensions (e.g. HMIPv6 and FMIPv6). Analysis shows that the effective data resumption time of TCP-CM is better that of existing approaches. Simulation results show that our proposed approach can yield better performance than the standard TCP in MIPv6.

The remainder of this paper is organized as follows. Section 2 analyzes the problems of TCP related to mobility using MIPv6-based approaches. Section 3 describes our proposed approach. Section 4 presents analytical model to demonstrate the benefits of our approach over existing approaches, followed by simulation results in Sect. 5. Finally, conclusions and future work are provided in Sect. 6.

PROBLEMS STATEMENT

The TCP protocol [25] is a general-purpose, reliable stream protocol. It works in different network environments, independent of medium, transmission rate, delay, and error rate. Standard TCP (TCP Reno) integrates support for acknowledgement, retransmission, congestion control, timing, maximum segment transmission mechanisms, and performs well in classic fixed networks. It primarily estimates retransmission timeout (*RTO*) based on measured round-trip time (*RTT*) and maintains two key parameters in the TCP sender: the congestion window (*CWND*) and the slow start threshold (*SSTHRESH*) for congestion control. In the case of MIPv6, when the MN moves and switches between the access networks, MIPv6 will perform the following procedures: movement detection, configuration of Care-of-Address (CoA) and binding registration. During these procedures, the communicating endpoints are not able to continue the communication between each other for some time (aka handover latency), which may however introduce the transport delay and packet loss, resulting in TCP congestion control mechanism in the following subtle ways. More details about MIPv6 and TCP can be found in [1, 17, 15, 24–26, 29].

Figure1 visualizes the segment transmission from the Correspondent Node (CN) to the MN during the handover, where we assume that segment 1 is the last successful sent segment through the previous access network. From this figure, we can see that when the MN moves out of its previous network at time*t*1 (i.e. The initiation of hand over), because the TCP layer in the CN is unaware of the occurrence, it may continue to send segments to the MN's previous network even if the MN has moved away (see Fig.1 from segment 2 to segment *k*). Thus, any segments sent out by the CN from time *t*0 (i.e. half RTT before the handover initiation) to time *t*2 (i.e. the termination of handover) will not reach the MN. This causes that if the

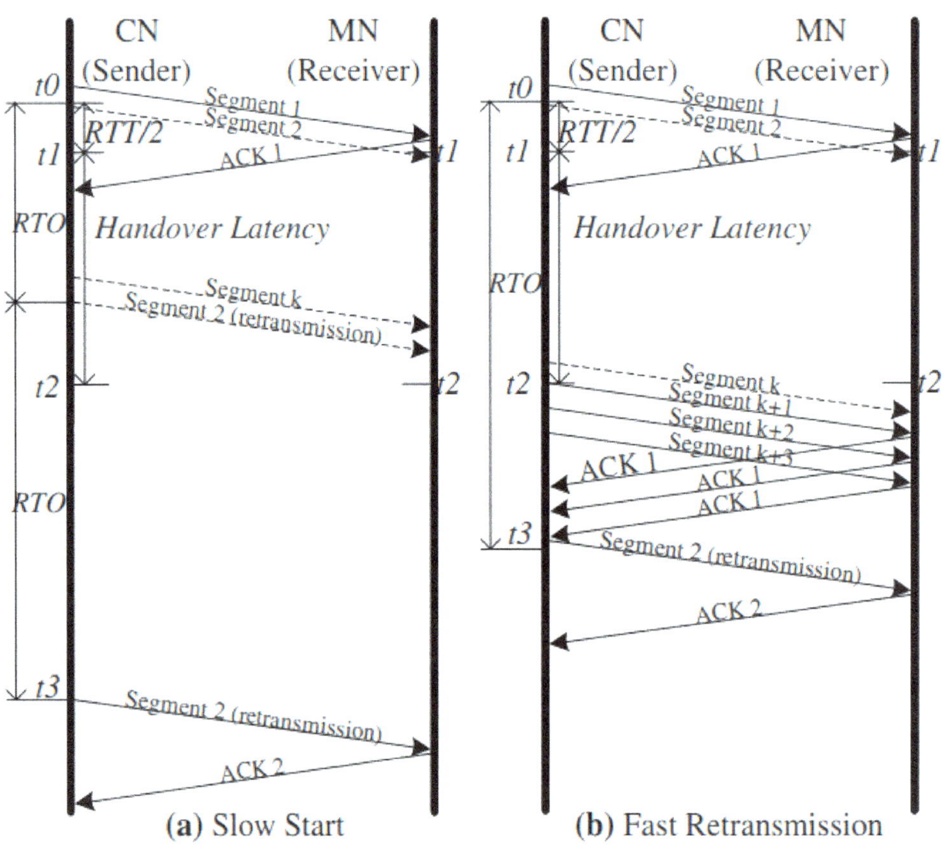

Figure 1: Visualization of segment transmission from the CN to the MN during a MIPv6 handover

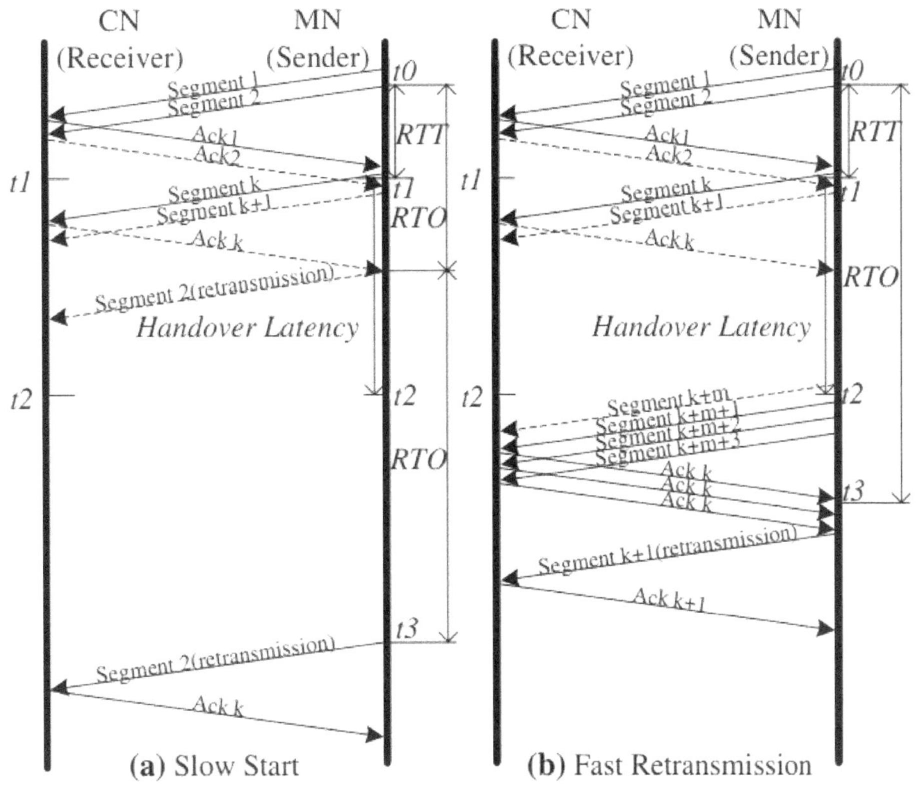

Figure 2: Visualization of segment transmission from the MN to the CN during a MIPv6 handover

timeout occurs before the handover termination, the slow start congestion control will be invoked (see Fig.1a), which will drop CWND to the size of one segment, and retransmits the earliest unacknowledged data segment (see Fig.1a retransmitted segment 2); if the timeout does not take place during the handover latency (i.e. from time *t*1 to *t*2), the MN may receive the new data segment from the CN after the termination of handover if allowed by the current window size of the CN (see Fig.1b from segment *k* +1 to segment *k* +3), which will incur the CN to retransmit the lost packets and begin the fast retransmission congestion control because of the consecutive duplicate acknowledgements from the MN (see Fig.1b retransmitted segment 2).

The case for segments sending from the MN to the CN (see Fig.2) is similar: when the MN moves out of its previous network, all the segments sent but not yet acknowledged in the sending window will not be acknowledged anymore because of the unreachability of the acknowledgement messages, which are responded from the CN to the MN's previous network (see Fig.2 from segment 2 to segment *k*). If the lost ACKs do result in a retransmission timeout, the MN will return to the slow start stage (see Fig.2a), which reduces its CWND to one segment and retransmit the earliest unacknowledged data segment (see Fig.2a retransmitted segment 2); if the timeout does not take place during the handover latency, the CN may receive the new segment from the MN after the handover termination if allowed by the current window size of the MN (see Fig. 2b segment $k + m + 1$). Then the packets sent before *t*1 (see Fig.2b from segment 2 to segment *k*) may be confirmed after the new acknowledgmentisdeliveredduetothecumulativefeatureofACKs.However,segmentssent during the time from time *t*1 to *t*2 (see Fig.2b from segment *k* +1 to segment *k* +*m*) may be discarded by the interface of the MN during the movement detection and CoA configuration procedures or by the Home Agent (HA) during the stage of binding update procedure, which may likewise incur duplicated ACKs retransmission or timeout retransmission (see Fig.2b).

In either scenario, the packet loss induces the timeout retransmission, finally causing a very small *CWND* and an equally small slow start threshold (SSTHRESH). Moreover, consecutive retransmission timeouts incur an artificial inflation of RTO due to Karn'sexponential timer back off algorithm [17]. As a result, the sender may waste long time to wait for a timeout before it can send new data continuingly (see Figs.1a and 2a from *t*2 to *t*3). In case of duplicate ACKs retransmission (see Figs.1b and 2b), even if SSTHRESH is dropped to half of their previous values, it may also be stale for the new network. For example, when the new network has a lower capability (i.e. Bandwidth Delay Product, *BDP*) than the previous network, the new setting may still be too high for the new network. The sender may inject more segments into the new network than necessary, which could cause the transmission queue at the bottleneck link's router to overflow. Conversely, if the new network is capable of carrying much more segments or is more lightly loaded than the previous network, the value of *SSTHRESH* will become irrelevant, and the increase of *CWND* is limited to one segment per *RTT*. It therefore will take multiple RTTs before reaching a reasonable throughput due to the slowness of additive increase. In addition, in MIPv6, when the MN moves from one network to another network, the route between communicating endpoints will change, causing the end-to-end transport delay to vary and the deviation of *RTT* increases quickly due to different network characteristics.

Therefore, the traditional RTO update algorithm [24] does not adapt to the old *RTT* and its deviation to estimate the *RTO* in the new network, which may result in TCP to mistakenly make the congestion control decision based on invalid *RTT/RTO*

A CROSS-LAYER APPROACH FOR IMPROVING TCP PERFORMANCE OVER MOBILE IPV6

The above analysis demonstrate that, in highly dynamic network environments, it is difficult to maintain high transport performance as well as extended mobility functionality via a single-layer solution. This motivates the design of a cross-layer approach, so-called TCP-MC, enabling TCP be more responsive to mobility behaviors in MIPv6. This approach introduces only modest changes to TCP and MIPv6 to enhance cooperation between them through inter-layer signalling interactions, allowing TCP to tune its behaviours reactively according to the procedures of the underlying mobility management protocol. The following subsections detail the key design rationales. We will see in Sect. 4 that the proposed approach is extensible for different MIPv6 extensions.

Enhancement of Inter-Layer Interaction between TCP and MIPv6

To enhance the cooperation between TCP and MIPv6, MIPv6 must monitor the mobility related procedures and is aware that subsequent procedures should be optimized by TCP. In addition, TCP also requires a way to learn of these mobility-related events of MIPv6.

We define two types of mobility events in MIPv6: Initiation_of_Handover (IoH) and Termination_of_Handover (ToH). The MN can easily identify its handover initiation through the Neighbour Unreachability Detection mechanism [22], through which the MN can actively probe the current access router using unicast Neighbor Solicitation messages to verify if the forwarding route is still working. If the MN receives a solicited Neighbor Advertisement with the Solicited flag set, it confirms that the current access router is reachable. Otherwise, it indicates the MN is moving out of its current network, causing the event of IoH. For the event of ToH, we can determine the event through the binding registration mechanism, through which the MN will receive the successful binding acknowledge when handover terminates. Following the mobility-related events, MIPv6 may signal TCP. The signalling may be delivered through the Interface Control Information (ICI) [32], which allows the IP module to notify mobility events and triggers corresponding TCP operations.

At the CN, an additional communication step is necessary to inform TCP of the events occurring at the other endpoint (i.e. the MN) of the connection. For the notification of IoH, it is necessary for thc MN to predict the reasonable time (i.e. a warning period) before which an actual Neighbour Unreachability occurs. Therefore, in our approach, after the notification of the completion of handover as described above, the MN forwards the signal of other events except the IoH over the new network to the CN. The signalling travels from the MN to the CN through the normal IP route and can choose either of the following two ways.

One way is to extend TCP, similar to the explicit notification mechanism by setting an Explicit Mobility Notification (EMN) flag [10] and introducing an Explicit Handover Notification bit [11,13] in a specially marked TCP acknowledgement segment, or using a special option in TCP [30]. This mode simplifies some of the inter-layer issues at the CN. However, it has several drawbacks: the mobility notification may have to be redone for other transport protocols, it becomes difficult to share messages between different transport protocols, and the connection-oriented state maintained by TCP may not easily extend to save events for long periods. In addition, the inter-layer interaction of the MN is still inevitable

Another way is to extend the IP layer to make use of the MIPv6 binding registration mechanism to transfer mobility-related events of the MN. We assume both the MN and the CN support the MIPv6 route optimization mode. hence this network-layer extension approach is preferred. By extending the Binding Update message with a ToH bit to indicate the CN's IP layer upon the detection of the mobility-related event, the event can then be transferred to TCP through an inter-layer interaction mechanism.

As soon as TCP is aware of the mobility-related events of the MN, it can react to mobility behaviors of the MN by tuning the state parameters. The adjustments of state parameters are derived from the mechanisms described in the following subsection.

TCP Activities to Mobility in MIPv6

When TCP is aware of the mobility-related behaviours of MIPv6, it invokes the following activities in response to mobility in MIPv6:

Congestion Avoidance: Prevent the invocation of congestion control while the MN moves. Since it is common that the communicating endpoints change their route while communicating with each other, it is not suitable that each time the route variation occurs, the window size starts from the slow start stage. Therefore, we need a way to optimize *CWND/SSTHRESH* values by quickly probing the route's available capacity. The optimal congestion window size can be determined based on the bandwidth estimation. In addition, updating the congestion window is based solely on ACKs for data sent through the new network.

Fast Recovery: Ignore the outstanding segments in the send window. These segments or their ACKs, which are sent before the handover initiation, are in flight. Although the retransmission timeout of these segments may not take place, they in fact cannot reach the receiver any more. Therefore, these segments should be retransmitted immediately.

Timing Re-initialization: Reset the bogus timing parameters. It is obvious that the RTT fluctuates significantly in mobile environments. Traditional TCP computes the smooth RTT by using previous values and RTO depends on the smooth RTT [24]. In order to avoid false timeout and artificial inflation of pause, the timing parameters RTT/RTO should be reinitiated to the value of RTT acquired from the new network.

In [7], it simply retransmits the earliest unacknowledged segment immediately followed by the congestion control (i.e. the timeout) back off. However, it cannot prevent from invoking the congestion control procedure. In our approach, the unnecessary congestion control can be avoided through the mechanism (1) and (3). Besides, when the MN moves and the handover takes place, the data of TCP connection may pass through different routes over heterogeneous networks (i.e. mobile packet routing). Due to the different network characteristics, the parameters of TCP connection (e.g. CWND, SSTHRESH, RTT and RTO) for the old network may not be suitable to the new network, so they need be tuned timely along with the changes of attachment networks. In [9], the Freeze TCP chokes all retransmission timers and enters persist mode without shrinking its CWND/SSTHRESH size through the Zero Window Advertisement (ZWA) when the handover takes place. When the MN moves into the new network, the TCP connection resumes through the Triplicate Re-connection ACKs (TR-ACKs) mechanism as suggested in [7]. This simple tuning of TCP parameters cannot match the new network characteristics, especially with frequent handovers. Our above congestion avoidance and timing re-initialization together are employed to handle differences in network characteristics upon mobility events by optimizing CWND/SSTHRESH values and re-initializing the timing variables RTT/RTO. Ho etal. [11] Simply does the mechanism (1) by employs Vegas to adjust the congestion window for congestion avoidance in the new network after handover. However, it does not consider stale data segments in retransmission queue and bogus timing. Thus, it is incapable of handling artificial inflation of RTO. Through the combination of above mechanisms, as also shown in the simulations in the following section, the adverse impacts on TCP introduced by the mobility features of MIPv6 can be overcome effectively in a timely fashion, which in turn prevent the degradation of transport performance.

After describing the changes for TCP, the following subsections describe the mechanisms required for different types of MIPv6 entities.

Activities for Mobile Node

Consider the scenario where the MN is sending data to the CN. When the MIPv6 of the MN determines that the current access router is no longer reachable through the neighbor unreachability detection mechanism, it notifies TCP of the event of IoH. TCP then stops sending segments to the IP layer and the counter of RTO is suspended immediately. When the MN receives a packet of valid Binding Acknowledgement, which indicates the handover completion, it notifies TCP of the event of ToH. Once TCP receives the signal of ToH, it resumes the *RTO* value and TCP may resume the transmission if allowed by the sending window size. If the usable sending window size is zero, the window size is increased by one segment, and then TCP goes on transmitting the segment to the CN. Therefore, the unacknowledged segments to the CN sent through the previous network may be confirmed after the new acknowledgment through the new network is delivered due to the cumulative feature of ACKs.

Activities for Correspondent Node

Consider the second scenario where the CN is sending data to the MN: as soon as TCP at the CN receives the signal of ToH, it clears the *SRTT* (i.e. the smoothed *RTT*) and *RTTVAR* (i.e. the *RTT* variation) and resets *RTO* according to the following formulas [24]

$$SRTT = R \tag{1}$$

$$RTTVAR = R/2 \tag{2}$$

Table 1: Effective data resumption time

Protocol	Effective data resumption time
FMIPv6	$4T_{CN-oAR} + 2T_{oAR-nAR}$
HMIPv6	Intra-domain: $4T_{CN-MAP} + 2T_{MAP-oAR} + 2T_{MAP-nAR}$
	Inter-domain: $2T_{CN-oAR} + 4T_{CN-MAP} + 12T_{MAP-nAR} + 4T_{HA-MAP}$
FR	$T_{CN-oAR} + 3T_{CN-nAR} + 2T_{HA-nAR}$
Freeze TCP	$T_{CN-oAR} + 3T_{CN-nAR} + 2T_{HA-nAR}$
Mobile TCP	$T_{CN-oAR} + 3T_{CN-nAR} + 2T_{HA-nAR}$
TCP-CM	$T_{CN-oAR} + 2T_{CN-nAR} + 2T_{HA-nAR}$
TCP-CM o.FMIPv6	$2T_{CN-oAR} + 2T_{oAR-nAR}$
TCP-CM o.HMIPv6	Intra-domain: $4T_{CN-MAP} + 2T_{MAP-oAR} + 2T_{MAP-nAR}$
	Inter-domain: $2T_{CN-oAR} + 2T_{MAP-nAR} + 2T_{MAP-nAR} + 2T_{HA-MAP}$

$$RTO = SRTT + \max(G, K * RTTVAR) \tag{3}$$

Where *R* is the value of RTT observed over the new network, *G* is the granularity of the clock, and *K* is a coefficient.

Since the outstanding segments in the retransmission queue travel through the previous network, and will never reach the MN, these segments are retransmitted, but not invocate the congestion control mechanism. Besides, the CN resets *CWND* and *SSTHRESH* according to the *BDP* of the new network, which represents a sender's TCP congestion window [29]. That is, the *SSTHRESH* is set equal to the available pipe size when connections are switched to the new network (i.e. $BWE * RTT_{new}$), and *CWND* is set to be equal to *SSTHRESH*. Then, the congestion avoidance takes over, where *BWE* means the estimated bandwidth and the value RTT_{new} is measured over the new network.

ANALYSIS AND COMPARISON

Our propose dapproach has several differences from previous approaches as discussed above. In the following sub sections, we will identify its advantages in terms of effective data resumption time introduced by mobility based on a mathematical model.

Considered Model

We firstly introduce a MIPv6 topology model used for analytical study. Figure 3 depicts the considered model. We assume that a Mobile Node (MN) is initially located at the old Access Router (oAR) and then moves from the oAR to a new AR (nAR). The mobile node receives the data packets sent from the Correspondent Node (CN). The MAP refers to the mobility anchor point employed by HMIPv6 for hierarchial mobility management [27]. The analysis will study the network-layer approaches (i.e.FMIPv6 [18] and HMIPv6 [27]), transport-layer approaches of FR [7], Mobile TCP [28] and Freeze TCP [9], and our proposed TCP-MC. The reason is that all of they can handle the packet loss during the effective data resumption time introduced by the handover as discussed in Sect. 2 (see Table 1).

Assumptions and Parameters

In order to compute the effective data resumption time we have to consider the delay introduced by packet transmission between nodes. The effective data resumption time will be analyzed considering the mobile node initiated handover case.

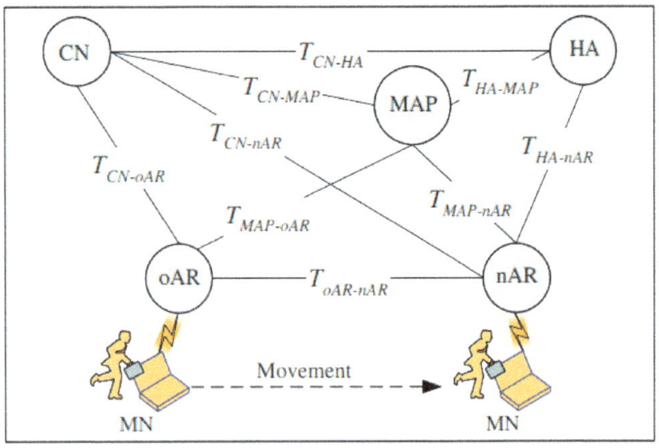

Figure 3: Considered scenario for numerical analysis

We assume that the processing delay are negligible compared to access to the channel and transmission delays. With respect to the parameters, we have the following assumptions.

- T_i–j only denotes the transmission delay of path between any two entities i and j. For example, T_{CN-oAR} is referred to as the time required by forwarding packets on the path from the CN to the oAR, or reversely. We suppose the same value for both uplink and downlink directions.
- The processing latency of local trigger in a mobile node's protocol stack is ignored. Thus, the period used to receive a moment hint with link-layer support is zero.
- The latency of wireless links between a mobile node and access router is ignored since it is much smaller than that of transmission delay on the global Internet.

Numerical Analysis

In MIPv6, if the network prefix has changed, the MN need configure a new temporary CoA. Then, the MN updates its location by advertising its new CoA with its HA and eventually it's CN(s). In general, the handover procedures can be regarded to comprise three blocks: the movement detection, the CoA configuration and the home (or correspondent) registration. During the handover procedures, there is a period when the MN is unable to receive TCP data segments both due to link switching and TCP/IP protocol operations, called TCP payload discontinuity for TCP connection. Therefore, we define the effective data resumption time of TCP connection as follows:

Definition of Effective Data Resumption Time The effective data resumption time of TCP connection refers to the time period between the receipt of the last TCP data segment on a previous access network and the resumption receipt of a new TCP data segment on the new access network when a communicating endpoint moves and changes its access networks.

Based on this definition, the overall effective data resumption time of TCP connection, which is represented by T, may thus be computed as the sum of the time steps as follows:

$$T = T_1 + T_2 + T_3 \quad (4)$$

Here T_1 refers to the transmission time that the TCP data segment is sent by the CN to the previous network of the MN before the handover initiation; T_2 refers to the time period that elapses during the handover procedures. It includes the movement detection time, CoA configuration time and binding update time; T_3 refers to the time of RTO for TCP data segment lost during the handover after the handover termination, or the time for duplicated TCP data segment transmission after the handover termination.

TCP-CM

Considering our proposed model described in Fig.3, the effective data resumption time of

T_{TCP}–CM will be the sum of the following contributions: (1) the time that any segment is sent by the CN to the old network of the MN before the handover initiation (i.e. T_{CN-oAR}); (2) the time the MN receive Unsolicited Router Advertisement (URA) messages from the access routers, which indicates that an MN entering a new link (i.e. the movement detection and handover initiation) [22]; (3) the time that the MN sends Router Associated (RA) messages for stateless address auto-configuration [31] over wireless link; (4) the time required to send the BU and receive the BA through the wireless medium for home registration; (5) the time required by the BU to reach the HA who forwards the packet to the MN (T_{HA-nAR}); (6) the time required by the BA to reach the current nAR (T_{HA-nAR}); (7) the time required to send the BU and receive the BA through the wireless medium for correspondent registration; (8) the time required by the BU to reach the CN who continuously sends the packet to the MN (T_{CN-nAR}), and (9) the time required by the BA to reach the current nAR (T_{CN-nAR}).

Based on our assumption, we ignore the part of local wireless link delay. Thus, the T_{TCP}–CM due to performing a handover from the oAR to the nAR in MIPv6canbecomputed through the following equation:

$$T_{TCP-CM} = T_{CN-oAR} + T_{HA-nAR} + T_{HA-nAR} + T_{CN-nAR} + T_{CN-nAR} \quad (5)$$

FR

Considering the case of FR [7], as soon as the connection is re-established through the nAR, the MN sends triplicate ACKs without waiting for a retransmission timeout for the last data segment which it received prior to the discontinuity through the oAR [7]. When the CN receives three copies of ACK, it will continuously transit TCP segments. Therefore, the effective data resumption time of T_{FR} for FR is composed of the following parts:

(1) the time required to successfully send the last TCP segment through oAR to the MN; (2) the time required by the MN performing the handover procedures, and (3) the time required by the TR-ACKs from the nAR to the CN.

Since the first two steps is calculated in Eq.5, the T_{FR} performing a handover from the oAR to the nAR in FR can be calculated through the following expression:

$$T_{FR} = T_{CN-oAR} + T_{HA-nAR} + T_{HA-nAR} + T_{CN-nAR} + T_{CN-nAR} + T_{CN-nAR} \quad (6)$$

Freeze TCP has the same effective data resumption time as that of FR because it employs the same TR-ACKs mechanism suggested in FR [9]. Therefore, we get the T_{TCP}–CM as follows:

$$T_{FreezeTCP} = T_{CN-oAR} + T_{HA-nAR} + T_{HA-nAR} + T_{CN-nAR} + T_{CN-nAR} + T_{CN-nAR} \quad (7)$$

Mobile TCP

Considering the case of Mobile TCP [28], when an MN changes its access routers (e.g. movement from the oAR to nAR), the icotl call corresponding to the downing of an interface or changing of its address is trapped by the kernel device control code. All the sockets that use the affected interface are notified, and a switch detection routine is invoked. This causes the MN to notify the CN via an option field in the next Mobile TCP packet [28]. Therefore, an additional timeliness is required to be considered beside the time required to successfully send the last TCP segment through the oAR to the MN and the time required by the MN performing the handover procedures, namely the time required for explicit handover notification through a special TCP option.

Thus, the $T_{MobileTCP}$ due to performing a handover from the oAR to the nAR in MIPv6 can be expressed through the following equation:

$$T_{MobileTCP} = T_{CN-oAR} + T_{HA-nAR} + T_{HA-nAR} + T_{CN-nAR} + T_{CN-nAR} + T_{CN-nAR} \quad (8)$$

FMIPv6 and HMIPv6

Based on above analysis, we find that the main contributions of effective data resumption time come from the MIPv6 handover latency. Next, we study the effective data resumption time under the FMIPv6 and HMIPv6, which tries to optimize the TCP performance through decreasing the handover latency of MIPv6 (see Sect. 3).

FMIPv6 allows a MN to configure a new CoA before it moves and connects to a new network. It also allows the MN to use the new CoA immediately upon connecting to the new network. Besides, the FMIPv6 provides a bi-directional tunnel between the old and new networks for packets forwarding while the BU procedures are being performed. Thus, compared to standard MIPv6 operations, the FMIPv6 not only can remove IPv6 configuration delay introduced by movement detection and CoA configuration, but also eliminate the delay involved during the MN's BU procedures. However, due to the features of timing and re-transmission of TCP, it is not necessary that the real effective data resumption time can be reduced in practice.

The RTO can be calculated through the algorithm of TCP's retransmission timer [24] when the MN is attached to the oAR as follows:

$$RTO \approx 2T_{CN-oAR} \quad (9)$$

During the handover, the data packets will be forwarded between the oAR and the nAR through IP tunneling in FMIPv6 [18], hence the data transmission time between the CN and

MN will be T_{CN}–$oAR + T_{oAN-nAR}$. Therefore, the inequality 2 $*(T_{CN}$–$oAR + T_{oAN-nAR}) > RTO$ will be satisfied according to Eq. 9, which causes re-transmission timeout events.

Because of the features of timing and re-transmission of TCP, when a timeout occurs, the TCP sender will retransmit data segments and reset the retransmission timer to backoff interval according to Karn's exponential backoff algorithm [17]. Therefore, the TCP sender (e.g. the CN) has to waits for yet another transmission of duplicate data segments before it can continuously transmit new data segments.

Thus, the T_{FMIPv6} due to performing a handover from the oAR to the nAR in FMIPv6 can be expressed through the following equation:

$$T_{FMIPv6} = T_{CN-oAR} + T_{CN-oAR} + T_{CN-oAR} + T_{oAR-nAR} + T_{oAR-nAR} + T_{CN-oAR} \quad (10)$$

HMIPv6 [27] is another extension to MIPv6 for optimizing the handover latency by introducing a new entity of Mobility Anchor Point (MAP) severing as a local home agent. When the MN moves within the MAP domain (called intra-domain mobility), mobility is handled by the MAP, and the MN only performs the BU procedures (i.e. local BU) to local MAP, which then eliminate the global BU procedures with its HA and CNs. However, when the MN moves between the MAP domains (called inter-domain mobility), the MN will have to perform both local BU procedures with the local MAP and global BU procedures with its HA and CNs. Therefore, the real effective data resumption time in HMIPv6 depends on the movement modes (i.e. intra-domain mobility and inter-domain mobility).

For intra-domain mobility, the handover latency ($T_{HO-Intra-Domain}$) will be the time required for the MN to send the local BU and receive the BA from the MAP:

$$T_{HO-Intra-Domain} = 2T_{MAP-nAR} \quad (11)$$

Similarly, we can calculate the RTO in MIPv6 before the handover initiation as follows:

$$RTO \approx 2T_{MAP-oAR} + 2T_{CN-MPA} \quad (12)$$

Since the MAP is located at the same local domain as the oAR and nAR, we assume that $(TMAP_{-oAR}) \approx (TMAP_{-nAR})$. Therefore, we can get $THO_{-Intra-Domain} < RTO$ based on Eqs. 11 and 12. Because the TCP segments send out by the CN during the handover are lost, the retransmission timeout will occurs and the lost TCP segment will be retransmitted. Thus, the effective data resumption time in the intra-domain mobility is calculated as follows:

$$T_{HMIPv6-Intra-Domain} = T_{CN-MAP} + T_{MAP-oAR} + T_{MAP-oAR} + T_{CN-MAP} + T_{CN-MAP} + T_{MAP-nAR} + T_{MAP-nAR} + T_{CN-MAP} \quad (13)$$

For inter-domain mobility, the handover latency ($T_{HO-Inter-Domain}$) will be the time required for the MN to send the local BU and receive the BA from the MAP, and plus the time required for global home registration and correspondent registration:

$$T_{HO-Inter-Domain} = T_{MAP-nAR} + T_{MAP-nAR} + T_{MAP-nAR} + T_{HA-MAP} + T_{HA-MAP} + T_{MAP-nAR} + T_{MAP-nAR} + T_{CN-MAP} + T_{CN-MAP} + T_{MAP-nAR} \quad (14)$$

From the Eqs. 12 and 14, we get $T_{HO\text{-}Inter\text{-}Domain} > RTO$, which causes that the handover risnotyet complete when the timeout occurs. Thus, during the and over, the retransmitted TCP segments are also lost, resulting consecutive timeouts. Based on multiple bakeoffs of the retransmission timer, the effective data resumption time will double the handover latency. Then, the effective data resumption time in the inter-domain mobility of HMIPv6 can be expressed as follows:

$$
\begin{aligned}
T_{HMIPv6-Inter-Domain} = 2 * (&T_{CN-oAR} + T_{MAP-nAR} \\
&+T_{MAP-nAR} + T_{MAP-nAR} \\
&+T_{HA-MAP} + T_{HA-MAP} \\
&+T_{MAP-nAR} + T_{MAP-nAR} \\
&+T_{CN-MAP} + T_{CN-MAP} \\
&+T_{MAP-nAR})
\end{aligned} \tag{15}
$$

TCP-CM over FMIPv6 and HMIPv6

TCP-CM is a flexible approach and as such it is also applicable for MIPv6 extensions such as FMIPv6 and HMIPv6. In the case of these extensions, TCP-CM also enjoys benefits of FMIPv6 and intra-domain HMIPv6 such as reduced handover time.

Following the above analysis, the effective data resumption time for TCP-CM over these MIPv6 extensions could be similarly derived as following equations:

$$T_{TCP-CM-FMIPv6} = T_{CN-oAR} + 2T_{oAR-nAR} + T_{CN-oAR} \tag{16}$$

$$T_{TCP-CM-HMIPv6-Intra-Domain} = 2T_{CN-MAP} + 3T_{MAP-nAR} \tag{17}$$

$$
\begin{aligned}
T_{TCP-CM-HMIPv6-Inter-Domain} = &2T_{CN-oAR} + 2T_{MAP-nAR} \\
&+2T_{MAP-nAR} + 2T_{HA-MAP}
\end{aligned} \tag{18}
$$

Numerical Results

To summarize the above analysis, the effective data resumption time introduced by TCP-CM, FR, Freeze TCP, Mobile TCP, FMIPv6 and HMIPv6, as well as TCP-CM over FMIPv6 and HMIPv6, are represented in Table 1.

First, we compare the effective data resumption time introduced by network-layer approaches (i.e. FMIPv6 and HMIPv6) with that of our proposed scheme. Then, the effective data resumption time caused by transport-layer approaches (FR, Freeze TCP and Mobile TCP) will be compared with that of our proposed scheme.

In the analytical model illustrated in Fig.3, we assume the paths between to entities, i.e. from the CN to the oAR, from the CN to the nAR, from the CN to the HA, from the CN to the MAP, from the oAR to the nAR, from the HA to the MAP, and from the HA to the nAR, are parts of the Internet since these entities are independent of each other from the geographical location point of view. Without the loss of generality, the transmission time of these pathes is considered the same as the delay of successively-transmitted Internet packets. We denote the total delay in the Internet as $T_{Internet}$ [16], so we can get the following expression:

$$
\begin{aligned}
T_{Internet} &\approx T_{CN-oAR} \approx T_{CN-nAR} \approx T_{CN-HA} \\
&\approx T_{CN-MAP} \approx T_{oAR-nAR} \approx T_{HA-MAP} \approx T_{HA-nAR}
\end{aligned} \tag{19}
$$

Besides, since the MAP is located at the same local domain as the oAR and nAR, we assume the following expression reasonably:

$$T_{Internet} > T_{MAP-oAR} \approx T_{MAP-nAR} \tag{20}$$

Thus, based on Eqs. 16 and 17, the effective data resumption time of FMIPv6 will be mainly produced by six times of the delay that the data packet passes through the Internet, i.e. $T_{FMIPv6} = 6 * T_{Internet}$. For HMIPv6, the effective data resumption time will be mainly produced by only over four times when the data packet passes through the Internet in this case of intra-domain mobility, i.e. $T_{HMI\ Pv6} > 4 * T_{Internet}$, but over ten times in this case of interdomain mobility, i.e. $T_{HMI\ Pv6} > 10 * T_{Internet}$. Comparatively, it takes only $5 * T_{Internet}$ for our TCP-CM approach, so it can be easily understood that TCP-CM has lower effective data resumption time compared with the network-layer approaches.

Considering the transport-layer approaches, it is very easy to get all of the FR, Freeze TCP and Mobile TCP will lead to at least $6 * T_{Internet}$ of effective data resumption time, which is also larger than that of our TCP-CM.

It is also shown that TCP-CM over MIPv6 extensions such as FMIPv6 and HMIPv6 performs better than the standard TCP-CM, FMIPv6 or HMIPv6 alone.

The above analytical results indicate that TCP-CM outperforms both the network-layer approaches and transport-layer approaches in terms of the effective data resumption time. In the next section, we will further evaluate our approach by means of simulation.

SIMULATION AND EVALUATION

In this section, we evaluate the effect of the proposed mobility control mechanisms through simulations performed using OPNET simulator [23]. The simulation models are built by extending the mobility control mechanisms from the standard MIPv6 and TCP model. The simulation results demonstrate how the enhancement of TCP performance can be achieved in our approach.

Model Modifications

Like the traditional TCP/IP model, in the OPNET model, the TCP and IP modules are not allowed to transmit control information except the data delivery. To enhance inter-signalling communication, we implemented a layered protocol interfacing between the TCP and IP modules through the Interface Control Information (ICI) mechanism. The ICI allows the IP module to notify mobility events and incurs corresponding mobility controls.

ICI Format Specification

An ICI format defines the structure of an ICI in terms of the attributes that it contains. An attribute's data type determines what sort of information it can store. In our ICI contents, we specify the following attribute names: handover_initiation and handover_terimination, where, handover_initiation means the event that the MN determines that the current access router is no longer reachable through Neighbour Unreachability Detection mechanism; handover_termination means the MN receives a packet with a valid Binding Acknowledgement. The format specifications for ICI between IP and TCP are shown in Fig.4.

ICI Mechanics and its Implementation

The sequences of operation for the TCP and IP modules are as follows:

We create a new ICI at the IP module using the Kernel Procedure (KP) op_ici_create(), and supply the ICI format specified in previous Subsection. Then we associate the created

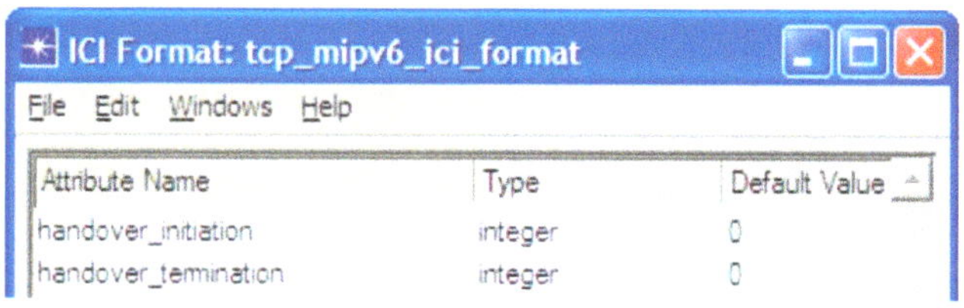

Figure 4: ICI format specification

ICI with the events of IoH and ToH respectively by calling the KP op_ici_install(). When the IP module is about to execute processes that will generate the events described above, and the ICI is to be associated with the corresponding events, it assigns correct values to the ICI's attributes by calling the KP op_ici_attr_set(). The following code for the indication and notification of IoH is implemented in the away state of the mipv6_mn process model:

node_objid = op_topo_parent (op_id_self ());

tcp_mod_objid = op_id_from_name (node_objid, OPC_OBJTYPE_PROC, "tcp"); if (aye_handover_ici_ptr == OPC_NIL)

{

aye_handover_ici_ptr = op_ici_create ("aye_handover_ici_format");

} endif op_ici_attr_set (aye_handover_ici_ptr, "aye_handover_initiation", 1); op_ima_obj_attr_get (compound_attr_objid, "Route Optimization", &aye_opt_flag); if (aye_opt_flag)

{

op_ici_attr_set (aye_handover_ici_ptr, "aye_route_optimization", 1);

} endif if (!L3_HO_HOME)

{

op_ici_attr_set (aye_handover_ici_ptr, "aye_l3_mobility_away", 1);

} endif op_ici_install (aye_handover_ici_ptr);

op_intrpt_force_remote(AYE_MOBILITY_CONTROL_INTRPT_CODE, tcp_mod_objid);

When the indicated events occur and result in an interrupt, the interrupted process in the TCP module obtains the ICI by calling the KP op_intrpt_ici(). TCP extracts information that it needs from the ICI by calling the KP op_ici_attr_get(), and responds correspondingly. Figure 5 shows the extended state diagram of the TCP process model. In this figure, we create a new state of MOB_CTR and define the condition of MOBILITY_CONTROL, which is associated with mobility-related events. As soon as the condition is true, it will transit to the MOB_CTR state, and perform corresponding Enter Executions as described below.

The following code is implemented in the extended MOB_CTR state of the TCP process model:

if (op_ici_attr_get (aye_ici_ptr, "handover_initiation", &tcp_

handover_initiation) == OPC_INITIATION_OF_HANDOVER)

{

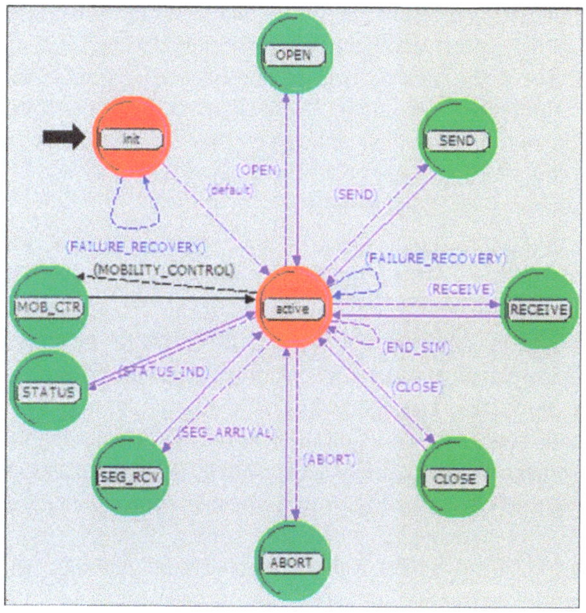

Figure 5: Extended TCP process model

current_time = op_sim_time (); avail_wnd = 0;

} endif

if (op_ici_attr_get (aye_ici_ptr, "handover_termination",

&tcp_handover_termination) == OPC_TERMINATION_OF_HANDOVER)

{

retrans_rtt = measured_rtt; retrans_rtt_dev = measured_rtt/2; current_rto = retrans_rtt + rtt_dev_coef*retrans_rtt_dev; ssthresh = (bw*retrans_rtt)/snd_mss cwnd = ssthresh snd_wnd = int(snd_wnd/snd_mss)*snd_mss if (snd_wnd<=snd_nxt-snd_una)

{

snd_wnd = snd_wnd+snd_mss

} endif next_timeout_time = next_timeout_time - (op_sim_time ()current_time);

avail_wnd = snd_una + total_wnd - snd_nxt;

} endif

Experimental Setup

To evaluate the effect of our approach, the TCP performance test is carried out in our mobile IPv6 test environments as illustrated in Fig.6. The simulation network topology model is composed by the IP cloud model, Access Network (AN) model, Access Routers (ARs), and communicating endpoints (i.e. the MNs and the CNs) etc. The IP cloud model represents the Internet through which the IP traffic can be modelled. The access network model represents the access network technology with different features and capacities. AR represents the wireless access base station of access networks. In addition, each AR consists of two interfaces, among which the wired interface is connected to the access network model through wired link and the wireless interface running IEEE802.11b with a coverage area of approximately 300m in radius provides wireless access for the MNs. The HA (i.e. home AR) represents the home agent with home agent function, while the AR2, AR3 and AR4 (i.e. foreign ARs) represent the foreign access router of AN2, AN3 and AN4 respectively. Each AR is positioned to be 250m apart with free space each other to ensure the overlapping distance. For simulation operation, we define a FTP file transfer application through the Application object and Profiles object. The concrete application configuration will be described in the following subsections in detail.

In this simulation scenario, the MNs (i.e. the MN_FTP_Server and MN_FTP_Client) and the CNs (i.e. the CN_FTP_Server and CN_FTP_Client) communicate to each other by running a ftp application as a source of TCP traffic. Initially, the MNs are placed at their corresponding home networks, and the MNs move in a counterclockwise trajectory roaming through all four access networks on the Internet, so that three general movement modes, namely movement from home network to foreign network (see Fig. 6 movement from access network1 to access network2), movement among foreign networks (see Fig. 6 movement from access network2 through access network3 to access network4), and movement from foreign network to home network (see Fig. 6 movement from access network4 to access network1), can be observed in simulations. The MNs are served by the home agent (e.g. the HA) when they moves into the foreign networks.

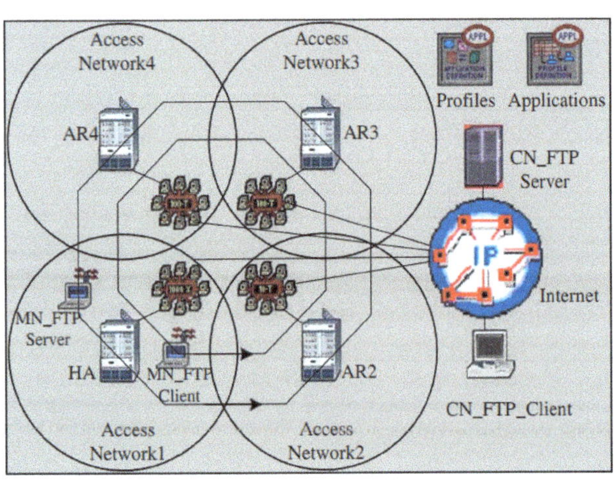

Figure 6: Simulation network model

Measurements and Results Evaluation

In this subsection, we measure the adaptive TCP behaviours for standard MIPv6, including the received sequence number, RTO, and CWND, to verify the correctness of our approach. Besides, we also investigate the performance of throughput to validate the effect of our approach.

TCP Parameter Measurements and Evaluation

First we measure the sequence number transmitted between the MN and CN by running a FTP application. In our simulation, we use the standard TCP Reno model and set the MN_FTP_Server as the FTP server and the CN_FTP_Client as the FTP client. The FTP server was sending a file of 16Mbytes to the CN_FTP_Client while it moves in the range of ANs. In addition, we set the buffer size of ARs larger than the send window of TCP in order to avoiding the packet loss due to the buffer overflow of link layer, which eliminates the impact of link layer factor.

Figure 7 depicts the sequence number of the received TCP segments versus time for one of the experiments, from which the comparison of sequence number progression in a connection using our MIPv6-aware TCP and a connection using unmodified TCP can be observed. It can be seen from the figure that our proposed approach recovers fast than the traditional TCP/MIPv6. This is because that we use the fast recovery mechanism, through which the TCP sender transmits data segments as fast as possible without waiting for an unnecessary timeout.

The retransmission timeout will be directly affected by the RTT, including SRTT and its deviation RTTVAR. In the standard MIPv6, the successive retransmission timeouts result in the artificial inflation of RTO. Figure 8 illustrates the growth and variety of RTO versus the simulation time when the endpoints are communicating with each other while the MN is roaming. From this figure, we can see that at about time 50s when the MN start switching

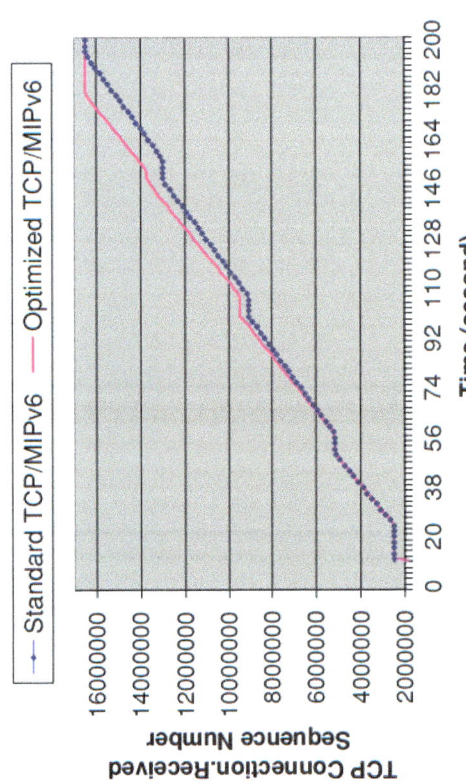

Figure 7: Optimized TCP sequence number in MIPv6

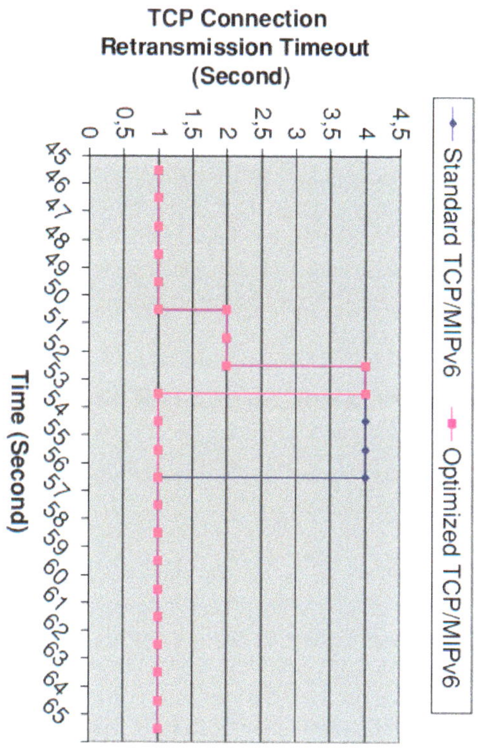

Figure 8: Optimized TCP RTO in MIPv6

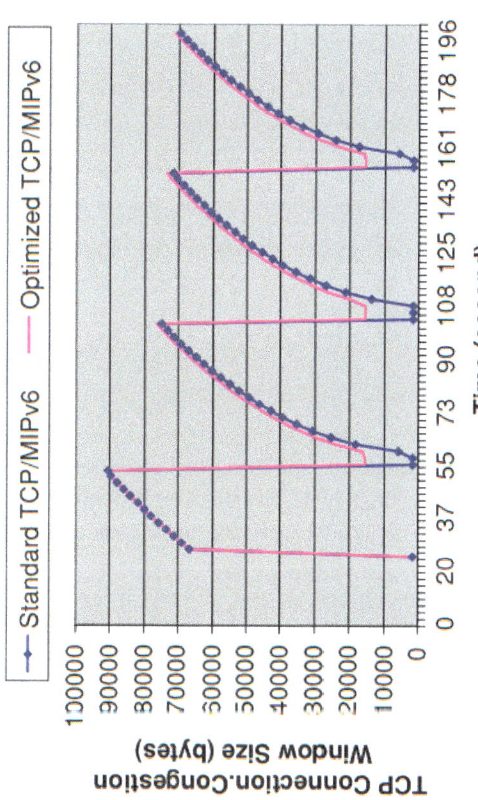

Figure 9: Optimized TCP CWND in MIPv6

To a new network (i.e. the handover initiation), the RTO grows exponentially due to consecutive retransmission timeouts and exponential timer back off algorithm [24]. When the MN completes the handover at about time 53s (i.e. the handover termination), with the standard TCP/MIPv6, the TCP has to wait for a long time before it could continue sending the segment due to the artificial inflation of RTO. However, with our approach proposed in above section, the RTO is reduced significantly compared to the standard TCP/MIPv6 as soon as the handover termination. This is because, in our approach, the TCP can be aware of the handover termination timely, and it then could reinitiate the SRTT and RTTVAR, which leads to RTO recovery quickly.

We also trace the congestion window depicted in Fig. 9. In this experiment, we set the receive buffer size larger than the maximum size of CWND in order to avoiding the effect of size of the remote advertised window on the send window.

Figure 9 plots the congestion window size with respect to the simulation time. From this figure, we can see that a tab out time 23swhenthe FTP server of CN (i.e. the CN_FTP_Server) starts to transmit the file, the size of CWND starts to increase exponentially. At time 26s, the CWND rises up to 65,535Bytes, which is the default value of SSTHRESH, and TCP enters the congestion avoidance stage, where CWND increases linearly. Whenever the connection is broken during the handover, the standard TCP decreases the congestion windows size by one segment due to the timeout (see the curve of standard TCP/MIPv6 in Fig. 9). However, our approach does not invoke the slow start during the mobility over MIPv6 (see the curve of optimized TCP/MIPv6 in Fig. 9). Each time the handover finishes, the TCP CWND is reset according to the BDP [29] and the SSTHRESH is equal to the CWND, which is usually higher than that of standard TCP. In this way, TCP prevents the invocation of any congestion control by estimating a suitable SSRESHTH and CWND, and the congestion avoidance takes over.

Performance Measurements and Evaluation

Series of simulation experiments are also performed in examining the optimized TCP through put performance for proposed TCP-CM. Figures 10 and 11 illustrate the effect of TCP-CM from the MN as TCP sender's and receiver's point of view respectively.

Figure 10 shows the average throughput when the CN_FTP_Server transfers a file to the MN_FTP_Client for performed experiments, from which the comparison of average throughput in a connection using our MIPv6-aware TCP and a connection using unmodified TCP can be observed. In this figure, the horizontal axis indicates the number of handover when the MN_FTP_Client moves around the wireless access networks, in which the endpoints are communicating with each other, and the vertical axis indicates the average throughput in terms of KBytes/Sec. The upper curve is the throughput of TCP with TCP-CM, and the bottom curve shows the throughput with standard TCP/MIPv6. As illustrated in this figure, when the number of handover increases, the TCP throughput will decline due to the frequent handovers. However, the overall throughput using TCP-CM is better than that of standard TCP/MIPv6. Moreover, the declining rate of TCP-CM is slower than that of standard TCP/MIPv6, which demonstrates that TCP-CM performs more outstanding when the handover occurs frequently. This is because that TCP-CM can detect the mobility events timely for any movement modes in frequent handovers, so that TCP-CM transmits data segments as fast as possible without waiting for an unnecessary timeout.

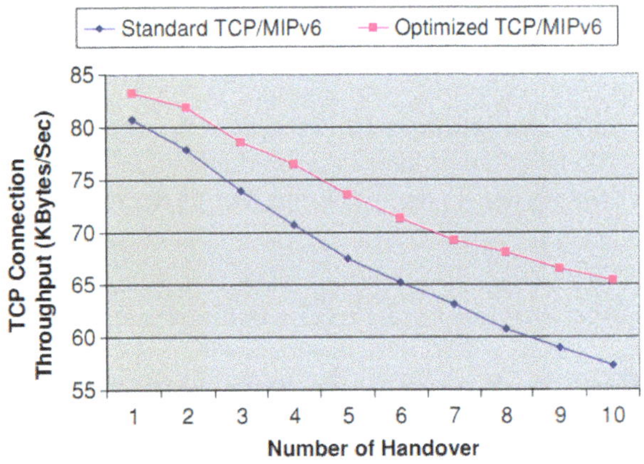

Figure 10: Average throughput of TCP from CN sender's perspective

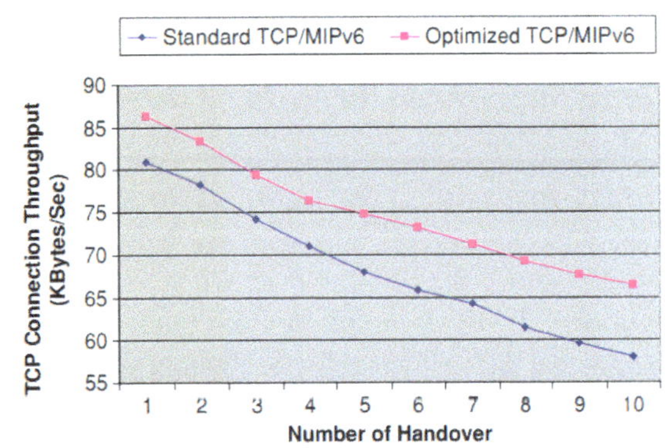

Figure 11: Average throughput of TCP from MN sender's perspective

We also investigate the improvements of the throughput when the mobile is the TCP sender. Figure 11 depicts the average throughput versus the number of handover when the MN_FTP_Server transfers a file to the CN_FTP_Client for performed experiments, from which we can observe that the average throughput of TCP-CM will be still better than that of standard TCP/MIPv6. This is because that TCP-CM differentiates the mobility issue from the congestion issue and prevents TCP from invoking the unnecessary congestion control procedures.

CONCLUSIONS AND FUTURE WORK

Like the network congestion, mobility is an important issue that TCP needs to care in mobile environments. In particular, the mobile Internet over MIPv6 suffers from the packet loss, deviation of transport delay and packet route changes, which are not due to the network congestion. We have shown how handover procedures of MIPv6 cause the packet loss and inflation of RTO and presented how the deviation of transport delay causes the RTT stale. We have also identified the factors that contribute to the performance degradation.

To adapt TCP to mobility better as well as improve the transport performance in networks over MIPv6, we have also proposed a new approach TCP-CM. The fast recovery and timing re-initialization mechanisms can eliminate the unnecessary waits for retransmission timeouts, and restore to a suitable network capability quickly; and the congestion avoidance mechanism prevents unnecessary invocation congestion control procedures after the MN moves to a new network, avoiding the degradation of throughput.

Our work makes clear that the need for TCP to deal with mobility control and congestion control respectively. Our results can be used to adapt TCP to the mobile Internet over MIPv6. Simulation results show that mobility control is significantly more robust than traditional TCP in the presence of mobile networks over MIPv6. Note TCP-CM is primarily designed as an extension to standard TCP (TCP Reno), which includes slow-start, congestion avoidance, fast retransmit and fast recovery. Clearly, although some other TCP variants such as TCP Tahoe (which lacks of fast recovery from TCP Reno) [14] may not be better suitable in mobile environments than TCP Reno, it is still possible to apply the TCP-CM mechanisms described in this paper to improve the transport performance.

While our approach does deal with many of the transport performance problems associated with MIPv6, we still have further work needed to be done. For example, we do not consider the effect of some wireless links due to the presence of bursty wireless channel error, which however is in existence in the real wireless networks. In addition, since the MIPv6 is still not the perfect mobility solution, it is necessary to investigate the integration of our TCP-CM with the variants of MIPv6. In the future, we will study the effect of our approach in the real wireless networks with the presence of bursty wireless channel error, and we will integrate our TCP-CM with the MIPv6 extensions (i.e. HMIPv6 and FMIPv6) for further expected performance improvement of TCP.

ACKNOWLEDGEMENTS

This work has been partially sponsored by the EU IST ENABLE research project. The use of OPNET Modeler in the research was facilitated through OPNET's university program. In addition, we would like to thank Fang-Chun Kuo, Ingo Juchem, Niklas Steinleitner, Jun Lei, and anonymous reviewers for their insightful comments to this paper

REFERENCES

1. Allman, M., Paxson, V., & Stevens, W. (1999). TCP congestion control. In RFC 2581.
2. Antoine, S., Wei, M., & Aghvami, A. (2003). Impact of Mobile IPv6 handover on the performance of TCP: An experimental testbed. ACM SIGMOBILE Mobile Computing and Communications Review, 7(1), 31–33.
3. Bakre, A., & Badrinath, B. (1995). I-TCP: Indirect TCP for mobile hosts. In Proceeding of 15th International Conference on Distributed Computing Systems (ICDCS'95).
4. Balakrishnan, H., Seshan, S., & Katz, R. (1995). Improving reliable transport and handoff performance in cellular wireless networks. Wireless Networks, 1(4), 469–481.
5. Border, J., Kojo, M., Griner, J., Montenegro, G., & Shelby, Z. (2001). Performance enhancing proxies intended to mitigate link-related degradations. In RFC 3135.
6. Brown, K., & Singh, S. (1997). M-TCP: TCP for mobile cellular networks. ACM SIGCOMM Computer Communication Review, 27(5), 19–43.

7. Caceres, R., & Iftode, L. (1995). Improving the performance of reliable transport protocols in mobile computing environments. IEEE Journal on Selected Areas in Communications, 13(5), 850–857.

8. Elaarag, H. (2002). Improving TCP performance over mobile networks. ACM Computing Surveys, 34(3), 357–374.

9. Goff, T., Moronski, J., Phatak, D., & Gupta, V. (2000). Freeze-TCP: A true end-to-end TCP enhancement mechanism for mobile environments. In Proceedings of Nineteenth Annual Joint Conference of the IEEE Computer and Communications Societies (INFOCOM'00).

10. Gurtov, A., & Korhonen, J. (2004). Effect of vertical handovers on performance of TCP-friendly rate control. ACM SIGMOBILE Mobile Computing and Communications Review, 8(3), 73–87.

11. Ho, C., Chan, Y., & Chen, Y. (2005). An efficient mechanism of TCP-Vegas on mobile IP networks. In Proceedings of 24th Annual Joint Conference of the IEEE Computer and Communications Societies (INFOCOM'05).

12. Hsieh, R., Seneviratne, A., Soliman, H., & El-Malki, K. (2002). Performance analysis on hierarchical Mobile IPv6 with fast-handoff over end-to-end TCP. In Proceedings of Global Telecommunications Conference (GLOBECOM'02).

13. Izumikawa, H., Yamaguchi, I., & Katto, J. (2004). An efficient TCP with explicit handover notification for mobile networks. In Proceedings of IEEE Wireless Communications and Networking Conference (WCNC'04).

14. Jacobson, V. (1988). Congestion avoidance and control. In ACM SIGCOMM'88 (pp. 314–329). Stanford, CA.

15. Johnson, D., Perkins, C., & Arkko, J. (2004). Mobility support in IPv6. In RFC 3775.

16. Kalman, M., & Girod, B. (2004). Modeling the delays of successively-transmitted Internet packets. In Proceedings of the IEEE International Conference on Multimedia and Expo (ICME'04).

17. Karn, P., & Partridge, C. (1991). Improving round-trip time estimates in reliable transport protocols. ACM Transactions on Computer Systems, 9(4), 364–373.

18. Koodli, R. (2005). Fast handovers for Mobile IPv6. In RFC 4068.

19. Le, D., Fu, X., & Hogrefe, D. (2006). A review of mobility support paradigms for the Internet. IEEE Communications Surveys and Tutorials, 8(1), 2–15.

20. Matsushita, Y., Matsuda, T., & Yamamoto, M. (2005). Network supported bandwidth control for TCP in hierarchical mobile Internet. IEICE Transactions on Communications, E88-B(1), 266–273.

21. McNair, J., Tugcu, T., Wang, W., & Xie, J. (2005). A survey of cross-layer performance enhancements for mobile IP networks. Computer Networks, 49(2), 119–146.

22. Narten, T., Nordmark, E., & Simpson, W. (1998). Neighbor discovery for IP version 6 (IPv6). In RFC 2461.

23. OPNET Technologies, I. (2007). OPNET Modeler. OPNET Technologies.

24. Paxson, V., & Allman, M. (2000). Computing TCP's retransmission timer. In RFC 2988.

25. Postel, J. (1981). Transmission control protocol. In RFC 793.

26. Soliman, H. (2004). Mobile IPv6: Mobility in a Wireless Internet. Addison-Wesley.

27. Soliman, H., Castelluccia, C., El-Malki, K., & Bellier, L. (2005). Hierarchical Mobile IPv6 mobility management (HMIPv6). In RFC 4140.

28. Stangel, M., & Bhargavan, V. (1998). Improving TCP performance in mobile computing environments. In Proceedings of International Conference on Communications (ICC'98).

29. Stevens, W. (1994). TCP/IP Illustrated, Vol. 1: The Protocols. Addison-Wesley.

30. Swami, Y., Le, K., & Eddy, W. (2006). Lightweight mobility detection and response (LMDR) algorithm for TCP. Draft swami-tcp-lmdr-07.txt, Internet Draft (work in progress).

31. Thomson, S., & Narten, T. (1998). IPv6 stateless address autoconfiguration. In RFC 2462.

32. Wehrle, K., Pählke, F., Ritter, H., Müller, D., & Bechler, M. (2004). The Linux Networking Architecture: Design and Implementation of Network Protocols in the Linux Kernel. Prentice Hall

CHAPTER 19

DETECTION OF TURBO-CODED OFDM SIGNALS

K. Vasudevan

Keywords: OFDM, Coherent detection, matched filtering, Turbo codes, Frequency selective Rayleigh fading, Channel capacity

INTRODUCTION

Future wireless communication standards aim to push the existing data-rates higher. This can only be achieved with the help of coherent communications, since they give the lowest bit-error-rate (BER) performance for a given signal-to-noise ratio (SNR). Conversely, they require the lowest SNR to attain a given BER, resulting in enhanced battery life. If we look at a mobile, it indicates a typical received signal strength equal to −100 dBm (10^{-10} mW). However this is not the signal-to-noise ratio! Therefore, the question is: What is the operating SNR of the mobiles? Would it be possible to achieve the same performance by transmitting at a lower power? The recent advances in cooperative communications has resulted in low complexity solutions, that are not necessarily power efficient [1, 2]. In fact, it is worth quoting the following from [3]:

1. The Myth: Sixty years of research following Shannon's pioneering paper has led to telecommunications solutions operating arbitrarily close to the channel capacity—"flawless telepresence" with zero error is available to anyone, anywhere, anytime across the globe.
2. The Reality: Once we leave home or the office, even top of the range iPhones and tablet computers fail to maintain "flawless telepresence" quality. They also fail to approach the theoretical performance predictions. The 1,000-fold throughput increase of the best third-generation (3G) phones over second-generation (2G) GSM phones and the 1,000-fold increased teletraffic predictions of the next decade require substantial further bandwidth expansion toward ever increasing carrier frequencies, expanding beyond the radio- frequency (RF) band to optical frequencies, where substantial bandwidths are available.

The transmitter and receiver algorithms proposed in this paper and in [4, 5] are well suited for implementation on a DSP processor or hardwired and may perhaps not require quantum computers, as mentioned in [3]. The reader is also referred to the brief commentary on channel estimation and synchronization in page 1351 and also to the noncoherent schemes in page 1353 of [1], which clearly state that cooperative communications avoid coherent receivers due to complexity.

Broadly speaking, the wireless communication device needs to have the following features:

1. Maximize the bit-rate
2. Minimize the bit-error-rate
3. Minimize transmit power
4. Minimize transmission bandwidth

A rather disturbing trend in the present day wireless communication systems is to make the physical layer very simple and implement it in hardware, and allot most of the computing resources to the application layer, e.g., for internet surfing, video conferencing etc. While hardware implementation of the physical layer is not an issue, in fact, it may even be preferred over software implementation in some situations, the real cause for concern is the tendency to make it "simple", at the cost of BER performance. Therefore, the questions are:

1. Was signal processing for coherent communications given a chance to prove itself, or was it ignored straightaway, due to "complexity" reasons?

Citation: K. Vasudevan, Detection of Turbo-Coded OFDM Signals, DOI 10.1007/s11277-015-2303-

2. Are the present day single antenna wireless transceivers, let alone multi-antenna systems, performing anywhere near channel capacity?

This paper demonstrates that coherent receivers need not be restricted to textbooks alone, in fact they can be implemented with linear (not exponential) complexity. The need of the hour is a paradigm shift in the way the wireless communication systems are implemented.

In this article, we dwell on coherent receivers based on orthogonal frequency division multiplexing (OFDM), since it has the ability to mitigate intersymbol interference (ISI) introduced by the frequency selective fading channel [6–8]. The "complexity" of coherent detection can be overcome by means of parallel processing, for which there is a large scope. We wish to emphasize that this article presents a proof-of-concept, and is hence not constrained by the existing standards in wireless communication. We begin by first outlining the tasks of a coherent receiver. Next, we scan the literature on each of these tasks to find out the state-of-the-art, and finally end this section with our contributions.

The basic tasks of the coherent receiver would be:

1. To correctly identify the start of the (OFDM) frame (SoF), such that the probability of false alarm (detecting an OFDM frame when it is not present) or equivalently the probability of erasure/miss (not detecting the OFDM frame when it is present) is minimized. We refer to this step as timing synchronization.
2. To estimate and compensate the carrier frequency offset (CFO), since OFDM is known to be sensitive to CFO. This task is referred to as carrier synchronization.
3. To estimate the channel impulse/frequency response
4. To perform (coherent) turbo decoding and recover the data.

To summarize, a coherent receiver at the physical layer ensures that the medium access control (MAC) is not burdened by frequent requests for retransmissions.

A robust timing and frequency synchronization for OFDM signals transmitted through frequency selective additive white Gaussian noise (AWGN) channels is presented in [9]. Timing synchronization in OFDM is addressed in [10–14]. Various methods of carrier frequency synchronization for OFDM are given in [15–21]. Joint timing and CFO estimation is discussed in [22–27].

Decision directed coherent detection of OFDM in the presence of Rayleigh fading is treated in [28]. A factor graph approach to the iterative (coherent) detection of OFDM in the presence of CFO and phase noise is presented in [29]. OFDM detection in the presence of intercarrier interference (ICI) using block whitening is discussed in [30]. In [31], a turbo receiver is proposed for detecting OFDM signals in the presence of ICI and inter antenna interference.

Most flavors of the channel estimation techniques discussed in the literature are done in the frequency domain, using pilot symbols at regular intervals in the time/frequency grid [32–36]. Iterative joint channel estimation and multiuser detection for multi-antenna OFDM is discussed in [37]. No coherent detection of coded OFDM in the *absence of frequency offset* and assuming that the channel frequency response to be constant over a block of symbols, is considered in [38]. Expectation maximization (EM)-based joint channel estimation and exploitation of the diversity gain from IQ imbalances is addressed in [39].

Detection of OFDM signals, in the context of spectrum sensing for cognitive radio, is considered in [40, 41]. However, in both these papers, the probability of false alarm is quite high (5 %).

In [42], discrete cosine transform (DCT) based OFDM is studied in the presence of frequency offset and noise, and its performance is compared with the discrete Fourier transform (DFT) based OFDM. It is further shown in [42] that the performance of DFT-OFDM is as good as DCT-OFDM, for small frequency offsets.

A low-power OFDM implementation for wireless local area networks (WLANs) is addressed in [43]. OFDM is a suggested modulation technique for digital video broadcasting [44, 45]. It has also been proposed for optical communications [46].

The novelty of this work lies in the use of a filter that is matched to the preamble, to acquire timing synchronization [47, 48] (start-of-frame (SoF) detection). Maximum likelihood (ML) channel estimation using the preamble is performed. This approach does not require any knowledge of the channel and noise statistics.

The main contributions of this paper are the following:

1. It is shown that, for a sufficiently long preamble, the variance of the channel estimator proposed in eq. (28) of [4] approaches zero.
2. A known postamble is used to accurately estimate the residual frequency offset for large data lengths, thereby increasing the throughput compared to [4, 5].
3. Turbo codes are used to attain BER performance closer to channel capacity compared to any other earlier work in the open literature, for channels having a uniform power delay profile (to the best of the authors knowledge, there is no similar work on the topic of this paper, other than [4, 5]).
4. A robust turbo decoder is proposed, which performs effectively over a wide range of SNR (0–30 dB).

In a multiuser scenario, the suggested technique is OFDM-TDMA. The uplink and downlink may be implemented using time division duplex (TDD) or frequency division duplex (FDD) modes.

This paper is organized as follows. Section 2 describes the system model. The enhanced frame structure is described in Sect. 3. The modifications in the turbo decoder in the presence of receive diversity and the variance of the channel estimation error, are presented in Sect. 4. The channel capacity is discussed in Sect. 5. The BER results from computer simulations are given in Sect. 6. Finally, in Sect. 7, we discuss the conclusions and future work.

SYSTEM MODEL

We assume 1st-order transmit diversity and Nth-order receive diversity. The data is organized into frames, as depicted in Fig. 1. The earlier frame structure considered in [4] is given in Fig. 2.

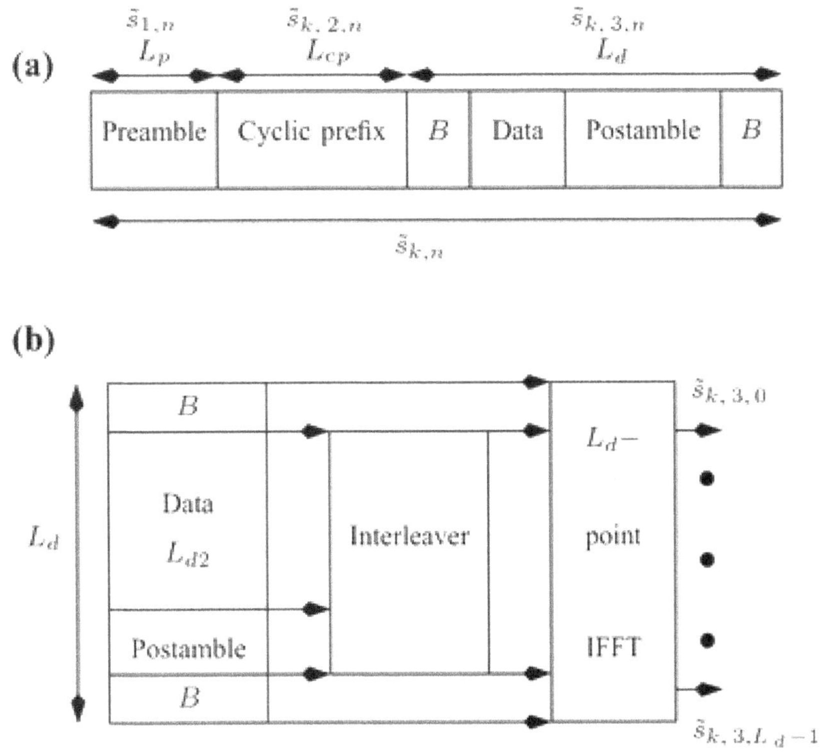

Figure 1: **a** Enhanced frame structure. **b** Processing of the data part at the transmitter

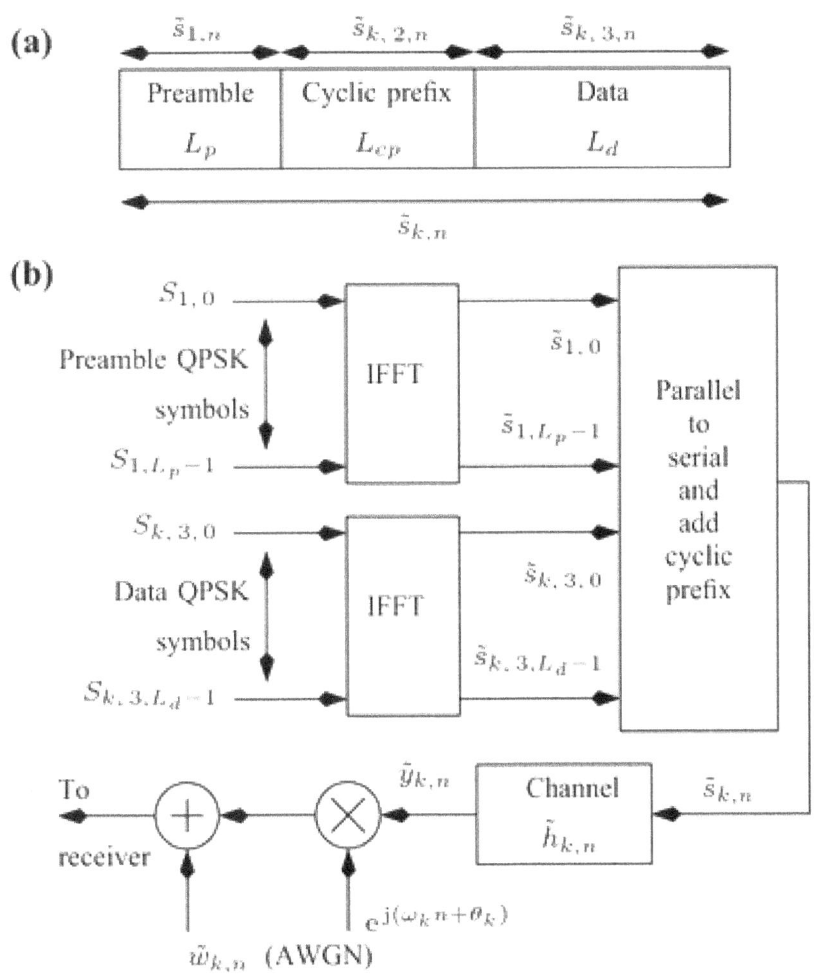

Figure 2: **a** The frame structure in [4]. b System model. k denotes the frame index and n denotes the time index in a given frame.

The received signal in each diversity arm (l) can be expressed as (see also eq. (5) in [4]):

$$\tilde{r}_{k,n,l} = \left(\tilde{s}_{k,n} \star \tilde{h}_{k,n,l}\right) e^{j(\omega_k n+\theta_{k,l})} + \tilde{w}_{k,n,l}$$
$$= \tilde{y}_{k,n,l} e^{j(\omega_k n+\theta_{k,l})} + \tilde{w}_{k,n,l} \quad (1)$$

for $1 \le l \le N$. The frequency offset is assumed to be identical for all the diversity arms, whereas the carrier phase and noise are assumed to be independent. The noise variance is same for all the diversity arms. Two extreme scenarios are considered in the simulations (a) identical channel and (b) independent channel in each diversity arm. The channel span is L_h and is assumed to have a uniform power delay profile.

The output of the FFT can be written as (for $0 \le i \le L_d - 1$):

$$\tilde{R}_{k,i,l} = \hat{H}_{k,i,l} S_{k,3,i} + \tilde{W}_{k,i,l} \quad (2)$$

for $1 \le l \le N$ diversity arms. The notation in (2) is self-explanatory and is based on eq. (36) in [4].

ENHANCED FRAME STRUCTURE

The basic motivation behind the enhanced frame structure, is to increase the throughput, which in turn, depends on L_d. The accuracy of the frequency offset estimate depends on the length of the preamble L_p. Increasing the number of frequency bins B_1 and B_2 [4] for a given L p, does not improve the accuracy. From Fig. 3 it can be seen that the RMS value of the fine frequency offset estimation error is about 2×10^{-4}, at an SNR per bit equal to 8 d_B. The subcarrier spacing with data length L_d = 4,096 is equal to $2\pi/4{,}096 = 1.534 \times 10^{-3}$ radians. Therefore, the residual frequency offset is 0.0002 × 100/0.001534 = 13 % of the subcarrier spacing, which is quite high and causes severe ICI. Note that the RMS frequency offset estimation error can be reduced by increasing the preamble length (L p), keeping the data length (L_d) fixed, which in turn reduces the throughput given by (for the frame structure in Fig. 2):

$$\mathcal{T} = \frac{L_{d1}}{L_p + L_{cp} + L_d} \quad (3)$$

Where L_{d1} is defined in Fig. 4, for the frame structure given in Fig. 2. Note that for a rate-1/2 turbo code $L_d = 2_{Ld1}$, whereas for a rate-1 turbo code, $L_d = L_{d1}$. This motivates us to look for an alternate frame structure which not only solves the frequency offset estimation problem, but also maintains the throughput at a reasonable value.

Consider the frame in Fig. 1a. In addition to the preamble, prefix and data, it contains "buffer" (dummy) symbols of length B and postamble of length L_o, all drawn from the QPSK constellation. In Fig. 1b we illustrate the processing of L_d symbols at the transmitter. Observe that only the data and postamble symbols are interleaved before the IFFT operation. After interleaving, the postamble gets randomly dispersed between the data symbols. The buffer symbols are sent directly to the IFFT, without interleaving. The preamble and the cyclic prefix continue to be processed according to Figure 1 in [4] and eq. (3) in [4]. We now explain the reason behind using this frame structure. In what follows, we assume that the SoF has been detected, fine frequency offset correction has been performed and the channel has been estimated.

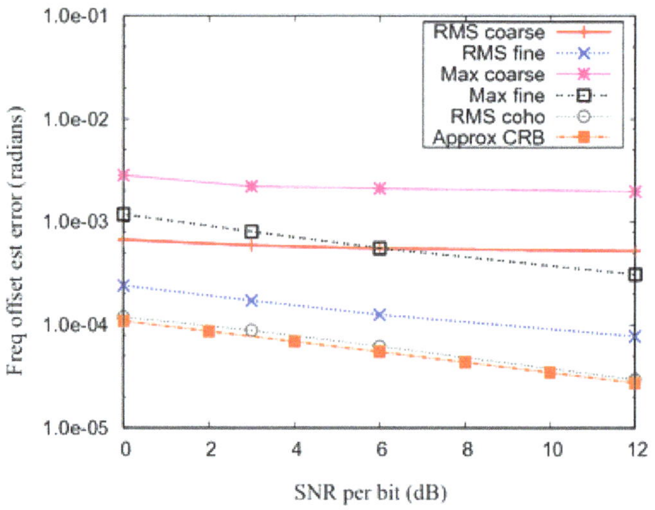

Figure 3: RMS and maximum frequency offset estimation error for $L_p = 512$

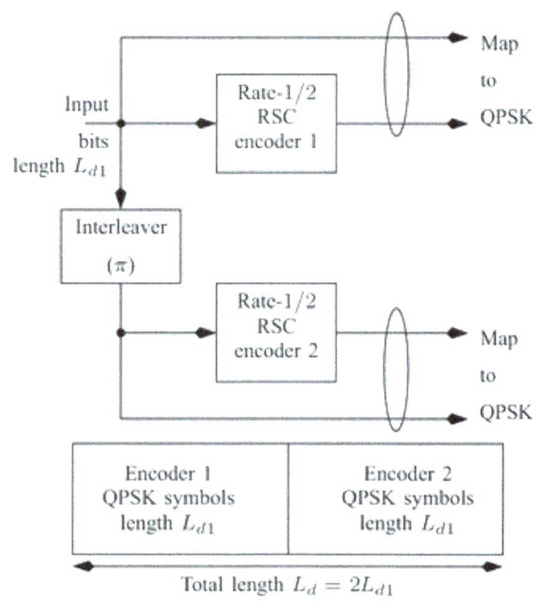

Figure 4: Encoder block diagram for the frame structure in Fig. 2. © 2013 IEEE. Reprinted, with permission, from [4]

We proceed by making the following observations:

1. Modulation in the time domain results in a shift in the frequency domain. Therefore, any residual frequency offset after fine frequency offset correction, results in a frequency shift at the output of the FFT operation at the receiver. Moreover, due to the presence of a cyclic prefix, the frequency shift is circular. Therefore, without the buffer symbols, there is a possibility that the first data symbol would be circularly shifted to the last data symbol or vice versa. This explains the use of buffer symbols at both ends in Fig. 1. In order to compute the number of buffer symbols (B), we have to know the maximum residual frequency offset, after fine frequency offset correction. Referring to Fig. 3, we find that the maximum error in fine frequency offset estimation at 0 dB SNR per bit is about $\pm 2\times 10^{-3}$ radians. With L^d = 4,096, the subcarrier spacing is

$2\pi/4{,}096 = 1.534 \times 10^{-3}$ radians. Hence, the residual frequency error would result in a shift of$\pm 2/1.534 = \pm 1.3$ subcarrier spacings. Therefore, while B = 2 would suffice, we have taken B = 4, to be on the safe side.

2. Since the frequency shift is not an integer multiple of the subcarrier spacing, we need to interpolate in between the subcarriers, to accurately estimate the shift. Interpolation can be achieved by zero-padding the data before the FFT operation. Thus we get a 2Ld–point FFT corresponding to an interpolation factor of two and so on. Other methods of interpolation between subcarriers is discussed in [49].

3. After the FFT operation, postamble matched filtering has to be done, since the postamble and $\hat{H}_k \approx \tilde{H}_k$ are known. The procedure for constructing the postamble matched filter is illustrated in Fig. 5. From simulations, it has been found that a postamble length $L_o = 128$ results in false peaks at the postamble matched filter output at 0 dB SNR per bit. Therefore we have taken Lo = 256. With these calculations, the length of the data works out as $L_{d2} = L_d - 2B - L_o = 4{,}096 - 8 - 256 = 3{,}832$ QPSK symbols. The throughput of the enhanced frame structure (with rate-1 turbo code) is

$$\begin{aligned}\mathcal{T} &= \frac{L_{d2}}{L_p + L_{cp} + L_d} \\ &= \frac{3{,}832}{512 + 18 + 4{,}096} \\ &= 82.84\,\%. \end{aligned} \tag{4}$$

The throughput comparison of various frame structures is summarized in Table 1.

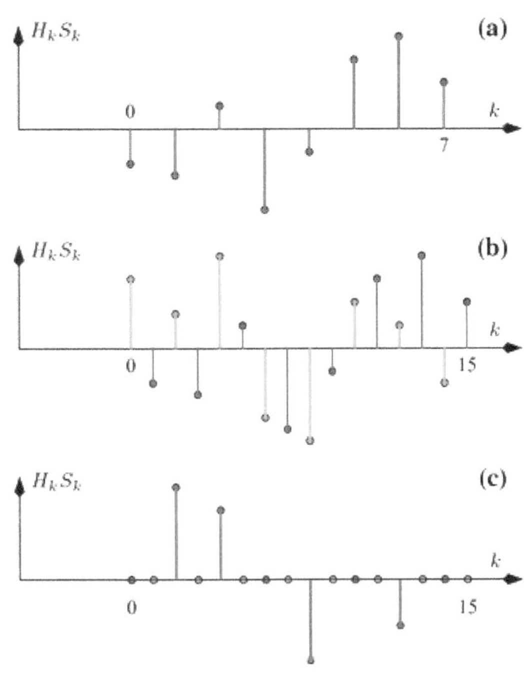

Figure 5: Obtaining the postamble matched filter for $L_d = 8$. Buffer symbols are not shown. The frequency offset (π/Ld) is half the subcarrier spacing ($2\pi/L_d$). H_k and S_k are assumed to be real-valued. Noise is absent. a Output of the L_d -point FFT in the absence of frequency offset. The red lines represent postamble and the blue lines represent data symbols. b Output of the 2_{Ld} -point FFT in the presence of frequency offset. Observe that the red and blue lines have shifted to the right by π/L_d . Green lines denote the output of the Ld -point FFT in the presence of frequency offset. c The postamble matched filter

Table 1: Throughput comparison of various frame structures with $L_p = L_{d1} = 512$, $L_{d2} = 3{,}832$, $L_{cp} = 18$

	Frame structure in Fig. 2 rate-1/2 turbo code [4] Eq. (3)	Frame structure in Fig. 2 rate-1 turbo code [5] Eq. (3)	Frame structure in Fig. 1 rate-1 turbo code (proposed) Eq. (4)
Throughput (%)	32.95	49.14	82.84

RECEIVER

The receiver algorithms for start-of-frame (SoF) detection, frequency offset, channel and noise variance estimation are already discussed in [4, 5], and apply also to the enhanced frame structure given above and receive diversity. In what follows, we describe the modifications required in the turbo decoder in the presence of receive diversity.

Turbo Decoding

In the turbo decoding operation, (for decoder 1, transition from state m to n, kth frame, N diversity arms, rate-1/2 turbo code, the enhanced frame structure in Fig. 1 and $0 \le i \le L_{d2}/2 - 1$), we have (assuming independent noise in all the diversity arms)

$$\gamma_{1,k,i,m,n} = \prod_{l=1}^{N} \gamma_{1,k,i,m,n,l} \tag{5}$$

where

$$\gamma_{1,k,i,m,n,l} = \exp\left[-\frac{\left(\tilde{R}_{k,i,l} - \hat{H}_{k,i,l} S_{m,n}\right)^2}{2L_d\hat{\sigma}_w^2}\right] \tag{6}$$

where $\hat{\sigma}_w^2$ is the average estimate of the noise variance over all the diversity arms and Sm, n is the QPSK symbol corresponding to the transition from state m to n

Similarly at decoder 2, for $0 \le i \le L_{d2}/2 - 1$, we have:

$$\gamma_{2,k,i,m,n} = \prod_{l=1}^{N} \gamma_{2,k,i,m,n,l} \tag{7}$$

where

$$\gamma_{2,k,i,m,n,l} = \exp\left[-\frac{\left(\tilde{R}_{k,j,l} - \hat{H}_{k,j,l} S_{m,n}\right)^2}{2L_d\hat{\sigma}_w^2}\right] \tag{8}$$

where

$$\begin{aligned} j &= L_{d3} + i \\ L_{d3} &= L_{d2}/2. \end{aligned} \tag{9}$$

For a rate-1 turbo code obtained by puncturing, alternate gammas have to be set to unity [5, 7]. The rest of the BCJR algorithm is described in [5, 7].

Robust Turbo Decoding

At high SNR, the term in the exponent (b is the exponent of e^b) of (6) and (8) becomes very large (typically b > 100) and it becomes unfeasible for the DSP processor or even a computer to calculate the gammas. We propose to solve this problem by normalizing the exponents. Observe that the exponents are real-valued and negative. Let $b_{1,j,i}$ denote an exponent at decoder 1 due to the jth symbol in the constellation ($1 \leq j \leq 4$ for QPSK) at time i. For notational convenience, we again assume a rate-1/2 turbo code. Let

$$\mathbf{b}_1 = \begin{bmatrix} b_{1,1,0} & \cdots & b_{1,1,L_{d3}-1} \\ \vdots & \vdots & \vdots \\ b_{1,4,0} & \cdots & b_{1,4,L_{d3}-1} \end{bmatrix} \tag{10}$$

denote the matrix of exponents for decoder 1. Let $b_{1,\max,i}$ denote the maximum exponent at time i, that is

$$b_{1,\max,i} = \max \begin{bmatrix} b_{1,1,i} \\ \vdots \\ b_{1,4,i} \end{bmatrix}. \tag{11}$$

Let

$$\mathbf{b}_{1,\max} = \left[b_{1,\max,0} \cdots b_{1,\max,L_{d3}-1} \right] \tag{12}$$

denote the vector containing the maximum exponents. Compute:

$$\mathbf{b}_1' = \mathbf{b}_1 - \begin{bmatrix} \mathbf{b}_{1,\max} \\ \vdots \\ \mathbf{b}_{1,\max} \end{bmatrix}. \tag{13}$$

Note that in (13), the vector $b_{1,\max}$ has to be repeated as many times as the number of symbols in the constellation.

If any element of $\mathbf{b}_1'$ is less than say, −30, then set it to −30. Thus we get a normalized exponent vector b_1, norm, whose elements lie in the range [0, −30]. It has been found from simulations that normalizing the exponents does not lead to any degradation in BER performance, on the contrary, it increases the operating SNR range of the turbo receiver. In practice, we could divide the range [0, −30] into a large number (e.g. 3,000) of levels and the exponentials (e^b) could be precomputed and stored in the DSP processor, and need not be computed in real-time. The choice of the minimum exponent (e.g. −30), would depend on the precision of the DSP processor or the computer.

Variance of the Channel Estimation Error

To see the effect of noise on the channel estimate in eq. (28) of [4], consider

$$\tilde{\mathbf{u}} = \left(\tilde{\mathbf{s}}_1^H \tilde{\mathbf{s}}_1\right)^{-1} \tilde{\mathbf{s}}_1^H \tilde{\mathbf{w}}_{k,m_1}. \tag{14}$$

When $m_0 = L_h - 1$ (eq. (24) of [4]), observe that

$$\hat{\mathbf{h}}_k = \tilde{\mathbf{h}}_k + \tilde{\mathbf{u}}. \tag{15}$$

Since $\tilde{s}_{1,n}$ is a zero-mean random sequence with good autocorrelation properties (approximately a Kronecker delta function weighted by $L_1\sigma_s^2$), it is reasonable to expect

$$\begin{aligned} &\tilde{\mathbf{s}}_1^H \tilde{\mathbf{s}}_1 = L_1\sigma_s^2 \mathbf{I}_{L_{hr}} \quad \text{for } L_p \gg L_{hr} \\ &\Rightarrow \left(\tilde{\mathbf{s}}_1^H \tilde{\mathbf{s}}_1\right)^{-1} = 1/(L_1\sigma_s^2)\mathbf{I}_{L_{hr}} \\ &\Rightarrow \tilde{\mathbf{u}} = 1/(L_1\sigma_s^2)\tilde{\mathbf{s}}_1^H \tilde{\mathbf{w}}_{k,m_1} \end{aligned} \tag{16}$$

where σ_s^2 is defined in eq. (4) of [4], L1 is defined in eq. (11) of [4], and I_{Lhr} is an $L_{hr} \times L_{hr}$ identity matrix. It can be shown that

$$E\left[\tilde{\mathbf{u}}\tilde{\mathbf{u}}^H\right] = \frac{2\sigma_w^2}{L_1\sigma_s^2}\mathbf{I}_{L_{hr}} = \frac{\sigma_w^2 L_d}{L_1}\mathbf{I}_{L_{hr}} \stackrel{\Delta}{=} 2\sigma_u^2\mathbf{I}_{L_{hr}}. \tag{17}$$

Therefore, the variance of the ML channel estimate (σ_u^2) tends to zero as $L_1 \to \infty$ and L_d is kept fixed. Conversely, when L_d is increased keeping L_1 fixed, there is noise enhancement.

THE CHANNEL CAPACITY

The communication system model under consideration is given by (2). The channel capacity is given by [50]:

$$C = \frac{1}{2}\log_2(1 + \text{SNR}) \quad \text{bits/transmission} \tag{18}$$

Per dimension (real-valued signals occupy a single dimension, complex-valued signals occupy two dimensions). The "SNR" in (18) denotes the minimum average signal-to-noise ratio per dimension, for error-free transmission. Observe that:

1. The sphere packing derivation of the channel capacity formula [50], does not require noise to be Gaussian. The only requirements are that the noise samples have to be independent, the signal and noise have to be independent, and both the signal and noise must have zero mean.
2. The channel capacity depends only on the SNR.
3. The average SNR per dimension in (18) is different from the average SNR per bit (or E_b/N_0), which is widely used in the literature. In fact, it can be shown that [7,50]:

   $$\text{SNR} = 2C \times \text{SNR per bit}. \tag{19}$$

4. It is customary to define the average SNR per bit (E_b/N_0) over two dimensions (complex signals). When the signal and noise statistics over both dimensions are identical, the average SNR per bit over two dimensions is identical to the average SNR per bit over one dimension. Therefore (19) is valid, even though the SNR is defined over one dimension and the SNR per bit is defined over two dimensions
5. The notation E_b/N_0 is usually used for continuous-time, passband analog signals [50– 52], whereas SNR per bit is used for discrete-time signals [7]. However, both definitions are equivalent. Note that passband signals are capable of carrying information over two dimensions, using sine and cosine carriers, inspite of the fact that passband signals are real-valued.

6. Each dimension corresponds to a separate and independent path between the transmitter and receiver.
7. The channel capacity is additive with respect to the number of dimensions. Thus, the total capacity over 2N real dimensions is equal to the sum of the capacity over each real dimension.
8. Each $S_{k,3,i}$ in (2) corresponds to one transmission (over two dimensions, since $S_{k,3,i}$ is complex-valued)
9. Transmission of L_{d2} data bits in Fig. 1 (for a rate-1 turbo code), results in N L_{d2} complex samples (2N L_{d2} real-valued samples) of $\tilde{R}_{k,i,l}$ in (2), for Nth-order receive diversity. Therefore, the channel capacity is

$$C = \frac{L_{d2}}{2NL_{d2}} = \frac{1}{2N} \quad \text{bits/transmission} \tag{20}$$

per dimension. In other words, (20) implies that in each transmission, one data bit is transmitted over 2N dimensions. Similarly, for a rate-1/2 turbo code with Nth-order receive diversity, transmission of $L_{d2}/2$ data bits results in N L_{d2} complex samples of $\tilde{R}_{k,i,l}$ in (2), and the channel capacity becomes:

$$C = \frac{L_{d2}}{4NL_{d2}} = \frac{1}{4N} \quad \text{bits/transmission} \tag{21}$$

per dimension. Substituting (20) and (21) in (18), and using (19) we get the minimum (threshold) average SNR per bit required for error-free transmission, for a given channel capacity. The minimum SNR per bit for various code rates and receiver diversity is presented in Table 2. Note that [50] the minimum E_b/N_0 for error-free transmission is −1.6 dB only when C → 0.

10. In the case of fading channels, it may not be possible to achieve the minimum possible SNR per bit. This is because, the SNR per bit of a given frame may be less than the threshold average SNR per bit. Such frames are said to be in outage. The frame SNR per bit can be defined as (for the kth frame and the lth diversity arm):

$$\text{SNR}_{k,l,\text{bit}} = \frac{1}{2C} \frac{\langle |\tilde{H}_{k,i,l} S_{k,3,i}|^2 \rangle}{\langle |\tilde{W}_{k,i,l}|^2 \rangle} \tag{22}$$

where $\langle \cdot \rangle$ denotes time average over the L_{d2} data symbols. Note that the frame SNR is different from the average SNR per bit, which is defined as [4]:

$$\text{SNR per bit} = \frac{1}{2C} \frac{E\left[\left|\tilde{H}_{k,i,l} S_{k,3,i}\right|^2\right]}{E\left[\left|\tilde{W}_{k,i,l}\right|^2\right]} = \frac{1}{2C} \frac{2L_h \sigma_f^2}{L_d \sigma_w^2}. \tag{23}$$

The kth OFDM frame is said to be in outage when:

$$\text{SNR}_{k,l,\text{bit}} < \text{minimum average SNR per bit} \tag{24}$$

for all l. The outage probability is given by:

$$P_{\text{out}} = \frac{\text{number of frames in outage}}{\text{total number of frames transmitted}}. \tag{25}$$

Table 2: The minimum SNR per bit for different code rates and receiver diversity

	Rate-1/2 turbo code 1st-order receive diversity	Rate-1 turbo code 1st-order receive diversity	Rate-1 turbo code 2nd-order receive diversity
Min avg SNR per bit (dB)	−0.817	0	−0.817

SIMULATION RESULTS

In this section, we present the simulation results for turbo-coded OFDM. In the simulations, the channel length L_h is equal to 10, hence L_{hr} = 19 [4]. The fade variance $\sigma_f^2 = 0.5$ [4]. Figures 6 and 7 depict the simulation results without and with data interleaving respectively, for the frame structure in Fig. 2a and the rate-1/2 turbo code.

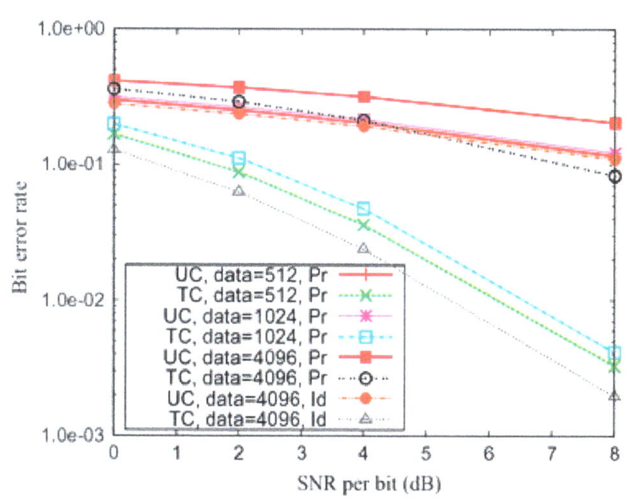

Figure 6: Simulation results without data interleaving, frame structure in Fig. 2a, rate-1/2 turbo code. © 2013 IEEE. Reprinted, with permission, from [4]

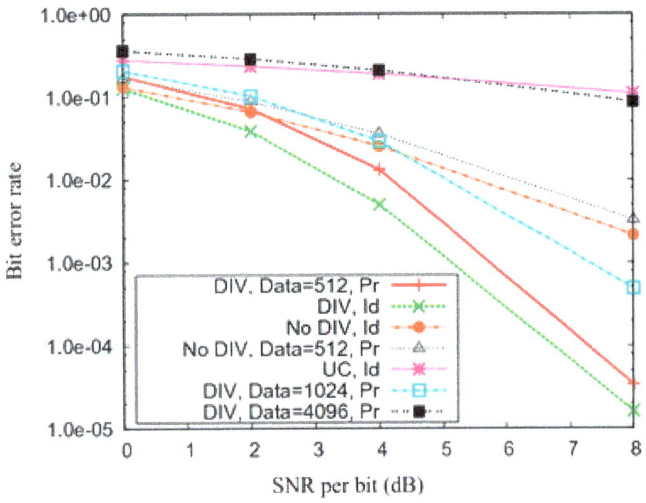

Figure 7: Simulation results with data interleaving, frame structure in Fig. 2a, rate-1/2 turbo code. © 2013 IEEE. Reprinted, with permission, from [4]

In Fig. 8, we present simulation results for the rate-1 turbo code, with enhanced frame structure, 1st-order receiver diversity and interpolation factors (ip) equal to 2, 4, 8, 16 and 32. We find that the performance of the practical receiver is as good as the ideal receiver for ip equal to 16 and 32. However, there is a 4 dB degradation in performance of the ideal receiver for the rate-1 turbo code, with respect to the ideal receiver for the rate-1/2 turbo code in Fig. 7, at a BER of 10^{-5}. This degradation in performance can be compensated by using receiver diversity, which is presented next

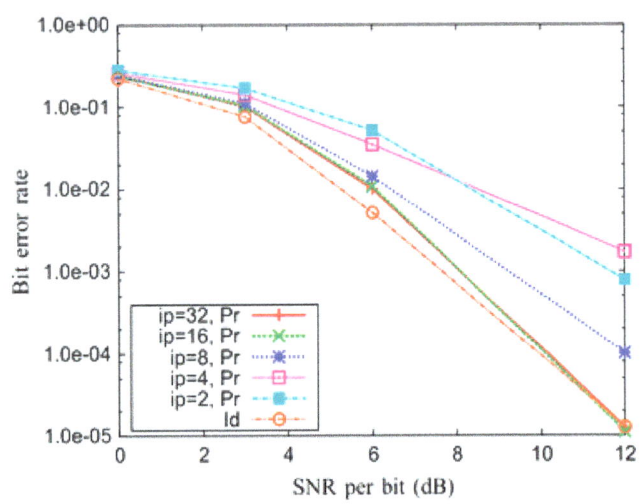

Figure 8: Simulation results with data interleaving, enhanced frame structure in Fig. 1a and rate-1 turbo code

In Fig. 9, we present simulation results for the rate-1 turbo code, with enhanced frame structure and 2nd -order receiver diversity. The channel in both diversity arms is assumed to be identical. However, noise in both the diversity arms is assumed to be independent. Comparing Figs. 8 and 9, we find that the ideal receiver with 2nd-order diversity is just 2 dB better than the one with 1st-order diversity, at a BER of 10^{-5}. Moreover, the practical receivers, with ip=32 have nearly identical performance. This is to be expected, since it is well known that diversity advantage is obtained only when the channels are independent.

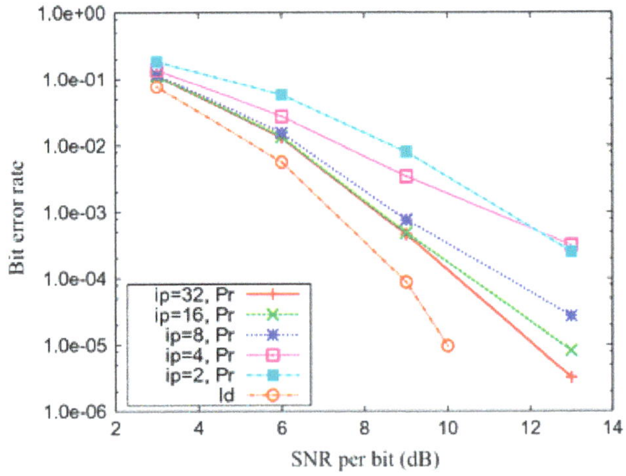

Figure 9: Simulation results with data interleaving, enhanced frame structure in Fig. 1a and rate-1 turbo code with 2nd order receive diversity. Identical channel on both diversity arms

In Fig. 10, we present simulation results for the rate-1 turbo code, with enhanced frame structure and 2nd-order receiver diversity. The channel and noise in both diversity arms are assumed to be independent. Comparing Figs. 8 and 10, we find that the ideal receiver with 2nd order diversity exhibits about 5 dB improvement over the one with 1st order diversity, at a BER of 10^{-5}. Moreover, the practical receiver with ip=16, 32 is just 1 dB inferior to the ideal receiver, at a BER of 10^{-5}.

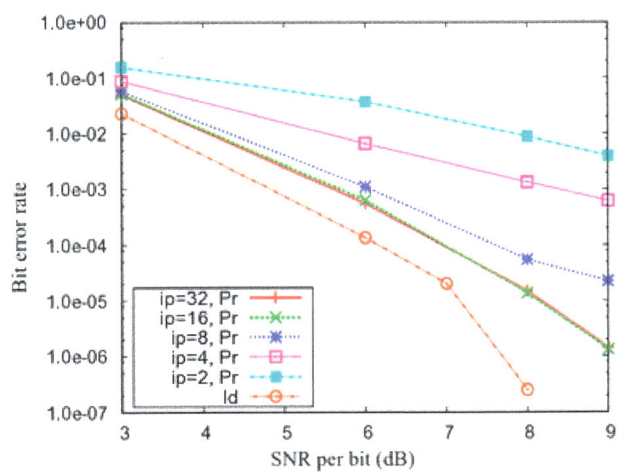

Figure 10: Simulation results with data interleaving, enhanced frame structure in Fig. 1a and rate-1 turbo code with 2nd order receive diversity. Independent channel on both diversity arms

Finally, in Fig. 11 we present the outage probability for the rate-1 turbo code with 1st and 2nd order receive diversity. The outage probability for 1st order receive diversity, at 6 dB SNR per bit is 3×10^{-4}. In other words, 3 out of 10^4 frames are in outage (no error correcting code can correct errors in such frames). Therefore, in the worst case, the number of bit errors for the frames in outage would be $0.5 \times 3 \times 3{,}832$ (assuming probability of error is 0.5). Let us also assume that for the remaining $(10{,}000 - 3 = 9{,}997)$ frames, all errors are corrected, using a sufficiently powerful error correcting code. Therefore, in the best case situation, the overall BER at 6 dB SNR per bit, with 1st order diversity would be $0.5 \times 3 \times 3{,}832/(10{,}000 \times 3{,}832) = 1.5 \times 10^{-4}$. However, from Fig. 8, even the ideal coherent receiver exhibits a BER as high as 10^{-2} at 6 dB SNR per bit. Therefore, there is large scope for improvement, using perhaps a more powerful error correcting code. However, it has been found from simulations that, increasing the number of trellis states does not result in a significant improvement in the BER performance. This is probably due to the fact that puncturing leads to loss of information.

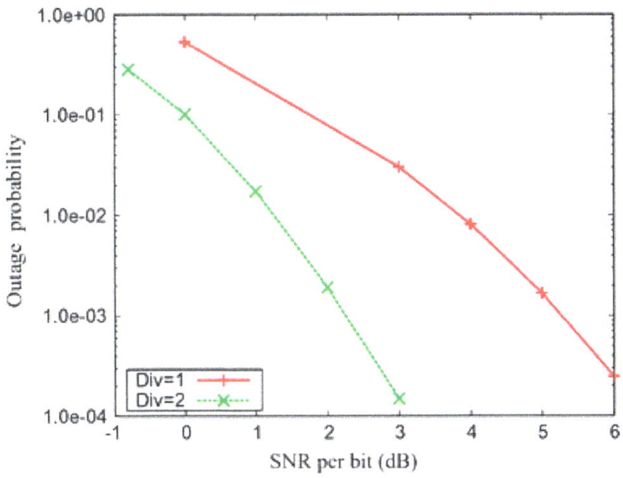

Figure 11: Simulation results for outage probability with data interleaving, enhanced frame structure in Fig. 1a and rate-1 turbo code with 1st and 2nd order receive diversity. Independent channel on both diversity arms

Similarly we observe from Fig. 11 that, with 2nd order receive diversity, the outage probability is 10^{-4} at 3 dB SNR per bit. This implies that 1 out of 10^4 frames is in outage. Using similar arguments, the best case overall BER at 3 dB SNR per bit would be $0.5 \times 3{,}832/(10{,}000 \times 3{,}832) = 0.5 \times 10^{-4}$. From Fig. 10, the ideal coherent receiver gives a BER of 2×10^{-2}, at 3 dB SNR per bit, once again suggesting that there is large scope for improvement.

CONCLUSIONS AND FUTURE WORK

This paper deals with linear complexity coherent detectors for turbo-coded OFDM signals transmitted over frequency selective Rayleigh fading channels. Simulation results show that it is possible to achieve a BER of 10^{-5} at an SNR per bit of 8 dB and throughput equal to 82.84 %, using a single transmit and two receive antennas.

With the rapid advances in VLSI technology, it is expected that coherent transceivers would drive the future wireless telecommunication systems.

It may be possible to further improve the performance, using a better code.

AN APPROXIMATE AND SIMPLE CRAMÉR–RAO BOUND ON THE VARIANCE OF THE FREQUENCY OFFSET ESTIMATION ERROR

Consider the signal model in eq. (5) of [4], which is repeated here for convenience (for notational simplicity, we drop the subscript k, assume $\theta_k = 0$ and $N - 1 = L_p - L_h + 1$):

$$\tilde{r}_n = \tilde{y}_n e^{j\omega n} + \tilde{w}_n \quad \text{for } 0 \le n \le N-1. \tag{26}$$

We assume that the channel is known, and hence $\tilde{y}_n$ is known at the receiver. Moreover, we consider only the steady-state preamble part of the received signal (note that time is suitably re-indexed, such that the first steady-state sample is considered as time zero, whereas, actually the first steady-state sample occurs at time $L_h - 1$). Define

$$\begin{aligned} \tilde{\mathbf{y}} &= \left[\tilde{y}_0 \ldots \tilde{y}_{N-1}\right] \\ \tilde{\mathbf{r}} &= \left[\tilde{r}_0 \ldots \tilde{r}_{N-1}\right]. \end{aligned} \tag{27}$$

The coherent maximum likelihood (ML) estimate of the frequency offset is obtained as follows: choose that value of $\hat{\omega}$ which maximizes the joint conditional pdf

$$\max_{\hat{\omega} \in [-\omega_{\max},\, \omega_{\max}]} p\left(\tilde{\mathbf{r}} | \tilde{\mathbf{y}},\, \hat{\omega}\right) \tag{28}$$

Where ω_{max} denotes the maximum possible frequency offset in radians. Substituting for the joint conditional pdf in (28), we obtain

$$\max_{\hat{\omega}} \frac{1}{(2\pi\sigma_w^2)^N} \exp\left(\frac{-\sum_{n=0}^{N-1}\left|\tilde{r}_n - \tilde{y}_n e^{j\hat{\omega} n}\right|^2}{2\sigma_w^2}\right) \tag{29}$$

which simplifies to

$$\max_{\hat{\omega}} \Re\left\{\sum_{n=0}^{N-1} \tilde{r}_n \tilde{y}_n^* e^{-j\hat{\omega} n}\right\}. \tag{30}$$

Observe that eq. (32) in [4] is the non-coherent ML frequency offset (and timing) estimator, whereas (30) is the coherent ML frequency offset estimator assuming timing is known. The root-mean-squared (RMS) frequency offset estimation error for the coherent detector is shown in Fig. 3 as "RMS coho".

Since ML estimators are unbiased, the variance of the frequency offset estimate is lower bounded by the Cramér–Rao bound (CRB):

$$E\left[(\hat{\omega} - \omega)^2\right] \ge 1 \Big/ E\left[\left(\frac{\partial}{\partial\omega} \ln p\left(\tilde{\mathbf{r}} | \tilde{\mathbf{y}},\, \omega\right)\right)^2\right] \tag{31}$$

since $\tilde{\mathbf{y}}$ is assumed to be known. It can be shown that

$$\frac{\partial}{\partial\omega} \ln p\left(\tilde{\mathbf{r}} | \tilde{\mathbf{y}},\, \omega\right) = \frac{j}{2\sigma_w^2} \sum_{n=0}^{N-1} \left[n\tilde{y}_n e^{j\omega n} \tilde{w}_n^* - n\tilde{y}_n^* e^{-j\omega n} \tilde{w}_n\right]. \tag{32}$$

Substituting (32) in (31) and assuming independent noise (the real and imaginary parts of noise are also assumed independent), we obtain:

$$E\left[\left(\frac{\partial}{\partial\omega} \ln p\left(\tilde{\mathbf{r}} | \tilde{\mathbf{y}},\, \omega\right)\right)^2\right] = \frac{1}{\sigma_w^2} \sum_{n=0}^{N-1} n^2 \left|\tilde{y}_n\right|^2 \tag{33}$$

and hence

$$E\left[(\hat{\omega} - \omega)^2\right] \ge \left[\frac{1}{\sigma_w^2} \sum_{n=0}^{N-1} n^2 \left|\tilde{y}_n\right|^2\right]^{-1} \tag{34}$$

when $\tilde{y}_n$ is known. When $\tilde{y}_n$ is a random variable, which is true in our case, then the right hand side of (34) needs to be further averaged over $\tilde{y}$ [53, 54]. In other words, we need to compute

$$E\left[\left(\frac{1}{\sigma_w^2}\sum_{n=0}^{N-1} n^2 |\tilde{y}_n|^2\right)^{-1}\right] = \int_{\tilde{\mathbf{y}}}\left[\frac{1}{\sigma_w^2}\sum_{n=0}^{N-1} n^2 |\tilde{y}_n|^2\right]^{-1} p(\tilde{\mathbf{y}})\, d\tilde{\mathbf{y}} \tag{35}$$

which is complicated. The purpose of this Appendix is to provide an alternate and a much simpler solution to (35), by assuming that $\tilde{y}_n$ is ergodic.

We claim that, for large values of N (in our case N = 504)

$$\begin{aligned}\sum_{n=0}^{N-1} n^2 |\tilde{y}_n|^2 &\approx \sum_{n=0}^{N-1} n^2 E\left[|\tilde{y}_n|^2\right] \\ &= \text{a constant.}\end{aligned} \tag{36}$$

Now

$$\tilde{y}_n = \sum_{i=0}^{L_h-1} \tilde{h}_i \tilde{s}_{n-i}. \tag{37}$$

Therefore

$$\begin{aligned}E\left[|\tilde{y}_n|^2\right] &= E\left[\sum_{i=0}^{L_h-1} \tilde{h}_i \tilde{s}_{n-i} \sum_{j=0}^{L_h-1} \tilde{h}_j^* \tilde{s}_{n-j}^*\right] \\ &= \sum_{i=0}^{L_h-1}\sum_{j=0}^{L_h-1} E\left[\tilde{h}_i \tilde{h}_j^*\right] E\left[\tilde{s}_{n-i}\tilde{s}_{n-j}^*\right]\end{aligned} \tag{38}$$

Where we have assumed

1. $\tilde{h}_n$ And $\tilde{s}_n$ to be independent
2. $\tilde{s}_n$ (The preamble) varies randomly from frame to frame and is not a constant.

Hence (38) can be rewritten as:

$$\begin{aligned}E\left[|\tilde{y}_n|^2\right] &= \sum_{i=0}^{L_h-1}\sum_{j=0}^{L_h-1} 2\sigma_f^2 \delta_K(i-j)\sigma_s^2 \delta_K(j-i) \\ &= 2\sigma_f^2\sigma_s^2 L_h.\end{aligned} \tag{39}$$

where σ_f^2 is defined in eq. (2) of [4], σ_s^2 is defined in eq. (2) of [4] and $\delta_K(\cdot)$ is the Kronecker delta function. With these developments (35) becomes

$$E\left[\left(\frac{1}{\sigma_w^2}\sum_{n=0}^{N-1} n^2 |\tilde{y}_n|^2\right)^{-1}\right] \approx \left[\frac{2\sigma_f^2\sigma_s^2 L_h}{\sigma_w^2}\sum_{n=0}^{N-1} n^2\right]^{-1}. \tag{40}$$

Therefore, the CRB on the variance of the frequency offset estimate is (assuming N −1 = M)

$$E\left[(\hat{\omega}-\omega)^2\right] \geq \left[\frac{2\sigma_f^2\sigma_s^2 L_h}{\sigma_w^2}\left(\frac{M^3}{3}+\frac{M^2}{2}+\frac{M}{6}\right)\right]^{-1}. \tag{41}$$

The approximate CRB is depicted in Fig. 3 as "Approx CRB".

REFERENCES

1. Hanzo, L., El-Hajjar, M., & Alamri, O. (2011). Near-capacity wireless transceivers and cooperative communications in the MIMO Era: Evolution of standards, waveform design, and future perspectives. *Proceedings of the IEEE*, *99*(8), 1343–1385.
2. Zhang, R., et al. (2013). Advances in base- and mobile-station aided cooperative wireless communications. *IEEE Vehicular Technology Magazine*, *8*(1), 57–69.
3. Hanzo, L., Haas, H., Imre, S., Brien, D. O., Rupp, M., & Gyongyosi, L. (2012). Wireless myths, realities, and futures: From 3G/4G to optical and quantum wireless. *Proceedings of the IEEE*, *100*(Special Centennial issue), 1853–1888.
4. Vasudevan, K. (2013). Coherent detection of turbo coded OFDM signals transmitted through frequency selective rayleigh fading channels. In *Proceedings of the IEEE ISPCC*. Shimla, India, Sept. 2013.
5. Samal, U. C., & Vasudevan, K. (2013). Bandwidth efficient turbo coded OFDM systems. In*Proceedings of the IEEE ITST* (pp. 490–495), Tampere, Finland, Nov. 2013.
6. Bingham, J. A. C. (1990). Multicarrier modulation for data transmission: An idea whose time has come. *IEEE Communications Magazine*, *28*(5), 5–14.
7. Vasudevan, K. (2010). *Digital communications and signal processing, (CDROM included), 2nd edn*. (India), Hyderabad: Universities Press. www.universitiespress.com.
8. Hanzo, L., & Keller, T. (2006). *OFDM and MC-CDMA: A primer*. New York: Wiley.
9. Schmidl, T. M., & Cox, D. C. (1997). Robust frequency and timing synchronization for OFDM. *IEEE Transactions on Communications*, *45*(12), 1613–1621.
10. van de Beek, J.-J., Sandell, M., Isaksson, M., & Börjesson, P. O. (1995). Low-complex frame synchronization in OFDM systems. In *Proceedings of the 4th IEEE international conference on universal personal communications*, Nov. 1995 (pp. 982–986). Tokyo, Japan.
11. Landström, D., Wilson, S. K., van de Beek, J.-J., Ödling, P., & Börjesson, P. O. (2002). Symbol time offset estimation in coherent OFDM systems. *IEEE Transactions on Communications*, *50*(4), 545–549.
12. Park, B., Cheon, H., Kang, C., & Hong, D. (2003). A novel timing estimation method for OFDM systems. *IEEE Communications Letters*, *7*(5), 239–241.
13. Ren, G., Chang, Y., Zhang, H., & Zhang, H. (2005). Synchronization method based on a new constant envelope preamble for OFDM systems. *IEEE Transactions on Broadcasting*,*51*(1), 139–143.
14. Kang, Y., Kim, S., Ahn, D., & Lee, H. (2008). Timing estimation for OFDM systems by using a correlation sequence of preamble. *IEEE Transactions on Consumer Electronics*,*54*(4), 1600–1608.
15. Baum, K. L. (1998). A synchronous coherent OFDM air interface concept for high data rate cellular systems. In *Proceedings of the IEEE VTS 48th vehicular technology conference*, May 1998, pp. 2222–2226.
16. Julia Fernández-Getino Garcia, M., Edfors, O., & Páez-Borrallo, J. M. (2001). Frequency offset correction for

coherent OFDM in wireless systems. *IEEE Transactions on Consumer Electronics*, *47*(1), 187–193.

17. Bradaric, I., & Petropulu, A. P. (2003). Blind estimation of the carrier frequency offset in OFDM systems. In *4th IEEE workshop on signal processing advances in wireless communications*, June 2003, (pp. 590–594).
18. Kuo, C., & Chang, J.-F. (2005). Generalized frequency offset estimation in ofdm systems using periodic training symbol. In *Proceedings of the IEEE international conference on communication*, May 2005, pp. 715–719.
19. Lin, J.-S., & Chen, C.-C. (2005). Hybrid maximum likelihood frequency offset estimation in coherent OFDM systems. *IEE Proceedings - Communications*, *152*(5), 587–592.
20. Ahn, S., Lee, C., Kim, S., Yoon, S. & Kim, S. Y. (2007). A novel scheme for frequency offset estimation in OFDM systems. In *9th international conference on advanced communication technology*, Feb. 2007, pp. 1632–1635.
21. Henkel, M., & Schroer, W. (2007). Pilot based synchronization strategy for a coherent OFDM receiver. In *IEEE wireless communications and networking conference (WCNC)*, March 2007, pp. 1984–1988.
22. Tufvesson, F., Edfors, O., & Faulkner, M. (1999). Time and frequency synchronization for OFDM using PN-sequence preambles. In *Proceedings of the IEEE VTS 50th vehicular technology conference*, Sept. 1999, pp. 2203–2207.
23. Minn, H., Bhargava, V. K., & Letaief, K. B. (2003). A robust timing and frequency synchronization for OFDM systems. *IEEE Transactions on Wireless Communication*, *2*(4), 822–839.
24. Zhang, Z., Long, K., Zhao, M., & Liu, Y. (2005). Joint frame synchronization and frequency offset estimation in OFDM systems. *IEEE Transactions on Broadcasting*, *51*(3), 389–394.
25. Abdzadeh-Ziabari, H., & Shayesteh, M. G. (2011). Robust timing and frequency synchronization for OFDM systems. *IEEE Transactions on Vehicular Technology*, *60*(8), 3646–3656.
26. Mattera, D., & Tanda, M. (2013). Blind symbol timing and CFO estimation for OFDM/OQAM systems. *IEEE Transactions on Wireless Communications*, *12*(1), 268–277.
27. López-Salcedo, J. A., Gutiérrez, E., Seco-Granados, G., & Swindlehurst, A. L. (2013). Unified framework for the synchronization of flexible multicarrier communication signals.*IEEE Transactions on Signal Processing*, *61*(4), 828–842.
28. Frenger, P. K., & Svensson, N. A. B. (1999). Decision-directed coherent detection in multicarrier systems on Rayleigh fading channels. *IEEE Transactions on Vehicular Technology*, *48*(2), 490–498.
29. Merli, F. Z., & Vitetta, G. M. (2008). A factor graph approach to the iterative detection of OFDM signals in the presence of carrier frequency offset and phase noise. *IEEE Transactions on Wireless Communications*, *7*(3), 868–877.
30. wei Wang, H., Lin, D. W., & Sang, T.-H. (2012). OFDM signal detection in doubly selective channels with blockwise whitening of residual intercarrier interference and noise. *IEEE Journal on Selected Areas in Communications*, *30*(4), 684–694.
31. Chen, C.-Y., Lan, Y.-Y., & Chiueh, T.-D. (2013). Turbo receiver with ICI-aware dual-list detection for mobile MIMO-OFDM systems. *IEEE Transactions on Wireless Communications*, *12*(1), 100–111.
32. van de Beek, J.-J., Edfors, O., Sandell, M., Wilson, S. K., & Börjesson, P. O. (1995). On channel estimation in OFDM systems. In *Proceedings of the IEEE VTS 45th vehicular technology conference*, July 1995, pp. 815–819.
33. Edfors, O., Sandell, M., van de Beek, J.-J., Wilson, S. K., & Börjesson, P. O. (1998). OFDM channel estimation by singular value decomposition. *IEEE Transactions on Communications*, *46*(7), 931–939.
34. Coleri, S., Ergen, M., Puri, A., & Bahai, A. (2002). Channel estimation techniques based on pilot arrangement in OFDM systems. *IEEE Transactions on Broadcasting*, *48*(3), 223–229.
35. Ribeiro, C., & Gameiro, A. (2008). An OFDM symbol design for reduced complexity MMSE channel estimation. *Journal of Communications*, *3*(4), 26–33.
36. Kinjo, S. (2008). An MMSE channel estimation algorithm based on the conjugate gradient method for OFDM systems. In *The 23rd international technical conference on circuits/systems, computers and communications (ITC-CSCC)*, July 2008, pp. 969–972.
37. Jiang, M., Akhtman, J., & Hanzo, L. (2007). Iterative joint channel estimation and multi-user detection for multiple-antenna aided OFDM systems. *IEEE Transactions on Wireless Communications*, *6*(8), 2904–2914.
38. Fischer, R. F. H., Lampe, L. H.-J., & Müller-Weinfurtner, S. H. (2001). Coded modulation for noncoherent reception with application to OFDM. *IEEE Transactions on Vehicular Technology*, *50*(4), 910–919.
39. Marey, M., Samir, M., & Dobre, O. A. (2012). EM-based joint channel estimation and IQ imbalances for OFDM systems. *IEEE Transactions on Broadcasting*, *58*(1), 106–113.
40. Kamalian, M., Tadaion, A. A., & Derakhtian, M. (2012). Invariant detection of orthogonal frequency division multiplexing signals with unknown parameters for cognitive radio applications. *IET Signal Processing*, *6*(3), 205–212.
41. Turunen, V., Kosunen, M., Vä ä rä kangas, M., & Ryynä nen, J. (2012). Correlation-based detection of OFDM signals in the angular domain. *IEEE Transactions on Vehicular Technology*, *61*(3), 951–958.
42. Tan, P., & Beaulieu, N. C. (2006). A comparison of DCT-based OFDM and DFT-based OFDM in frequency offset and fading channels. *IEEE Transactions on Communications*,*54*(11), 2113–2125.
43. Yu, C., Sung, C.-H., Kuo, C.-H., Yen, M.-H., & Chen, S.-J. (2012). Design and implementation of a low-power OFDM receiver for wireless communications. *IEEE Transactions on Consumer Electronics*, *58*(3), 739–745.

44. Sari, H., Karam, G., & Jeanclaude, I. (1995). Transmission techniques for digital terrestrial TV broadcasting. *IEEE Communications Magazine*, *33*(2), 100–109.

45. Reimers, U. (1998). Digital video broadcasting. *IEEE Communications Magazine*, *36*(6), 104–110.

46. Takahashi, H. (2009). Coherent OFDM transmission with high spectral efficiency. In *35th European conference on optical communication*, Sept. 2009, (pp. 1–4).

47. Vasudevan, K. (2008). Synchronization of bursty offset QPSK signals in the presence of frequency offset and noise. In *Proceedings of the IEEE TENCON*, Hyderabad, India, Nov. 2008.

48. Vasudevan, K. (2012). Iterative detection of turbo coded offset QPSK in the presence of frequency and clock offsets and AWGN. *Signal, Image and Video Processing*, *6*(4), 557–567.

49. Adakane, D. V., & Vasudevan, K. (2013). An efficient pilot pattern design for channel estimation in OFDM systems. In *Proceedings of the IEEE ISPCC*, Shimla, India, Sept. 2013.

50. Proakis, J. G., & Salehi, M. (2005). *Fundamentals of communication systems*. Pearson Education Inc.

51. Haykin, S. (2001). *Communication systems* (4th ed.). New York: Wiley Eastern.

52. Proakis, J. G. (1995). *Digital communications* (3rd ed.). New York: McGraw Hill.

53. Morelli, M., & Mengali, U. (2000). Carrier-frequency estimation for transmissions over selective channels. *IEEE Transactions on Communications*, *48*(9), 1580–1589.

54. Li, Y., Minn, H., & Zeng, J. (2010). An average Cramer–Rao bound for frequency offset estimation in frequency-selective fading channels. *IEEE Transactions on Wireless Communications*, *9*(3), 871–875.

CHAPTER 20

ACOUSTIC INFORMATION LEAKAGE THROUGH FIBER OPTIC COMMUNICATIONS

Vladimir V. Grishachev

Keywords: Fiber Optic Communications, Acoustic (Speech) Information Leakage Channel, Protection of Acoustic (Speech) Information

INTRODUCTION

Modern technologies of remote and local cable communication systems are based on optical data transmission systems due to the advantages of fiber optic cable over electrical cable as the transport medium. One of the main directions of development is to ensure the broadband subscriber access which is based on optical networks, completely passive (PON) in the future. Technologies such as fiber to the building/home/office/desk (FTTB/FTTH/FTTO/FTTD) lead to the fact that the fiber replaces wire technology in near environment of user [1]. In addition to communications technology fiber is actively used in measuring and security systems. Fiber optic distributed measuring network can control all the basic physical fields in real time with high sensitivity and accuracy [2, 3]. One of the most active trends of fibers in the security systems is the application of optical interfaces for extension of special lines in Closed Circuit Television (CCTV), perimeter security system object.

Advancement of optical structured cable systems (SCS) closer to the man create new threats to information security circulating in building, office, workplace. One of the risks associated with possibility eavesdropping on confidential conversations, using influence of acoustic fields on transmission of light in the fiber. Optical fiber is successfully used to create sensors and distributed measuring networks. Hence regular optic structured cabling system in building is nothing short of a distributed measurement network, which can be used to measure various physical fields, including acoustic field.

Thus, in commercial and government buildings is necessary to protect confidential negotiations in office manager, office space, meeting rooms and other allocated areas of acoustic (speech) information leakage through the optic structured cabling systems. This problem is new, understudied in connection with what is very dangerous.

Citation: V. V. Grishachev, " Acoustic Information Leakage Through Fiber Optic Communications," Journal of Information Security, Vol. 3 No. 2, 2012, pp. 149-155. doi: 10.4236/jis.2012.32017.

PHYSICAL PRINCIPLES EAVESDROPPING

Covert obtain of acoustic (speech) information by using regular fiber optic communications for various purposes is one of the new methods of acoustic intelligence, which is called an acousto-optic (fiber) information leakage channel [4,5]. Forming leakage channel due to the fact that acoustic field from holder information affects the fiber of regular cable systems and causes a modulation of light passing through optical fiber, passive or active elements of optical equipment by acoustic frequencies, as well as reflection from a heterogeneities in them (Figure 1). Modulation of light in the optical fiber can take place in amplitude, phase, polarization and frequency of emission due to exposure on optical fiber of acoustic field. Light which has been modulated by sound can come out far beyond the protected premises through regular fiber optic communications. Then, attacker can gain access to functioning in the establishment of confidential information by demodulating.

On the principles of acousto-optic modulation implemented fiber optic sensor of acoustic field in sonar [6,7], vibration sensors [8,9], and other devices [2,3]. For example, in fiber optic perimeter security systems by vibration acoustic effects of the intruder on fiber is registered penetration of object.

Also, for a long time being developed fiber optic hydrophones, whose operation is similar to vibration sensors with various types of optical schemes. Thus, our research are supported by numerous practical work in neighboring areas. The intruder using various schemes to connect to fiber optic communications is able to conduct tapping conversations on securable.

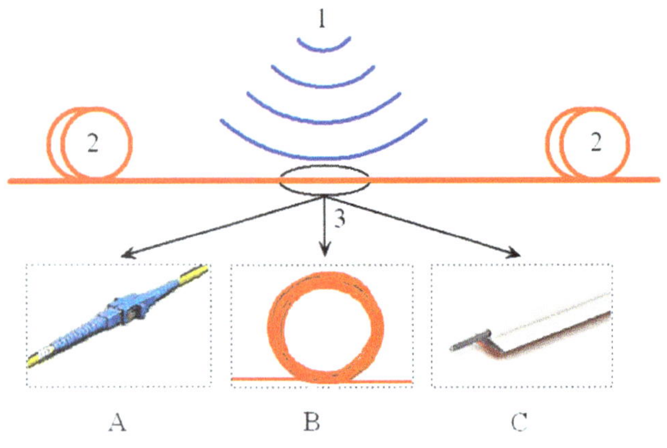

Figure 1: Model of acousto-optic (fiber) information leakage channels. 1: sound source, 2: optical cable, 3: element of optic cable system is subject to sound influence including optical plug contact (A), free optical cable (B), cable with vibro-acoustic contact with designs building (C).

The bases of channels leakage are light beams in an optical cable lines. All light beams can be divided into regular (legal), related to the physical implementation of data transmission protocol, and irregular (illegal), specially generated by an attacker to gain unauthorized output of the speech information. Regular light beams that are formed digital transmission techniques can create a leakage channel without disrupting the entire system, since level of acoustic action on a regular light beam reduces the signal/ noise is negligible. By irregular flows will be assigned any radiation, which generated by light sources with unauthorized connections to fiber optic communications.

Research on the effectiveness of speech leakage carried by the articulator method, which defines speech intelligibility W (%) as the number of words correctly understood at channel output to the number of words spoken at entrance channels leakage. They showed a high risk of new method of eavesdropping. Estimation of efficiency made for amplitude modulation of light passing flows in communication line, containing the basic elements of passive optical networks—fiber optic cable free and attached to the building design, detachable connections, attenuators, etc. This research on shared standard equipment was shown possibility eavesdropping of voice over fiber optic communications with the sound pressure level (SPL) of 60 dB in intelligibility of W up to 80%. The modulation depth of intensity of transmitted light reached saturation into 0.3% at an SPL of 90 dB in the surrounding space.

SCRIPT EAVESDROPPING THREATS

We discuss the overall sequence of actions infringer to obtain acoustic information through fiber optic communications and give a general description used by special technical facilities (Figure 2). Formation of an acousto-optic

(Fiber) information leakage channel is virtually impossible without physical access to optic cable that passes through selected rooms. Cabling must be free of active optical equipment on site between the infringer and the source of acoustical information, which is associated with recovery of regular shape signal and suppressing noise components of radiation in active equipment. Between the infringer and the source of acoustical information must be placed only passive optical elements, which do not change significantly modulation of light. To passive optical elements, except the optical cable relate sockets, adapters, splitters, couplers, attenuators. It should be noted that such a structure of the optical cable network is the most promising for subscriber access and rapidly developing as technology of passive optical networks.

Figure 2: Generalized script of acoustic (speech) information leakage through fiber optic communications (the model of acousto-optic (fiber) leakage channel). I: allocated room; II: secondary rooms. 1: location of confidential negotiations; 2: workplace (workstation); 3: fiber optic communications; 4: regular active optical network equipment; 5: location of leakage channels such as A, B, C; 6: technical means of acoustic intelligence: a (source) and b (receiver) for leakage channel on passing of light; a or b (source and receiver in one location) for leakage channel in reflected light.

Implementation of leakage channel requires applying technical facilities to connect to the cable and recording optical emission. Connection is implemented through regular plug connections, which are used to connect parts of network among themselves and to attach to optical line (OLT) and optical network (ONT) terminals. Connection is dropped and into gap is inserted insertion with input of probing radiation and outlet of part. Another method of connection is to apply coupler radiation on macrobends optic cable. All the proposed methods do not require special technical facilities, distribution of which is regulated by normative documents, such devices are used for installation the optical network. Another method is using cable break to insert the coupler by welding fibers.

Optical scheme of eavesdropping can be accomplished in several ways (Figure 3). First, it can be applied the special probe light sources which are not provided regular network.

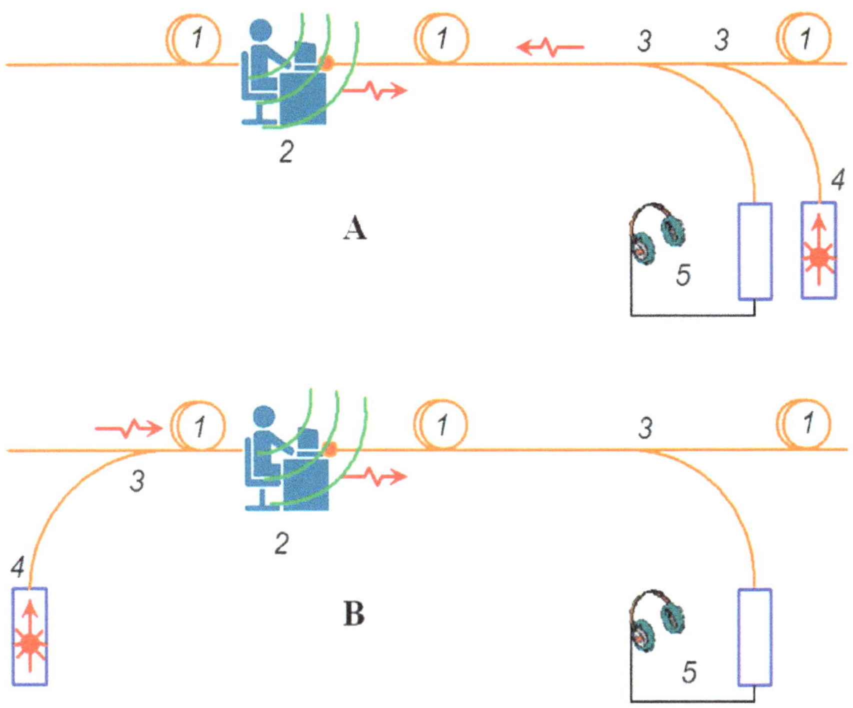

Figure 3: Structure of acousto-optic (fiber) leakage channel in scheme at reflection (A) and scheme for passage (B). 1: optical cable network; 2: confidential source (speech) information with sensitive to vibro-acoustic influence of optical cable; 3: optical coupler; 4: light source (laser); 5: analog optical detector with acoustic demodulator and headphones.

Probing by light can be produced by reflection or by passing from place of modulation. In this case it is possible to combine transmitting and receiving radiation. Second, for eavesdropping can be used regular radiation which applies for traffic within the network.

The danger leakage channel is determined by efficiency of acoustic modulation of light in location of sound source. Acoustic field causes various kinds of modulating light in optical fiber, by choosing parameters demodulation (amplitude, phase, polarization and frequency) is always possible to achieve very high efficiency of leakage channel acoustic (speech) information. Another danger associated with availability of installation equipment that can be used as special technical facilities of acoustic intelligence. For example, for voice communication between installers network uses fiber optic phone, which allows for direct connection to fiber carry voice communications over a distance of 200 km. Fiber optic phone can connect to fiber optic cable without it breaking through macrobend fiber. On the same principle joining works detection of optical signal in fiber, which allows establishing direction of optical signals in coated with 250 micron, 900 microns, as well as in standard optical cords up to 3 mm without gap. Level meters backscatter is designed for monitoring quality of polishing of single-mode fiber optic connectors and measure level of backscatter from other components of communication lines can be used for eavesdropping, also. Still has great potential optical time domain reflectometer (OTDR)—basic device condition monitoring optical. The above instruments are widely available commonly used for installation of optical cable systems, which increases their use in the channel leakage [10].

PREVENTING EAVESDROPPING THREATS

All of the major ways to counter of speech information leakage through waveguide channels can be divided into the following types:

- Soundproofing channel environment, the passive method is to reduce influence of acoustic field on channel environment;
- Filtration of data carrier in transmission channel, the method consists in not passing through channel of irregular signals and modulations with confidential speech information;
- Masking data carrier in transmission channel, the method consists in concealment by addition of a special mask signal and modulations;
- Moisy channel environment, the active method consisting in creating synthetic interference and noise on acoustic frequencies [11, 12].

Each method has its advantages and disadvantages, but the overall effectiveness of any security depends largely on the technical capabilities of detecting threats to information security [13]. Technical means to detect the fact of eavesdropping or preparing equipment to implement it will undoubtedly increase the reliability of the protection system. In the case of fiber optic communications should take into account physical characteristics of fiber optic communication channel, such as small size, direction of radiation and absence of side light beams that go beyond the channel.

Features fiber optic channel allows us to offer a simple and effective way to detect unauthorized output of information (eavesdropping) by monitoring the current in channel of light beams. Any attack on the security system via fiber optic channel for accessing acoustic (speech) information associated with light beams in fiber. Monitoring parameters of light beams in the channel allows identifying any possibility of unauthorized output. This will require registration of radiations pass through fiber optic elements, the allocation authorized by the data carriers (regular radiation), to identify unauthorized flows (irregular radiation) and modulation of acoustic frequencies in any of them. Irregular emission (from external sources) may have a spectral composition as the crossover with regular radiation and does not intersect with it, which is modulated by an external acoustic signal contains confidential information.

Preventing eavesdropping is achieved by performing the following rules. First, regular light beams should not be modulated on audio frequencies. Second, irregular flows that are not provided by the physical implementation of data transmission protocol of network must be absent, and when available, they should not be modulated sound. These simple rules make it possible to detect an attack on a security system and neutralize it. Thus, the degree of risk of acoustic (speech) information leakage is determined by the following features:

1) Irregular light beams is detected in the channel information transmission;
2) Regular light beams is modulated by one of the parameters of optical radiation (amplitude, phase, polarization, frequency) and/or simultaneously on several parameters by external acoustic signal;
3) Abnormal light beams which are separated from spectrum are modulated by one of the parameters of optical radiation (amplitude, phase, polarization, frequency) and/or simultaneously on several parameters by external acoustic signal on a given optical wavelength.

That at least one of these conditions is sufficient for the formation of acoustic (speech) information leakage and can be used to estimate threats to information security.

PRACTICAL IMPLEMENTATION PROTECTION

The problem of detecting possibility of speech information leakage through regular fiber optic communication is solved by installing special equipment, registering light beams in transmission channel information. Implementation can be carried out based on standard or specially created items, which include photodetector, connected to the fiber optic link, also, optical, electronic and optoelectronic analytical element for allocation of acoustic oscillations parameters detected optical radiation. Protection device can be done in two structural decisions: as a separate unit, which has its own alarm system threats, or block the built-in active equipment, which has informational link with the main equipment. Let us discuss possible implementation of devices and their features function.

The external indicating device threat (Figure 4) includes in an optical channel with standard optical connectors and closes the communication link without a significant impact on passing traffic. The main risk associated with probing (illegal) radiation, which is separated from regular light by beam splitters. Incident radiation at regular wavelength is reflected, and irregular radiation at otherwise wavelengths passes through it and is registered a photodetector. Part passing through beam splitters of regular radiation is allocated and is recorded by another photodetector. Received at photodetector output signals are analyzed on existence of probing (illegal) radiation and on possible amplitude modulation. The decision about danger is taken from obtained data, which consists in answer to questions—what is the danger level? And on which side of indicator is a threat comes? The highest danger level corresponds to existence of irregular radiation or amplitude modulation of regular radiation. The average level corresponds to existence of amplitude modulation of regular radiation at the level noise in optical channel. Absence of irregular emissions and any irregular modulation corresponds to the safe mode.

Although the device registers only amplitude modulation of optical radiation and does not register other modulation types, but given that other types of modulation can be effectively observed only when using probing (illegal) radiation, we can assert control of all types of modulation by detecting irregular radiation. Another possibility inherent in this device is that it acts as a filter irregular optical radiation since the regular radiation passes indicator threats and irregular do not pass. Such property greatly limits the application optical design on the passage, which is more effective scheme for reflection. Reflected signal is always weaker than direct probing radiation. Scheme on reflection to the infringer demands more intense radiation to reach an acceptable echo. However for threat indicator any probe signal is a direct that goes either left or right of him so its check will be much safer than the infringer, recording only reflected signal.

The internal indicating device threat (Figure 5) can be integrated directly into the active optical network equipment. It can be integrated into equipment or join a removable modules—transceivers. In the last case, the physical changes to basic equipment may be required and to change the driver itself transceiver.

The main problem of this conversion is to place additional optical elements in required form factor of transceiver. Discuss the structure and operation of indicator threats in integrated dual-port active equipment with separate fiber optic input (the channel receiver) and output (the channel transmitter). Own monitoring system be put on each fiber in the form of additional sensor. In the port of receiver signal is divided into regular and irregular radiation through an optical circulator. Information signal from the regular receiver is processed by conventional means for transmission channel, and arrives at the integrator output which generates an analog signal from the regular radiation that can be modulated by an acoustic frequencies. Selection of irregular radiation is recorded and converted to its own receiver to an analog signal. In the port of transmitter there is no incoming radiation, thus the channel is required separation of radiation on regular and irregular, they are divided by propagation direction. Photodetector is connected to transmitter through coupler and the signal from it is also integrated. Thus, the control system has three receivers with integrated units that make up three analog signals, by which concludes that existence of a threat eavesdropping. Analysis is performed on the existence of irregular radiation and the presence typical components of speech in signal spectrum. On this basis is generated danger signal.

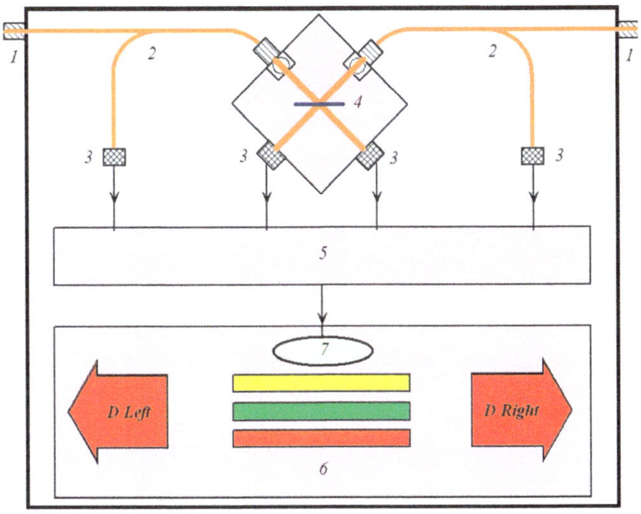

Figure 4: External indicator attacks. 1: optical inputs; 2: optic couplers; 3: photodetectors; 4: beam splitters; 5: spectrum analyzer; 6: display unit; 7: sound indicator attacks. D Left—LED attack on the left, D Right—LED attack on the right, the color risk indicators.

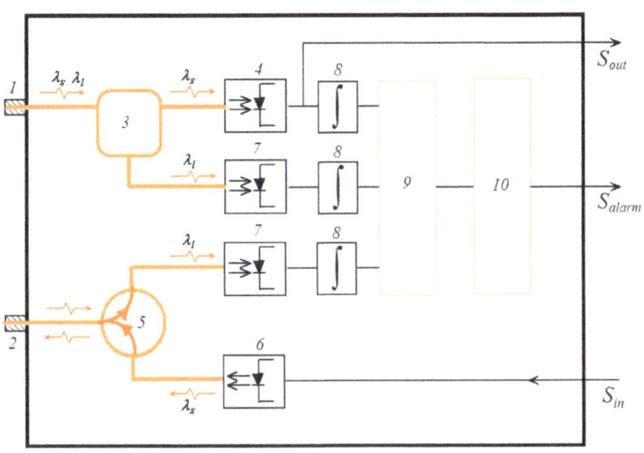

Figure 5: Integrated indicator attacks. 1: optical input; 2: optical output; 3: multiplexer which separates data signal at wavelength l_s from probing light beam at wavelength l_l; 4: regular photodetector with input information signal S_{out}, 5: coupler; 6: regular transmitter with output information signal S_{in}; 7: additional monitoring photodetectors; 8: integrating elements; 9: spectrum analyzer; 10: analog-digital converter, generating danger signal.

ADVICE ON PROTECTION

At present the implementation of methods described above to identify threats to speech information leakage does not exist. As can be seen from a general description of functioning principles of protection devices, to develop working models for detection of eavesdropping by fiber-optic communications is possible based on standard equipment. The main element of protection system is optical detector with an amplifier at sound frequencies, which is present in any analog fiber-optic phone. Standard analog fiber optic phone has high sensitivity which allows recording very small fluctuations in intensity and detect attempts at eavesdropping. However, it has significant disadvantages for the systems security one of them being sensitivity shift in the infrared spectrum, which does not register with highly reliable probing emission of visible spectrum. At distances of several hundred meters, total optical loss will amount to several dB at wavelengths in the visible range in a standard quartz fiber that cannot reliably detect weak optical radiation regular photodetectors.

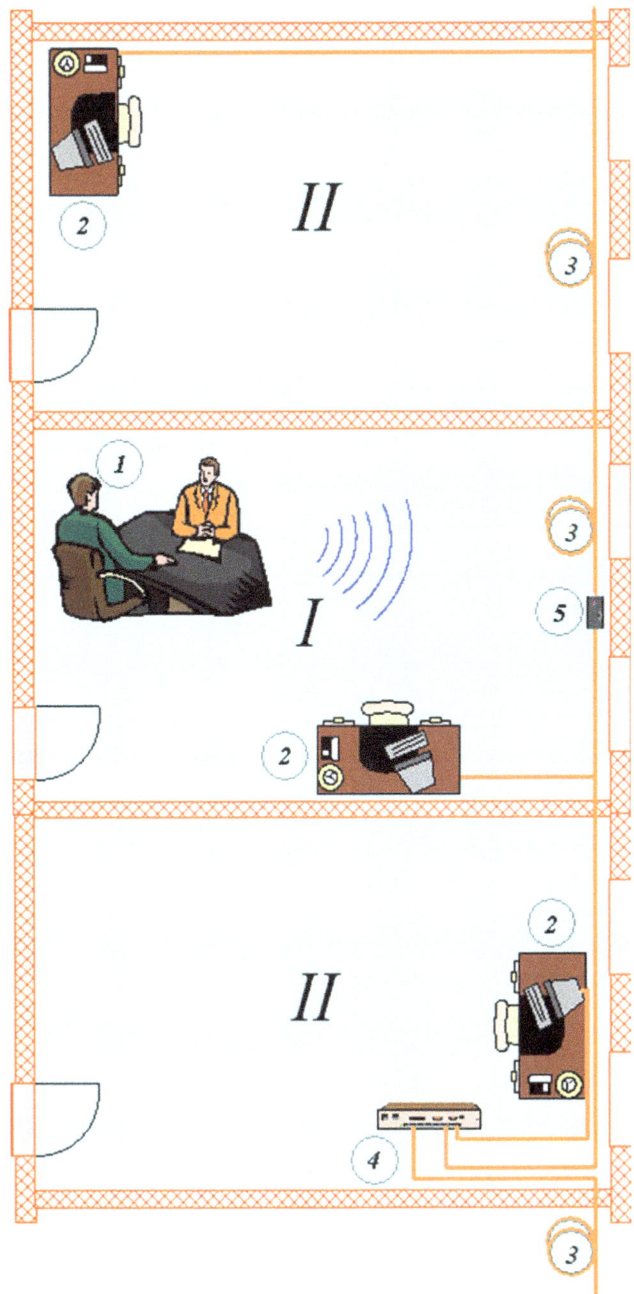

Figure 6: Schematic diagram of protection against leakage information through fiber optic communications, based on detector of attack. I: allocated room; II: secondary rooms. 1: location of confidential negotiations; 2: workplace (workstation); 3: fiber optic communications; 4: regular active optical network equipment; 5: inclusions detector attack.

Another disadvantage is need for additional fiber optic components to register modulation by polarization, frequency and phase. But in any case, a fiber optic phone is closest by principles of functioning for use in protection from eavesdropping through fiber optic communications.

Protection against eavesdropping of selected rooms through a fiber optic communications can be represented as follows (Figure 6). Optic cable passes by the selected room and connected to a computer at workplace. The whole cable with connecting elements inside the room appears as a system which is subject to acoustic effects, formed a speech of confidential information carriers. Light output is modulated a speech and exceeds the selected rooms and may be registered by an attacker. Dangerous for connecting technical reconnaissance attackers are all parts of network in the selected room from one active equipment to another. Defining dangerous section, establish the device detecting attacks in the selected room, way optical insert.

We conducted a trial listening to fiber optic communication lines consisting of optical cable with dual fibers of length exceeding 25 m and a thickness of 3 mm each. Light output was formed an optical tester or a heliumneon laser and registered a fiber-optic phone with analog modulation. Acoustic effects produced locally using computer speakers, acting directly at cable and network elements. The acoustical signal was very noisy, but the words be recognized speech at the hearing.

The presented modeling studies confirmed possibility of implementing such schemes identify attacks even with the help not profile equipment. Production of specialized equipment can more reliably resolve the problem identification of eavesdropping and the security services to help protect speech information in modern rapid proliferation of fiber optic communications technologies.

CONCLUSION

This work is a development patent for invention [13], which proposes a solution to ensure information security talks in selected areas by identifying potential threats on formation leakage channels of acoustic (speech) information through fiber optic communication systems and proposed for use in systems to protect confidential voice communications. Detection of acoustic (speech) information leakage channel is carried out by monitoring optical emissions in regular fiber optic communications. Appearance of any irregular light emissions or modulation on acoustical frequencies of regular light streams creates a potential threat of acoustic information leakage.

REFERENCES

1. F. L. Cedric, "Passive Optical Networks: Principles and Practice," Elsevier, San Diego, 2007.
2. R. Hui and M. S. O'Sullivan, "Fiber Optic Measurement Techniques," Elsevier Academic Press, Waltham, 2009.
3. E. Udd and W. B. Spillman Jr., "Fiber Optic Sensors: An Introduction for Engineers and Scientists," 2nd Edition, John Wiley & Sons, Hoboken, 2011.
4. V. V. Grishachev, D. B. Khalyapin, N. A. Shevchenko and V. G. Merzlikin, "New Channels of Leakage of Confidential Information over the Voice of Fiber Optic Subsystems SCS," Special'naya Tehnika, No. 2, 2009, pp. 2- 9.
5. V. V. Grishachev and O. A. Kosenko, "The Practical Estimation of Convert Audio (Voice) Channel Efficiency from Fiber-Optic Communications," Voprosy Zashity Informacii, No. 2, 2010, pp. 18-25.
6. J. A. Bucaro and T. R. Hickman, "Measurement of Sensitivity of Optical Fibers for Acoustic Detection," Applied Optics, Vol. 18, No. 6, 1979, pp. 938-940. doi:10.1364/AO.18.000938
7. B. Culshaw, D. E. N. Davies and S. A. Kingsley, "Acoustic Sensitivity of Optical Fiber Waveguides," Electronics Letters, Vol. 13, No. 25, 1977, pp. 760-761. doi:10.1049/el:19770537
8. J. C. Juarez and H. F. Taylor, "Field Test of a Distributed Fiber-Optic Intrusion Sensor System for Long Perimeters," Applied Optics, Vol. 46, No. 11. 2007, pp. 1968-1971.doi:10.1364/AO.46.001968
9. J. C. Juarez, E. W. Maier, K. N. Choi and H. F. Taylor, "Distributed Fiber-Optic Intrusion Sensor System," Journal of Lightwave Technology, Vol. 23, No. 6. 2005, pp. 2081-2087.doi:10.1109/JLT.2005.849924
10. Fiber Optic Devices Ltd. (FOD), "Products," http://www.fods.com
11. V. V. Grishachev, D. B. Khalyapin and N. A. Shevchenko, "Method and Device for Actively Protecting Confidential Spoken Information from Leaking over an Acousto-Optic Fibre Channel by Using External Optical Noise Masking," 2010. http://www.wipo.int/patentscope/search/en/WO2010126400
12. V. V. Grishachev, D. B. Khalyapin and N. A. Shevchenko, "Methods and Devices for Actively Protecting Spoken Information against Eavesdropping via an Acousto-Optic Fibre Leakage Channel," 2010. http://www.wipo.int/patentscope/search/en/WO2010126401
13. V. V. Grishachev, "Fiber-Optic Detector for Detecting Threats of Verbal Information Leaks," 2011. http://www.wipo.int/patentscope/search/en/WO2011031186

INDEX

A

Access Network (AN) 220
Access Routers (ARs) 220
adaptive antenna systems (AAS) 23
Adaptive frequency hopping (AFH) 100
additive white Gaussian noise (AWGN) 174, 226
Additive white Gaussian noise (AWGN) 101
adjacent frequency leakage ratio (AFLR) 60
Admission Control (AC) 2
antenna access point (AAP) 167
asynchronous connection-oriented links (ACL) 173

B

Base Station Controller (BSC) 74
Binary Integer Linear Programming (BILP) 160
Binary Integer Linear Programs (BILP) 160
bit error rate (BER) 21, 23
bit-error-rate (BER) 225
Bit Error Rate (BER) 211
bit error ratio (BER) 174
Bluetooth (BT) 173
body area network (BAN) 173
brain-computer interface (BCI) 137
Branch and Bound (B&B) 151
Branch and Cut (B&C) 151
broadband (BB) 124

C

Call Session Control Function (CSCF) 3
carrier frequency offset (CFO) 226
channel quality indicator (CQI) 23, 32
channel state information (CSI) 113
Closed Circuit Television (CCTV) 237
Common Control Channel (CCCH) 77
configuration coordination 170
Convex functions (DC) 152
coplanar strip (CPS) 120
Core Network (CN) 2
correctly identify the start of the (OFDM) 226
Correspondent Node (CN) 212, 215
Cutting Plane (CP) 151
cyclic shift diversity (CSD) 22

D

DC Algorithms (DCA) 152
Directional Scheduled MAC 106
Direct sequence code division multiple access (DS-CDMA) 113
discrete cosine transform (DCT) 226
Discrete multitone (DMT) 156
Dynamic Spectrum Management (DSM) 156

E

encapsulant with a dielectric constant of 4 (ENC4) 130
enhanced data rates (EDR) 173
European Telecommunications Standards Institute (ETSI) 74
Evolved Packet Core (EPC) 2, 3
extended cell (EC) 169

F

Fast Handovers for Mobile IPv6 (FMIPv6) 211
Federal Communications Commission (FCC) 61
fibre-to-the-cabinet (FTTC) 201
fibre-to-the-premises (FTTP) 201
finite impulse response (FIR) 187, 188
fourth generation (4G) 57
frame error rate (FER) 32
frequency division duplex (FDD) 226
frequency hopping (FH) 102

G

gateway mobile switching centre (GMSC) 74
Gaussian frequency shift keying (GFSK) 173
general packet radio service (GPRS) 2
geometric mean decomposition (GMD) 21
Global System for Mobile Communication (GSM) 73

H

handovers (HO) 169
hard-threshold decision (HTD) 193
Hierarchical Mobile IPv6 (HMIPv6) 211

high-definition television (HDTV) 165
High Frequency Structure Simulator (HFSS) 120
High Pass Filter (HPF) 139
High Speed Packet Access (HSPA) 59
home communications controller (HCC) 167
Home Location Register (HLR) 74
human-computer interface (HCI) 137

I

Industrial, Scientific and Medical (ISM) 173
information abstraction 170
Integrated circuits (ICs) 130
International Mobile Telecommunication (IMT) 1
International Telecommunication Union-Radiocommunication Sector (ITU-R) 57
Internet protocol television (IPTV) 165, 202
inter-symbol interference (ISI) 129
inverse fast Fourier transform (IFFT) 62

J

joint diversity and multiplexing (JDM) 25

K

Kernel Procedure (KP) 219

L

light emitting diodes (LEDs) 145
line of sight (LOS) 145
local area networks (WLANs) 226
local loop unbundling (LLU) 201
location register (LR) 5
long term evolution (LTE) 22, 23
Long Term Evolution (LTE) 1, 7
Long-Term Evolution (LTE) 57
low pass filter (LPF) 146
Low Pass Filter (LPF) 139
LTE-Advanced (LTE-A) 57

M

Maximum likelihood (ML) 226
maximum-likelihood sequence detection (MLSD) 193
medium access control (MAC) 113, 226
minimum mean-square error (MMSE) 62
Mixed Integer Linear Programming (MILP) 188, 190, 195
Mixed Integer Programming (MIP) 190
Mixed Integer Quadratic Programming (MIQP) 190
Mobile Node (MN) 211, 215
mobile station (MS) 169
Mobility Anchor Point (MAP) 217
Mobility Management Entity (MME) 6
multi-band (MB) 124
multi-packet reception (MPR) 113
multiple code word (MCW) 23
multiple-input and multiple-output (MIMO) 9
multiple-input multiple-output (MIMO) 145
Multiple-input multiple-output (MIMO) 21

N

next-generation networks (NGN) 201
Non Access Stratum (NAS) 2

O

optical line (OLT) 238
optical network (ONT) 239
optical time domain reflectometer (OTDR) 239
Optimal Spectrum Balancing (OSB) 156
Organisation for Economic Co-operation and Development (OECD) 201
orthogonal frequency division multiplexing (OFDM) 226
orthogonal frequency-division multiplexing (OFDM) 21, 187

P

packet error rate (PER) 174
Partitioning-Hub Location-Routing Problem (PHLRP) 160
personal area network (PAN) 173
physical layer (PHY) 166
physical (PHY) 60
plane wave (PW) 188
power spectral density (PSD) 156
pulse-amplitude modulation (PAM) 193

Q

Quadratic programs (QP) 161
quality of-service (QoS) 158

R

Radio Access Networks (RAN) 1
radio access technology (RAT) 58
Radio Network Controller (RNC) 74
radio resource allocation (RRA) 59
Radio Resource Control (RRC) 2
radio resource management (RRM) 60
Radio Resource Management (RRM) 2
Received signal strength indication (RSSI) 100
Retransmission timeout (RTO) 212
Right-Leg-Driven(RLD) 138
Round-trip time (RTT) 212

S

second-generation (2G) 225
serial port profile (SPP) 175
Serving Gateway (SGW) 6
signal-to-interference-plus-noise-ratio (SINR) 159
signal-to-noise ratio (SNR) 102, 225
single-input single-output (SISO) 21
Singular value decomposition (SVD) 26
sound pressure level (SPL) 238
space-time codes (STCs) 21
structured cable systems (SCS) 237
subscriber stations (SS) 23
symbol error rate (SER) 31
System Architecture Evolution\" (SAE) 2

T

time division duplex (TDD) 226
Traffic Flow Templates (TFTs) 3
transfer matrix method (TMM) 188
Triplicate Re-connection ACKs (TR-ACKs) 215

TV White Spaces (TVWS) 59

U

uninformed transmitter (UT) 28
Universal Mobile Telecommunications System (UMTS) 74
user equipment (UE) 2

V

vector quantization (VQ) 31
vertical Bell Laboratories Layered Space-Time (VBLAST) 23
Visible Light Communication (VLC) 145
Visitor Location Register (VLR) 74
Visual Networking Index (VNI) 202, 208

W

White Space Devices (WSD) 60
wideband code-division multiple access (WCDMA) 33
Wireless Application Protocol (WAP) 174
wireless local area networks (WLAN) 165
wireless personal-area networks (WPAN) 129
Wireless Sensor Networks (WSNs) 105
Worldwide Interoperability for Microwave Access (WiMAX) 57

Y

year-over-year (YoY) 202

Z

Zero Window Advertisement (ZWA) 215